BLOOD EVIDENCE

How DNA Is Revolutionizing
the Way We Solve Crimes

Henry C. Lee, Ph.D.
Frank Tirnady

PERSEUS
PUBLISHING
A Member of the Perseus Books Group

This book is dedicated
to the memory of Bill Schaaf.

Copyright © 2003 by Frank Tirnady and Henry C. Lee, Ph.D.

Library of Congress Control Number: 2002105970
ISBN 0-7382-0602-4

Perseus Publishing is a member of the Perseus Books Group.

Find us on the World Wide Web at http://www.perseuspublishing.com

Perseus Publishing books are available at special discounts for bulk purchases in the United States by corporations, institutions, and other organizations. For more information, please contact the Special Markets Department at the Perseus Books Group, 11 Cambridge Center, Cambridge, MA 02141 or call (800) 255-1514 or (617) 252-5298, or email j.mccrary@perseusbooks.com

Text design by Trish Wilkinson
Set in 11-point Perpetua

First printing, May, 2003
1 2 3 4 5 6 7 8 9 10—04 03

Contents

Acknowledgments

First and foremost, I want to thank my co-author Frank Tirnady. His hard work, persistence and his deep interest in forensic application of DNA make this project possible.

Perseus Publishing has been extremely professional throughout this entire project. Rebecca Marks, Beth Buschmann-Kelly, and John Rodzvilla are outstanding editors. They have provided this project with their excellent insights in the subject area. I want to thank all the others at Perseus who have given this book their considerable professionalism and expertise. I also want to thank Linda Greenspan Regan for her interest to initiate this project.

A special thanks to Dr. Carll Ladd, Dr. Michael Bourke, Dr. Michael Adamowicz, Dr. Heather Miller Coyle and Nicholas Yang for their assistance in reading the manuscript and giving us their critical evaluation and suggestions. I also want to thank the members of the Connecticut State Forensic Science Laboratory and the University of New Haven for their interest and support. A special thanks you to Valerie Shook, Timothy Palmbach, Joseph Sudol and Elaine Pagliaro.

Over the past 45 years I have been involved in forensic research and casework. During this time, I have had the honor and pleasure to meet many of the major players in the forensic DNA field. I want to thank them for their advice and friendship. Also, I want to thank many excellent attorneys, scientists and judges for their help in this project and for their effort in bringing justice through blood evidence.

Finally, I want to thank those closest to me for their friendship and support, especially my mother An-Fu Lee, my sister Sylvia Lee-Huang, my wife Margaret, my daughter Sherry Lee Hersey and my son Dr. Stanley Lee for their patience, love and understanding.

Henry C. Lee

A great many people deserve thanks for their efforts in making this book possible. Enormous gratitude is owed to Raymond R. Smollen, whose continued interest, assistance, and unrivaled computer skills greatly facilitated this effort. The contributions of Elise Peterson cannot be overstated, as she generously lent her expertise in both science and writing to this project. Much appreciation is also due Edward "Spooky" Johnson, whose keen interest in science and energetic research efforts made life a lot easier. A very special thanks is owed to writer Donald Honig, who generously offered his counsel regarding the mysteries and imponderables of publishing. The encouragement and guidance provided by University of Connecticut Professor Emeritus George F. Cole were, in many ways, responsible for this book. And it will never be possible to quantify the contributions of Yasutaka "Yaz" Maruki, whose passion for language and writing are truly inspiring.

Many professional people generously gave their time to provide insight and assistance during the writing of this book. These include, most notably, Attorney Robert L. Curzan, University of Connecticut School of Law Professor Todd Fernow, University of Connecticut School of Law Professor Timothy Everett, Senior Assistant State's Attorney John Malone, Attorney Jeffrey Van Kirk, Superior Court Judge Joseph Purtill, Attorney Carl P. Fortuna, and Court Reporter Barbara Cutter. A particular note of thanks is due the people at the Office of Special Investigations, Criminal Division, U.S. Department of Justice.

A number of people deserve special mention for their advice, assistance, and encouragement; these include Rick Brown, Michael Kubica, Brian King, Syed Nadeem Ahmed, Mildred Beattie, Dante Aiudi, Qu Joh, Robert Smith, Harriet Greenberg, Superior Court Judge William Shaughnessy, Phil Lodato, Attorney Wesley Spears, James Brown, Robin Walters, Norman Slen, Patty Hayes, Betty Wheelock, and Aaron David King.

Finally, the efforts of those at Perseus are greatly appreciated. Especially noteworthy are the heroic efforts by John Rodzvilla, whose calm managerial style helped the process enormously, and the exceptional editorial skills demonstrated by Norman MacAfee and Beth Bushman-Kelly. Needless to say, too, there would be no book if not for the solicitous oversight of Publisher David Goehring, Managing Editor Chris Coffin, and Senior Project Editor Rebecca Marks.

Frank Tirnady

Introduction:
The New DNA Paradigm

DNA EVIDENCE CAN BE INFURIATING, PARTICULARLY IF YOU'RE A CRIMINAL defendant. This was surely the case for 46-year-old William Scott, on trial for raping an elderly woman in Queens, New York, during a 1999 burglary. The trial wasn't going well for the defense, causing friction between Scott and his court-appointed lawyer. The prosecution was about to call its final witness, a forensic scientist who would testify that Scott's DNA matched the DNA from a semen stain found on the victim. Scott's attorney, Harold Ehrentrew, had been urging Scott to plead guilty in exchange for a 20-year sentence rather than let the trial proceed to a jury verdict and face additional time in jail. Instead, Scott asked the judge for another lawyer, but the judge refused. As the jury was entering the courtroom to hear the DNA testimony, Scott threw the trial into an uproar by sucker-punching his lawyer, knocking him to the ground. Court officers tackled Scott, and the judge allowed the jurors to escape through a back door. Much to Scott's consternation, the stunt did not result in a mistrial. Worse still for Scott, the jury convicted him the very next day, and he was sentenced to 35 years.[1]

Since its introduction in 1985, forensic DNA analysis has been conducted literally millions of times in the United States, in Europe, and around the world. Forensic DNA tests are now routinely used in criminal cases, civil cases, family matters, the investigations of war crimes and disasters, and to address a wide range of socially significant issues. In the wake of the terrorist attacks against the United States on September 11, 2001, forensic investigators performed—for this single event—hundreds of thousands of DNA tests to identify the remains of the thousands killed in the collapse of the twin towers at the World Trade Center.

To reach a point where it could even begin to tackle a catastrophe like the September 11 attacks, forensic DNA analysis had to evolve. By the turn of the century, DNA analysis had already gone through several generations of development that took it from a laborious, weeks- or months-long process for a single test to one that can be conducted on an industrial scale that takes mere hours or days for each individual test. Throughout its brief lifetime, forensic DNA analysis has been subjected to an intense level of scrutiny in the courts, in the scientific and legal literature, and in the popular press. Yet it has survived and promises to remain a mainstay in forensic investigations for a long time to come. How does it work? What can it do? What are its limitations? These questions are the focus of this book.

The achievements of DNA analysis in its relatively short time have been impressive. DNA science has solved crimes considered otherwise unsolvable, abbreviated the careers of serial rapists and serial killers, identified the remains of soldiers missing in action or of historical figures, established paternity in hundreds of thousands of instances, assisted medical detectives in the tracking of diseases, helped federal authorities in controlling problems with the food supply, resolved contentious public debates, and illuminated countless other controversies involving biological issues. DNA analysis has demonstrated an extraordinary ability as an investigative and crime-solving tool for one principal reason: It permits the differentiation of one person from another on a genetic level. It can also distinguish one plant from another, one animal from another, and even one viral or bacterial strain from another. The impact of forensic DNA analysis has been widely hailed as equivalent in significance to the revelation more than a century ago that everybody's fingerprints are different. But that kind of praise doesn't even begin to tell the story.

DNA-identification evidence is predicated upon the genetic uniqueness of each individual (excepting identical twins, triplets, or clones), a uniqueness driven by the engine of sex. As will be discussed later in the book, the human body's production of eggs or sperm and the eventual combination of these so-called germ cells is, at bottom, nature's way of making every individual truly an individual. Theoretically, DNA typing could distinguish not only between all people who are alive in the world today but between all humans who've ever lived. It's that good. Very often, too, the principle holds true for animals and plants, but even when organisms do not reproduce sexually, there is generally some genetic clue to their identity.

Forensic DNA analysis is not a single technique. It is a swarm of techniques that have been coaxed out of the vast expanse of the human genome. In addition to the earliest techniques of Restriction Fragment Length Polymorphism (RFLP), human leucocyte antigen analysis (HLA DQAI), and amplification fragment length polymorphism (AFLP, such as DIS80), the various techniques now target, for example, mitochondrial DNA (mtDNA), short tandem repeats (STRs), Y chromosome markers, X chromosome markers, Alu repeat sequences and single nucleotide polymorphisms (SNPs, which include, but are by no means limited to HLA DQAI). With the exception of RFLP, all of the forensic techniques are generally powered by polymerase chain reaction, or PCR, a revolutionary process that permits the faithful reproduction of small, sometimes even vanishingly small, amounts of DNA. The process produces large amounts (identical copies) of DNA and makes DNA typing possible because a certain minimum volume of DNA is necessary to conduct the tests. It is PCR that allows investigators to obtain a DNA profile from postage stamps, coffee cups, and even single hairs.

The challenge to any new forensic technique is that it has to survive on two inter-related fronts: the intellectual battlefield of peer-reviewed scientific journals and, ultimately, in the crucible of the courtroom. Both of these hurdles can be smashmouth, of course, but forensic DNA analysis, despite occasional setbacks, has managed to stay squarely on its feet. The vetting process is an ongoing one, however, journal by journal and jurisdiction by jurisdiction, because forensic DNA analysis has been characterized by, if nothing else, a continual stream of innovations. For the most part, the innovations are arising out of the greater understanding scientists are gaining about the human genome, which is being unraveled a little bit more each day. Unquestionably, there will be new and improved versions of DNA tests ahead and the inevitable fights that DNA typing engenders. But there is little danger that forensic DNA analysis will one day be revealed to be the forensic equivalent of cold fusion. The stuff works. The real weakness of forensic DNA analysis, just as with any other legitimate forensic technique, is not *what* is being done, but *how* it is being done and by whom.

Indeed, in recent years, the reliability of fingerprints—fingerprints!—has been vigorously challenged in a dozen or more criminal cases around the country (and elsewhere in the world), notwithstanding a legal pedigree that dates to 1911. There is, the renowned legal scholar Edward J. Imwinkelried told *The New York Times*, "a widespread public perception that fingerprint testimony is infallible."[2] Fingerprint examiners, in fact, are trained to testify that a fingerprint match amounts to "absolute certainty" that an identification has been made. Yet several cases have demonstrated there may be cracks in the fingerprint edifice. After serving two years of a life sentence for a 1998 murder conviction in Pennsylvania, Richard Jackson was freed after prosecutors finally agreed that the three fingerprint experts who had testified at his trial had actually been wrong when they concluded there had been a fingerprint match linking Jackson to the crime. A Scottish police officer likewise was cleared in a murder investigation when two new fingerprint analysts discovered that the initial fingerprint "match" linking the officer to the crime was questionable. The ultimate issue, it is clear, is the competence of the fingerprint examiners and the uneven nature of their training. A 1995 proficiency test of 156 fingerprint examiners found, suggestively, that 20 percent of them made at least one false match in the course of the test. Although no case has yet to result in fingerprint-identification evidence being declared per se inadmissible, Philadelphia U.S. District Judge Louis H. Pollak was inspired by defense counsel arguments in January 2002 to rule that FBI fingerprint examiners could not testify that a defendant's fingerprints were a definitive match to fingerprints recovered from a crime scene in a capital murder case because fingerprint evidence, the judge explained, had never been subjected to rigorous scientific testing and consequently did not meet the U.S. Supreme Court's standard for valid scientific testimony. Two months later, however, Judge Pollak was persuaded by prosecutors to reverse his earlier ruling and allow the match testimony, after all. In any event, the U.S. Department of Justice is taking the legal challenges very seriously. The National Institute of Justice (NIJ, the DOJ's research arm) is currently looking to fund studies that will devise standardized and statistically validated procedures for examining fingerprints.[3]

The NIJ recently concluded in a report on DNA evidence: "DNA has had an intensity of scrutiny far greater than other methods of criminal investigation, such as ballistics, handwriting, lie detection, eyewitnesses, even fingerprinting. It has passed the test. The scientific foundations of DNA are solid. Any weaknesses are not at the technical level, but are in possible human errors, breaks in the chain of evidence, and laboratory failures." In fact, the NIJ authors suggest, the baptism of fire that DNA evidence has withstood is inspiring a thorough re-examination of other methods of identification, including fingerprinting.[4]

More than that, however, the success of DNA evidence, or "DNA fingerprinting," as it is commonly known, has inspired investigators to think big with respect to other forensic-identification methods, and even strive to invent some new ones. Veteran experts in the field of firearms examinations and ballistics, for example, who, among other things, endeavor to link spent bullets with the guns that fired them, are finding their discipline reinvented. It is certainly true that recent technological advances have greatly improved the ability to declare such matches, but the improvements are a matter of degree. Nevertheless, to speak in terms of a ballistics match (which is to say, a match based on the markings created on the bullet or bullet fragment as it passed through the rifled barrel of the gun) appears to be passé. It is now de rigueur to speak in terms of "gun DNA" and "ballistic fingerprinting" and to envision a state or federal computer database that records the markings that all guns make on the bullets they fire. In fact, "gun DNA" and "ballistic fingerprint" laws have recently been passed in New York and Maryland (requiring the manufacturers to provide the data) and Massachusetts is considering such a law. Although the efficiency of such databases for guns is controversial, the idea is that a bullet extracted from a corpse, for example, can be instantaneously traced to the gun of origin.[5]

The new state laws are really designed to piggyback, and greatly expand, a pre-existing federal database, the National Integrated Ballistic Information Network (NIBIN), that catalogues information about thousands of gun crimes committed across the country since 1993 and has already been instrumental in solving countless crimes through ballistics matches. By 2003, all state and local authorities will have access to NIBIN through specially designed computer software.[6] The new state laws are intended to compile ballistics data on weapons *before* they are used in crimes to facilitate the apprehension of the perpetrators. (Note that both Maryland and New

York's laws apply only to handguns.) Since "ballistic fingerprinting" laws require, in effect, that the government maintain a list of all gun owners, a de facto gun registry, critics have challenged the laws as an unconstitutional abridgement of the Second Amendment and a violation of the 1968 federal Gun Control Act. Moreover, opponents argue that the new laws will likely be ineffectual since the distinctive markings made by a gun can be altered by simply replacing a gun barrel or firing pin, for example, and will, in any event, be gradually altered over time with repeated use of the firearm. Nevertheless, as the Washington, D.C.,-area sniper attacks were unfolding, largely in Maryland and Virginia, in the fall of 2002—attacks that ultimately claimed ten lives—proponents of a national ballistic fingerprint database grew increasingly vocal and appeared to be winning converts. Even after suspects John Allen Muhammad and John Lee Malvo were apprehended and the fatal shootings stopped—and it was determined that the rifle apparently used in the shootings was not obtained legally (that is, traceably)—advocates of the new laws continued to press their case. Among other major newspapers, the editors of the *New York Times, the New York Daily News, The Boston Globe, The Hartford Courant,* and *USA TODAY* endorsed the idea of creating a national ballistic-fingerprinting system, contingent upon—the editors of *The New York Times* hastened to add—a carefully controlled, scientific demonstration that such a system actually works. (Early reports indicate, incidentally, that there is DNA evidence linking John Lee Malvo to some of the crime scenes.)[7]

A similar effort is currently underway to devise a surefire method of "diamond fingerprinting," aimed primarily at distinguishing so-called blood diamonds, or conflict diamonds, from diamonds mined under legitimate circumstances. Blood diamonds come from the mines of West African war zones, mines that have been seized by rebel groups or corrupt governments as war booty and used to fund an illegal international arms trade, terrorism, or the ceaseless and remorselessly cruel civil wars of the region. In fact, the Al Qaeda terrorist network inspired by Osama bin Laden has profited from cheaply purchased blood diamonds originating in Sierra Leone, according to intelligence officials, and diamonds may even have been stockpiled by Al Qaeda operatives to thwart a global crackdown on terrorist-linked bank accounts. Other terrorist groups have also been linked to the illegal trade.[8] Facing international outrage, trade sanctions, and the real prospect of a consumer boycott of all diamonds, the legitimate diamond industry is now actively seeking a method to differentiate between lawfully mined diamonds and blood diamonds. In November 2001, a global pact was reached as a first step to require certificates of origin for all shipments of rough stones and warranties to accompany finished diamonds sold by legitimate diamond merchants. At the same time, efforts are ongoing to come up with a diamond-identification process worthy of the name "diamond fingerprinting."[9] Some methods focus on tracing a given diamond to its geologic source based on individualizing characteristics of the diamond. Other methods being developed may ultimately help investigators trace the specific origin of individual diamonds and facilitate the return of stolen property, such as producing microscopic engravings on diamonds using lasers or ion beams. One company, Gemprint of Toronto, uses a laser beam directed at a diamond's largest facet to generate a unique pattern of reflected light and then stores the image in a computer database for future reference. The information can then be accessed by jewelers or law enforcement, and recutting the diamonds does not appear to affect the test.[10]

Another variation on the theme of DNA fingerprinting is "brain fingerprinting," developed, at least in part, by Dr. Lawrence Farwell (supported by funds from the Central Intelligence Agency) and patented in the mid-1990s. Using a headband embedded with electronic sensors, the brain fingerprint test is really an elaborate EEG, or electroencephalograph, that seeks to register a subject's neurological responses to various images flashed on a screen to determine whether the subject reacts with "guilty knowledge" or not. The test is based on a well-known EEG spike, called P300, that occurs when the brain notices something significant (such as an image related to a crime). Unfortunately for Dr. Farwell, the first court to consider the results of a brain fingerprint test (purporting to show that a convicted murderer, jailed for life in 1978, was actually innocent of the crime) found it unpersuasive. Despite the setback, Dr. Farwell continues to believe in his system. Independently, a half dozen or so other laboratories around the country are also working on similar brain-fingerprinting protocols.[11]

Although brain fingerprinting may well languish in obscurity, the field of biometrics is a growth industry both in the United States and abroad. Biometrics is the science of identifying people based on unique physiological (or physical) characteristics. Both dermal fingerprinting and "DNA fingerprinting" are biometric systems, and they seem to have inspired a spate of identification technologies that are becoming increasingly widespread. Some of the variations include palm-print systems, eye-identification systems (focusing on the pupil, iris, or retina), and face-recognition technology. Face-recognition technology, in particular, sparks fierce public debates over the issues of unreasonable searches, privacy, freedom of assembly, freedom of association, and the right to travel. The more likely it is that the digitalized information created by the system will end up in a centralized computer database, the more vitriolic the debates. Face-recognition technology is one of the more controversial of the new biometric technologies—or, at least it was until September 11, 2001. A number of companies have developed systems that scan facial features to measure, for example, the distance between pre-selected nodal points and examine key facial landmarks such as eye sockets, cheekbones, and chins. The information is reduced to a mathematical code, then digitalized, and can be cross-checked against a database. The database may include the face-prints of customers, authorized personnel, or, in the case of law enforcement systems, fugitives, criminals, missing persons, or terrorists. The systems can be easily integrated into pre-existing networks of surveillance cameras or used (as is done in West Virginia) to prevent people from obtaining duplicate driver's licenses under false names. While face-recognition software has been installed without great fanfare in some nationwide ATM systems, a national furor erupted after police in Tampa, Florida, used the technology at Super Bowl XXXV in January 2001 and then later integrated it into a system of three dozen surveillance cameras trained on passersby in the entertainment district of Ybor City. When it was revealed that about 100,000 people attending the Super Bowl had been secretly scanned as they entered the stadium (a process that identified 19 petty criminals but no one worth arresting), the operation was jointly denounced as the "Snooper Bowl" by the American Civil Liberties Union and Republican Representative and House majority leader Richard Armey, who called for congressional hearings on the technology. Even though the Ybor City face-recognition effort didn't become operational until signs were posted alerting the public to the presence of the cameras, Dick Armey was not mollified. "This is a full-scale surveillance system," he told Reuters. "Do we really want a society where one cannot walk down the street without Big Brother tracking our every move?" Others, including business owners in Ybor City, welcomed the technology as a crime-prevention measure, however. Tampa police officials noted that no scanned images are retained if they fail to match an image in the database. Additionally, the currently available software is far from foolproof, although efforts are ongoing to improve the technology. Efforts are also underway to devise behavioral-recognition technology that will alert authorities when specific conduct is observed by unmanned cameras.

The system used in Ybor City is called FaceIt software and is manufactured by Visionics Corporation of Jersey City, New Jersey. Visionics software creates a facial map based on 80 potential reference points; a hit is registered if a dozen or more points match an image in the database. Beginning in 1998, FaceIt began operating in various locations throughout England (starting with high-crime areas) where there are currently an estimated 2.5 million closed-circuit TV (or CCTV) surveillance cameras. The FaceIt software is only used in a few locations, however, although one locale, the London borough of Newham, has reportedly installed 300 face-recognition cameras. The cameras throughout Britain were installed in the 1990s, at least initially, in response to terrorist bombings by the Irish Republican Army. Over time, however, they came to be recognized as a crime-control measure, one that most Britons appear to welcome. Whether America follows the British lead on surveillance cameras, as it appears to be doing with the idea of a national DNA database, remains to be seen. Immediately after the terrorist attacks on America of September 11, 2001, while the stock market generally crashed, the stock price of Visionics more than tripled. Within a month or so of the attacks, officials at Logan Airport in Boston, where two of the four hijacked planes originated, agreed to install the face-recognition technology of two rival companies (Visionics and Viisage Technology, Inc., of Littleton, Massachusetts) at various checkpoints at the airport for a trial run. Other airports around the country were soon making similar plans. Less than a year later, controversy had erupted over the efficacy and practicality of some of the systems installed around the country.[12]

On the other hand, America has a deeply ingrained tradition that puts a premium on individual freedoms, as the oft-quoted words of Benjamin Franklin convey: "Those who desire to give up liberty in order to gain security, will not have, nor do they deserve, either one." Commenting in *The New York Times*, William Safire characterized the expanding network of surveillance cameras this way: "It is the pervasive, inescapable feeling of being unfree."[13]

Despite such sentiments, in the aftermath of September 11, 2001, a network of closed-circuit TV cameras (one already planned in the wake of the 1995 Oklahoma City bombing) was more rapidly installed at various national monuments in Washington, D.C., under the jurisdiction of the United States Park Police of the National Park Service, including the Washington Monument, the Lincoln Memorial, the Jefferson Memorial, the Franklin D. Roosevelt Memorial, and the Korean War Veterans and Vietnam Veterans Memorials. In addition, the Washington, D.C., City Council began debating the extent to which surveillance cameras should be deployed at various locations throughout the city. Neither the city authorities nor the Park Police had any immediate plans to equip their systems with face-recognition technology. Cameras equipped with the face-recognition software of Visionics were used, however, during the 2002 Winter Olympics in Utah to scan the faces of the hundreds of thousands of people entering the main arena known as the "E" center. By the summer of 2002, Virginia Beach, Virginia, authorities had likewise installed a surveillance network equipped with face-recognition software throughout the popular vacation spot. By this time, such cameras had also been installed, with perhaps a tinge of sad irony, at the ferry dock in Manhattan where tourists board the ferry to visit the Statue of Liberty on Liberty Island.[14]

DNA testing, in all its myriad forms, has not only affected detective work but has become ingrained in the popular culture and has wormed its way into virtually every facet of modern life. Armed with DNA analysis, scientific sleuths have waded into some of our era's most contentious social controversies. On most levels, DNA testing seems a godsend, particularly when a violent criminal is apprehended and convicted based on DNA analysis, but there are increasing instances when DNA testing raises troubling issues. One flashpoint involves the creation of a national DNA database that catalogues and computerizes the DNA profiles of an ever-increasing number of people (and interlinks all similar state DNA databanks). The impact that such a system will have on individual privacy is being closely monitored and fiercely debated, particularly in light of calls for a universal DNA database to which everyone may be required to contribute a DNA sample.

Although forensic DNA analysis raises some disconcerting social issues, it has been, by almost every measure, a resounding success. Two developments combined to so radically alter the forensic landscape as to make almost everything that came before DNA fingerprinting seem quaint and vaguely prehistoric: first, the understanding by Sir Alec Jeffreys that certain aspects of the human genome could be examined to differentiate between individuals and, second, the epiphany by Kary Mullis that the natural processes of DNA replication could be harnessed to produce unlimited quantities of whatever segment of the genome one cared to investigate. Jeffreys called his invention "DNA fingerprinting" and the name stuck (although, in a very technical sense, the term only refers to the specific process developed by Jeffreys). Mullis called his invention polymerase chain reaction (PCR) and the name stuck, also, except that it is commonly referred to as DNA amplification. Over time the two inventions became commingled in practice, so that now when people speak of "DNA fingerprinting," they mean a combination of the two. Most often DNA is first amplified and then examined for individualizing characteristics. (Increasingly, a third process, DNA sequencing, also known as the Sanger method, is being employed to unravel specific DNA segments letter by letter [base by base] so they can be compared to other DNA segments that have also been sequenced. The development of the Sanger method—named for its inventor, Frederick Sanger—predates the work of Jeffreys and Mullis, but the automation of the process occurred subsequently and now drives advances in virtually all branches of biotechnology. As significant as the method has been to genetic research and the Human Genome Project, DNA sequencing was not the impetus of forensic DNA analysis, however. Nevertheless, the Sanger method currently plays an indispensable role in the process.) To say, as many have, that DNA fingerprinting and PCR are "revolutionary" contributions to forensic science is surely to understate their significance. In the forensic world, their impact has been felt as profoundly as the discovery of fire and the invention of the Gutenberg press. When early humans first harnessed fire, the fossil record suggests they were happy just to cook with it for awhile (perhaps a million years or so). Eventually, however, the primitive hearth became the stove, the furnace, the smelter, the locomotive, and the power plant, not to mention the oil lamp and Bic lighter. Likewise, the Gutenberg press gradually spawned daily newspapers, more readily available textbooks, Thomas Paine's "Common Sense," the Encyclopedia Britannica, *Tom Sawyer,* and *Playboy* magazine. The progress has been much the same with forensic DNA analysis since Jeffreys and Mullis laid the groundwork in the early and mid-1980s, except that the innovations and new applications have occurred with breathtaking speed. The burgeoning use of forensic DNA analysis (and the proliferation of the necessary DNA machines and experts) closely mirrors the contemporaneous revolution in personal computers. Personal computers are now everywhere and are globally interconnected. Forensic DNA analysis is similarly everywhere and is rapidly becoming a globally interconnected enterprise. (Indeed, the United States and United Kingdom envision the joint use of their national DNA databases to solve crimes of international import.) The rise of DNA and that of computers are not merely parallel phenomena, of course. Computers and DNA are rapidly becoming intertwined on a fundamental level, well beyond the current use of computers to store analogues of DNA information and transmit it over the Internet as part of the national DNA database. Advances in miniaturization and microchip technologies are being combined with the well-established analytic techniques of forensic DNA analysis. The imminent fusion of these two seemingly disparate technologies is promising a new DNA fingerprinting revolution that will operate at lightning speed. By using photolithography, chemical etching, or spotting techniques, various academic laboratories and private companies are currently experimenting with devices that are essentially high-speed, miniature instruments that perform all of the analytical steps currently conducted by laboratory technicians using sophisticated machinery, but do so in a tiny fraction of the time. Currently, for example, a DNA sample can be amplified and then separated into a variety of identifiable short repeating segments (STRs) in several hours. Miniaturization technologies are demonstrating that the entire procedure can be done instead in just a few minutes.[15]

Even without the propulsion of miniaturization and microchips, however, forensic DNA analysis has gone far. Although it was invented fairly recently, it seems as if eons have passed since "DNA fingerprinting" was simply a novel forensic technique called into service in some faraway rape or murder case. The techniques of forensic DNA analysis are now the commonplace items of contemporary life. For those people interested in tracing their family roots, for example, there are user-friendly home kits for securing your DNA so it can be analyzed at a laboratory and positioned with greater certitude (although this is controversial) on your family's genealogical chart. The service is made available by companies like FamilyTreeDNA of Texas and GeneTree of California; the Y chromosome can be analyzed to trace paternal lineages and mitochondrial DNA can be used to trace maternal lineages. (Warning: The resulting DNA tests may provide unwanted information, like, for example, your real father is your father's best friend, Ray.)[16] For the safety conscious, there are child-identification kits that are available to record vital information about children, including a DNA sample, in case they are lost or kidnapped. For some families, the techniques of forensic DNA analysis can help resolve or bring closure to some unthinkable family catastrophe, such as the death of a family member in the terrorist attacks of September 11, 2001, or perhaps even some long-ago calamity that swept away a relative or family member.

In May 2001, in fact, three bodies were exhumed from numbered graves in Halifax, Nova Scotia, to help three families find answers through DNA tests about relatives lost on April 15, 1912, when the *Titanic* sank off the coast of Newfoundland. About 150 of

the *Titanic*'s 1,500 victims are buried in Halifax, and 44 of them have never been iden-
tified. Ongoing DNA tests are being conducted by the Paleo-DNA Laboratory at
Lakehead University in Thunder Bay, Ontario, under the direction of Dr. Ryan Parr.[17]
On November 6, 2002, Dr. Parr and his colleagues announced the first DNA identification of a victim of the Titanic tragedy. The remains were those from a grave marked
simply "Unknown Child," who mtDNA analysis demonstrated was 13-month-old Eino
Panula from Finland. Eino perished along with his mother and four brothers when the
Titanic sank. Mrs. Panula, who reportedly declined a seat in a lifeboat because she had
become separated from some of her sons, was taking her children to join her husband,
who had gone ahead to prepare for the family's resettlement in America in a small
coal-mining community in Pennsylvania.[18]

For another family, the Toma family of Sulyanbokor, Hungary, the upheaval of
World War II had marched a brother, like so many countless others, into oblivion. Incredibly, DNA tests revealed in the summer of 2000 that the brother was still alive,
having been held in Russia since 1944 after being captured by the Soviet army during
a battle with Nazi-led Hungarian troops. Upon his capture, Andras Toma was sent to
a prisoner-of-war camp in Siberia but was transferred to a mental hospital in 1947,
where he was forgotten. During his captivity, Andras learned very little Russian so he
remained in virtual isolation. In August 2000, the 75-year-old Toma, whose existence
was discovered only after an accidental encounter with a Russian who spoke Hungarian, was returned to Hungary, where a search uncovered surviving half-siblings.
Their DNA established Toma as the longest-held POW of World War II.[19]

Certainly one of the greatest contributions of forensic DNA analysis (as will be
examined more fully in a later chapter) has been the use of the technology since 1989
to free the wrongly imprisoned. By late 2002, more than 120 men had been released
from jail (death row in some cases) after DNA tests had excluded them as contributors of biological evidence linked to the crimes for which they had been convicted.
(As will be seen later, however, some of these exonerations are extremely controversial, since some people argue that some of the DNA tests themselves are misleading.)
In this context, New York attorney Peter Neufeld—famous both for his role as defense counsel for O. J. Simpson and for championing the causes of convicts who
maintain their innocence—has called forensic DNA analysis "the gold standard of innocence or guilt."[20]

At the same time, if the behavior of actual criminals is any gauge, forensic DNA
evidence can be a major impediment to pursuing a life of crime. In fact, police have
for years seen a rash of cases where criminals have attempted to foil DNA detectives
by, for example, wearing gloves and condoms during rapes, forcing rape victims to
bathe or otherwise destroy biological evidence, scattering the blood or semen of
other people around crime scenes, and even sending imposters in to take their DNA
tests. While some of these efforts have surely succeeded in hopelessly frustrating

investigators, the ever increasing sensitivity and sophistication of DNA tests have managed to stay one giant step ahead of criminals who think they can outsmart DNA investigators. (Fortunately, too, much of the information obtained by criminals on the street or in prison is misinformation.) In fact, it is now virtually impossible to commit a criminal act, particularly a crime against another person, without leaving an identifiable DNA calling card behind. As Salt Lake County, Utah, Sheriff Aaron Kennard told *USA TODAY,* "These days you need to wear a spacesuit to commit a crime without leaving any [DNA] behind."[21] Nevertheless, authorities are well-advised to remain vigilant for the latest ploy; some criminals are very inventive. For example, when Anthony Turner, who was suspected in three rapes in and around Milwaukee, Wisconsin, argued to authorities in 1999 that the real rapist just happened to have the exact same DNA profile as he did, police were, to say the least, unpersuaded. He was arrested, tried, and convicted in March 1999. Shortly afterward, while Turner remained in jail awaiting sentencing, police were astonished to learn that the DNA sample from a new rape case had come back identifying Turner as the rapist, even though he couldn't possibly have done it. It was a DNA match that called into question the whole enterprise of forensic DNA analysis. If two unrelated people in Milwaukee, or the United States for that matter, had the exact same DNA profile (assuming a sufficient number of markers were examined), the criminal justice system had big problems. Investigators soon learned, however, that it was Turner who had the problem, not DNA science. In a clever scheme, Turner had smuggled his own semen out of jail in a ketchup packet and had family members bribe a woman to stage a rape scene complete with Turner's smuggled semen. The plot failed and Turner got 120 years.[22]

Prosecutors in the United States submit an estimated 10,000 biological samples for DNA testing and try thousands of cases with DNA evidence every year.[23] About 25 to 30 percent of the samples submitted for testing come back excluding the prime suspect. Most of the rest inspire guilty pleas. The laboratories that conduct forensic DNA tests are a combination of state, federal, and commercial laboratories. It was commercial laboratories, in fact, that first brought DNA evidence into the courtroom in criminal cases, but over time state and federal laboratories caught up with the technology and absorbed a greater and greater proportion of DNA criminal casework. The private laboratories continue to thrive, however, because (as will be discussed later) there is a burgeoning public demand for forensic DNA tests to resolve a variety of legal disputes, such as paternity issues. Private laboratories continue to do criminal casework, too, for defense attorneys and prosecutors alike, because their resources occasionally outstrip that of state police laboratories. Increasingly, too, forensic DNA tests are being used, even devised, to address questions that are best described as social emergencies, or potential social emergencies. The spread of contagious diseases is always high on the list of escalating emergencies that can confront officials. New DNA typing techniques can now reliably identify and trace unusual cases of infection back to a source. In early 2000, for example, disease detectives concluded through sophisticated DNA tests (using RFLP analysis) that a Baltimore, Maryland, mortician probably contracted tuberculosis, which became active, from a cadaver during an embalming procedure. The genetic signature of the TB from the mortician was identical to the TB samples taken from the cadaver.[24] Very often, genetic research in other fields can facilitate DNA-identification techniques and pay big dividends for doctors or scientists concerned with diagnosing patients or tracking disease transmissions. In recent years, of course, there has been a frenzy of research among geneticists to map the genomes of important organisms, including everything from the human genome to the banana genome. As will be examined later, such undertakings are a great deal more than academic exercises. They can have profound social implications and even assist forensic investigators in solving crimes. For example, when, in October 2001, the United States was confronted with its first experience of wide-scale bioterrorism and anthrax-laden letters began turning up in the offices of high-profile media companies, the mail rooms and offices of all three branches of the federal government, as well as U.S. Postal Service stations, the level of understanding that science had accumulated about potential biological weapons became crucial. This will be examined later in the book.

As significant as DNA profiling has been to investigations of grave national and international importance, the true mark of its success is that it has quietly and inexorably filtered down to the level of the common, everyday dispute. Of late, forensic DNA analysis has been called upon to solve lower-level, even petty crimes and even referee run-of-the-mill interpersonal disputes. A principal reason for the "everyman DNA boom" is that DNA tests have become a lot cheaper. Given this, the feeling often is, "Why not?" In some cases, it may well seem, it gives an unfair advantage to those who "Think DNA" when something doesn't seem quite right. In Cape Town, South Africa, for example, a dispute over the ownership of a dog—a German shepherd claimed by a local woman after she spotted him with a Taiwanese crew aboard a cargo ship, who insisted the dog was their mascot—remained deadlocked until DNA tests were carried out by local veterinarians to determine if the dog came from a breeder identified by the Taiwanese crew as the dog's original owner.[25] Common burglars, too, are finding themselves implicated by DNA analysis far more often, something made all too easy for investigators when the burglar has had the misfortune of cutting himself during the commission of the burglary.[26] Little or no police investigation is necessary when the culprit has the double misfortune of having his DNA on file in a state or federal DNA database.[27] But it's turning out that even transgressions of a noncriminal variety are facing expiation by genotyping. In a first of its kind, a Palm Beach County, Florida, Circuit Court judge admitted DNA tests into evidence in a divorce case in December 2000 that tended to support a claim of marital infidelity. In this case, proof of infidelity triggered a clause in the couple's prenuptial agreement—as amended in 1999—that would permit the heartsick spouse, in this case the wife of a wealthy retired businessman, to collect millions of dollars. (In Florida, infidelity is not a specific ground for divorce, so the infidelity was only relevant to the extent that it bore on the prenuptial agreement.) Here, Nanette Sexton found clues, including stained bed sheets, at her husband's Vermont horse farm that indicated he was having an affair in 1999. She locked the items in a vault in Vermont and then later sent them to a laboratory in Denver for DNA analysis. The laboratory reported back that the stain on the sheet contained a mixture of male and female components. The male DNA was Nanette's husband's, 74-year-old Richard Bailey; the female DNA did not match Nanette's. As the case approached a trial date, Richard Bailey's attorney was arguing, among other things, that his client had lost his mind by the time he signed the amendment to the prenuptial agreement (as a result of Alzheimer's). The case was subsequently put on hold—for at least three years—while the judge investigates Bailey's health and mental condition.[28] A few months after the Florida decision, a court of appeals in Louisiana made another unprecedented decision and required that a defendant in a sexual harassment suit provide a blood sample for DNA analysis. The higher court ruled that the same legal test for compelling a DNA sample in a paternity case applied in the sexual harassment case; it was particularly appropriate here, the court ruled, because the alleged victim, who'd been fired from her job, demonstrated that the DNA profile of the defendant would likely match a semen stain on her skirt that resulted from a non-consensual encounter with the defendant. Significantly, the defendant had consistently maintained that nothing *of any kind* had occurred between them. A Virginia employment law newsletter warned employers: "[The decision] is important to you because it signifies a break from the historic opposition to introducing DNA evidence in civil cases. Employees who possess similar damaging evidence will certainly waste no time in trying to convince courts across America that DNA testing is the best way to end the swearing contest between the alleged victim and alleged harasser."[29]

For most of its relatively short history, DNA evidence in criminal cases was reserved for the most heinous crimes: rape, murder, domestic or foreign terrorism, cop killings, genocide. The penalties for such crimes, of course, are severe and may even include the death penalty. Gradually, however, DNA-typing expertise proliferated, the process was streamlined, and the costs of doing the tests dropped precipitously. Now DNA evidence is increasingly being wielded against minor offenders, offenders who were disinclined to "Think DNA," as others reflexively do. For example, when Indiana State Trooper Dan Jones was handed his chicken club sandwich at a Burger King fast-food restaurant not too long ago, he felt there was something really odd about the way the kid at Burger King said "Enjoy," but it was just a feeling. "He had a real stupid grin on his face," Officer Jones later told a reporter, "like a deviant kind of grin that said, 'I just spit on your sandwich.'" Sure enough, upon inspection, Officer Jones found a roughly one-inch diameter pool of spit underneath the bun of his sandwich. With this, Officer Jones made a general announcement to the Burger King staff that everybody in the joint was under investigation for consumer-product tampering, whereupon the bun was carefully preserved as evidence and transported to the state police forensic laboratory for DNA analysis. A judge later found sufficient grounds to issue a warrant to secure blood samples for DNA comparison from three kitchen workers who were on duty during the incident. When the DNA came back as a match for a 16-year-old grill cook, he was fired by Burger King and arrested. In the end, however, the boy received only 40 hours of community service and apologized to Officer Jones. He got off easy compared to two former Burger King employees elsewhere in Indiana, who both received 30-day jail sentences for aggravated assault after spitting on the burger of another police officer. Such cases, it turns out, are far from unusual. Across the country, police officers are finding all manner of hideous contaminants in their fast-food orders, so much so that many officers prefer to pack a lunch or eat where they can observe the food being served. Aside from the gross-out factor, there is the ever-present fear of contracting a contagious disease, perhaps a very serious one. (In the second Indiana case, in fact, both of the employees later tested positive for exposure to hepatitis, but fortunately the officer did not contract the disease.) One episode in Rochester, New York, sparked two arrests and a $13.5 million dollar lawsuit against Burger King, its franchisee, and two employees, brought by a Monroe County sheriff who discovered that his burger contained, according to the allegations, urine and oven cleaner. The phenomenon is by no means limited to Burger King restaurants. North Carolina highway patrolman Chris T. Phillips used to enjoy eating at Taco Bell until he discovered what turned out to be phlegm, according to laboratory tests, on one of his nachos. The miscreant was identified without extensive DNA tests on this occasion and fired. Much to the dismay of Officer Phillips, however, North Carolina has no criminal statute that makes it a crime to spit on food. The reasons for such behavior on the part of teen-aged fast-food workers is murky and perhaps varied. One of the malefactors apprehended doing it later explained that he resented the cops harassing him all the time about his skateboarding.[30]

As one writer recently put it, "DNA has gone decidedly mainstream."[31] It's a phenomenon driven largely by the ubiquity of forensic DNA analysis—solving recent crimes, solving older crimes (so-called "cold cases"), freeing the innocent, determining paternity, tracking diseases, monitoring problems in the food supply, weighing in on historical debates and resolving momentous scientific debates. To be sure, too, crime is a singular obsession in America. "Crime may be down across America," *The Wall Street Journal* noted in 2001, "but on television, it is up—way up."[32] In a July 2002 preview of the fall television line-up, *USA TODAY* noted in a headline: "CBS to

Launch TV Crime Wave."[33] Featured regularly on the many television dramas and documentaries about crime, of course, is "DNA fingerprinting." NBC's *Law & Order* dramas routinely confront DNA evidence, for example, as do CBS's popular *CSI: Crime Scene Investigation* and the more recently introduced *CSI: Miami, Without a Trace,* and *Robbery Homicide Division.* How accurate are the shows? Los Angeles County Sheriff's Department criminalist Elizabeth Devine, who is the technical advisor for *CSI,* conceded to *Popular Science* in 2001 that there were some inaccuracies on the details of the forensic science portrayed in the show. DNA results are obtained nearly instantaneously, for example, to speed the drama along; the TV stars don't wear gloves in the laboratory as they should; and the multi-colored bottles in the background of laboratory scenes are chosen exclusively for their color, rather than the chemicals they're supposed to contain.[34] But it is certainly significant that such shows have highly qualified technical advisers. Even Fox TV's *The X-Files,* which clearly falls under the rubric of science fiction (we're pretty sure), has long had a scientist on the payroll to consult in matters related to biology and DNA analysis. Reality-based documentaries abound on television as well, often featuring crimes that were solved with the help of DNA. The Discovery Channel has been dedicating an entire evening of late (Tuesdays) to true-crime shows focusing on murder and mayhem: an alternating schedule includes, variously, *The Prosecutors, The New Detectives, The FBI Files,* and the *Justice Files.* The Learning Channel carries *Scene of the Crime, The Secrets of Forensic Science,* and *Medical Detectives,* among others, and Court TV, which features wall-to-wall coverage of all of the available trials du jour, recently added *The Forensic Files, Dominick Dunne's Power, Privilege, and Justice,* and *I, Detective* to its programming. A&E, the Arts & Entertainment Channel, also has several fact-based shows on crime, including *American Justice, Investigative Reports, Cold Case Files,* and *City Confidential.* And there have been, of course, any number of movies featuring DNA tests, such as CBS's miniseries *The Third Twin,* based on Ken Follett's novel by the same name.

There is a considerable market for true-crime books as well, of course, which necessarily examine DNA evidence, at least in passing, when the crimes involved are of recent vintage. Perhaps more importantly, however, is the degree to which DNA evidence is woven into the fabric of contemporary crime and detective fiction (not to mention legal thrillers), including, for instance, the novels of Linda Fairstein, Kathy Reichs (*Deadly Decisions,* 2000), Patricia Cornwell, Donald E. Westlake (*Bad News,* 2001), Dean Koontz (*Fear Nothing,* 1998), and Jeffrey Deaver (*The Coffin Dancer,* 1998). But how could it be otherwise? What should contemporary writers write about if not the latest breakaway forensic advance? In fact, there is nothing at all new in this. The genesis of the crime and detective genre, beginning with Edgar Allan Poe's story "Murders in the Rue Morgue," published in 1842, is directly traceable to the ascendancy of forensic science as a serious discipline. A persuasive case for this theory, in any event, is made by English professor Ronald R.

Thomas of Trinity College in Hartford, Connecticut, in his book, *Detective Fiction and the Rise of Forensic Science*.[35] As forensic science evolved, so, too, did the crime and detective novel, but the reverse is also true to some degree. Edward Rothstein of *The New York Times* calls this interrelationship "the double helix of crime fiction and science."[36] Just as Sir Arthur Conan Doyle, who wrote the stories featuring Sherlock Holmes, was intrigued by the early days of mug shots, lie detectors, and fingerprinting, so, too, have criminal investigators been fascinated by the deductive and observational powers of Sherlock Holmes, to name one example. Indeed, teachers of forensic science have been known to recommend the works of Conan Doyle (among others) to their students.[37] Certainly, the novelists and investigators of today continue to share a fascination with assembling mental jigsaw puzzles. The singular contribution of forensic DNA analysis is that it has made everything biological a potential clue, and this adds an entirely new dimension to figuring stuff out.

For some observers, America's obsession with crime seems unhealthy, however. "Distressed cultural critics call it 'serial killer fetishism' or 'forensic chic,'" commented Thomas Doherty in *The Boston Sunday Globe*, noting, with some horror, a "serial killer marathon" that ran on Court TV in 1999.[38] What fuels the phenomenon? There's every indication that the answer, at least partly, has something to do with a more generalized interest in science, as opposed to being simply another social indicator pointing to the downfall of Western Civilization. "Outbreak!" *The New York Post* warned in 1998, "Suddenly 'science' is *tres* chic."[39] A more recent headline in *The Wall Street Journal* may have captured it best: "How Science Got Cool."[40] The fashion industry even flirted with "lab chic" in recent years, according to *The New York Times* (although the short-lived craze may have been driven as much by angst over germs as anything else).[41] An indication that the trend is continuing is found in the November 2001 launch of the new science magazine *Seed*, distributed by AOL Time Warner. Billed as a magazine of "science couture," the magazine combines the sex appeal of more mainstream publications with the heady topics of the day in science. According to one reviewer, "*Seed* is a mark of the growing influence of science on culture."[42] That science "got cool" is due, in no small measure, to the unraveling of the DNA molecule that occurred over the last half of the 20th century and continues into the 21st.

In a number of surveys taken around the turn of the century, the discovery of the structure of DNA by James Watson and Francis Crick in 1953 ranked at, or near, the top of the accomplishments of the 20th century (its chief rival was the discovery of penicillin).[43] When the completion of a first draft of the human genome was jointly announced in June 2000 by the federally funded Human Genome Project (under the auspices of the National Institutes of Health) and the privately owned Celera Genomics, Francis Collins, who led the federal side of the effort, described the genetic map as "a glimpse of an instruction book previously known only to God."[44] From this feat, a writer for *The New York Times Magazine* predicted, there would flow "Biotech Chic," as the magazine's headline called it: "Soon, each person may be able to have the three billion DNA units of his or her unique genetic sequence encoded on a computer chip, for perfect identification in matters medical, official, criminal or otherwise." A major dilemma facing the fashion conscious, the writer suggested, was how to wear "the ultimate 21st-century accessory" (so the magazine solicited suggestions).[45]

Just the very *idea* of DNA is so powerful that it has inspired people in a wide range of professions to "Think DNA" and attempt to incorporate it, on some level, into their disciplines. Until Watson and Crick figured it out, nobody really knew what the specific mechanism of inheritance was, although Gregor Mendel had brilliantly observed the operation of mysterious inheritance "factors" some 90 years before. Indeed, even before Mendel, Darwin had jolted the world into acknowledging there was some unseen force driving evolution. Watson and Crick demonstrated that the language of DNA is written in four letters or bases (A, T, G, and C) and is inscribed along the coiled double-helix of the chromosomes. These four letters, it turns out, are—quite remarkably—enough to produce every living thing on earth. Others wondered, What else could the four letters do? Beginning in 1994 with the experiments of Leonard Adleman of the University of Southern California, a number of researchers have attempted to pirate the four-letter code of DNA for use in enhancing computer power. DNA computing, in fact, is seen by some people as the possible salvation of the computer industry. Computers store information based on a binary code, ones and zeroes, and computer capacity and sophistication are fast approaching the limits of miniaturization for the silicon chip, upon which computers depend. The elegant complexity of DNA may offer a solution. About a dozen research teams in the United States are working with DNA or RNA molecules on the problem, and scientists have even gone so far as to organize the Consortium of Biomolecular Computing.[46] In November 2001, an Israeli team at the Weizmann Institute, led by Professor Ehud Shapiro, announced that they had built the first programmable DNA computer in which all functions (including software and hardware) were performed by biomolecules. The DNA computer they created was so tiny, according to Shapiro, that a trillion of them could fit into a single test tube capable of performing a billion functions.[47] Other computer scientists have taken a step in a different direction and joined forces with musicians to develop software programs that translate DNA sequences into music. The music of DNA now comes in a variety of programs that can be downloaded from the Web. Molecular scientist Linda Long of the University of Exeter in England, for example, offers "molecular music" based upon the sounds of various human proteins, as derived from the "notes" of their DNA coding. Assistant piano professor Brent Hugh of Missouri Western State College wrote a DNA software program and even released a CD entitled *Music from the Human Genome*. "The opening from chromosome 23 really caught my interest," he told *The New York Times*. "It had an interesting sweep to it, and it was almost like a fugue."[48]

It is not merely the hard sciences that are capitalizing on the DNA idea, although many disciplines do so only in the broadest sense. A new form of literary criticism has arisen, for example, called "genetic criticism," which, according to D. T. Max, "focuses on the evolution of literary manuscripts from drafts to published form, taking into account the inevitable impact of editors and publishers."[49] Inspired, perhaps, by the staying power of evolutionary biology (which has spawned, among other things, evolutionary psychology), the neo-Darwinist school that presupposes a biological imperative for most social interactions and behaviors, virtually every field of study is now influenced by DNA to one degree or another. "Every academic department, it seems," writes Emily Eakin in *The New York Times*, "has its biology-enamored theorist," noting the recent ascension of "biohistory."[50]

One of the more colorful adaptations of the new DNA paradigm has occurred in the field of theoretical anthropology. In particular, the study of cultural evolution has spawned not one, but at least two intellectual camps (rival camps, in fact) that embrace various biological models to describe the manner in which cultural traits and ideas are transmitted within populations and across generations. The mainstream model is called cultural selectionism, which borrows heavily from the field of population genetics, the hard science dedicated to observing and predicting (lots of math here) the frequency with which genetic traits (and even genetic information unrelated to physical traits) occur within human populations. Physical traits that are favored, naturally, are selected more often in decisions involving reproduction. Population genetics figures prominently in the story of forensic DNA analysis, so it will be discussed more fully later on. Suffice it to say that cultural selectionism tracks cultural phenomena in a population using a methodology similar to the way population geneticists track genetic phenomena in a population. They can make only generalized predictions. Cultural selectionists, for example, have monitored the spread of Roman Catholicism through the Philippines, but they wouldn't try to predict whether a particular family will convert. At the same time, however, cultural selectionists recognize that cultural ideas and traits are not genes. Thus, the model has significant limitations. An upstart rival camp to the selectionists, on the other hand, known as memeticism, appears to embrace its own biological model with far fewer inhibitions than the selectionists. Memetics is the study of memes, or "self-replicating units of cultural information," as one proponent defines them.[51] Memes, of course, are the cultural analog of genes, and memeticism is a way of conceptualizing the transmission of cultural information in biological terms. "Just as genes are the basic units of biological evolution," one memeticist explains, "so memes are the basic units of cultural evolution."[52] Memes are not seen as merely popular ideas, however; they are, apparently, imbued with an inherent, self-animating mission to reproduce themselves. They have been likened to infectious agents or parasites. The meme movement is traced to zoologist and author Richard Dawkins of the University of Oxford, who first proposed the existence of memes in his 1976 book *The Selfish Gene,* citing, as examples, "'tunes, ideas, catch-phrases, clothes fashions, ways of making pots or of building arches.'"[53] Others have suggested that memes include, among other things, ideas in people's heads, commonly repeated words, or even cultural artifacts such as the wheel.[54] Whatever memes are, various theories for their persistence have been advanced. Memeticist Susan Blackmore, a psychologist at the University of the West of England in Bristol, suggests, for example, that memes, whose survival is dependent upon imitation and repetition, have thrived because imitative behavior in humans is a favored trait and that imitators, down through the ages, have mated with imitators and begotten many more little imitators.[55] Memeticist John Aunger suggests, instead, that memes simply came along and exploited what genes had produced, namely the loosely configured neuronal networks of the human brain. The brain, according to Aunger, necessarily stores information redundantly to avoid the loss of knowledge crucial to survival. Aunger posits a brain backup system composed of "troids" and "trions" (terms he coined, by his own admission, to give the theory a biomedical edge) that are, essentially, physical manifestations of memes. After ricocheting around in the brain for awhile, meme entities, apparently, occasionally pop out of the brain and seek to inhabit the unwitting brains of others. Other memeticists speak in terms of memetic determinism (akin to genetic determinism) and being a slave to your memes.[56] It's all very biological.

All of this is very spooky, of course, and may well be taking the cultural imperative to "Think DNA" a logistical bridge too far. But, who knows? Maybe it's all true. If it is, then, ironically, one of the most successful memes in recent years is DNA itself, which, of course, was the inspiration for memetics in the first place. Spurred on by developments in medicine, genomics, and forensic science, the DNA theme (meme?) has so permeated modern culture and language as to be simply another integral part of the landscape. DNA has, in fact, actually altered the physical landscape. So many biotechnology companies sprouted up in Cambridge, Massachusetts, for example, that the town became known as Genetown.[57] Similarly, a 15-mile corridor along Interstate 270 in Montgomery County, Maryland, is referred to as DNA Alley because it now boasts the largest collection of private genomics companies in the world, including Celera Genomics, Human Genome Sciences, and Gene Logic, Inc., not to mention The Institute for Genomic Research (TIGR), a not-for-profit institution, and the National Institutes of Health (NIH) in Bethesda, Maryland. Finally, a frenzy of laboratory building has occurred throughout the United States over the past ten years as the federal government and almost every state have endeavored to bring their forensic science laboratories into the 21st century with, among other things, the ability to properly perform the most sophisticated DNA tests available.

The singular impact of DNA science on modern life is everywhere reflected in the language. Certainly one measure of the cultural significance of any idea or thing is the degree to which the language adopts it. That the United States is a religious society is easily demonstrated by pointing to such common expressions as "Thank God," godsend, godforsaken, God complex, "God Bless America," and "In God We Trust." Indeed, the oath administered in court cases ends with "so help me God." The influence of baseball is reflected in the common use of such phrases as "curve ball," "foul ball," "7th inning stretch," "pinch hitter," and "couldn't get to first base." William Shakespeare's works introduced so many memorable expressions into the English language that many people who use them have no idea that they are quoting Shakespeare. These include, to cite a few, "to thine own self be true," "sound and fury," "pound of flesh," "band of brothers," "brave new world," and the "nature/nurture" dichotomy.

In recent years, the term "DNA" has become, perhaps inexplicably, embroidered into the English language, far beyond usages describing the technical aspects of genetics, "DNA fingerprinting," or "DNA tests." While there are, of course, biotechnology companies or research foundations that employ a DNA-esque title, such as DNA Sciences, DoubleTwist, Inc., and DNAX Research Institute (pioneer biotechnology giant Genentech Inc. managed to get dibs on the stock symbol "DNA"), there are also any number of outfits using the DNA moniker even though their businesses have nothing whatsoever to do with genetics. There are, for example, DNA New Media Group (a marketing and advertising company), DNA Creative (an ad agency), and just plain DNA (a high-fashion modeling agency). In addition to company names, there are a variety of products that seek to exploit DNA's cachet. One photo-editing software program, for example, goes by the name "PhotoGenetics."[58]

For years, the Hewlett-Packard computer company ran ads promoting itself and its founders as "The DNA of Silicon Valley" and asking "Do you have the hp gene?"[59] Motorola, Inc., trademarked the advertising slogan "Digital DNA" to promote its products and characterize its innovative computer-systems designers (the "Digital DNA team"). There is even a Web site, www.digitaldna.com.[60] For civil libertarians, however, the idea of "digital DNA" conjures up Big Brother or, at least, a lot of snooping little brothers crawling through the inescapable digital trails left by modern devices. One's travels, medical status, personal habits, and political beliefs are continually being registered and stored somewhere by someone for reference by who knows whom. *Popular Science* magazine editor-in-chief Scott Mowbray warned readers in an editorial in July 2002 entitled "Your Other DNA": "From our genetic code to our shopping habits, from Google searches to the words we speak and hear during the day, we are each a database." The cover of the magazine explains, "Privacy in Peril: Who's Scanning Your Digital DNA?"[61]

The Internet, of course, has introduced countless new terms into the language. Some have been entirely invented, such as cyberspace or hyperlink, but many have arisen after being commandeered by the Internet's pioneers from more common usages, like bookmark, mouse, domain, surfing, and virtual reality. When the discussion turns to the basics of how all of this computer stuff actually works, naturally the instruction manuals carry such names as "Microsoft DNA Technology" and a "Step-By-Step Windows DNA Web Application Tool Kit." When promoting computer software for industrial automation, Microsoft describes the product as "DNA for Manufacturing." (None of this is related to the computer programs that actually are designed to convert DNA information into digital codes or computer graphics. When analyzing STR markers in criminal casework, forensic science laboratories use, for example, a software program called Genescanner produced by Applied Biosystems, Inc.).

DNA chic goes beyond the product lines of the computer industry, of course. A popular piano trio goes by the name Triple Helix, for instance.[62] An early entrant to the DNA craze was fragrance designer Bijan, which introduced DNA Perfume in 1993. Available at Saks Fifth Avenue, the perfume came in bottles that looked suspiciously like glass-blown double helixes. In an abundance of caution, perhaps, ads for "DNA" carried an asterisk after the product name, noting in small print that "DNA fragrances do not contain deoxyribonucleic acid (DNA) except as included in the ingredient list on product packaging."[63] (Wait a minute. Does that mean that it has no DNA in it, or that it does?) When O. J. Simpson attorney Robert Shapiro began giving DNA Cologne away as an in-your-face gag in 1994, months before the infamous criminal trial started, he was roundly criticized by the press for being insensitive. On the eve of the announcement in June 2000 that a rough map of the human genome had been jointly completed by Celera Genomics of the PE Corporation and the federally funded Human Genome Project, *The New York Times* advised its readership, which it assumed had already mastered computer speak, that "it's time to start learning a new dialect," that of "genomese." The paper listed a few examples, including phenotype, genetic marker, and ATGC.[64]

Of late, DNA also seems to be driving the plots and themes of Hollywood blockbusters. *USA TODAY* suggested in 2002, in fact, that DNA seems to figure into the formula for success of current films, noting, in particular, the success that year of *Spider-Man* and *Star Wars: Attack of the Clones*. "Once a staple of paranoia cinema, radioactive monsters no longer rule the silver screen," the paper explained. "DNA now makes the creatures, and the heroes, in Hollywood."[65] When Sony Pictures released a new *Godzilla* in 1998, according to the paper, the film did not generate the box-office receipts expected by the movie's producers, and certainly not those of the three *Jurassic Park* movies. *Jurassic Park*, of course, blended dinosaur mania with DNA technology to resurrect the creatures in the movie. *Godzilla*, on the other hand, stuck with the original explanation for the monster's existence, exposure to radiation. In the original 1962 comic-book version of *Spider-Man*, the hero is bitten by a radioactive spider. For the 2002 incarnation of *Spider-Man*, however, scriptwriter James Cameron changed the radioactive spider into a spider that had been genetically engineered.[66]

DNA is now a mind-set, even a worldview. It is an analytical tool for studying, well, anything. One brand agency (which goes by the lower-case name nickandpaul) actually went so far as to trademark the notion of "Brand DNA" to describe its approach to assessing the health and marketability of its clients' products. As Paul Bennett of nickandpaul explained to a reporter, "A good child needs good DNA to succeed, and so does a product."[67] Think tank Walker Digital Corp. of Stamford, Connecticut, was pursuing a patent in 2000 for a business idea known as "Retail DNA," described by *The Wall Street Journal* as "a software-driven method for adding incremental sales of french fries, shakes and the like at fast-food restaurants."[68] There is in architectural design, too, an essential "design DNA," according to Pilar Viladas, writing in *The New York Times Magazine*.[69] It seems clear that the father of memes, Richard Dawkins, would wholeheartedly agree, having cited architecture as an example of "the extended phenotype," or DNA's "effects on the world at large."[70] Emeritus professor of art history at Yale University George Hersey likes to "call a blueprint a genome," intentionally reversing the standard description of the genome as a blueprint.[71] Pondering the health of capitalist economies around the world, *The Wall Street Journal* concluded that the degree to which capitalism flourishes or languishes "can be traced to the different traditions that emerged in England and France in the 12th century and spread throughout their colonies." (English common law versus French civil law.) What really mattered, as the *Journal*'s headline described it, was "The Legal DNA of Good Economics."[72]

In recent years, "DNA" has similarly become a convenient shorthand for the "essential nature" or "basics" of something. Speaking of baseball, for example, journalist and author David Halberstam commented in 2001, "It's in the DNA of the country."[73] The magazine *Piano Today* carried a cover story in 1998 promising that the writer would be "uncovering music's DNA."[74] "Main Street. Route 66. TV dinners. Playboy. Elvis," writes Rob Owen of *The Pittsburgh Post-Gazette*. These are among the "cultural icons that have become part of this country's DNA."[75] William Shakespeare's works, of course, are commonly referred to as "the DNA of English literature." DNA has even been dragged into the gender wars, as suggested by the title of an op-ed piece by columnist Maureen Dowd, "Decoding Male DNA: Whassup?"[76]

On occasion, of course, a DNA comparison actually has some substance, if only to the extent that it may be a reference to a process with a forensic application. There is, for example, according to book reviewer Ellen Clegg of *The Boston Globe*, a "DNA of the written word" or "a sort of literary DNA" arising from the idiosyncratic manner in which writers write. The differences in penmanship, syntax, spelling, and punctuation are so distinctive, in fact, as Clegg's book review of *Author Unknown: On the Trail of Anonymous*[77] by Don Foster makes clear, that the study of the written word has given rise to a new forensic discipline, forensic text analysis, which combines literary criticism and computer-aided text comparisons. Indeed, Foster is one of the discipline's pioneers and was contacted by the FBI to assist in their analysis of the Unabomber manifesto written, it was later learned, by Ted Kaczynski.[78] When the manifesto was published in *The Washington Post,* it was, in fact, Kaczynski's brother, David, who quickly recognized the text as that of his misanthropic brother (whose prosecution, incidentally, was shored up by, among other things, actual DNA analysis).

1 The DNA of DNA

THE VERY FIRST TIME THAT DNA TYPING WAS USED IN A CRIMINAL INVESTIGA-
tion was at the request of Scotland Yard. The year was 1986. It was during the investigation of two brutal strangulation murders (two-and-a-half years apart) that had
occurred in two neighboring villages about a hundred miles north of London. The
victims, Lynda Mann and Dawn Ashworth, were 15-year-old girls who had also been
raped in the attacks. The savagery of the killings sparked a sense of community outrage that is rare in a world that has grown numb to horrific news.

Early in the investigation into the second slaying, on August 8, 1986, the police arrested a 17-year-old kitchen porter at a local psychiatric hospital who had a history
of aberrant sexual behavior. After a long interrogation, Richard Buckland made a
graphic confession, providing details apparently unavailable in the newspapers. The
second case, at least, seemed open and shut. In an effort to link Buckland to the earlier killing, preserved semen samples from both cases, along with a blood sample
from Buckland, were sent to Dr. Alec Jeffreys at Leicester University, who had developed a process called "DNA fingerprinting" (which soon gave birth to Cellmark Diagnostics Corporation in the UK and the United States) using "multi-locus" probes.
The technique had become well known in England after Jeffreys had used it in a
highly publicized immigration case where a question arose about an individual's relationship to a family in England. Dr. Jeffreys devised, in essence, an entirely new
method for establishing familial relationships (paternity) using DNA. Dr. Jeffreys's
laboratory was less than ten miles from where the murders took place.

Jeffreys stunned everyone when he reported back that Buckland was *not* the
rapist/murderer. At the same time, Jeffreys reported his conclusion that both girls
had indeed been raped by the same man. Buckland then became the first person in
the world to be cleared with the help of DNA typing. (Some investigators now believe that Buckland simply stumbled upon the crime scene of the second victim after
the killing, which accounts for his knowledge of certain unpublicized details.)

Frustrated by a long series of dead ends in their investigation, the police finally announced a campaign of voluntary blood testing for all the men in the two communities between the ages of 13 and 34. A total of 5,511 men eventually came forward to (more or less) voluntarily give blood samples. Only one of those called upon to give a sample stood on his rights and refused to do so, but the police had already ruled him out as a suspect.

When he received notice of the blood-testing campaign in 1987, Colin Pitchfork (true name), who worked at a local bakery, expressed to his wife his reluctance to comply with the request because he had a police record for exposing himself in public. As luck would have it, Pitchfork was able to convince a very gullible co-worker, Ian Kelly, who lived outside the range of the blood-testing dragnet, to submit blood in his stead using a false identity card. Consequently, Pitchfork's blood type (the authorities were led to believe) didn't match, and he was excluded early on. Only those samples with the right blood type—Type A—went on to be subjected to DNA analysis. Unfortunately for Pitchfork, Kelly eventually bragged about the blood switch while out drinking, and a woman who overheard the boast called police. Upon his arrest in 1987, Pitchfork readily confessed to both killings, and subsequent DNA tests tied him to the crimes.[1] For his contributions to science, Dr. Jeffreys was knighted in 1994 by Queen Elizabeth.

Since its introduction by Jeffreys, "DNA fingerprinting" has evolved. (Note that many scientists continue to prefer such terms as DNA typing, DNA profiling, DNA analysis, or DNA matching to avoid any undue association with the methodologies used in fingerprint identification.) It has been refined to improve its power of discrimination, increase the chances that analysts can retrieve usable DNA, and speed the entire process (through automation). The essential idea, however, is that one person can be distinguished from all others by examining repetitive DNA sequences, has not changed. As the technology evolved and scientists came to understand more about the genome, the repetitive segments grew shorter, and a variety of new and interesting genetic sequences were uncovered that made DNA identifications not only more certain but more and more accessible to analysts, no matter what shape the biological sample was in. In the early days of DNA analysis, samples that were either small or old or had been left to the mercy of the elements for too long (in other words, degraded) were beyond the reach of DNA-typing techniques. Over time, however, the addition of new technologies and refinements in retrieval and processing have made the untestable sample very much the exception. Nowadays, the saliva found on a coffee cup, or on a cigarette butt, or even on the back of a licked postage stamp, is likely to provide more than enough genetic material to perform even the most sophisticated and discriminating of DNA tests.

Chromosomes—which are housed in the nuclei of cells—carry the genes. Each chromosome carries thousands of genes, which determine the physical characteristics of an individual. The position or location of a gene on a chromosome is called a gene locus. Because chromosomes occur in pairs, the genes that they carry are likewise paired. Thus, every individual has a pair of genes determining each inherited genetic characteristic.

DNA (deoxyribonucleic acid) is the blueprint of life. The entire DNA molecule is coiled tightly in little bundles that make up the chromosomes. Every normal human being has 46 chromosomes, 1 pair of sex chromosomes (two X chromosomes for females; an X and a Y for males), and 22 pairs of autosomes (the term used to describe any chromosome that is not a sex chromosome). All humans receive 22 autosomes plus one X chromosome from their mothers and 22 autosomes and one X or Y chromosome from their fathers. Of the estimated ten trillion cells in the human body, DNA is contained in the nucleus of all of them. Some cells, however, have no nucleus, such as red blood cells, and thus contain no DNA. (All cells that have a nucleus are called nucleated cells.)

The DNA in the nucleus of any single cell contains all the genetic information necessary to build the entire human being from top to bottom. The reason for this is that from the moment of conception the 46 original chromosomes are faithfully reproduced each time cell division occurs, whether the cell division is aimed at growing to maturity or the replacement of exhausted tissue. We are, thus, genetically identical from head to toe. As with any hard and fast rule, however, there are, of course, exceptions: Sperm cells only have half the normal complement of DNA (that is, 23 chromosomes). The story is more complicated for eggs, but they, too, ultimately contribute only 23 chromosomes to the next generation. The reason that DNA typing can successfully be performed on sperm cells is that when many sperm are analyzed together, the entire genetic makeup is represented.

It is estimated that if fully stretched out, the DNA molecule in humans would be roughly six feet in length. Since the DNA molecule is identical in every nucleated cell in a person's body, each cell contains a perfect copy of the master blueprint for the whole body. Significantly, each copy of the blueprint will remain unchanged throughout a person's life, being the same in every detail at birth as it is at death (except for rare mutations or cancer cells).

Although DNA was first detected in 1868, the fundamental structure of DNA wasn't identified until James Watson and Francis Crick (with the help of exquisite X-ray photographs of crystallized DNA produced by Rosalind Franklin) did so in 1953, for which they received the Nobel Prize. The structure they identified is that of a double helix resembling a coiled rope ladder, the supporting sides of which consist of deoxyribose sugar and phosphate. Between the sugar-phosphate supports are four alternating organic bases, adenine (A), guanine (G), cytosine (C), and thymine (T). These are the rungs of the ladder and are formed when two of the four building-block bases, or nucleotides, chemically bond to form a single base pair. Due to their chemical nature, A can only pair with T, and C can only pair with G. There are thus only four possible combinations of base pairs: A-T, T-A, C-G, and G-C. These combinations repeat endlessly as the rungs of the seemingly infinite rope ladder.

Although the affinity of the nucleotides for one another exists, the attachment is weak. Consequently, the two sides of the DNA molecule can be easily loosened by, for example, simply raising the temperature to near the boiling point of water or immersing the DNA in an alkaline solution. This is known as "denaturing" the DNA molecule and amounts to the unzipping of the genetic code, like cutting the rope ladder right down the middle. This is what occurs naturally each time a cell divides: The DNA strand comes apart and free-floating nucleotides attach to their complements (that is, A's with T's and G's with C's), thus creating two sets of identical strands— one destined for each daughter cell.

The entire set of genetic instructions for humans is referred to as the human genome and consists of approximately 6 billion base pairs, and only about 0.1 percent of that (approximately 6 million on average) might differ between any two individuals.[2] The objective of forensic DNA analysis is to detect and compare the similarities and/or differences between two DNA samples. With the exception of identical twins, each person's DNA in theory has identifiable differences. The various genes or DNA sequences randomly punctuate the chromosomes at predictable locations, or loci (pronounced "low-sigh," from the Latin word for "places"). Some loci are well known, such as the locus for cystic fibrosis on chromosome 7, or that for sickle-cell anemia (the disease resulting from a single nucleotide substitution in the beta-globin gene) on chromosome 11. Since human chromosomes come in pairs (except for the sex chromosomes in men), there will be two copies of any given gene or DNA sequence at a particular locus in normal people. Altogether, there are tens of thousands of genes occurring along the expansive human genome, although the exact number remains a subject of considerable debate. When the completion of a rough map of the human genome was jointly announced in June 2000 by the federally funded Human Genome Project and the private company Celera Genomics, it was concluded by both teams that the human genome consisted of between 30,000 and 35,000 genes, far fewer than many scientists had predicted. Since the announcement, a number of prominent scientists (among them, William Hazeltine, the chief executive officer of the biotechnology company Human Genome Sciences) have argued that the intense race to complete the mapping of the human genome left tens of thousands of genes unaccounted for. These scientists have called for, in effect, a recount. The true number of human genes, according to these scientists, may well be two, three, or perhaps even four times as many as reported by the federal project and Celera.[3]

Scientists refer to genetic variations at a particular locus as alleles. Each individual has two alleles at a given locus (the same place on each chromosome in a pair), one inherited from each parent, and the alleles may be identical or entirely different. If they are identical (meaning the same exact DNA sequence was inherited from both parents), the locus is called homozygous. If different DNA sequences make up the locus, it's called heterozygous. .

Before Alec Jeffreys introduced his novel DNA-fingerprinting technique, there were a number of other genetically based identification techniques available to forensic scientists. But these sought to isolate the expression of the gene, known as the phenotype or gene product, rather than examine the nucleotide sequence that dictated that particular phenotype. The nucleotide sequence for a specific phenotype is known as its genotype. But it gets a little trickier. In some cases, for example, one of the alleles a person inherits is dominant over the other one. This means that when a dominant allele is paired with a recessive allele, only the characteristic dictated by the dominant allele will be expressed. Conventional serology can only demonstrate that the expressed trait is present.

In one of the most significant forensic tests ever developed, the ABO blood-grouping test, there is an even subtler complication. In the ABO blood-grouping system, there are three different genetic characteristics: A, B, and O. Since each of us has two alleles for blood type, one allele inherited from our father and the other from our mother, there are six possible combinations of alleles: AA, AO, BB, BO, AB, and OO. The blood groups are, in fact, co-dominant (not dominant/recessive) and their detection by conventional serology focuses on the antigens that are phenotypically expressed. With Types O and AB, this is straightforward. With the remaining blood types, however, it gets confusing. The genes for A and B add a chemical to pre-existing "H" antigens that are on all blood cells; the gene for O does not add anything. This makes it impossible to detect the presence of Type O in the presence of Types A or B. In practice, therefore, it is impossible to distinguish between Types AA and AO, for example, by traditional serological blood-typing methods; both will be identified as Type A. The O of the AO will be missed.

DNA analysis targets the genotype rather than the phenotype, of course, thus avoiding the shortcomings of the traditional serological methods. This superior ability of DNA analysis prompted some people in the law enforcement community to question the utility of even doing traditional ABO blood grouping now that DNA typing is readily available and will avoid equivocal results. With serious budget constraints and the crushing backlog of cases that many state and local forensic science laboratories face around the country, the argument was advanced that it is a waste of resources to continue doing traditional ABO testing. At first, this was extremely controversial because ABO testing is a long-standing, reliable, and legally unassailable method. During the furor, the Broward County (Florida) Sheriff's Office crime laboratory, among others, embraced the slogan "Just say no to ABO." When it decided in 1996 to discontinue traditional ABO testing in favor of DNA testing.[4]

Genetic variations that are widely distributed in the population can now be detected at their most fundamental level. Many different alleles can exist within the general population and be potential contributors to a particular locus. The resulting genetic differences among humans that stem from those loci are called polymorphisms. These are what make people different and DNA typing possible. Scientists have identified thousands of polymorphic regions amenable to DNA analysis.

Highly polymorphic loci (those genetic sites that differ the most among people) are usually the general target areas of DNA-typing tests. DNA typing endeavors to isolate certain highly polymorphic loci (or aggregate of loci that, collectively, show great variation) and distinguish among the alleles that can exist there. Since the vast majority of the human genome is exactly alike for any two individuals, these tiny differences are sought by scientists to set one individual apart from all the rest. This aim is complicated by the fact that even unrelated people can share the same alleles at polymorphic sites. It is for this reason that scientists look at a number of loci before attempting to assess just how different one person is from another.

As noted, the protein-coding regions of the genome are the genes. But it is the non-coding regions that generally interest forensic scientists performing current methods of DNA analysis. Most DNA tests focus on vast non-coding regions of the DNA molecule; these areas are sometimes referred to as "junk DNA" because they serve no known biochemical role. (One of the few exceptions to this is PCR-based HLA DQA1 typing, which is also used in medicine as a method for detecting specific tissue types related to immunity.)

The breakthrough in DNA technology, the DNA fingerprinting technique pioneered by Sir Alec Jeffreys, analyzes regions of the genome known as variable number tandem repeats (VNTRs), which constitute one component of "junk DNA." Discovered in 1980, VNTRs are identical bursts of DNA interspersed throughout the genome that repeat in tandem a varying number of times (which creates the allelic variation). A particular DNA segment may occur once along the genome or it may repeat multiple times. The number of base pairs associated with different VNTRs varies considerably, and they are classed accordingly. One type of VNTR is known as a hypervariable minisatellite region; these VNTRs range from several hundred to several thousand base pairs in length. The core repeating unit is typically 15 to 35 base pairs long (although the boundaries can range from 8 to 80 base pairs[5]). Another class of VNTR is known as variable microsatellites, so called because the core repeating segments are only a few (two to seven) bases long. This latter class of VNTR is commonly referred to as short tandem repeats (STRs). STRs (the progeny of Jeffreys's discovery) have several advantages over RFLP. For one thing, they are smaller (usually less than a few hundred bases long) and are thus amenable to DNA amplification (PCR). In practical terms, this means analysts can work with far smaller biological samples (such as tiny blood spots or even single hairs with an intact hair root). STRs are also more numerous than their longer counterparts.

A repeating segment is like a line of identical boxcars on a railroad track, each boxcar constructed of however many bases it takes to manufacture that particular model. The number of times the boxcar is repeated depends upon the person; it might be a few times, or it might be a few hundred times. This wide variation of repeats accounts for the size differences among different alleles, which is how RFLP or STR analysis differentiates among individuals. In other words, the forensic scientist is looking at a "size allele," or repeat-sequence allele, when performing RFLP or STR analysis. Single-locus polymorphic sites are only at a single locus on a pair of homologous, or matched, chromosomes, while multi-locus sites are at loci scattered throughout the chromosomes. (Multi-locus analysis has historically been most widely used in paternity determinations, although many commercial DNA testing laboratories have recently switched to STRs.)

PCR technology is a sophisticated chemical manipulation that takes advantage of DNA's natural ability to replicate itself. PCR can faithfully reproduce millions of copies of specific genetic segments. The singular importance of the technique is that it enables forensic scientists to genetically analyze extremely small evidence samples or samples that are highly degraded (i.e., old). PCR provides the raw genetic material for a variety of forensic techniques. PCR has been aptly described as "molecular photocopying" since it amplifies the available DNA until there is sufficient genetic material for analysis. PCR also permits the generation of sufficient material for any number of retests; such confirmatory tests can be performed at another law enforcement laboratory or by an independent laboratory hired by a defendant. It is axiomatic that the hallmark of reliability in DNA typing, as with any scientific analysis, is reproducibility. After an extensive study of forensic DNA typing, the Office of Technology Assessment (a research arm of Congress) wrote in 1990 in a published report, *Genetic Witness: Forensic Uses of DNA Tests:* "Reliable tests must perform reproducibly within a laboratory, across many laboratories, and in the hands of different practitioners."[6]

Some PCR-based tests are more like traditional genetic or blood tests in that they can simply give yes/no responses to specific questions (that is, are the targeted alleles present?). The earliest PCR-based tests that were developed targeted genetic information with far fewer alleles than occur in VNTRs. Consequently, many people share the alleles. For that reason, these tests have much less discriminating power than tandem-repeat tests. More recently, however, scientists have targeted segments of the genome with a greater ability for differentiation, including STRs and D1S80. Enhanced tests of the yes/no variety have also been introduced, such as the Polymarker test. One serious drawback of PCR is that the potential for contamination is ever present with the technique. Conceived by Nobel–prize winner Kary Mullis and developed by Cetus Corporation, PCR was first brought into the courtroom by Forensic Science Associates, a private laboratory.

The underlying principles of DNA typing have reached not merely general acceptance but virtual universality of acceptance. "There is nothing controversial about the theory underlying DNA typing," early detractors of DNA typing William Thompson and Simon Ford admitted as early as 1989. "Indeed, this theory is so well accepted that its accuracy is unlikely even to be raised as an issue in hearings on the admissibility of the new tests."[7] "Given the variation in DNA sequence among individuals," reported the Office of Technology Assessment in *Genetic Witness* in 1990, "no scientific doubt exists that technologies available today accurately detect genetic differences." In 1996 a report published by the National Research Council (a research arm of the National Academy of Sciences) entitled *The Evaluation of Forensic DNA Evidence* forcefully echoed earlier reports about DNA typing in general: "The technology of DNA profiling and the methods for estimating frequencies and related statistics have progressed to the point where the reliability and validity of properly collected and analyzed DNA data should not be in doubt."[8] As a result of such assessments, DNA typing has skyrocketed.

It is fitting—given the inspiration for forensic DNA typing—that DNA paternity testing has burgeoned in the years since Alec Jeffreys's pioneering work. Paternity testing is one of the principal uses of forensic DNA analysis. By 1998, people living in about 25 major cities in the United States probably couldn't help noticing the distinctive billboards of Houston-based Identigene, Inc., a company specializing in paternity tests. The ads asked, "Who's the father?" in large pink letters set against a black background, and offered the curious the telephone number 1-800-DNA-TYPE. The ads, which also ran in newspapers and magazines, started appearing in 1995 just as O. J. Simpson began his trial on charges of double-murder (a lucky break, the company insists) and attracted a good deal of attention. (In New York City, the ads drove around Manhattan on top of taxicabs.) The company logo, a thumbprint with a sperm's tail, is elegantly clear about the company's mission.[9]

While genetic-marker tests have been available in paternity disputes for decades, there has been a dramatic surge in the number of paternity tests conducted in the United States since the more refined DNA tests became available. Paternity tests shot up from about 88,000 in 1988 to roughly 247,000 in 1998 alone.[10] In 1999, according to the American Association of Blood Banks, 280,000 paternity tests were performed in the United States and established that 28 percent of the men tested were not the fathers.[11] (A reported 300,000 paternity tests were conducted in 2000.[12]) The reasons for the upswing are varied. Since 1988, DNA paternity testing has evolved right along with forensic DNA testing, and now the testing procedures are virtually identical when automated STR systems are used. (Some paternity-testing laboratories, however, such as DNA Diagnostics Center of Fairfield, Ohio, have opted to continue using RFLP-based testing because of its greater discriminatory power, rather than switch to PCR-based STRs.[13]) DNA tests are now much more precise, more automated, and far faster than the earliest versions. As a consequence, paternity tests are far less expensive than they once were. Moreover, welfare laws require DNA tests when issues of paternity arise, and it is difficult to imagine a paternity suit getting very far without one, even though it has not been that long since judges or juries were actually expected to render paternity decisions by, among other things, comparing the facial features of a defendant and the child in question, who was sometimes literally held up before the jury box for inspection. There has, of course, also been a great deal of advertising and consciousness raising.

The most fundamental reason that paternity testing is a growth industry, however, is the accuracy of DNA testing. In all but a tiny minority of cases, DNA test results are deemed the final word on "Who's the Daddy?" There are, of course, exceptions. When Ralph von Weingarten was sued for paternity in 1992, his brother Ross, who happened to be his identical twin, testified that he had also had sex with the plaintiff, a claim that, even though the woman ultimately prevailed against Ralph, could not be satisfactorily resolved with current DNA tests in a paternity action. In the future, highly refined DNA tests (focusing on single nucleotide polymorphisms, SNPs) might be able to determine which of the twins had fathered the child—since mutations that differentiate the two could theoretically be detected by sequencing the DNA—but determining paternity in such a case with currently available DNA testing methods would be next to impossible. Such problems with twins are likely to recur because fertility treatments for women have spiked the number of twins born yearly in the United States, both fraternal and—for reasons that are not entirely clear—identical twins. And the von Weingarten twins are not the only pair of identical twins prone to mischief, identity swapping, and blaming the other sibling for a misdeed.[14]

High-profile cases involving paternity (and sometimes maternity) issues fill the headlines and run the gamut from the straightforward to the bizarre. Sometimes the DNA tests simply exclude the individual. Such was the case—although a bit circuitously—when then-President Bill Clinton was accused in 1999 of having fathered a boy, by this time 13, in Little Rock, Arkansas, after allegedly visiting a prostitute in the mid-1980s while he was serving as the governor of that state. The timing of the story was particularly bad for President Clinton, coming as it did directly on the heels of the Monica Lewinsky scandal involving perhaps the most famous DNA tests ever performed. "CLINTON PATERNITY BOMBSHELL," screamed the front page of *The New York Post,* "DNA test will determine if prez is father of teen." The story detailed the long-standing allegations of the mother.[15] As things turned out, however, President Clinton never had to submit to a round of paternity tests, and he was never sued. By this time, the highly sophisticated DNA tests conducted by the FBI laboratory in the Monica Lewinsky scandal were so well publicized that he didn't have to. It was only necessary that the boy and his mother submit to tests and then compare their results with those of the president, results contained in the report submitted to Congress by Independent Counsel Kenneth Starr. Such a comparison was arranged by *Star* magazine, a national tabloid, and undertaken at a Tennessee laboratory hired by the magazine. The information in the Starr Report was sufficient to exclude President Clinton as the father of the boy.[16]

Because of the increasing sophistication of DNA-typing technology, there now exist a number of DNA-typing methods to examine paternity questions short of the standard DNA paternity test. In fact, Charles Lindbergh was excluded in 2000 as the father of a man, Robert Dolfen, who had believed for more than 60 years that he was, quite possibly, Lindbergh's kidnapped son who had somehow survived the kidnapping ordeal and wound up being raised by Glendora Dolfen in Ohio. Glendora apparently believed—and led her son to believe—that a bizarre mix-up had occurred on a trip she had taken to New Jersey with her then-four-year-old son around the time of the kidnapping. For a number of reasons, including physical similarities between the Lindbergh baby and Glendora Dolfen's boy, similarities the neighbors thought uncanny, New Jersey State Police actually investigated the claim in 1936, dismissing it, however, when they found that the Dolfen boy's fingerprints didn't match those retrieved from toys belonging to the Lindbergh baby. Even so, questions about Robert's parentage lingered until the end of the century. Thus, in the year 2000, an analysis was performed on mitochondrial DNA (which is maternally inherited and is packaged in the body's cells separately from the nuclear DNA). Robert's mitochondrial DNA was compared with that of his niece (Glendora had died by this time), and it was determined that the maternal lineage was unbroken.[17] It was not necessary to exhume Glendora for the DNA analysis because of the nature of mitochondrial DNA, which is faithfully inherited from woman to child. (There is some controversy about this, however.) The maternal line rules for this type of DNA. The mitochondrial DNA of Robert and his niece would not have matched, as they did, if Robert were not Glendora's son; the mitochondrial DNA inherited by Robert's sister, and thus by her daughter (Robert's niece), would have been identical to Robert's only if Robert were his mother's son.

In an even more unusual case, both the science and the underlying facts were complicated. When the seaplane of American multimillionaire Larry Lee Hillblom, age 52, crashed into the Pacific Ocean near Saipan in 1995, his death set in motion, as *The Wall Street Journal* described it five years later, "one of the world's most tangled paternity cases."[18] Mr. Hillblom had been a co-founder of air-courier giant DHL Corp. (the "H" stands for Hillblom) and had amassed a $600 million fortune in business holdings. He also had a penchant for having sex with teen-aged Asian girls (virgins especially) and within weeks of his death young women in Vietnam, the Philippines, and the islands of Micronesia produced six children they claimed were fathered by Hillblom. Although Hillblom had drafted a will leaving almost his entire fortune to charity, specifically a trust to further medical research at his alma mater, the University of California, he neglected to specifically disinherit anybody, which is a requirement under the laws of California, from whence Hillblom hailed. He had never formally acknowledged any of the children in question but reportedly assisted financially with the schooling of one Palaun boy, born in 1984; the boy's birth certificate identifies him as "Junior Larry Hillblom," on the insistence of the boy's mother.

The ensuing paternity actions stalled in the Saipan courts when no usable biological material from Larry Hillblom, Sr., could be found for DNA testing to determine which children, if any, were entitled to a share of Hillblom's estate. Hillblom's body was never recovered, and by the time investigators arrived at his Saipan mansion to look for hair or blood or skin they found the place antiseptically purged of all traces and his clothing and personal effects (including combs and toothbrushes) buried in the backyard, which had destroyed their evidentiary value. Initially, both Hillblom's charitable trust and DHL fiercely contested the paternity claims, and the years of legal caterwauling eventually drained $40 million in legal fees from the estate. For a time, since none of Hillblom's living relatives would cooperate with the testing, all hopes for a DNA sample resided in a preserved mole taken off Hillblom's face years earlier at a hospital in San Francisco. Unfortunately, the hospital balked at producing it and then later produced a mole that they said was not Hillblom's after all, due to some sort of mix-up.[19] The judge in Saipan grew impatient and called in an unusual consultant, Dr. Charles Brenner, a Ph.D. in number theory, who calls himself a forensic mathematician. By this time, the DNA profiles of the children had been compiled; there just wasn't any DNA from Hillblom to compare them with. To compensate, Dr. Brenner devised a computer model to compare the children's genetic profiles with one another and he concluded that four of the children had to have the same father, a finding that clearly put the advantage in these children's corner.[20] Later, to cement the findings, a deal was struck with Hillblom's mother to obtain a sample of her blood for comparative purposes in exchange for $1 million and a French villa. The final DNA results brought a settlement that made a four-year-old Vietnamese boy, two Philippine toddlers, and Junior Larry Hillblom very rich, with each receiving $50 million in cash and property. (The charity still got $200 million, and the rest went to taxes.) The windfall made Junior Larry Hillblom a very confused 15-year-old, one who, like his father, was very much preoccupied, according to *The Wall Street Journal,* with teen-aged girls.[21] One hopes he grows out of it.

In the United States a question of paternity for a celebrity or high-profile civic leader is a bonanza for television and newspaper journalists. Even before Lucas Maurice Morad Jagger was born in New York City on May 18, 1999, to 29-year-old

Brazilian model Luciana Morad, the story was a big one both here and abroad. A Brazilian gossip magazine, *Caras,* was the first to publish exclusive photos of mother and child a few months later,[22] just as Luciana's New York attorney, Raoul Felder, was shifting into high gear in the paternity suit he had filed against the putative father, Mick Jagger. For Jagger, 56, the year 1999 was a drag. Although his band, The Rolling Stones, was by 1999 in its 37th year of continuous success and was the top-grossing concert band in the world that year, earning a reported $89.2 million, this was only about half of what the band had earned the year before.[23] Furthermore, Jagger's 22-year relationship with Texas model Jerry Hall—with whom he had four children—ended because of the Luciana Morad pregnancy. "IT'S ALL OVER NOW," read the cover of *People* magazine in February 1999, featuring a picture of Hall and Jagger in happier times.[24] "GIMME $50M," blared the front page of *The New York Daily News* on January 16, 1999, announcing the split of Hall and Jagger. Appropriately enough, perhaps, an influential poll of 700 people in the music industry conducted in 1999 by VH1 named 1965's "(I Can't Get No) Satisfaction" by The Rolling Stones the greatest rock 'n' roll song of all time.[25] Reportedly, Jagger, a British subject, maintained that he was not the father of the boy and vowed, even before the child's birth, to take a DNA test to prove it.[26] Felder was happy to take him up on it. "Except for the O. J. Simpson case, DNA has never let us down," Felder was quoted as saying.[27] Ultimately, a DNA paternity test was anything but a downer for Luciana Morad and her son, Lucas. And quite sensibly, Mick Jagger was persuaded by it when the DNA tests confirmed that he was the father of Lucas. On March 15, 2000, Jagger responded simply "Yes" and "I admit" to the paternity questions of Manhattan Family Court hearing officer David Kirchblum, saying further that he was "satisfied with the blood test results" and would forgo a trial on the matter.[28] By the time of Lucas's second birthday a final settlement had been reached and a sense of international calm had returned to rock 'n' roll. Mick Jagger, in fact, according to an agreeable Luciana Morad, was now insisting that Lucas (a U.S. citizen because of his birth in New York) receive his education in England.[29]

With DNA paternity testing, there is no wiggle room, since the standard tests now demonstrate to a near statistical certainty that the individual in question is, or is not, that child's father. "With DNA, we can prove that he either is the father or he's not," Lisa McDaniel of DNA Diagnostic Center told ABC News on *PrimeTime Thursday,* "It is completely definitive."[30] (By convention, however, the probability of paternity is typically stated in paternity actions, when deemed proven, as 99.9 percent, not 100 percent.) Nevertheless, as with Mick Jagger, a great deal can be made in the news of an "admission" of paternity following DNA tests, even though there would be little point in denying it, barring some sort of fraud. The question immediately becomes a matter of support. But even this pressing question can be swallowed up in the social ramifications of the tests. In some high-profile paternity cases, the political consequences can be severe, and not only for the philanderer. Moreover, there are even run-of-the-mill paternity cases that can shake up the legal landscape.

By the time the Reverend Jesse Jackson admitted in mid-January 2001 that an extramarital affair had produced a daughter, 20 months old at the time, the primary questions were political, but they got worse. The DNA results had long been known to Jackson and his family, and he was reportedly already paying more than $3,000 a month in voluntary child support to the child's mother, Karin Stanford, the former executive director of Jackson's Citizenship Education Fund (CEF), a nonprofit, tax-exempt organization. The CEF is one of four tax-exempt organizations founded by Jackson, which also include the Rainbow/PUSH Coalition, Push for Excellence, and People United to Serve Humanity. "This is no time for evasions, denials or alibis," Jackson sanely said in a prepared statement,[31] timed to preempt a story about the child within days of appearing in the weekly tabloid *The National Enquirer* ("JESSE JACKSON'S LOVE CHILD," was the headline). But political questions can easily mutate into legal ones, and it was no different for Jackson, who had "long been hailed as the pre-eminent political and social voice of black America," as *Newsweek* put it,[32] and had twice sought the Democratic nomination for president. Questions arising from the scandal led to greater scrutiny of Jackson's overall relationship with corporate America and calls for tax audits of his organizations.[33] Financial disclosures followed and amended tax returns were filed,[34] all on the heels of the birth of a child. In the end, then, the Reverend Jackson's DNA results triggered an avalanche of problems.

Karin Stanford filed a paternity suit against the Reverend Jackson in an effort to formalize their child's support and visitation arrangements.[35] At least she won't have the burden of proving paternity. "The Jesse Jackson story has made one thing clear," wrote Jill Porter, a columnist for *The Philadelphia Daily News,* "Diamonds are no longer a girl's best friend. DNA is."[36]

DNA testing has been called into service in countless high-profile controversies, but it has also engendered a few. Few controversies over DNA paternity testing have been as heated as when the DNA reveals that "Dad" is not the biological father, after all.

With DNA testing so prevalent, an ever increasing number of men are finding out (usually after a divorce or over a child-support dispute) that they are not the actual fathers of the children they've been raising or supporting. Dubbed "paternity fraud" by Carnell Smith of Georgia, a man who found himself in this predicament, it is an issue that has caused nationwide turmoil. Some men, like Smith, are finding judges willing to rule in their favor and terminate their support obligations. (Smith, who was inspired to get a DNA test by a "Who's the Daddy?" billboard after his former girlfriend sought to raise his financial obligations, subsequently filed a lawsuit against the mother, alleging fraud in an effort to recoup the $80,000 he had paid out over the years in child support.)[37] But more often men in this position are finding that the

DNA revelation is not enough to relieve them of pre-existing child-support obligations. Why? Well, it depends, in part, on the circumstances of the child's birth.

When the children in question were born while the parents were married, many
courts in the country have cited an ancient rule of English common law known as
the presumption of paternity, which provides that children born during a marriage
are presumed to be the husband's for all legal purposes. The only way to overcome
the presumption is for the husband to demonstrate that he is sterile or impotent or
was beyond "the four seas bordering the kingdom" (in the words of William Black-
stone) when the child was conceived. Originating under the Romans, the rule was
embraced in England 500 years ago.[38] In keeping with this doctrine, the U.S.
Supreme Court ruled (five to four) in 1989 that California's presumption of paternity was enough to prevent a man from trying to prove through DNA tests that he
was the biological father of a child born to a married woman with whom he'd had
an affair, thus squelching the man's efforts to gain visitation and other rights regarding the child. To rule otherwise, Justice Antonin Scalia said, would violate the state's
policy precluding "inquiries into the child's paternity that would be destructive of
family integrity and privacy."[39] The presumption-of-paternity rule, however, has
now become the basis for rejecting the arguments of ex-husbands that continued
child support is grossly unfair given the DNA tests they have obtained. Many courts
won't consider the DNA evidence and sometimes add additional reasons for denying the claims. A New York judge, for example, ruled that an ex-husband, who was
prepared to demonstrate nonpaternity genetically, must continue to pay because
terminating support would "stigmatize the child as illegitimate" and "reward the defendant for the poor judgment of the plaintiff at the expense of the innocent
child."[40] Ultimately, most family courts are deciding that maintaining the status quo
is in the "best interests of the child," which remains the guiding principle in such
family matters. The issue is complicated by a patchwork of state laws that set time
limits on how long a man may challenge paternity, which generally range from two
to five years.[41]

For men supporting children born out of wedlock, the best-interests-of-the-child
doctrine and the time limits for challenging paternity are the chief obstacles they
must overcome to be relieved of their child-support obligations. And a DNA test
proving nonpaternity is generally not enough. In April 2001, for example, the
Supreme Judicial Court of Massachusetts ruled—consistent with rulings in most
other states—that both the time limits (one year in Massachusetts) and the best-
interests-of-the-child principle trumped the DNA results the man obtained in 1999
that established nonpaternity. Significantly, the court recommended in a footnote
that the state require all unmarried men to undergo DNA paternity tests before signing paternity agreements,[42] something the plaintiff in the Massachusetts case had declined to do in 1993, because, he said, he believed the mother and couldn't afford the

test, anyway. In 1998, however, he explained, the mother made a statement that suggested that he wasn't the real father after all and that he was, in fact, a "sucker." This prompted the 1999 DNA test.[43]

The controversy has pitted men's groups against women's groups and proponents of traditional families against supporters of nontraditional ones. Intense lobbying efforts began in a number of states to change the laws that set time limits on offering DNA proof of nonpaternity. And there have been successes. In Ohio, on the heels of a lawsuit challenging the presumption of paternity on constitutional grounds, a law was enacted that relieved men of child support when DNA established nonpaternity. In answer to a flurry of legal challenges in Maryland, the state Court of Appeals ruled in 2000, four to three, that there is no longer any time limit in Maryland for a man to challenge paternity through DNA testing and, further, that the best-interests-of-the-child doctrine need not be paramount in paternity determinations. (The ruling upheld a 1995 Maryland law that permitted courts to modify paternity orders when DNA tests reveal nonpaternity.)[44] In late 2002, the Michigan House of Representatives overwhelmingly passed a law targeting cases of "paternity fraud," although the legislation must clear additional hurdles before becoming law.[45] In January 2001, the South Dakota Supreme Court went even further and ruled that a man who obtained DNA proof of nonpaternity could seek reimbursement from his ex-girlfriend for the child support he had paid her.[46] (As of late 2002, less than a dozen states, either through legislation or as a result of legal decisions, allowed men to terminate child-support payments on the basis of DNA tests proving nonpaternity once they were beyond the governing statutes of limitation.) Carnell Smith went on to found the group U.S. Citizens Against Paternity Fraud, which successfully lobbied for a change in the law in Georgia. The group has a Web site, www.paternityfraud.com, which displays the slogan: "If the genes don't fit, you must acquit."[47]

But these are currently the exceptions. "In Genetic Testing for Paternity," a *New York Times* page-one headline read in 2001, "Law Often Lags Behind Science." The article featured a divorced Texas man who discovered, after a genetic test for cystic fibrosis had alerted his doctor to the fact that he could not have fathered one of his children who has the disease, that three of his four children were not biologically his. In Texas, a court ruled he must continue to pay child support for all four children. In November 2002, the Nebraska Supreme Court came to the opposite conclusion reached by the South Dakota Supreme Court a year earlier, refusing to allow an ex-husband to sue his former wife to recoup child support payments. The issue is not likely to go away soon because some estimates put the number of children not related to marital fathers as high as 10 percent.[48]

DNA paternity testing has gained a solid reputation in criminal cases for solving crimes or, at a minimum, identifying the victims of crimes. When human remains are found, even if the remains are partial, DNA tests have often been instrumental in identifying the victim. A reverse-paternity analysis can be performed allowing the unknown DNA (of the remains) to be compared with the DNA of the likely parents or other living relatives to confirm an identification. Even in cases where bodies have never been recovered, DNA tests of this kind have routinely resulted in convictions for murder (assuming there was a big enough bloodstain left behind to indicate a possible death occurred). There are other uses for DNA paternity tests in the criminal context, as well. When an eight-year-old girl surfaced in Puerto Rico in 1997 in the custody of a woman who was being investigated for child abuse, it was subsequently determined that the woman had falsified the child's birth certificate. DNA paternity tests ultimately reunited the girl with her parents, who had not seen her since she was abducted in 1990 as a toddler from their California home.[49] And in what has been called a "dumpster baby" epidemic around the country, DNA paternity tests are being used to identify abandoned infants and their mothers, who are sometimes criminally charged. (The epidemic has led about half the states to pass "safe haven" laws, allowing distraught parents to legally desert the infants at designated places, such as hospitals.[50])

In October 1998, a 24-year-old patient in a Massachusetts nursing home, who had been in a drug-induced coma since 1995, gave birth to a premature baby girl. The child was unquestionably the product of rape, but the police had no suspect because, based on the developmental stage of the baby, it had occurred some five months earlier.[51] Use of a conventional rape kit would have been pointless since any semen deposit would have long ago disappeared. Only a paternity analysis could identify who had raped the 24-year-old patient. Police immediately made clear their plans to request DNA samples from all male employees of the nursing home, something that immediately caught the eye of civil libertarians.[52]

By this time, DNA dragnets, as they are known, were becoming increasingly common in the United States. As in the first DNA case, the Colin Pitchfork case, in England, police initiate a DNA dragnet by identifying an ever widening circle of male suspects and "requesting" a DNA sample for comparison with the DNA evidence they have. The suspects can provide DNA either through a blood sample or, more commonly, a swabbing taken from the inner cheek. Such dragnets are routine in several European countries, such as England, Belgium, and Germany, where police solved the rape and murder of an 11-year-old girl in 1998 after securing DNA samples from 16,4000 area men. In the United States, one of the most expansive DNA dragnets accumulated more than 2,000 DNA samples from men in south Florida as the police endeavored to solve a string of murders of prostitutes. (Although the killer was eventually caught, it was not as a result of the DNA dragnet.)[53]

Although Britain allows authorities to secure a DNA sample from someone based solely on a policeman's suspicions, in the United States a lawful demand for DNA can only be made by warrant after a judge is persuaded by police that the individual is a legitimate suspect. Consequently, without a warrant or court order consistent with the strictures of the Fourth Amendment, police can only obtain a DNA sample from someone if he or she provided it voluntarily (and, typically, a consent form is signed). This is true in Germany, too. But U.S. critics of the practice, and not a few DNA donors, contend that the police pressure people into signing consent forms and, on occasion, unfairly target minorities. Moreover, fierce legal battles have erupted over what is to become of the DNA samples after the cases are solved. When police in Ann Arbor, Michigan, announced their intention to retain for 30 years the DNA samples obtained from a dragnet, only a lawsuit filed on behalf of some of the donors secured the return of the samples. The subsequent assurances of other police departments that samples from their DNA dragnets will be destroyed once the case is solved have not mollified many of the critics, who remain distrustful of the police and deeply concerned over the privacy implications of the process.[54] "You could have a hundred reasons why you don't want to give your DNA that doesn't relate to this particular case," explained attorney William B. Moffitt, a former president of the National Association of Criminal Defense Lawyers, on CBS's *60 Minutes* in 2001, such as "I may have committed another crime that I don't want you to know about."[55]

For Massachusetts police investigating the rape of the comatose patient, a modest DNA dragnet seemed the obvious way to eliminate a long list of potential suspects and eventually unmask the rapist. To this end, they collected DNA samples from the victim's father and 20 male nursing-home employees, which were then compared to that of the infant's and mother's DNA.[56] Eventually, another dozen men also provided samples as the dragnet broadened. When the DNA paternity tests were completed in January 1999, state-certified nurse's aide Israel Moret was identified as the father of the infant and the woman's attacker. Moret was promptly arrested and fired by the nursing home.[57] The DNA tests were performed by Cellmark Diagnostics of Maryland employing the same DNA-typing methods (STRs) used for other criminal cases.[58] Moret had volunteered his DNA sample and continued to deny the crime when confronted by the DNA results. "This case, it is no secret, is turning on DNA," Essex County District Attorney Kevin M. Burke told the press,[59] and his office remained confident of a conviction based solely on the DNA analysis, whatever Moret's position.[60] After a year of thinking about the DNA results, however, Moret pleaded guilty to raping the coma victim and received an 11-year prison term, to be followed by 10 years of strict supervision.[61] "It is often hard to argue against success," William Moffitt remarked about the case. "I suggest to you that if we ran a police state, we would be much more successful at solving crime."[62] A few months after Moret's plea, in May 2000, the comatose woman died. Her severely disabled baby daughter remained on life support.[63]

DNA paternity tests can also help untangle some extraordinarily tangled situations that arise due to scientific advances, but it cannot necessarily resolve them. In

1999, for example, DNA tests revealed that an embryo mix-up had occurred during an in vitro fertilization procedure at a Manhattan fertility clinic, IVF New York, resulting in a white Staten Island woman's giving birth to twins of different races. When the woman, Donna Fasano, gave birth to two boys, one white and one black, the evident foul-up triggered years of litigation for the parents involved. DNA tests revealed that one of the embryos that had been implanted into Mrs. Fasano actually belonged to a black couple, Robert and Deborah Rogers, who immediately filed a lawsuit seeking custody of the child.[64] Unfortunately, the case of the "scrambled eggs," as some in the press were quick to refer to it, soon bogged down in legal problems. When the Fasanos turned over custody of four-month-old Akeil to the Rogerses in May 1999, they asked the Rogerses to sign an agreement that called for the two boys to be raised, in effect, as brothers who would enjoy regular visitation. Before long, the Rogerses found the visits with the Fasanos onerous and were dismissive of Mrs. Fasano's characterization of Akiel as her "son" and the two boys as "brothers."[65] As the Rogerses pressed to get out of the visitation altogether, Mrs. Fasano countered through her attorney that she was Akiel's "gestational parent" and was entitled to the visits.[66] A New York appellate court, however, disagreed with Mrs. Fasano and terminated the visitation in October 2000, saying that Akiel should have been given over to the Rogerses at birth.[67] Similar cases have been reported elsewhere, sometimes involving the fertilization of one couple's egg with the sperm of another couple. In England, dozens of couples were reportedly undergoing DNA tests in late 2000 after it was uncovered that a series of mistakes at two fertility clinics there may have really bollixed things up and in July 2002, it was revealed that a white couple from London had given birth to black twins after an IVF mix-up at a National Health Service clinic. It soon became apparent that the mix-up also involved a black couple who had also sought treatment at the clinic. An initial round of DNA tests soon established that although the white woman was indeed the biological mother of the twins, her husband was not the biological father.[68] What about cases that are less apparent than the Rogers heartbreak or the London case? And how could you possibly sort it all out without DNA tests? And what should be done when a baby mix-up isn't determined until years later?

The latter question, of course, has already been confronted when it has been revealed that babies were somehow mixed up at the hospital and the kids went home with the wrong parents. Perhaps the most famous example is the saga of Kimberly Mays, who learned at age nine that she had gone home with the wrong parents in 1978 after identification tags were accidentally switched at a rural hospital in Wauchula, Florida. Kimberly was born to Regina and Ernest Twigg but went home with Robert and Barbara Mays, whose own daughter, Arlena, went home with the Twiggs. After Arlena developed heart disease in the late 1980s, from which she died in 1988, blood tests revealed that Arlena was not the biological daughter of the Twiggs.

An investigation determined that the Twiggs' biological daughter was apparently living with single-parent Robert Mays (whose wife Barbara died of cancer in 1981 and whose second marriage had ended in divorce). In 1989, Mays agreed to DNA tests, but only on the condition that the Twiggs seek only visitation of Kimberly, not custody, should the tests prove Kimberly was the biological daughter of the Twiggs. When the tests demonstrated that Kimberly was the Twiggs' daughter, a visitation schedule was hammered out, but only five visits with the Twiggs and their seven other children took place before Robert Mays put a stop to them, arguing that the visits were detrimental to Kimberly's performance in school. The Twiggs countered by pressing their visitation rights and eventually seeking custody, which was bitterly contested. As the Twiggs continued to pursue at least visitation, a 14-year-old Kimberly responded by getting a lawyer and seeking to have the Twiggs' parental rights completely terminated. After a high-profile trial in 1993, Robert Mays was awarded custody and Kimberly was allowed to sever all ties with the Twiggs. Things were just warming up, though, because Kimberly later ran away from the Mays household (Mays had remarried), accusing Mays of being monstrously strict and even of molesting her—allegations she later admitted were entirely false. Kimberly went first to live in a youth shelter, then she went to live with the Twiggs, but she ran away from the Twiggs several times and even did a spell in reform school. (The Twiggs themselves separated in 1996.) Eventually, Kimberly managed to patch things up with Robert Mays, and she moved back home, but only long enough to get married to Jeremy Weeks and have her own baby, son Devon, in October 1997, an event that triggered a new series of problems for Kimberly. By all accounts, the marriage (which nevertheless survived) was rocky at the start. At one point, Jeremy obtained a restraining order against Kimberly, and in April 1999 the State of Florida took custody of Devon. Custody of Devon was eventually restored by a Florida judge to Kimberly and Jeremy in April 2000, however, and things were looking up for the young family.[69]

But Kimberly Mays's tumultuous story is not the only one. In 1998, Paula Johnson and her ex-boyfriend, Carlton Conley, underwent DNA paternity tests prior to settling a child-support dispute initiated by Johnson. The tests revealed something startling, that neither Conley, nor Johnson herself, were biologically related to Paula's three-year-old daughter, Callie. In fact, subsequent DNA tests confirmed that Johnson's actual baby girl left the University of Virginia Medical Center in 1995 with (unbeknownst to them) Whitney Rogers and Kevin Chittum and was being raised as Rebecca Chittum. Conley was, in fact, Rebecca's father. Tragically, as tests were being conducted, Rogers and Chittum were killed in a car crash, whereupon Rebecca went to live with Chittum's parents.[70] "In the end," *Newsweek*'s T. Trent Gegax wrote, "the case of the switched babies is another national drama that hinges on a DNA test."[71] But the end would be years away in this drama, all the result of a seemingly routine DNA test ordered by the judge in Johnson's child-support case.

Lawsuits were filed against the state of Virginia, the hospital, its staff, and the company that manufactured the identification bracelets used for keeping track of babies at the hospital. Rebecca's caretakers reached a settlement worth millions for Rebecca in February 1999.[72] In mid-2000, however, a custody agreement was finally reached allowing the children to stay put while permitting ongoing visitation.[73] Eventually, Johnson reached a settlement with the state of Virginia on behalf of Callie similar to that awarded Rebecca.[74] A judge's ruling allowed Carlton Conley to avoid paying child support for Callie because, well, Callie wasn't really his child.[75]

While the Virginia story is a modern story because of scientific advances, it is also an old one, as Janna Malamud Smith noted in a reflection on the case written for *The New York Times:* "How could I not be glad to know my birth?" declared Oedipus in the classic play written 2,500 years ago. But DNA adds new wrinkles to the drama. "Whatever else, it seems that we have some hard thinking to do about DNA testing," wrote Smith. "Science offers us DNA testing, cloning and in vitro fertilization. But as often as these techniques lift us out of one human dilemma, they drop us straight into another."[76]

Even before *Sports Illustrated* ran a special report entitled "Paternity Ward" in 1998, and posed the question on its cover to the world of professional sports, "Where's Daddy?", it was clear that DNA analysis would play at least a peripheral role in the world of athletes.[77] If nothing else, forensic DNA analysis would be used in court against fallen athletes (and even sportscasters) accused of major crimes. O. J. Simpson's criminal and civil cases, in particular, made that resoundingly clear. But, as things were turning out, the role of DNA, in all its myriad forms, was destined for greater things and might even be called upon one day to define what a human athlete is, or is not.

The *Sports Illustrated* story demonstrated that many pro athletes were behaving just like a lot of other celebrities, only more so. Larry Johnson of the Los Angeles Lakers, for example, whose 1995 marriage produced two children, was revealed over a period of years by paternity actions and DNA tests to have fathered an additional three children by three other women. But Johnson is far from alone. There is, according to *Sports Illustrated,* an "NBA All-Paternity team of players who have had children out of wedlock and have subsequently been the subject of paternity-related lawsuits"— namely Larry Bird, Patrick Ewing, Juwan Howard, Shawn Kemp, Jason Kidd, Stephon Marbury, Hakeem Olajuwon, Gary Payton, Scottie Pippen, and Isiah Thomas.[78] Shawn Kemp of the Cleveland Cavaliers has fathered at least seven children around the country, according to one of his attorneys.[79] Since this kind of conduct can be expensive, one NFL player reportedly tried to foil a court-ordered DNA paternity test by sending another athlete in to take the blood test. The plot fell apart, however, when the laboratory took a photograph of the imposter and demanded, not only blood, but fingerprints. The real subject of the test appeared at the lab five minutes later and fessed up.[80] These ongoing paternity scandals involving highly paid athletes have forced the courts to devise child-support payment schedules that do not bankrupt the more profligate athletes after the first three or four children, as happened to baseball's Kevin Mitchell in 1996, even though he was earning $4.5 million a year.[81]

There are glorious moments of sport, too, of course. And, significantly, DNA analysis is beginning to play just as much of a role in the exalted moments as it has been playing in the low. When an Oregon collector of baseball memorabilia put a baseball on the auction block in early 1999 that he claimed was the ball Mickey Mantle hit for his 500th home run (one the man had purchased for $25,000 two years earlier), Mantle's 66-year-old widow, Merlyn Mantle, came forward to say that she was, in fact, in possession of Mantle's 500th home-run ball. This scotched the sale of the Oregon ball by the auction house and left the collector fuming. "I feel sorry for the man who bought that ball," said Merlyn Mantle, who had been married to Mickey for 43 years, "but I also resent having to prove that I have the ball. In fact, I'm amazed that anyone would question it."[82] Unfortunately, the Oregon ball appeared to have a legitimate pedigree (having been sold to the Oregon collector by the Little League Museum in Baxter, Kansas), so neither side backed down. The Oregon collector offered to pay for forensic examinations of the two balls, but Merlyn wasn't budging.[83] Even if such examinations were undertaken, however, how could the examiner be sure? For Mantle's milestone homer, there may be a way to identify the real ball and, then again, there may not be. Nowadays, DNA analysis can take out the guesswork. And there can be a lot at stake with sports memorabilia. In November 1998, an auction record was set when an anonymous collector paid $126,500 for the baseball smacked by Babe Ruth in 1923 on Yankee Stadium's inaugural day; it was the stadium's first home run. In 2000, the most highly prized baseball card in the world, referred to with scientific precision as the 1909 Gretzky T206 Honus Wagner (to distinguish it from an inferior 50 or so others known to exist, 49 of which do not bear a value-enhancing Piedmont cigarette ad on the back as does the Gretzky card), was sold at auction on eBay for $1.26 million. Between 1909 and 1911, the American Tobacco Company produced the T206 series of baseball cards depicting both major league and minor league players. When Pittsburgh Pirate shortstop Honus Wagner got into a dispute with the company, his cards were pulled from the series, making the Wagner cards rare. In 1936, Wagner was in the first class of inductees into the Baseball Hall of Fame. At one point, the near pristine baseball card sold on eBay was owned by National Hockey League legend Wayne Gretzky, which explains why it is identified as the 1909 Gretzky T206 Honus Wagner.[84]

So far, though, the most expensive sports collectible arises out of the 1998 home-run duel between Mark McGwire of the St. Louis Cardinals and Sammy Sosa of the Chicago Cubs. Not since the 1961 single-season home-run race between New York Yankees Roger Maris and Mickey Mantle, when Maris finally beat out the more popular Mantle with 61 homers and simultaneously broke the seemingly inviolate 60-homer record of Babe Ruth set back in 1927, had their been such excitement over a baseball rivalry. And, of course, controversy. The Maris record is controversial because it took Maris more games than Ruth to reach and surpass Ruth's record. And, despite Maris's heroic efforts, Mantle garnered most of the media attention. During the 1998 McGwire-Sosa contest, it was revealed that McGwire was regularly using the dietary supplement androstenedione (among others), popularly known as andro, which is reputed to raise testosterone levels, promote healing, and help build lean muscle mass. Although the National Collegiate Athletic Association, the NFL, and the International Olympic Committee (IOC) include andro on their lists of banned substances, the substance is not regulated by the Food and Drug Administration and is not an illegal substance under baseball rules (nor those of the NHL or the NBA, for that matter), although baseball began actively studying it after the McGwire controversy. The issue grew more complicated early in the 1999 baseball season when White House drug czar Barry McCaffrey began issuing warnings that some andro on the market was adulterated and contained testosterone, which would be illegal since testosterone is a steroid that is only available by prescription. It was at this point that McGwire voluntarily stopped using andro.[85] Whatever the effects of andro on McGwire, he certainly was not slowed down by the controversy in the 1998 season or in the andro-free years to come. By the end of the 2000 baseball season, McGwire's overall ability to hit one home run for every 11 times at bat (actually 10.83)[86] inspired many commentators to call him the greatest home-run producer in history, although some might disagree.[87]

On August 5, 1999, McGwire hit his 500th career home run and became the 16th player in baseball history to do so, and it took fewer times at bat for McGwire than any of his predecessors.[88] In a surprise to many, however, McGwire retired from baseball at the end of the 2001 season.

Mark McGwire's most dramatic contribution to baseball was hitting 70 home runs by the end of the 1998 season to best both Roger Maris and his 1998 rival Sammy Sosa, who ended the year with 66 home runs. The McGwire-Sosa home-run race was listed by the Associated Press as no. 2 on its list of the top ten stories of 1998, second only to the Bill Clinton/Monica Lewinsky scandal.[89] *Time* magazine named McGwire "Hero of the Year" for 1998,[90] and *Sports Illustrated* named him and Sosa "Sportsmen of the Year" for their good-natured competition throughout the 1998 season. One of the biggest winners of the year, however, was 26-year-old research scientist Philip Ozersky, who caught McGuire's 70th homer in St. Louis.[91] Initial reports suggested that the value of the ball might reach $1 million and may not even be worth the tax headaches that would befall its possessor. But Ozersky shrewdly held on to his prize, and it was not long before it was being referred to as the Hope Diamond of sports collectibles.[92] Ozersky did more than hold on to the ball, however, he took measures to protect it. With help from an agent, he contacted Professional Sports Authenticators of Newport Beach, California. Although the ball has other distinctive markings (one of which is visible only under ultraviolet light), the company sought to guarantee the ball's future authentication by marking it with a specially synthesized DNA smear, the profile of which is a carefully guarded secret. (The original batch of DNA was destroyed.)[93] Ozersky decided he wanted to auction the ball off and consigned it to Guernsey's, the Manhattan auction house, and it was scheduled for auction on January 12, 1999. (This was the same auction from which Mickey Mantle's putative 500th homer ball was pulled just before bidding began. The Oregon collector's ball, by the way, was eventually authenticated by Professional Sports Authenticators, tagged with DNA, and sold at auction.) By the time the bidding was over, Mark McGwire's 70th home-run ball had been sold for a record-smashing $2.7 million (plus another $305,000 in commissions) to comic book entrepreneur Todd McFarlane.[94] The following day, *The New York Times* carried a photograph of Philip Ozersky standing next to the pricey baseball. Ozersky was wearing a big grin and sporting a necktie bearing the distinctive pattern of the DNA molecule.[95]

Barry Bonds of the San Francisco Giants surpassed McGwire's home-run record on October 7, 2001, when he hit his 73rd home run of the 2001 season at Pacific Bell Park in San Francisco. For Bonds it was perhaps the ultimate reward for a career characterized by consistently excellent baseball playing. For others it was both a lesson in the fleeting nature of sports records and the enduring nature of disputes over sports memorabilia. The very instant a fan emerged from the roiling mob attempting to claim the ball, soon estimated to be worth more than $1 million, there was controversy. Recorded by television cameras, the ball dispute began when baseball fan Alex Popov appeared to catch the ball out of the air with a baseball glove, but he was immediately swallowed up by a crush of eager baseball fans scrambling for a piece of baseball history. Security guards were still laboring to untangle the human pile-up when, lo and behold, another fan, Patrick Hayashi, emerged from the swarm holding the coveted baseball. Popov sued Hayashi, claiming, among other things, that he, Popov, was, basically, mugged by the crowd. Hayashi just somehow managed to penetrate the mob to snatch the ball. A Superior Court judge in San Francisco subsequently seized the ball, pending a resolution of the dispute.[96] Since neither side would back down from claiming sole ownership of the ball, a trial was scheduled for October 2002 to determine the ball's fate. As the trial date approached, it was estimated that Todd MacFarlane's investment (McGwire's 70th) had declined in value by about 75 percent, almost guaranteeing that Bonds's 73d home-run ball, would become the most valuable item of sports memorabilia in history.[97] Evidently torn himself over the dispute, Judge Kevin McCarthy finally ruled in December 2002 that Popov and Hayashi were both equally entitled to the historic ball. He ordered that the ball be sold and that the proceeds be split, in Solomonic fashion, equally between Popov and Hayashi.

While the controversy continues over whether Mark McGwire's use of andro during the 1998 season somehow diminishes his home-run milestone (just as it does over the extent to which steroids and other performance-enhancing drugs are used in professional baseball generally), at least there will be no controversy in 30 years over who has the real 70th home-run ball. Given all this, it was not long before the DNA identification technology used in his case caught fire in the world of sports. It has become so pervasive, in fact, that it is almost certainly vulnerable to an entirely new racket: DNA counterfeiting. Developed by DNA Technologies, Inc., of Los Angeles, which licenses it to, among others, PSA/DNA Authentication Services (named by its developer, Professional Sports Authenticators), DNA authentication tags come in a few varieties and are now used for an ever increasing number of purposes. In addition to McGuire's 70th, PSA/DNA used the system on Hank Aaron's 715th home-run ball and bat (the milestone, reached on April 8, 1974, was significant because it broke Babe Ruth's record) and the bases and infield balls used in the 1999 and subsequent World Series games. Sometimes the DNA tag is made of an artificial, or synthetic, piece of DNA that is about 20 or more DNA letters, or bases, in length. Once the sequence of DNA letters is established (which would be next to impossible to crack by guessing the order of the letters that are derived from the four-letter DNA code), batches of the strand are reproduced by PCR for insertion onto the item. This was the type of DNA tag used for all Super Bowl game footballs in January 2000 with Super Bowl XXXIV.[98] Given FBI estimates that between 70 and 80 percent of memorabilia is fake, DNA tags have been well-received as an anti-counterfeiting measure.[99]

Another type of DNA tag uses a snippet of DNA from a real person (which, again, is made plentiful by PCR). Artist Thomas Kinkade, for instance, signs his artwork with a special DNA Technologies' ink that has been laced with a piece of his own DNA. To assure the authenticity of autographs, athletes and celebrities have also begun using their own personalized DNA ink.[100]

DNA Technologies got a major boost when the Sydney Olympics organizing committee adopted the DNA tags as a way to authenticate the official merchandise and souvenirs of the summer games in 2000. In this instance, ink impregnated with a DNA strand of an unidentified Australian athlete (perhaps more than one) was applied directly to a product or inscribed on a tag attached to it. A spot check can be made with an ultraviolet light to verify the item is so tagged or, if a further dispute develops, the product can be sent for a DNA analysis to DNA Technologies. The U.S. Customs Service estimates that the U.S. economy loses $200 billion every year from counterfeit merchandise, so the DNA tags may well catch on.[101] Organizers of the Winter Olympics 2002 in Salt Lake were also inspired to examine using the DNA tags, and Warner Brothers studios has used the tags on certain movie props to discourage theft.[102]

Unfortunately, the wide-scale use of DNA tags on millions of items, as was done in Sydney, virtually guarantees there will be efforts to identify the encoded DNA tags in order to make counterfeit goods. It is one thing to use a one-of-a-kind DNA tag on a unique item of sports memorabilia, such as Mark McGwire's home-run ball, because that item will not be in circulation. It is quite another to use the same DNA strand (or even several) to mark millions of products. What enterprising chemist, cursed with the genes of villainy, would not be tempted to take up the challenge of deciphering the DNA tags? Once the DNA snippet is identified and amplified by PCR, a counterfeiter could make a fortune. Anticipating this, of course, DNA Technologies has taken measures to prevent it, including specially treating the DNA in the tags to make it difficult to amplify with PCR and including other snippets of DNA in the DNA ink to confound would-be counterfeiters.[103] Nevertheless, it still seems awfully tempting. After all, if DNA Technologies can manage to amplify the mystery DNA in its laboratory to authenticate a souvenir (very easy for them, to be sure), why can't a bio-crook do the same? However, a more fundamental problem threatens the burgeoning DNA tagging industry: DNA degradation. When DNA is exposed to either the natural environment or chemical insults, such as when an article of clothing is washed with detergent, the bonds that hold the string of letters together break down and the original strand of DNA fragments into smaller pieces. Depending on the degree to which such degradation has occurred, this may well hopelessly frustrate an authentication.

Whether DNA tags survive the test of time or not, DNA analysis promises to play an increasingly important role in future Olympics. The reason is the persistent use by athletes of performance-enhancing drugs that many believe often go undetected. Dozens of athletes at the 2000 Sydney Olympics were expelled from the games, including the entire Bulgarian weight-lifting team, after testing positive for banned substances.[104] Some athletes had their medals, even gold medals, stripped. It is an old problem for the IOC, which continued to lag behind in drug-testing efficacy. "Drugs Taint Games," read the page-one headline of *USA TODAY* in late September 2000 as the games were closing out.[105] "Athletes' Drug Use Outpaces Testing," read the front-page of *The Boston Globe* a few days earlier.[106] Just as the games ended, the U.S. Anti-Doping Agency (USADA) began operations, wresting control of all domestic drug testing from the U.S. Olympic Committee and promising to greatly expand testing. Its counterpart, the World Anti-Doping Agency (WADA), was beginning its operations and would soon take over all testing from the IOC.[107] While some of the doping problems stemmed from the sophistication of many athletes who have learned to finesse drug screenings, some substances, like human growth hormone, could be taken in such a way as to evade current tests. The challenges of the future for the Anti-Doping Agencies, however, were about to get a great deal more daunting.

Olympic doping scandals have a long, even tragic, history. One of the most egregious examples is the pervasive administration of steroids to young athletes in the former East Germany during the 1970s and 80s. Many, if not most, of the athletes had no idea what they were being given by their doctors and coaches. As Steven

Ungerleider recounted in *Faust's Gold*,[108] the doping in East Germany was a system-atic, state-sponsored program that was conducted largely by medical doctors. Its purpose was to demonstrate to the world that a socialist system was better than a capitalist one. All told, 10,000 athletes were fed or injected with steroids during the 1970s and 80s, resulting in catastrophic physiological changes for some of them as they entered adulthood, including cancers, ovarian cysts, liver ailments, excessive hair growth, and permanently deepened voices. The changes later prompted one of the female athletes to undergo a sex-change operation. Some reported birth defects in their children as well, which they ascribe to the doping.[109] During the period of abuse, East Germany racked up a wildly disproportionate 160 gold medals.[110] After the fall of the Berlin Wall, an investigation led to the indictment of more than 400 people who participated in the doping program, which resulted in a series of trials, including those of former high-ranking sports officials, coaches, and doctors, that ended in 2000. During one of the trials, a lawyer barked at a former senior doctor, "You are the Josef Mengele of the [East German] doping system."[111]

As the East German episode reveals, few competitors who take performance-enhancing drugs are ever caught, even where suspicions and accusations abound. While authorities have sought to improve testing for known substances of abuse, new drugs or formulations are continually being thrown into the mix. Until new, more sensitive tests were developed in 2000 for EPO, or erythopoiten, for example, use of the drug simply eluded the IOC's radar. (The tests are still imperfect.) The medical use for EPO is to treat anemia (it regulates red blood cell production), but it is con-sidered a favorite of distance runners and cyclists because it enhances endurance. The general perception remains that the IOC's drug-testing policies and techniques were a failure.[112] This led to the creation of WADA and USADA and a commitment to re-double screening efforts. Unfortunately, this new resolve may well be swallowed up by a new nemesis, one spawned by the age of genetics.

"Gene doping will be the next issue," warns Dr. Patrick Schamasch, medical direc-tor of the IOC's WADA. "Injections of synthetic EPO and human growth hormone will be nothing by comparison. If somebody increases the production of hormones by direct genetic manipulation—what are you going to do?"[113]

Penn State professor Charles Yesalis voices similar concerns about professional baseball, which has yet to fully put to rest its steroid controversy. Yesalis is among sev-eral health experts who have urged those in professional baseball to accept outside, in-dependent monitoring of illegal drug use among players to help restore confidence in the fairness of the game. In fact, in 2002, the U.S. Anti-Doping Agency offered to become that independent drug-testing agency, but professional baseball elected to po-lice itself. "Two, four, six years from now," Yesalis told *USA TODAY,* "[the] whole con-versation [about steroid abuse] could be passé due to genetic engineering."[114]

To confront the apparent imminence of genetically engineered athletes, Dr. Schamasch has sought information from biotech companies working on the cutting edges of genomics and genetic therapies. The IOC and WADA convened several meetings on the subject in 2001, fully recognizing they were entering a legal, ethical, and technical minefield. Genetics experts were among those invited to the conferences and have taken positions on WADA's various committees. Questions addressed included some very basic ones: Should gene enhancement be illegal at all? How safe will it be? What if an athlete is being legitimately treated for injury or disease? And what about the athlete whose genes were altered by his parents when he was an embryo? (This kind of genetic change is known as a germ-line genetic change because it is permanent and will be passed on to future generations; specifically, the germ-line refers to the sperm and the egg. Such enhancements at birth are not currently allowed, but are considered imminent. Interestingly, germ-line genetic modifications have resulted from in vitro fertilization procedures, but they were not induced for purposes of genetic enhancement.) And, as *The New York Times* put it, "Ultimately, at the heart of the issue will be a profound question: what is a human athlete?"[115]

Doctors and pediatricians are already being pressured by parents to make their children taller and more athletic by administering human growth hormone, or to give them steroids and improve their chances of winning college athletic scholarships.[116] The challenge for drug testers will depend on the method of gene enhancement, whether by gene therapy or germ-line genetic changes. Although the promise of gene therapy in humans has been elusive, experiments in mice have shown impressive results. For example, gene therapy experiments, which (usually) use redesigned viruses to introduce genes into a subject, allowed scientists working with a growth factor to transform aging mice into what the scientists subsequently dubbed "Schwarzenegger mice," an effect that lasted for months. Moreover, one experiment with a growth hormone showed that oral antibiotics could be used to turn the gene off or on, a very troubling prospect for human drug testers. To catch a gene doper, highly specific genetic tests will have to be developed for a class of suspect genes, and a tester will have to know where in the body the genes were inserted. To make matters worse, an examination of this sort would seem to require a biopsy, something that would meet with such fierce resistance as to be untenable. Consequently, only noninvasive methods would be practicable, such as a test for a specific chemical marker or a genetic chip designed to signal the presence of an illegal gene sequence.[117] Identifying a germ-line alteration, however, would present additional and perhaps insurmountable problems. The only way to unmask such an athlete would be to cross-reference the athlete's DNA with that of both parents to see if anything was added. So, in the future, should the parents of athletes be required to register their DNA with the IOC for comparative analyses? Should deceased parents be exhumed to get their DNA? Whatever the answers to these questions, many scientists are taking the prospect of genetically enhanced athletes very seriously. In fact, the American Museum of Natural History dedicated part of its 2001 exhibition, "The Genomic Revolution," to considering these very questions. A full-page ad for the exhibit in *The New York Times* offered this glimpse of the future: "Some kids dream of growing up to be just like their favorite player. With the mapping of the human genome, they just might. Genetic enhancement could, in fact, result in a breed of super athletes. Imagine: 150 home runs in a season; the 2-minute mile. But think, just because we can do it, should we?"[118]

2　"No Doubt at All"

O N NOVEMBER 30, 1987, LISA GRAY[1] GOT HOME FROM WORK AT ABOUT four o'clock. A single parent in her mid-30s, she worked as a renewal ratings analyst for one of the many insurance companies in Hartford, Connecticut. Shortly afterward, her live-in fiancé got home, too, and the couple began a discussion that turned into an argument.

"To clear my head," as she put it, she decided to take a walk around the block of her suburban Bristol, Connecticut, neighborhood at about 5:20 P.M.[2]

By then, it was already pitch dark outside. About two thirds of the way around her route, Lisa was startled by a man running up behind her who said "Hi," prompting her to say "Hi" in return. She got a fleeting look at the man as she turned around and sensed she was in trouble.

Suddenly, Lisa was knocked off her feet, which sent her glasses flying (she is very near-sighted), and the man pounced on top of her, pinning her to the ground. He held her there while he looked around suspiciously, then flipped her onto her stomach, bound her hands behind her back, shoved a dirty rag into her mouth, and dragged her to her feet.

Before she knew what was happening, Lisa was being rushed across the street and through a short stretch of woods. She was led across a track field of the local public high school and into a parking lot adjacent to some tennis courts. The lot was nearly vacant, but Lisa saw a car coming. Gagged, she couldn't cry out. As the car neared, the abductor slowed their pace and put his arm around Lisa's shoulder to make it appear the two were lovers out for a stroll. The car never stopped.

After a few minutes, they arrived at a late-model sedan. Fearful of doing otherwise, Lisa confessed through the gag that the rope around her wrists had come undone. Looking around nervously, the man opened the door and hustled Lisa into the front seat of the car.

He forced her to put her head across his lap as he started up the car, pushing her face down whenever she endeavored to look upward. Each reflexive glance, in any event, revealed nothing more than the silhouettes of treetops, telephone wires, and buildings flickering by in an unsettling whirl. Lisa had no idea where she was. After about 10 or 15 minutes of driving, according to her estimate, the car hit a bumpy road and Lisa heard what sounded like branches scraping the car. She remembered that the man looked at a watch dangling from the gear shift, just as he was turning the car onto the bumpy road.

The man stopped the car and instructed Lisa to remove her winter coat and she obeyed. He then ordered her to perform fellatio upon him, unbuckling his pants to prepare himself. He forced her head into his groin with such force that Lisa slashed her ear on his opened zipper. Then he commanded that she remove one leg from her blue jeans and underwear and lie still while he penetrated her. He then crawled out of the car to fix his pants. After it was over, the rapist drove a short distance, all the while threatening to kill Lisa and her son—whom she'd mentioned while pleading with him not to rape her—if she ever breathed a word of this to anyone. Lisa later recalled that the interior of the car was "very messy" and that in the back was a baby seat with tears in it and stuffing coming out.

"I can't believe this man could have a kid," she thought to herself. "He can't be a father doing this to me."

Pulling the car over, the man forced Lisa out the passenger side and gestured toward a main road. He wiped her eyeglasses clean of fingerprints, then hurled them into the dirt where Lisa was standing and drove away with his lights off. Badly shaken, and in unfamiliar territory, Lisa ran to the nearest house and phoned for help. By seven o'clock she was at the Bristol Police Station with her fiancé. The two were quickly ushered into a small room directly behind the front desk, but the boyfriend was immediately asked to leave so Lisa could speak freely. Although she was visibly shaken, between her sobs Lisa regained enough composure to give a detailed description of her attacker.

After the interview, Lisa was taken to the hospital where she underwent a thorough medical examination with a doctor who used a Johnson's Rape Kit, a commercially produced package of medical materials designed to collect forensic evidence in rape cases. The doctor described her appearance that night as "very anxious, disheveled, teary-eyed, very upset." As part of the ordeal at the hospital, Lisa had to leave most of her clothes behind as evidence. After the examination, with her life now completely upside down, Lisa went home.

The next day, Lisa went early to the Bristol Police Department and, with the help of Detective Roy Bredefield, put together a composite sketch of her attacker from a series of pre-drawn overlays. The sketch was published in the local newspaper, *The Bristol Press,* in the ensuing days. A police flyer bearing the sketch described the assailant as "a white male approx. 20 years of age, slender build, approx. 5ft. 7in. tall, weight approx. 140 lbs., having brown hair and a thin mustache." It also noted that the "vehicle involved is an older model chevy impala, dark in color, vehicle was a mess inside."

Approximately a week and a half after the rape, on December 11, two young women, Beth Swanson of Bristol and Tina Hagerman of nearby Terryville, came to the Bristol Police Department with the name of a person who fit the description Lisa gave almost to a tee. It was someone they knew personally, they said, who was acting "strange" and "weird" and had become obsessed with accounts of the rape in the newspaper, and the composite sketch.

The name they gave was Ricky Hammond, a 24-year-old local man. As it turned out, Hammond lived a mere quarter of a mile from where Lisa Gray was freed. As it turned out, too, Hammond was no stranger to the Bristol police.

"We were familiar with the name the two ladies had given us," Detective Brede-field later remembered. The two women told police that Hammond, who was married, had been having an affair with Valerie Hilton and that the police should talk with her. On December 16, the detectives went to Hilton's home. She admitted her sexual involvement with Hammond and explained that she had broken off the affair because of his strange reaction to the newspaper accounts of the Bristol rape. She told police that the day after the rape, Hammond showed her a newspaper story about the incident and denied his involvement, while at the same time openly worrying about his uncanny resemblance to the composite sketch that ran with the story. She said the stories in the paper inexplicably rattled Hammond and that he talked about little else. On one occasion when she and Ricky were walking together down Union Street in Bristol, she told police, Hammond insisted they duck into the woods to avoid approaching police officers. Ricky was certain the police were looking for him, she said. Moreover, she was convinced, Hammond was capable of committing the rape they were investigating. Hilton also related some intimate details that further confirmed police suspicions about Hammond. For example, Hammond always had dirty rags on hand, she said, because he often worked on cars. Additionally, she said, Hammond had taken Hilton a couple of times to a dirt road in Burlington, Connecticut, known to be a lovers' lane, just a few minutes from her place in Bristol.

Even before being approached by police, Hilton explained, she had herself tried to verify an alibi Hammond had given her. Although he replied he wasn't at first certain of his exact whereabouts, at one point Hammond told Valerie that he had been with a mutual friend, Andy Neil, on the evening of the rape, but Neil denied this.

The police found a car parked at the apartment complex where Hammond lived with his wife and four-year-old daughter, which, according to the search warrant subsequently obtained, "matched the description of the vehicle used in the rape," "was messy inside and had a child's seat in the back seat." The car was registered to

JoAnne Hammond, Ricky's wife. There were what appeared to be newly made hori-
zontal scratches on one of the front fenders, and there was mud caked above the tires
in the wheel wells. Hammond denied to police having driven the car in over a
month, since, he pointed out, the registration on the car was only temporary and had
expired in late November. Tina Hagerman, another friend of Hammond's, however,
said that just wasn't true.

Hammond was "very evasive about answering questions" when the police first
showed up at his house, according to one of the warrants issued in the case. He reluc-
tantly agreed to an appointment at the Bristol Police Department to answer a few
questions and pose for a few pictures, but he never showed up. At this first encounter, a detective noted that Hammond had about a two-week-old beard. When
the police came to seize the car, a greenish-blue 1976 Chevrolet Malibu, Hammond
blocked the car in with another vehicle in an effort to prevent the police from taking
it, which didn't work. The police found that the interior of the car contained items
described by the victim, plus several human head hairs that were submitted for
examination.

In mid-December, Lisa was shown two photo-arrays at her home by Bristol Detective Mark Schulz that had been put together by Detective Bredefield after his investigation had settled on a suspect. She picked out Hammond's picture, but, for a
variety of reasons, was reluctant to commit fully. In early March, she was shown
a third photo-array with a more recent photograph. Now, she said, she was sure. After a three-month investigation, Hammond was arrested and charged with kidnapping and two counts of rape.

In an effort to get definitive proof before going to trial, Senior Assistant State's Attorney John Malone had availed himself of the most sophisticated forensic technology
at the disposal of the law enforcement community, DNA typing. The enthusiasm of
law enforcement for the technology is clear from the comments of John W. Hicks,
deputy assistant director of the FBI Laboratory Division, Washington, D.C., writing
in 1989 in *The Scientist* magazine: "[T]oday, the application of DNA typing technology
to criminal investigations is perhaps the most significant forensic breakthrough of the
century." When the results of the tests came back excluding Ricky Hammond as a
possible contributor to the stains on Lisa Gray's clothing, however, Malone said, "I
was devastated."

The very best that forensic evidence had to offer had just excluded the prime suspect. Science said Ricky Hammond was not the guy who contributed the semen. But
Malone—based on Lisa Gray's identification—was absolutely convinced Ricky Hammond did it.

"When the test came back," recalled Victim's Advocate Noreen Wilcox, who
worked out of the State's Attorneys Office in Hartford with Malone at the time, "[one
prosecutor] wanted to flush the case because he said the guy's not guilty. When they

told the victim, she went crazy, saying, 'You can't do that! He's the guy!' So . . . [Malone] picked up the ball and said, 'I don't care. I'll try it.'"

Prior to the DNA test, traditional blood-typing tests had been performed a couple of times on seminal samples taken from Lisa's clothing, and these, too, had excluded the man Lisa maintained attacked her, Ricky Hammond. As a result of the earlier blood tests, Hammond's bond had been reduced from $100,000 to a Promise to Appear, and he was released from jail. He'd been in jail nearly a year by that point.

In the face of the DNA-typing evidence that said it was somebody else who had raped Lisa Gray, the state's case against Hammond seemed hopeless, particularly since the DNA tests were performed by the FBI's DNA Analysis Unit in Washington, D.C., one of the most proficient laboratories in the world for conducting such tests. Nothing in Malone's 13 years as a prosecutor prepared him for a curve ball like this one.

Still, Malone believed Lisa Gray.

Moreover, there were other aspects of the case that seemed to point to the conclusion that Hammond was the right guy. For one thing, Malone discovered there was an "extremely high" rate of exclusion by the FBI in samples tested from around the country in rape cases. The FBI, Malone learned, was encountering a roughly 37 percent exclusion rate with DNA typing. This was an extraordinarily high percentage of exclusions, many believed, since it meant that more than a third of the people that the police were arresting for rape were wrongly arrested.

Clearly, *something* was wrong, Malone believed.

The 1990s saw an increasingly aggressive approach throughout the United States regarding sex offenses. In 1994, seven-year-old Megan Kanka was raped and killed by a neighbor who had been twice convicted of sex crimes, unbeknownst to Megan's parents. In response, New Jersey passed a novel law designed to monitor the whereabouts of convicted sex offenders and alert local communities to their presence. The new law, which came to be known as Megan's Law, eventually served as a model for similar laws passed in all 50 states, all of which continue to be referred to as Megan's Laws. The states have adopted various schemes for the registration of sex offenders, the notification of local residents about their presence in the neighborhood, and even the publication of the names and addresses of the sex offenders. By 2002, about a dozen states had launched Internet sites that identify the names, addresses, and other personal data about registered sex offenders. Although the laws have been vigorously challenged on a number of grounds, in most jurisdictions they have either withstood court challenges or been modified to accommodate judicial rulings. Several state high courts have ruled the laws unconstitutional, however. The U.S. Supreme Court is therefore poised to rule in 2003 on a number of constitutional issues arising from the Megan's Laws of several states, including whether the registry laws can apply to sex offenses committed before the laws were passed.[3]

Another legal strategy to combat sex offenses was also inaugurated in 1994, when Kansas passed a first-of-its-kind Sexually Violent Predator Act. Under the Kansas law, state authorities can continue to confine convicted sexual predators after their jail terms have expired by means of civil commitments to designated mental institutions. Such civil commitments are reserved for those who are deemed "unable to control their dangerousness" and are likely to offend again, as the U.S. Supreme Court explained it in a 1997 ruling that upheld the Kansas law. By 2002, about a dozen and a half other states had passed similar laws, most of which permit the institutionalization of sex offenders (who might otherwise be granted freedom) on the basis of "a mental abnormality" that appears to incline them toward the commission of future sex crimes. In January 2002, by which time approximately 1,200 people across the country were being detained under these laws, the U.S. Supreme Court revisited the Kansas law and narrowed its reach to individuals the state can prove have "serious difficulty in controlling behavior," as Justice Stephen G. Breyer defined the standard for the 7-2 majority. Since their inception, the sexual predator laws have been the subject of ferocious court challenges; the Supreme Court's 2002 ruling assures that the challenges will continue, but also seems to fix sexual predator legislation more securely in the firmament of American law.[4] Even more controversial laws aimed at repeat sex offenders have been passed, however: castration laws. By mid-2002, the laws of California, Georgia, Montana, Florida, and Louisiana authorized chemical castration for certain repeat sex offenders, and Texas allowed both chemical and, in some instances, surgical castration.[5]

When Ricky Hammond was first jailed, he was charged with one count of Kidnapping in the First Degree and two counts of Sexual Assault in the First Degree, one count for each alleged sexual act. He faced a total of 65 years in prison if he got convicted on all three counts and had consecutive sentences imposed. This figure tends to be, in fact is meant to be, more menacing than any realistic assessment of the type of sentence a defendant will usually draw upon conviction. But this assumes the defendant is not tried and sentenced by a judge inclined toward imposing the maximum sentence.

Since the scientific tests pointed away from Hammond (with one possible exception), Malone was confronted with a dilemma that a great many prosecutors around the country—based on the FBI's exclusion rate—were also starting to face. Deputy District Attorney of San Mateo County Kathryn Yolken, for example, faced the same dilemma in her efforts to prosecute Rigoberto Chavez in 1988 for the brutal rape and kidnapping of a young California woman, although in the Chavez case it was the defense attorney who had secured the testing from one of the commercial laboratories that also perform DNA tests. Yolken had a very persuasive witness/victim and a belief in the guilt of Chavez. "I'm sure she'll agonize over this for a while," said Dr. Edward T. Blake, the forensic scientist who performed the DNA tests in the Chavez case, "and I'm sure she'll ultimately do the only thing she can do. She'll dismiss the charges. Because to proceed with prosecuting the defendant in this case, in the face of the genetic evidence, would be nothing short of malicious prosecution."[6]

Ultimately, Yolken did dismiss the charges against Chavez. But Malone was not willing to let go so easily. Instead, he sought a middle ground. He offered Hammond a deal. In exchange for a guilty plea to the much lesser offense of Unlawful Restraint in the First Degree, Malone said he would agree to giving Hammond a five-year suspended sentence and five years of probation.

After consulting with his attorney, Jeffrey Van Kirk, however, Hammond rejected the plea. So Malone decided to take the road less traveled and prosecute Hammond in the teeth of the forensic evidence. This decision, however, posed a whole new set of problems for the prosecutor, since prior to this time Malone had been very much an advocate of DNA typing because of its reputation for establishing once and for all that a suspect had done it.

"It was my idea to have the DNA done in [the Hammond] case," he conceded. "I was in favor of anything we could do with DNA."

How could Malone address such a glaring flaw in his case against Ricky Hammond at trial? The idea of attempting to thoroughly debunk the science seemed, to say the least, counterproductive. Yet, to date, the only challenges that had been attempted in a courtroom had inclined toward the thoroughgoing variety. Malone's solution was, again, to seek a middle ground. "I didn't want it to be used to acquit some guy who I thought was guilty," he recalled.

Genetic evidence has had the biggest impact on rape cases, as compared to other violent crimes, because rapists often leave a substantial biological trail. Typically, DNA evidence is offered by the prosecution to show the jury that the rapist is in fact the guy they say it is, and no other. The Hammond case, however, Connecticut's first case to put DNA-typing evidence before a jury (although a number of guilty pleas had by that time been credited to the technology), completely reversed the roles of the adversaries.

"The sword that they brought to battle is being used against them," Van Kirk proclaimed to the press during the trial.

Malone had but one weapon in his arsenal, the eyewitness testimony of victim Lisa Gray. Her pretrial identifications of Hammond, and her anticipated in-court identification of him, remained the most crushing evidence against Ricky Hammond. Malone labeled the whole affair as "simply a case of either accepting the eyewitness' testimony or accepting scientific evidence."

Lisa Gray proved to be about as convincing an eyewitness as a prosecutor could hope for, repeating unwaveringly from the beginning that she had "no doubt at all" and was "absolutely sure" Hammond was her attacker.

"She was an excellent witness," Malone recalled later. "She was as good as any witness I've had." Superior Court Judge Joseph J. Purtill, who presided over the trial, agreed that in his decade or so on the bench he had seen "none better."

Beyond Lisa Gray, Malone had considerable circumstantial evidence (and that alone is sufficient for a conviction), including some weak but unchallenged expert testimony on hair comparison, and the very damning (but perhaps explainable) "consciousness-of-guilt" behavior on the part of Ricky Hammond.

Eyewitness-identification testimony is controversial evidence. It is for this reason that scientific evidence is so compelling in the courtroom. There's something reassuringly concrete about it.

Concern over eyewitness identification is something of a time-honored legal controversy, but its detractors are growing. As long ago as 1927, Felix Frankfurter, who later became a distinguished U.S. Supreme Court Justice, wrote in his book *The Case of Sacco and Vanzetti:* "The identification of strangers is proverbially untrustworthy. The hazards of such testimony are established by a formidable number of instances in the records of English and American trials."[7]

It wasn't for another 40 years, however, that the United States Supreme Court formally acknowledged the inherent difficulties with eyewitness testimony in the 1967 case of *United States v. Wade:* "The vagaries of eyewitness identification are well-known," wrote Justice William Brennan for the majority of the court. "[T]he annals of criminal law are rife with instances of mistaken identification," Justice Brennan explained, adding: "We do not assume [the] risks are the result of police procedures intentionally designed to prejudice an accused. Rather we assume they derive from the dangers inherent in the context of the pretrial identification."[8]

Many judges remain confident that juries can figure out for themselves the difference between good eyewitness-identification evidence and bad: "We are content to rely upon the good sense and judgment of American juries," wrote Justice Harry Blackmun for the majority of the United States Supreme Court in *Manson v. Brathwaite* in 1977, "for evidence with some element of untrustworthiness is customary grist for the jury mill. Juries are not so susceptible that they cannot measure intelligently the weight of identification testimony that has some questionable feature."[9]

Some people believe, however, that jurors are ill-equipped to distinguish between the good and the bad, not because they are stupid, but because, being human, they are inclined to follow their intuition and that, this viewpoint holds, will inevitably lead to errors in judgment.

The police were "familiar with the name" Ricky Hammond, because of an encounter he had with them a couple of years back when his brother, Marty Hammond, was arrested for the rape and murder of another Bristol woman, Peggy Krull. Ricky testified against his brother at the probable cause hearing (which Connecticut holds in murder cases instead of the more familiar grand jury hearing).

Peggy Krull, a divorced mother of two, was kidnapped and strangled to death by Marty Hammond, then 24, in Casey Field in Bristol on August 8, 1985. He then made off with her jewelry, according to police. Krull's decomposed body wasn't found for eight days, so it was impossible for forensic scientists to say for sure whether or not Krull had been raped before she was killed. If that could have been established, Marty Hammond could have been charged with capital felony murder, punishable by death.

According to Marty Hammond, Peggy Krull seduced him after she'd had an argument at a local bar with her boyfriend. She then threatened to tell police that Marty had raped her. Since Marty was already on probation for the 1982 rape of a 15-year-old Watertown girl, he got nervous.

"I panicked," says Marty Hammond's sworn statement. "I had a flashback to the last time and knew if I got busted again I was going to be put away for a long time. I freaked out. I killed her with my bare hands."

Marty pleaded guilty to the kidnapping and murder of Peggy Krull and, in accordance with a plea bargaining agreement, got 25 years in jail, the absolute minimum sentence for murder in Connecticut.

Ricky Hammond was interrogated by police in the Krull investigation, but he was never charged with anything. Ricky conceded being with his brother near the time of the murder and, as a result, played a role in the investigation and hearings that led to the guilty plea by Marty. Needless to say, this is an inauspicious way for the police in your town to get familiar with your name. And now it was something of an unfortunate backdrop to the current case against Ricky.

Before the trial began, Judge Purtill ruled, after a brief *Frye* (or admissibility) hearing, that the FBI's DNA-typing methodology met the "general acceptance" test for novel scientific evidence and would be admitted into evidence. (Connecticut has since adopted the federal *Daubert* standard for admitting novel scientific evidence, giving judges greater flexibility.[10]) What was odd, perhaps, was that it was the defense offering the evidence. But the DNA wasn't the only evidence brought by the defense. Ricky Hammond had an alibi for November 30, 1987. Late in the afternoon on the day of the rape, Hammond met with Chris Petruzzi, a stranger, and put a deposit on a truck Petruzzi had for sale. Hammond even got a receipt. The timing was close.

Petruzzi told the jury that, at the time of the transaction, Hammond was driving a car other than the one allegedly used in the rape. Petruzzi's description of it matches the description of Hammond's second car, a "shit-box Pinto," as Petruzzi called it in an affidavit. Moreover, he said the transaction took place "anywhere from 4:30 to 5:00, quarter after five." On cross-examination, however, he said the transaction only took between ten and fifteen minutes, and it was pointed out that the meeting was only a short distance away from the scene of Lisa Gray's initial attack.

Van Kirk maintained that for his client to have pulled this off (that is, meeting with Petruzzi, driving home, switching cars, driving to the high school parking lot, and running through the woods to where Lisa Gray was walking) would have required "commando-type timing."

Valerie Hilton testified at the trial about Hammond's "weird and paranoid" behavior around the time of the rape. Andy Neil, with whom Hammond said he'd been on the night of the rape, testified for the state at the trial, too, refuting Hammond's claim that the two were together on November 30, 1987. Neil said he couldn't remember whether they were together on the evening of the rape. Neil also told the jury that Hammond had a habit of taking his wristwatch off and dangling it from the gear shift. He said he remembered this specifically because he had admired the watch so much. Another friend, however, contradicted this.

Diminutive and attractive, wearing business-like attire, Lisa Gray told her story on the stand with compelling matter-of-factness. There were no emotional outbursts, yet there was emotion.

Lisa had been shown the third photo-array—one she had requested—more than three months after the assault on March 5, 1988. She picked Hammond out of all three of the photo-arrays she was shown but had balked, according to Sergeant Schulz, at committing herself completely to the first two arrays because of apparent discrepancies in the age and facial hair of the person depicted in the photos.

Van Kirk argued unsuccessfully that the three photo-arrays were suggestive in that Hammond was the only individual repeated in all three displays: Thus, he maintained, the entire identification was tainted. He would later argue the same point to the jury. Judge Purtill conceded "there may be a certain amount of suggestibility" in the third array, but overruled Van Kirk's motion to suppress the out-of-court identifications based on the "totality of the circumstances."

Lisa recounted for the jury the ritual of viewing the first two sets of pictures alone with Bristol Detective Sergeant Mark Schulz in her kitchen, while her son and fiancé waited in the living room. She told the jury that the picture she chose seemed a younger version of the man who raped her.

But was she certain it was him? Malone asked her in court.

"I was absolutely sure," she said confidently of her first choice.

The main difference in the second set of photographs she was shown that same night, she explained, was that the man had a beard in the photograph, but not on the night of the rape.

"Did you have any difficulty picking him out in either of those two arrays?" Malone asked.

"No," she said.

"And would you tell the ladies and gentlemen of the jury what your level of certainty was with respect to the second array?" Malone continued.

"I was sure it was him," she said.

"And did you indicate anything to Sergeant Schulz about wanting to see any other photographs?"

"I just asked if they had any with, you know, a clean shaven face. You know, more recent, but with his face not with a beard on it."

"And did you have any doubt in your own mind, however, at the time that the person who did this to you was included in those arrays?"

"No," Lisa said. "I didn't have any doubt. I knew that was him."

The third set of photographs were shown to Lisa at the police station several months later on March 5, 1988. Hammond had not yet been arrested.

"I had no doubt," she told the jury about the picture she selected from this last series. "I knew it was him."

Malone made sure to bring out, while he was asking the questions, the fact that Lisa had gone to Ricky's arraignment. This had the effect of blunting any attempt by Van Kirk to suggest that seeing him in custody this way may have altered her memory of the actual rape.

"I recognized him when he walked in," Lisa recalled of seeing Ricky at his arraignment.

"And was there anything different about his appearance the time that you saw him in Court and recognized him as opposed to when you'd seen him on November 30th, or when you'd seen the pictures?" Malone inquired, posing the question in a manner Van Kirk certainly would not have.

"He had a pair of glasses on."

"And was he bearded at that time, or not? Do you recall?"

"I don't know," Lisa responded. "I don't think so."

"Before November 30th of 1987," Malone went on, "had you ever seen ever before in your life that person who kidnapped you and raped you?"

"No, I never did."

"Do you see in Court today the person who kidnapped you and raped you on November 30th of 1987?"

"Yes, I do."

"And would you point him out for ladies and gentlemen of the jury, please, for the record?"

"He's over there, sitting at that table," she said, gesturing at Ricky Hammond.

"Is there any doubt whatsoever in your mind that that is the person who kidnapped and raped you on November 30th?" Malone concluded.

"No doubt at all," she said.

Van Kirk countered Lisa's eyewitness account by pointing out, both on cross-examination of Lisa and again in his closing argument to the jury, that Lisa's opportunity to observe her attacker was far from ideal. The whole encounter, he stressed, took place under conditions of unbroken darkness, from the moment of her abduction until the assailant released her on a darkened street and drove away with his headlights off. Moreover, he emphasized, Lisa is very nearsighted, with 20/375 vision, and she only had "a split second" to see the man as he approached her before her glasses were knocked from her face. Even during that split second, he insisted, the attacker's face must have been in shadow, since distant streetlights were positioned at his back.

"So her opportunity to view this individual plainly was very, very limited," Van Kirk told the jury.

He also reminded the jury that Sergeant Shultz seemed to feel Lisa's identification of Hammond from the first two photo-arrays, despite her testimony to the contrary, was a tentative one.

"[S]he was not 100 percent sure," Shultz had said. "The beard confused her." What Van Kirk did not do, perhaps because he felt current law militated against it, was attempt to explore the less obvious difficulties with eyewitness identification evidence.

"[C]ross-examination," wrote Justice Brennan in his dissent in Watkins v. Sowders, "is both an ineffective and a wrong tool for purging inadmissible identification evidence from the jurors' minds. It is an ineffective tool because all the scientific evidence suggests that much eyewitness identification testimony has an unduly powerful effect on jurors."[11]

The Connecticut Supreme Court had by this time occasionally expressed concern over the reliability of eyewitness testimony, too. "[M]emories fade over time," the court wrote in State v. Kemp four years before Hammond went to trial, "people under severe stress do not acquire information as well as alert persons not under stress, and people tend unconsciously to resolve apparent inconsistencies between their memories and after-acquired facts."[12] In State v. Vaughn ten years before that, the Connecticut high court had warned: "It is a familiar point to courts and legal scholars that sex cases create a special need for an evaluation of credibility. . . . [O]ften the charge incites sympathy for the prosecutrix and prejudice against the defendant."[13]

Despite the apparent sentiment, the Connecticut Supreme Court, to this point, had consistently deflected efforts by defense attorneys to introduce at trial expert testimony on the trouble with eyewitness identifications. In the Kemp decision itself, in fact, the court ruled that such expert testimony "invades the province of the jury to determine what weight or effect it wishes to give eyewitness testimony."[14] Such testimony "is ordinarily superfluous," the court said a year later in State v. Boscarino, going on to quote from Kemp, "since '[t]he weaknesses of identifications can be explored on cross-examination and during counsel's final arguments to the jury.'"[15]

But are the problems with eyewitness identifications self-evident to a jury?

In 1972 the United States Supreme Court outlined in Neil v. Biggers what have become the key factors to be considered in evaluating the likelihood of eyewitness misidentification. These include "the opportunity of the witness to view the criminal at the time of the crime, the witness's degree of attention, the accuracy of the witness's prior description of the criminal, the level of certainty demonstrated by the witness at the confrontation, and the length of time between the crime and the confrontation."[16]

In the ensuing decades, however, a number of reviewing courts began to conclude that juries can't possibly distinguish between a good and bad eyewitness account because of the sheer complexity of the task. This is a matter of grave concern, said the United States Court of Appeals for the Second Circuit (which includes Connecticut) in the 1983 decision of *Kampshoff v. Smith,* because "eyewitness testimony is among the most influential, even as it is among the least reliable, forms of proof. . . . Moreover, because eyewitness testimony is such powerful stuff and can decide a case on its own strength, it can blind a jury to other exculpatory evidence or inferences. Once loosed, improperly suggestive eyewitness testimony may taint a whole trial."[17] (Federal courts now routinely permit expert testimony concerning the hazards of eyewitness identification.) Also in 1983, the Arizona Supreme Court became the first state appellate court to permit experts to testify in criminal cases regarding the reliability of eyewitness identifications, ruling that it was an abuse of discretion for the trial judge to have excluded the expert testimony offered by the defense.[18] That case, *State of Arizona v. Chapple,* came to define the new trend. California followed suit the next year.[19]

"Forensic science," the *Kampshoff* court explained, "has so thoroughly impeached the reliability of eyewitness identification that studies now focus not on the fact of eyewitness inaccuracy but on the causes of the extraordinary gap between a criminal event and human perception and recall of it. . . . But too often the doubts cast in the laboratory have not found their way into the courtroom and the judgment of juries."[20]

The court was referring specifically to a growing body of psychological research aimed at clinically determining the ability, or inability, of people to accurately remember and describe significant events or individuals. The court noted in particular the work of Dr. Elizabeth Loftus, a psychology professor at the University of Washington and one of the country's foremost experts on the subject of eyewitness identification. When courts do permit expert testimony concerning eyewitness evidence, the expert is often Dr. Loftus.

"Juries have been known to accept eyewitness testimony pointing to guilt," Loftus writes in her book, *Eyewitness Testimony,* "even when it is *far* outweighed by evidence of innocence."[21]

There are a number of reasons for eyewitness mistakes, according to Loftus. Stress, for example, is commonly believed to enhance recall of a person or event. Yet the opposite is borne out by clinical experiments. There is also the phenomenon of "unconscious transfer," a phenomenon where a witness inadvertently substitutes the face of an individual seen later, perhaps in an identification procedure, for that of the actual criminal. Witnesses can also incorporate facts learned subsequent to an event into their recitals of the event. In this way, a description becomes irretrievably flawed.

Loftus has testified, further, that, contrary to popular belief, a witness's degree of confidence about an identification is completely unrelated to the accuracy of that identification. Her studies have shown that a diffident witness is just as likely to be correct about something as someone who is "absolutely positive." At the same time, a confident person is just as likely to be wrong.

"[E]yewitness testimony is likely to be believed by jurors," Loftus writes in her book, "especially when it is offered with a high level of confidence, even though the accuracy of an eyewitness and the confidence of that witness may not be related to one another at all. All the evidence points rather strikingly to the conclusion that there is almost *nothing more convincing* than a live human who takes the stand, points a finger at the defendant, and says 'That's the one!'"[22]

But the jury in the trial of Ricky Hammond remained oblivious to these concerns, since no expert testimony was presented on the issue. And Lisa Gray was the very first witness.

In the decade following Hammond's trial, a dramatic shift occurred across the country regarding the sanctity of eyewitness testimony, especially when it is the centerpiece of a case. The great confidence the U.S. Supreme Court expressed in *Manson v. Brathwaite* that juries could sort out good from bad eyewitness evidence is no longer the prevailing view. It has been replaced by a generalized acknowledgment that juries may well need expert help in understanding the subtle influences that can undermine an eyewitness account or identification, even a supremely confident one. In May 2001, the New York Court of Appeals, the state's highest court, joined the federal courts and the majority of states in the nation that now allow, at least under some circumstances, expert testimony to assist a jury in evaluating the reliability of eyewitness evidence. Although "juries may be familiar from their own experience with factors relevant to the reliability of eyewitness observation and identification," the court wrote, "it cannot be said that psychological studies regarding the accuracy of an identification are within the ken of the typical juror."[23] The court stopped short of requiring such expert testimony when it is offered, however, leaving it to the trial judge to determine whether the testimony is warranted.

"The research is becoming increasingly compelling," Iowa State University professor of psychology Gary L. Wells, one of about 100 experts who regularly testify around the country on the reliability of eyewitness evidence, told *The Wall Street Journal*, "and states are beginning to understand that experts have a great deal to offer."[24]

Prosecutors, however, continue to worry that juries' assessments of the credibility of eyewitnesses are being replaced by the assessments of highly paid psychological experts.[25]

While the Supreme Court of Connecticut did condemn one trial judge in 1996 for excluding expert testimony on the unreliability of an eyewitness's perception and memory, primarily because the witness was under the influence of cocaine at the time she observed the crime, the court in that case did not consider the exclusion of the testimony so egregious as to justify a new trial for the defendant since there were other corroborating witnesses.[26] More recently, the Connecticut high court made clear it prefers to stick with the more traditionalist view that juries can figure out for themselves what constitutes reliable eyewitness testimony.

"We decline to follow *Chapple* and continue to follow the rationale of *State v. Kemp*," the court ruled in *State of Connecticut v. McClendon* in 1999, a case that was co-incidentally prosecuted by John Malone and involved an eyewitness identification that was crucial to the conviction.[27] In the *McClendon* case (a double-murder that resulted in the defendant's being sentenced to 140 years), the defense offered the testimony of a psychologist who was prepared to underscore, among other things, that stress inhibits the memory of an event, rather than enhances it, as many believe; that confident eyewitness assertions are just as likely to be erroneous as less certain ones, based on various studies; that "unconscious transference" may produce false memories; and that the identification of a human voice heard only once by a witness is particularly unreliable, an important element of the case because the witness confidently claimed that she could identify the voice of the perpetrator long after the incident. After noting that the expert conceded the paucity of research on voice identifications and that the eyewitness in the case had viewed the perpetrator at least part of the time while not under stress (just before the incident), the court agreed with the trial judge that the general influences on eyewitness evidence that the expert was prepared to testify about "should come as no surprise to the average juror." Quoting from the *Kemp* decision, the court concluded it was sufficient for defense counsel to explore the weaknesses of eyewitness identifications through cross-examination and during final arguments to the jury, as McClendon's attorney had endeavored to do.

A strongly worded dissent upbraided the majority of the court for its ruling because, the judge wrote, "this case turned entirely on the seemingly credible identification testimony of [one eyewitness], whose identification of the defendant was made two years after the event and was not supported by any corroborating evidence," namely, "[n]o credible forensic evidence connected the defendant to the murder." The case was particularly troubling to the dissenting judge because it involved a cross-racial identification, which research indicates is a genuine problem, and an extremely confident eyewitness identification, even though the witness viewed eight photo-arrays with the defendant's picture in it before positively identifying him.[28]

3 "Does Not Match"

HAMMOND AND HIS ATTORNEY BROUGHT SOME POWERFUL EVIDENCE OF their own to the trial: DNA that indicated he wasn't guilty. Moreover, sitting in their corner was something most criminal defense attorneys only dream about—the FBI.

In fact, this was the first time an FBI DNA scientist had ever testified on behalf of a defendant in a case where DNA-typing evidence had been introduced. At that time, the FBI had testified for the prosecution in approximately 75 cases around the country. Special Agent Lawrence A. Presley, assigned to the DNA Analysis Unit of the FBI Laboratory in Washington, D.C., painstakingly explained to the jury the science behind the DNA technology and the carefully controlled steps the laboratory had taken to arrive at its conclusions. Samples from the victim's clothing were analyzed and compared with DNA derived from the blood of the victim, her boyfriend, and Ricky Hammond.

It was here that Van Kirk focused his energies, patiently coaxing Presley through the basic steps of the DNA test. Van Kirk's goal was to build to a forensic crescendo with Presley and end with a clash of cymbals that would ring in the ears of the jurors throughout their deliberations.

Presley explained the steps of RFLP analysis in very basic terms. He analogized, for example, the sought-after repeat sequences of DNA (the VNTRs) to the repetition of common names, such as Smith, occurring in a telephone book. These, he explained, are the highly variable, noncoding segments of the genome sometimes called "junk DNA."

"No one knows exactly the function of this repeat DNA," Presley explained, "but we know that most humans—almost all humans as far we've been able to measure so far, have this repeat DNA. Everyone has it. That DNA is in fact what we look at."

RFLP is one of the DNA-typing techniques that focuses on the size differences of certain genetic locations. It involves several steps. First, the genetic material is literally cut to pieces with the help of special enzymes, called restriction enzymes, which operate like molecular scissors to snip the DNA molecule wherever certain letter sequences are recognized by the enzyme. The genetic fragments thus created are called restriction fragments. Restriction enzymes create many thousands of DNA fragments ranging in size from a few to tens of thousands of base pairs.

Next, gel electrophoresis is used to allow the scientist to sort out (according to size) the DNA fragments that are produced by the restriction enzymes. The mixture of DNA fragments is loaded into a specific well of a gel, and exposed to an electric field. This process also has the effect of denaturing the DNA. Because the chemical makeup of DNA gives it a net negative charge, the DNA fragments travel through the pores in the gel toward the positive electrode. The gel acts as a sieve, with large DNA fragments moving more slowly than small ones. Thus, the mixture is separated, or resolved, according to size. In RFLP analysis of human DNA, the numerous fragments are laddered continuously along the entire length of the gel, but no single DNA fragment can be visualized without further effort. After a period of time, the electrophoresis is stopped and the DNA transferred out of the gel onto a nylon membrane in a process called Southern transfer. The nylon membrane, or Southern blot, retains the DNA in the orientation obtained in the gel after electrophoresis.[1]

Radioactively tagged probes are then applied to the membrane to seek out and bind to predetermined DNA segments (specific VNTRs) that are complementary to the nucleotide sequence of the probes (the DNA, at this point, has been denatured); this process is called hybridization. Finally, X-rays are taken; the probes are radioactively tagged so that when the nylon membrane is exposed to an X-ray film, the location of the responding DNA segments can be visualized as dark bands. The X-ray photographs are called autoradiographs, or autorads for short. The determination of size is made by comparing the test samples to pieces of DNA of known length—known as molecular size markers—that are run side by side on the same gel. The series is compared—visually and/or by computer imaging—to determine whether there is agreement between the individuals. Heterozygosity is typically the result since VNTRs have many alleles; therefore, a single-locus DNA profile will emerge in the form of two distinct bands on the autorad in the lane corresponding to the well into which the DNA was injected. The two bands identify the two alleles of the individual at that locus. (If, as is rare, both parents contributed an identically sized allele, the individual would be homozygous at that locus and only one band would appear in the lane.)

Once this analysis is completed, the radioactive probe material is cleansed from the nylon membrane and another radioactive probe is added seeking a different VNTR. Three to five VNTRs are ultimately detected on average, the whole process taking several weeks to complete. Five VNTRs were detected in the Hammond case.

Laboratory protocol dictates what constitutes a match between two bands being compared. This procedure is necessary because it is not possible in VNTR analysis to determine precisely the actual size of the DNA fragments being examined.

To give meaning to a DNA profile that is deemed a match, statistics and population genetics (a specialized branch of genetics) are employed to calculate the probability that a particular DNA profile will occur in the general population. Frequencies of the alleles are determined from random population samplings. A standard match window is developed by adding and subtracting 2.5 percent of the evidence sample's measurement value. The numerical interpretations that arise from these calculations were one of the most fiercely contested aspects of DNA typing in criminal cases because getting to the final numbers requires making certain mathematical assumptions. When comparing the DNA profile from a suspect with that of biological evidence retrieved from a crime scene, the point is this: The lower the probability of a match (one in a zillion, say, versus one in a hundred), the more likely it is that the suspect contributed the DNA to the crime-scene sample.

Successful forensic RFLP analysis depends, of course, on what happened to the evidence before the test is performed. If strict guidelines for proper sample collection were ignored, or the evidence was stored haphazardly, the results of the DNA test are suspect. Moreover, environmental effects can act on the DNA in evidentiary stains and cause degradation, or breakdown, of the DNA molecule. (This does not, however, change the DNA into somebody else's.) And if a forensic scientist doesn't receive a sufficiently large, intact (or so-called high molecular weight) DNA sample from case specimens, there's little hope of a successful RFLP test.

Before being superseded by advances in DNA technology, the RFLP technique was performed by the FBI and most state and local forensic science laboratories, as well as a number of commercial DNA-testing laboratories. In 2001, only one government DNA laboratory reported using RFLP in criminal casework, down from 43 a few years earlier.[2] Nevertheless, RFLP remains the method of choice for a few private laboratories performing DNA paternity testing and continues to be called into service on rare occasion in some cases that pose unique forensic questions. In the Hammond case, the FBI used the RFLP technique targeting single-locus probes (by far the most common approach). A single-locus probe, as its name implies, is designed to lock onto a highly polymorphic DNA segment that occurs only once along the DNA molecule, producing two readable bands on the autoradiograph with heterozygous DNA samples and one band with homozygous DNA samples. A multi-locus probe, on the other hand, targets polymorphic segments that are repeated any number of times along the DNA chain, usually producing about 15 or more bands.

The standard formula for computing the probability of a match involves multiplying the frequencies of the single loci together using the straightforward product (or multiplication) rule. The ultimate question is, what is the probability that any person tested at random would have all the DNA markers in his genetic blueprint that were found to exist in the genetic pattern of the felonious DNA.

"Now, if this technique really works," Presley explained to the Hammond jury, demonstrating with the X-rays, "then I should be able to look at the known blood stain from the victim, which is here, and here. And that these should line up with each other. And I think you can see that fairly clearly here, if I go straight across, that this band lines up this band, as it should. And that this band here, lines up with this one, as it should, because it's her DNA, and her DNA in her blood, and DNA in her vaginal epithelial cells, or her DNA in any other cell, should be exactly the same. And in this case, you can see I think visually, that it is."

The male DNA components extracted from the underpants, however, did not line up. The analysis Presley described for the jury, which showed Lisa Gray's DNA bands so distinctly, stubbornly resisted producing anything even remotely resembling Hammond's "DNA fingerprints." Presley stressed that the unknown male DNA came from a single individual. It just didn't match Lisa's boyfriend, Lisa, or Ricky Hammond.

"Let me ask you this," Van Kirk said triumphantly. "To a reasonable degree of scientific certainty, what does that mean to you, or what is your opinion concerning what you see on the screen as it has relevance to this case?"

"Basically," Presley answered, "that the profile—neither the suspect nor the boyfriend, but primarily the profile of the suspect—does not match the profile that was found in the panty crotch area from the victim."

"Does that to a reasonable degree of scientific certainty exclude the suspect in this—"

"In my opinion," Presley said, "yes, it does."

The cymbals clashed. According to the FBI Laboratory and the DNA tests, the assailant had to be someone else.

Van Kirk exploited the high note by following it with the rhythmic, repetitive testimony of Presley as he explained to the jury how each of the five probes had, in turn, been successfully deployed. Assuming that Hammond was their target, all five probes had failed to find their mark.

There are myriad ways for things to go awry in DNA analysis. These include things known as partial enzyme digestion, band shifting, cross-hybridization, star activity, loading errors, incomplete stripping of membrane before re-hybridization, and loss of small fragments during electrophoresis. There is also the problem of cross-contamination, such as when extraneous DNA comes in contact with an evidentiary sample. Consequently, scientists employ a series of laboratory controls to help recognize when things have gone wrong. A control, then, is a flare. Controls are the key to calling a DNA test valid. According to Presley, no flares went up in the Hammond DNA test.

Presley was emphatic, in fact, that his controls did not indicate any problems. One such control was a known sample of DNA, having nothing to do with the case except that it was run side by side with the known and questioned samples of the Hammond case.

"That's a cell-line control; that's a DNA that we know will show up exactly the same way every time, if the procedure has worked correctly," he said. "And we run those on every gel that we run."

"In this particular case," he explained, "the cell line appeared exactly where it should, with two bands exactly where they should be. So it indicated to me that the system was operating correctly."

Malone's strategy was two-pronged. The first prong (the contamination theory) involved the "suggestion" (Malone's own word) that somehow something had gone wrong in this case and, through no fault of the FBI, the articles of clothing secured from Lisa Gray on the night of her rape had been sullied with some foreign biological material that had led to the false exclusion of Ricky Hammond.

The contamination theory didn't arise out of the presentation of any specific, direct evidence that a particular mishap could conclusively account for the wrong results of the DNA test. Instead, the theory pivoted on whether a mishap was in fact possible. In his questioning of the expert witnesses who appeared at the trial, Malone repeated a number of themes intended to show the contamination theory had a few teeth.

Malone focused especially on the following: that it was impossible for scientists to say precisely when the stains on Lisa Gray's underwear were deposited; that the scientists could not definitively say whether it was in fact semen that was tested (although Presley stressed that the chemical tests he performed were designed to detect semen); that it was within the realm of possibility that some foreign biological material was accidentally transferred to the panties after the crime; and that the best evidence, the sealed vaginal swab, which is not vulnerable to contamination in the same way articles of clothing are, was never tested.

Malone's argument was that the samples from the patches of clothing were contaminated somehow *before* Presley received them for analysis, and even before the Connecticut toxicology lab was able to conduct the ABO blood-typing tests that had excluded Hammond very early in the investigation (although the contamination did not necessarily occur before state toxicologists had received the clothing, according to Malone's theory). There was, then, so far as the jury knew, only a very narrow window of opportunity for a mishap.

Malone suggested through Presley that any number of body fluids, from any number of sources, might account for the DNA that had ultimately found its way to the X-rays the jury was seeing. In particular, Presley acknowledged that DNA is contained in blood, sweat, and tears. Presley testified, though, that the most likely scenario—given the careful separation step—was that his DNA tests had typed the DNA of sperm, not something else.

"But there are other possibilities?" persisted Malone.

"[There] could be other possibilities," admitted Presley, referring to some kind of additional deposit or secondary transfer event.

Significantly, blood-grouping tests performed early in the investigation excluded Hammond's blood type. The day after Lisa Gray was raped, the rape kit and clothing were delivered to the Connecticut State Toxicological Laboratory in Hartford. About a week later, the kit was removed from the laboratory's vault where it had been stored and a series of tests were performed by Dr. Sanders Hawkins, who later testified for both sides at Hammond's trial. Dr. Hawkins performed an acid phosphatase test on the vaginal swab and the stains found on the clothing, and the test indicated the possible presence of semen. Acid phosphatase is an enzyme occurring in semen as well as other body fluids and some plant sources. Under a microscope, he further observed that sperm was present on the slides contained in the kit, which were smeared at the hospital with a second vaginal swab. Dr. Hawkins also conducted tests on Lisa Gray's blue jeans and discovered A and H antigens, thus indicating that a secretor with type A blood had contributed to the stain. Roughly 80 percent of the population are secretors, and their body fluids are typable by means of the ABO blood-typing system. The remaining 20 percent are non-secretors, who do not have soluble A or B antigens in their body fluids. Subsequent tests performed by Dr. Hawkins and forensic scientist Elaine Pagliaro at the Connecticut State Police Forensic Science Laboratory confirmed the presence of type A antigens on the swab, the underpants and blue jeans. Arguably, the DNA tests were further confirmation of Hammond's exclusion.

Ricky Hammond has type O blood. So does Lisa Gray and her boyfriend, so nobody seems to account for the A antigens. There was incriminating forensic evidence presented against Ricky Hammond, too. Although Dr. Hawkins was only qualified as a toxicologist and serologist, he nevertheless testified regarding hair-comparison examinations he performed on the hair samples recovered from Hammond's car in December 1987 pursuant to the search warrant and head hair taken from Lisa Gray. It was made clear in the course of the testimony that hair comparison is not a positive means of identification (in fact, Dr. Hawkins called it "the weakest link" in the array of laboratory tests available to him). But Dr. Hawkins's testimony had a certain aura about it that may have been misleading to the jury. Dr. Hawkins employed terminology with an impressive ring to it ("macroscopic features" and "cortical pigmentation," for example) and he concluded that the hair was "similar" and "very similar."

Van Kirk had arranged for a forensic expert in hair comparisons to counter Dr. Hawkins's testimony, but the expert, Dr. Peter R. DeForest, was not available on the day his testimony was to be presented, a Thursday. A motion for a continuance until the upcoming Monday was denied by Judge Purtill, who said that his "recollection of one thing Dr. Hawkins did say was that hair—his testimony, this was after cross examination—the hair method or the hair comparison procedure is a very, very poor method of identification." Whether the jury shared this assessment—certainly the correct one—is another question.

A letter sent to Van Kirk from DeForest, who had received the hair evidence during the trial, lambasted Dr. Hawkins's methodology for hair comparisons. After noting that the hair was "mounted in an idiosyncratic and unacceptable fashion," he went on to say:

> The reports [of Dr. Hawkins] are also the cause for some concern. . . . Much of the terminology used conveys little information and could be misleading. The use of such jargon as "flattened imbricate scales" may seem to lend an air of authority to the reports but means little in human hair comparison. Flattened imbricate scales are characteristic of human hair, and thus this is not a term for a criterion that is of any value in distinguishing among humans. . . .
>
> I have seen a limited number of hair samples prepared by the State of Connecticut's Toxicology Laboratory in Hartford. In addition, I have seen a somewhat larger number of hair comparison reports issued by this laboratory. From this exposure, it is clear to me that this laboratory should not be doing hair comparisons. I have reviewed hair cases from forensic science laboratories all over the country. I have never seen hair mounted in the extremely naive fashion described above. . . .[3]

The jury heard only the testimony on the hair comparison conducted; they never learned of DeForest's opinion about it.

Van Kirk attempted to dismiss the state's DNA contamination theory out of hand in his closing argument:

> So I guess the State's—the Bristol Police Department and the State of Connecticut's idea is this: If you go to [the] first lab, and you don't get the results you want, try it again. If you get the same results from that lab, and you still don't get what you want, send it to another lab. A better lab. If you still don't get the [results you want], let's try a new test, DNA. And lo and behold, what do we have? It still isn't what they want. What else do they say now? Contamination. Must be something wrong with the test. What Mr. Malone fails to do here is give you any contamination.

The second prong of Malone's argument, which crystallized during his closing argument, challenged the jury not to blindly follow science, since, Malone argued, history is replete with examples of scientific blunders.

Referring to Van Kirk's just concluded closing argument, Malone said:

Mr. Van Kirk had talked to you about the [DNA] test, and said to you "Isn't science wonderful?" Let me suggest to you that anyone who would simply blindly accept scientific results without the kind of analysis that I suggested, against Miss [Gray's] testimony, against all the corroborating evidence, is fooling him- or herself. The question is an old question, whether it's a man or a machine which is more important. I would suggest to you, and it's clear through many examples in history, that science and technology has to be the handmaiden of man, and not man or woman—mankind generally being the servant of science. And Mr. Van Kirk would have us blindly accept these scientific results and be led by them. I suggest to you a more reliable way to the truth is through Miss [Gray], and not through the science. Examples? Sure. Back in the 19th century, do you remember Mr. Nobel, who started the prize? He invented nitroglycerine.

An objection was overruled. Malone continued,

He invented it to end war, but didn't end war at all. Science was not an answer in that case. Science and technology produced for us an unsinkable ship called the Titanic in 1912, which didn't make it across the Atlantic the first time. That time science didn't have the whole answer. Science also, even more recently, tragically, as good as we are in space and science—it's science that gave us the Challenger, which didn't make it, and very tragically. Science—look back a few years. It's science in medicine that gave us, for women who have problems in the early part of pregnancy with pain, a medicine called Thalidomide. What happened with science there? We have families still scarred by that tragedy. How many of you have—remember the swine flu vaccine that science had warned us about? Well, more people got it from the vaccine. There wasn't any swine flu. The only problem was with the swine flu vaccine.

Science should not be blindly applied and followed. And I'm not saying to disregard it. You may decide to disregard the tests as I suggested, for the reasons that I suggested, the contamination. You may decide to consider the test, and yet to reject it in the light of Miss [Gray's] testimony. But whatever you do, do it smart. Don't just—just because they're tests doesn't mean to toss out the case. I don't think that that's a fair, proper, or really the right way to approach this case. If you think that Miss [Gray] was wrong because of a number of things, then as I said to you, I don't think that you can convict him, unless you believe Miss [Gray]. But [what] I suggest to you is you should use the science; if you decide to consider it, consider it. But in the end, I would suggest that you should not accept the results in light of the overwhelming weight and persuasiveness of Miss [Gray's] testimony, corroborated by the details, corroborated by the vehicle, and corroborated by the [guilty-seeming] behavior of the defendant.

Beyond the "consciousness-of-guilt" evidence the jury had heard about, Ricky Hammond's general appearance in court left a great deal to be desired. He was disheveled (but not, perhaps, for lack of trying to be otherwise), and he had a furtive air about him. Occasionally, too, he was disruptive in court.

"Ricky Hammond looked like," even Judge Purtill frankly recalled after the trial, "if you're going to get a guy to play the part of the depraved rapist, that would be Ricky Hammond."

In stark contrast to Hammond, Lisa Gray had a pleasant appearance and was composed, respectful, and believable (in addition to being steadfast in her testimony) throughout the trial. Malone endeavored to capitalize on this imbalance in his closing argument to the jury.

Hammond did not testify in his own defense at the trial. Van Kirk explains this by saying that, at the time, he felt things were going well and, consequently, "[Ricky] could only hurt us."

In 1893, in the case of *Wilson v. United States,* the United States Supreme Court tackled the question of "those who might prefer to rely upon the presumption of innocence which the law gives to every one, and not wish to be witnesses." In doing so, the Court took a moment to explain the reasons behind this fundamental constitutional safeguard:

> It is not every one who can safely venture on the witness stand, though entirely innocent of the charge against him. Excessive timidity, nervousness when facing others and attempting to explain transactions of a suspicious character, and offenses charged against him, will often confuse and embarrass him to such a degree as to increase rather than remove prejudices against him. It is not everyone, however honest, who would therefore willingly be placed on the witness stand.[4]

One criminal judge in the same division of the court as Judge Purtill told a roomful of newly recruited jurors at their jury indoctrination to forget what they had learned from their fathers when they were growing up. No doubt their fathers had taught them, said the judge, that if someone doesn't have anything to say for themselves when accused of a misdeed, they're probably guilty.

"That's not the law," the judge instructed them.

Later, coming down from the podium and out of earshot of any impressionable jurors, the judge—unconnected with the Hammond case—reflected on his speech.

"Forget what I said to the jurors up there about the presumption of innocence," he said with a wave of his hand. "If a defendant doesn't testify at trial, he's dead. Dead."

The jury returned a verdict of guilty on all three counts. Both Hammond and Van Kirk collapsed in their chairs as the verdict was being rendered. Hammond sobbed openly. Van Kirk was nonplussed. It was the first time in U.S. legal history that someone was convicted of a crime despite DNA evidence that appeared to exclude him.

"The victim's testimony was definitely a major part of it," said Dorothy Donaghue, one of the six jurors in the case, as she left the courthouse and met the press. "After going over all the evidence, I know the DNA was a strong defense there but we didn't think it was enough."

"We really had to look at the entire evidence in the entire case," said another juror, Dieter Zinsmeister, to *The Hartford Courant*, "and the DNA portion just did not carry that much weight in our decision because of the other evidence, particularly the victim's testimony."

"What [the case] means," Van Kirk commented, "is there is not a logical limit to a jury's ability to ignore good evidence and accept sympathy and emotion as a substitute to hard evidence."

"What we had, Your Honor," Malone admitted at Hammond's sentencing, where Judge Purtill imposed a sentence of 25 years, suspended after 23 years, "was simply a case of either accepting an eyewitness' testimony or accepting scientific evidence. . . . As far as I was concerned, without that DNA and scientific evidence this case could have been decided in favor of the state, that is with a guilty verdict, without the jury having even left the room. It was that strong."

Malone had urged the court to impose a sentence approaching the maximum, a sentence that would "keep him in jail for the rest of his fertile life." Moreover, Malone asked the court to consider one aspect of the case that, although perhaps obvious, would have been better left unsaid. Malone asked Judge Purtill to "compare this defendant with the victim. You've seen her testify," he said. "You have seen her conduct herself around the courthouse. You have seen some of her in the pre-sentence investigation. And I would like you just to focus on one aspect, because it would seem to me that anybody who saw these two people for longer than five minutes would make an obvious preference as to who the better human being is. . . ."

Van Kirk snapped back.

"It's just such comparison of the value of a particular human life over another that led to—that still leads up until today—to atrocities against particular ethnic and social groups by totalitarian societies."

Malone was duly castigated by the court. Yet Malone's remarks, as inappropriate as they were, go a long way toward explaining some of the dynamics of the trial of Ricky Hammond.

At Hammond's sentencing, Judge Purtill denied, in the interests of finality, Van Kirk's motion for a new trial and a separate motion for post-trial blood and DNA typing of the vaginal swab and smears, since the court felt it would have been cumulative in light of the jury's verdict. Malone argued that Van Kirk could have had the materials tested himself, but chose not to for tactical reasons. The issue was complicated by the fact that Van Kirk insisted Dr. Hawkins told him that the biological material on the swab had been entirely consumed by the earlier testing, making it impossible to do any further testing. Van Kirk later learned otherwise.

Hammond's appeal was taken up by the criminal law clinic of the University of Connecticut School of Law, headed jointly by Professors Todd Fernow and Timothy Everett. In an appellate brief (which went straight to the Connecticut Supreme Court after the court opted to take the appeal directly, bypassing the intermediate appellate court), the clinic argued that Judge Purtill improperly denied Hammond's motion for a new trial because the jury had rendered a verdict clearly against the weight of the evidence and disregarded the judge's instruction on reasonable doubt; the judge had improperly denied a defense post-trial motion for DNA typing and blood testing of the vaginal swabs secured in the Johnson Rape Kit (which had never been analyzed); and finally, the prosecutor had committed misconduct in his final argument to the jury, thus denying Hammond a fair trial.[5]

One of the unanswered questions raised by the Hammond trial was, If the evidence had become contaminated, how did it get that way? No definitive explanation was ever offered at trial. Van Kirk volunteered after the verdict that, in his view, it was almost impossible to imagine a scenario in which the stains became contaminated. That argument would become central to Hammond's appellate argument. What turned out to be a key footnote in the brief filed by the University of Connecticut's legal clinic read:

Wholly apart from the absence of any direct evidence establishing actual contamination in this case, any imagined contamination theory would require all of the following unlikely and preposterous events to have taken place: First, the contaminant would have had to have been applied only within the crayon outline of the original stain on the clothing which was made by Dr. Hawkins the day after the assault. Second, the contaminant would have had to have been semen and nothing else, because the male and female fluids were chemically separated by breaking open the male sperm heads in order to segregate the male DNA, and if the additional fluid had been something other than semen, it would not have been isolated from the victim's DNA by the process described. In any event, however, there would have been 4 bands of DNA in the victim's lane and *none* in the unknown male's lane—a fact which the prosecutor expressly noted an unwillingness to dispute (T. 915–16). Third, the same exact semen would have had to have been applied to both the blue jeans and the underwear via the same source, since they both indicated that a[n] antigen secretor with Type A blood deposited the stain. Fourth, the semen would have had to have been added in the laboratory or when the samples were in police custody, because no one else had access to the clothing. Fifth, since the male portion of the stain came from only one individual, the defendant's DNA and antigen secretion, under the state's theory, would have had to have been obliterated to the point that neither showed up in any of the test results *at all*. Sixth, in obliterating the defendant's contribution to the stain, the contaminant would also have had to simultaneously leave Miss X's contribution to the stain *completely* intact, so as to yield female profiles on all five autoradiographs clearly matching the victim's DNA.[6]

The Connecticut Supreme Court was very troubled by the Hammond case. Hammond's lawyers were arguing, in effect, that the DNA evidence overpowered the other evidence presented at the trial, even though the other evidence, standing alone, was more than enough to convict their client. The court certainly agreed that Malone put on a "strong case."[7] But for the blood and DNA evidence pointing away from Hammond, his cause would have been lost. The court wrote,

> Absent strong exculpatory evidence, a reasonable juror could have found the defendant's guilt established beyond a reasonable doubt by the victim's consistent positive identification of the defendant, her substantially accurate description of his car and its contents, the scratches on his car consistent with its being driven in the woods as described by the victim, the numerous actions by the defendant that appeared to display a consciousness of guilt, and the testimony of one former friend of the defendant that the defendant had a habit of hanging his watch on the gear shift of his car, as the assailant had done.[8]

The appeal turned, then, on the phrase "absent strong exculpatory evidence." Was the defense's blood-typing and DNA evidence strong exculpatory evidence? Or had the prosecution demolished it? On appeal, the defense's position was that there was nothing solid in the trial record to justify the jury's outright rejection of the forensic DNA analysis and the blood-typing tests. The court thought long and hard about the facts and circumstances of the case before the court; as Malone put it, "punted."

The court quickly focused on the prosecution's theory of contamination. "It is physically impossible," the court reasoned, "for the defendant to have been the man who sexually assaulted the victim unless *both* the blood typing tests *and* the DNA tests made on the samples taken from the victim's clothing are for some reason unreliable," meaning either the testing procedures were inherently flawed or the items had been contaminated. By this time, the court noted, ABO blood typing tests, for one, were deemed so reliable that if exclusionary ABO blood tests were offered in a run-of-the-mill paternity case (something which was routinely done), a jury simply wouldn't be allowed to bring back a finding that the man was the child's father. And that is with a much lower standard of proof in a case with much less at stake. Whatever problems there may or may not be with forensic DNA analysis did not appear to be relevant to this case, the court felt, because the DNA analysis of the stains produced an exact match to the victim's DNA, if nothing else, strongly suggesting the DNA testing procedures were working fine.[9]

If contamination was the only way to account for the exculpatory results in Hammond's case, the court saw two problems. First, there was no specific evidence in the trial record of contamination occurring (in fact, the record made it seem "improbable"[10]). Second (and here the court was guided by the defendant's brief to the court), any conceivable theory of contamination in this case didn't make any sense at all. "Such contamination had to take place in the laboratory or when the samples were in police custody," the justices wrote, echoing the defendant's brief. "Moreover, both contaminants would have to have been bodily fluids from a secretor with type A blood, and the contaminants would have to have been carefully applied only within the crayon outline of the original stain on the clothing that was made by Hawkins when he first examined the clothing."[11] Furthermore, "[t]he state's theory also fails to explain how the DNA test could have falsely exculpated the defendant. Even if someone else's blood, semen, or saliva had been added to the small patch of clothing that had been circled with a red crayon, the state has not suggested how such bodily fluids could have destroyed the DNA that was already present in the stain."[12]

The court concluded that the state's theory of contamination was "untenable."[13] Established legal precedent suggested that a verdict based on the "physically impossible" was also untenable, and the defendant's motion for a new trial should have been granted.[14]

Oddly, perhaps, the court did not take the final leap and simply reverse the case, possibly because the court saw this as a particularly "hard case."[15] Just as it seemed unlikely that Hammond was guilty, the court explained, it also seemed unlikely that he was innocent. "While . . . certain improbable speculations are necessary to support a theory of guilt," the court wrote, "equally improbable speculations are necessary to support a theory of innocence."[16] How could the jury simply disregard all the other evidence? the court wondered. It would have been extremely difficult.

In the end, Hammond was saved by a fundamental legal precept that says that a draw goes to the defendant. Since only "equally improbable speculations" tended to support either theory of the case, that Hammond was innocent or that he was guilty, the presumption of innocence was the defendant's trump card. "It follows," concluded the court, "that if the hypothesis consistent with guilt and that consistent with innocence are equally reasonable, because they are equally unreasonable, then the defendant's conviction cannot stand."[17]

The mechanism for reversing Judge Purtill's decision not to grant Hammond a new trial was a determination that the judge had "abused his discretion" in making the decision. The court was not prepared to say, however, that Judge Purtill had really done anything wrong. He was simply underinformed. "The logical inconsistencies that we have found in the prosecution's theory of the case were not brought to [Judge Purtill's] attention by counsel for the defense," the court concluded. "It is possible that, fully informed, the trial court might have ruled differently on the motion for a new trial."[18] Similarly, Judge Purtill was underinformed about the "logical inconsistencies" when he made his decision to deny Van Kirk's motion for post-trial blood and DNA tests.[19] Consequently, the court decided to give Judge Purtill another crack at these two motions and remanded the case back to him for further proceedings "consistent with this opinion."[20] Although the court felt that some of Malone's statements during his closing argument were improper, they did not feel his comments were so egregious as to undermine the essential fairness of the trial. At the same time, the court strongly suggested that the prosecution's failure to test the vaginal swabs on its own initiative, notwithstanding the state's claim that before and during the trial, they, too, were under the impression there was insufficient material on the swabs for further testing, was a breach of the prosecution's ethical duty to lay all evidence before the court tending to aid in the ascertainment of the truth and may even have been a violation of the state's duty to disclose exculpatory evidence to the defense.[21] Despite these observations, the court took no direct action on these matters. Instead, these criticisms of the state appeared to serve as simply a strongly worded subliminal message embedded in the decision. Even without that subtext, however, the decision was pretty confusing.

"Having gone so far in [our] direction," Todd Fernow remarked after the decision was announced in February 1992, "I can't figure out why they just didn't do it themselves," adding, "I don't know what this means. I feel sort of empty. I've never seen a remand like this."

At the first hearing in the case following the high court's ruling, Judge Purtill also wondered why the Supreme Court hadn't simply reversed him. It was unclear to everyone just how to proceed, a fact that dragged out the case. Leaving the decision for another day whether to grant Hammond a new trial, Judge Purtill reversed his earlier ruling and granted the defense's motion for post-trial blood and DNA testing. The basis of his decision to now grant the motion, he told the lawyers, "is that the Supreme Court says I have to."[22] With that, arrangements were made for the rape kit to be sent anew to the FBI laboratory for further tests, tests that would take months. One thing did become clear at the hearing. John Malone was not about to back down. Hammond remained in jail.

Six months later Judge Purtill finally revisited the motion for a new trial that he had earlier denied. He finally agreed to grant Ricky Hammond a new trial, noting, "I'm bound by the ruling of the Supreme Court."[23] The case now went back on the pre-trial docket. It was as if Hammond had just been arrested. The FBI had not yet reported back its new DNA findings. Malone made clear his intention to retry the case and Hammond remained in jail.

About a week later, the FBI laboratory reported back that it had succeeded in analyzing the vaginal swabs with a new, less discriminating DNA technique, the PCR-based HLA DQA1 test. There was insufficient material to conduct additional RFLP tests. The newer test actually looks at a specific gene sequence that varies within the population and is related to tissue type. Like conventional blood tests, DQA1 can only narrow the field by identifying specific genotypes that range from rare to reasonably common. At the same time, DQA1 focuses on genotypes that occur in much narrower slivers of the population than blood-typing tests. The most common DQA1 genotype, for example, is shared by only about 10 percent of the population. Such tests were most often used in cases where the available biological sample was too small for the more discriminating VNTR analyses available at the time or as a quick way to determine whether police had focused on the right suspect. (These were the first DNA tests conducted in the O. J. Simpson case, for example.) The FBI produced DQA1 types for Lisa Gray, her boyfriend, Ricky Hammond, and the unknown male contributor to the material preserved on the vaginal swabs and slides. Most significantly, the unknown male profile did not match Hammond's DQA1 type. At the same time, the sample did match the type of Lisa Gray's boyfriend, so the boyfriend was included in the sizable percentage of the population that could have contributed to that sample. At a hearing convened within days of the DNA news to consider Hammond's bond status (his bail remained too high for him to post), Malone declared before a pre-trial judge, Richard A. Damiani, that the new DNA testing "doesn't change anything," since it might only mean that the semen on the swab was residual from consensual relations the night before the attack. Perhaps the rapist did not ejaculate, he suggested. Despite the ambiguity, the judge released Hammond without bail, subject to strict supervision by the authorities. By this time, Hammond had spent more than three years in jail, including his time in jail before his original trial.

The case ended, somewhat oddly, on June 15, 1993, when Ricky Hammond pleaded no contest to two misdemeanor charges, sexual assault in the fourth degree and unlawful restraint in the second degree, a plea taken before Judge Damiani. Another prosecutor, Christopher Morano (now Chief State's Attorney), stood in for Malone. As the plea was being formally entered in open court, Hammond responded "no contest" to the charges for the record. The judge then recited the facts of the case leading up to Hammond's original arrest and asked Hammond whether he understood that he was pleading guilty to the facts as recited. Hammond responded forcefully, "I'm under the impression that I didn't plead guilty at all." "Strike the word guilty," the judge corrected. The sentence imposed was two years, less than the time that Hammond had spent in jail, so he was processed out of jail that day.

Todd Fernow, whose clinic had won Hammond his new day in court and who handled the plea agreement for Hammond, characterized the resolution of the case as "peace with honor."[24]

4 The Improbable
Origins of PCR

POLYMERASE CHAIN REACTION (PCR), OR DNA AMPLIFICATION AS IT IS commonly known, is a technology that has vastly expanded the ability of forensic scientists to type the DNA of an individual. While RFLP analysis is limited to substantial amounts of DNA, PCR-based tests can work on quantities as small as DNA from a single nucleated cell. Although it is possible the jury in the Ricky Hammond case thought otherwise, the hair-comparison evidence added little to the case against Hammond. But if the hair found in Ricky Hammond's car had been genetically typed with the help of PCR—as it is possible to do today—it could have been pivotal.

Forensic PCR work has been around from the beginning. In fact, PCR was used in the very first DNA case ever in the United States, *Pennsylvania v. Pestinikas*[1] in 1986—the same year Alec Jeffreys introduced his "DNA fingerprinting" technique. The case involved the DQA1 typing of preserved human organs believed to be from two separate autopsies. The organs had been stored in formaldehyde and the DNA had degraded well beyond the point where RFLP analysis, as pioneered by Alec Jeffreys only months before, was possible. A PCR analysis, however, indicated that the chances that the two sets of organs were from two different people, as had been believed, was only between 1 and 2 percent, since both sets exhibited the very same homologous subtype, DQA1 1.1, 1.1. It was therefore concluded that both organs belonged to Pestinikas.

So it was from the very beginning that the utility of PCR-DNA typing was recognized. Nevertheless, the FBI (which only began using PCR-based tests in mid-1992) and the courts approached the technology with a certain caution, primarily because of the ever-present risk of contamination with PCR.

PCR is not in fact DNA typing at all. It is, instead, a kind of genetic jump-start for forensic DNA typing, as well as a host of other genetic tests. In other words, PCR explodes the amount of DNA available for a DNA test by targeting a pre-selected DNA segment—available at the outset from as little genetic material as a single cell—and reproducing the segment in a carefully controlled chain reaction until there are millions, even billions, of copies of it. PCR has often been referred to as "genetic Xeroxing," but scientists are only replicating a DNA sequence of interest, not the entire genome. Once the DNA has been multiplied, scientists are then free to perform any number of genetic tests on the sample without worrying about exhausting the supply of DNA (which, of course, can have serious repercussions in criminal cases).

PCR "is to genes," proclaimed *U.S. News & World Report* in 1990, "what Gutenberg's printing press was to the written word, and it promises to be no less revolutionary."[2]

PCR harnesses the natural ability of DNA to replicate itself, except that with PCR, scientists jolt nature into working harder and faster than it would otherwise be inclined to do. Once a crime-scene sample has been determined (with chemical tests) to be blood or some other DNA-containing tissue, the very first thing a forensic scientist does in preparing to perform PCR (or any DNA test, for that matter) is to soak a portion of a crime-scene specimen, such as a swatch of material or a cotton swab, in a test tube or some other container to release the cellular material into solution. Ideally this procedure releases approximately 90 to 95 percent of the cells attached to the fibers of the material. The cellular material in solution is then reduced to pellet form by spinning the test tube in a centrifuge until a defined sediment of cells collects at the bottom of the test tube. If the specimen is blood, the scientist targets the white blood cells in the sediment since red blood cells are devoid of DNA. When blood is put in a centrifuge, the red blood cells settle out faster and collect as a bottom sedimentary layer in the test tube. Immediately above the red blood cells, the lighter white blood cells form a definable sediment.

In rape cases, the scientist initially looks under the microscope for sperm. Epithelial and sperm cells are separated by a process known as differential extraction. If sperm are detected under the microscope, their concentration (quantity of sperm) may be noted. Next, a DNA extraction buffer is used, which is a solution containing salts, a buffer material, a detergent (SDS, or sodium dodecyl sulfate), and the enzyme proteinase K. These ingredients would be added whether or not there are sperm in the specimen. The detergent operates to dissolve the cell membrane, and the proteinase K chews up proteins that are present. The mixture is then incubated in an oven for an hour, during which time the DNA is released from the interior of the cells' nuclei due chiefly to the action of proteinase K. If sperm cells were contained in the mixture, however, they will survive this stage and remain intact because the outer sperm membrane is chemically resistant to such treatment.

In sexual assault cases, the released epithelial DNA will then be collected and saved apart from the unaffected sperm cells. To do this, the digested epithelial DNA and the intact sperm cells are separated in a centrifuge, where the heavier undigested sperm cells settle to the bottom of the test tube. This separation process, where sperm and non-sperm cells are differentiated, involves a critical washing step where the scientist endeavors to cleanse the collected sperm cells of any remaining epithelial cell DNA. The washing is done several times. The washing procedure has its limits, however, and the separation is rarely considered perfect, so an analyst must consider the possibility that the ultimate typing results reflect a mixture.

Once the separation is deemed complete, a new extraction buffer solution is added to the isolated sperm cells for digestion, but this time an additional chemical is added; called a reducing agent, the most commonly used chemical is DTT, or dithiothreitol. DTT opens the sturdy sperm membrane and permits the detergent SDS and proteinase K to invade the cells and release their DNA into solution. The DNA is extracted from this solution by the addition of more organic chemicals, which dissolve any other remaining cellular material and remove the digestive proteins. Next, the isolated, purified DNA is concentrated and washed.

The first question that arises at this point is how much, if any, DNA has been preserved? And secondly, if there is any DNA, is it high molecular weight DNA or is the sample degraded?

The molecular weight can be estimated by electrophoresis. Electrophoresis is performed by placing a small percentage of the DNA material, perhaps 10 or 20 percent of the extracted DNA, on an agarose gel and applying an electrical charge. DNA will separate out based on size and the charge of the DNA molecule. DNA can be detected on the gel because DNA has the ability to chemically bind to ethidium bromide, a chemical that fluoresces when exposed to ultraviolet light. The quantity of DNA is estimated by the intensity of the fluorescence of the sample DNA compared to several control samples with known amounts of DNA.

The amplification stage has been enormously simplified by the development of pre-packaged amplification cocktails, originally developed by Cetus Corporation. The prepared cocktails contain the following: a buffer (most commonly, diluted magnesium chloride, which serves to maintain the proper PH balance); the primers, or short pieces of DNA that recognize their denatured gene complement and bind to it, thus bracketing off the region of interest in the targeted DNA region, the four building-block bases of DNA (which when stimulated to do so by the enzyme polymerase will assemble themselves along the denatured strands during the actual amplification process); and the enzyme DNA Taq polymerase, which is naturally responsible for the replication, or copying, of genes. Heating and cooling the mixture of reagents and the selected DNA in the proper sequence will achieve the desired result.

PCR provokes DNA to produce exact copies of itself, in the same way that it naturally does during cell replication. The first step is the denaturing, or separation of, the strands of the DNA molecule. This is achieved artificially in the PCR process by simply turning up the heat to approximately 94 degrees Celsius (near the boiling point of water). Each strand of the denatured DNA can now serve as a template for a new, complementary strand of DNA that will be induced by Taq to assemble itself along the template.

All of this, incidentally, takes place automatically in a machine called a thermal cycler that takes the reaction tubes placed inside through the paces of the amplification process. In practical terms, PCR amplification has been made extraordinarily simple for the analyst.

In the next step, oligonucleotide (multiple nucleotide) primers bind to their complementary sequences in each strand of the target DNA when the temperature is lowered (to, for example, 55 degrees C). The primers bind, or anneal, to their complementary sequences in such a manner as to flank, or mark off, the region of DNA to be amplified. However, the primers are designed to attach themselves just outside the region of interest; they are not part of the polymorphic (genetically variable) region. This accounts for why the primers work in everybody's DNA, since all humans share a sequence of DNA complementary to designed primers.

Next, DNA polymerase induces the reaction that causes the new strands of DNA to form. The natural functions of the DNA polymerase enzyme are the repair of DNA and DNA replication. The enzyme that proved most efficient for PCR is referred to formally as native Taq DNA polymerase. Nicknamed Taq, the enzyme was originally isolated from a thermophilic bacterium known as *Thermus acquaticus,* which somehow manages to survive in the hot springs and geysers of Yellowstone National Park. As it turned out, the enzyme that the bacterium uses to copy its own DNA is the same kind of enzyme humans require for copying DNA when a cell divides. Moreover, the enzyme is not fazed by high temperatures. Taq is capable of surviving essentially undamaged through the heat denaturation stage of the PCR cycle.

Taq and a plentiful supply of unattached nucleotides (the necessary building blocks of DNA) in the cocktail cause a new DNA strand to form along the template beginning at the primer and extending across the target sequence, thereby making copies of the target DNA. The process is repeated dozens of times (or cycles) to produce an adequate supply of the target DNA. With each cycle, the DNA segment is multiplied exponentially, increasing the amount of DNA from the previous cycle. In a matter of hours the reaction tube is teeming with identical copies of a genetic sequence scientists want to analyze.

PCR has a singular advantage over RFLP, one that is particularly important for forensic science: PCR can be successful with DNA that is considerably older or has been damaged somehow. When biological samples are exposed to a variety of environmental insults, such as sunlight, moisture, and bacteria, the DNA breaks apart, or degrades, into smaller and smaller sections. At a certain point, RFLP analysis is no longer possible because the DNA molecule has become too fragmented and disjointed for the test to yield interpretable results. PCR, on the other hand, requires less DNA at the start and is more forgiving of fragmented DNA. The shorter repeating sections of the genome, or STRs (short-tandem repeats), were seen as an ideal complement to the PCR process because they can often be detected in degraded DNA samples.

PCR can also be done very quickly. RFLP typically takes about six to eight weeks, while PCR tests can be run in a matter of hours. Many PCR-based DNA tests, however, have an important disadvantage when compared to RFLP tests: The types of analyses that can be done are far less discriminating than RFLP tests, which analyze highly polymorphic regions of human DNA. In contrast, many PCR tests are aimed at regions of the genome that do not vary nearly as much. (This includes STRs, so scientists must analyze more STR sites in the typical case.) The effect would be the same if a certain percentage of the population shared identical fingerprints (which they don't), so that finding a particular set of fingerprints around the throat of a victim would only tell investigators that an individual in the 3 percent of the population that has that set of fingerprints strangled the victim; the prints in the hypothetical case could not positively identify the murderer, they can only narrow the field.

PCR is the brainchild of Dr. Kary Mullis, who earned his Ph.D. in biochemistry at the University of California at Berkeley and remains one of the most colorful scientific figures of the 20th century. Mullis conceived the idea for PCR in 1983 while working at Cetus Corporation, one of the pioneer biotechnology companies established in the 1970s to take advantage of the newly available tools for studying DNA. The idea for PCR came to him while he was driving along a mountainous highway on the way to his weekend cabin in Mendocino, California, thinking about an experiment to examine a single allele of the gene that causes sickle-cell anemia. Although a group of rival scientists announced the first diagnostic test for sickle-cell anemia just months before, Cetus executives were determined to develop a simpler and commercially viable diagnostic test for the disease.[3] Mullis's ruminations naturally turned to this objective. Suddenly, he was seized by the idea of repeating a step of DNA replication; a particular stretch of DNA, he recognized, could be reproduced ad infinitum if the step was repeated over and over again, at least theoretically.[4] The idea that seized him on the way to Mendocino was so compelling that Mullis pulled the car off the road several times to scribble down calculations and stayed up the entire night working on the idea. Upon his return to work on Monday, Mullis began a search of the scientific literature to see if the idea had already been thought of, but he turned up nothing.

After a year of experimental setbacks with various genetic sequences, Mullis returned to the sickle-cell gene (a beta-globin mutation) that originally inspired him and succeeded in amplifying it in an experiment that would, to some degree anyway, withstand scrutiny from his colleagues. (Mullis was perceived by some co-workers as suffering from a lack of scientific rigor.) His efforts to promote his success at the 1984 Cetus annual meeting in Monterey, California, however, degenerated into a late-night brawl with another scientist over the viability of the process that had become known as PCR. Hotel security was called at one point to help Mullis walk off some steam.[5]

Now, Mullis's job was on the line and there was considerable pressure to fire him, but the potential of PCR proved his salvation, although he was removed as head of his lab and put on probation.[6] He was specifically given the opportunity to work on PCR for another year. Over the next few months, senior scientists at Cetus adjudged Mullis's experiments, to date, substandard (at least insofar as they believed the work would fail to survive the peer-review of journal submission), but they remained intrigued.[7]

A few months after Mullis's brush with unemployment, a unit at Cetus was formed under the direction of Dr. Henry Erlich referred to as "the PCR group" and staffed by talented scientists.[8] Using beta globin as the model, the PCR group succeeded over the next six months in demonstrating—to everyone's satisfaction— that PCR was indeed amplifying specifically targeted DNA. The first PCR patent applications were filed on March 28, 1985. By the end of that year, Cetus had entered into a joint venture with Perkin-Elmer Corporation to develop PCR instrument systems and reagents for the research market. This venture would produce the thermal cyclers that automated the PCR process and are now found in virtually every laboratory engaged in molecular biology. (The first prototype of the machine was dubbed "Mr. Cycle."[9])

Thermal cyclers are essentially heating blocks that are regulated by a computer; a technician places test tubes full of genetic material and chemical reagents in the block, programs the computer, and the machine runs the test tubes through the various heating and cooling cycles automatically.

Although there continued to be friction between Mullis and some of his colleagues, the general progress of PCR development was very promising. It was therefore agreed that publication of PCR should be done as quickly as possible. Publication was to be in the form of two papers, one fundamental and theoretical and written by Mullis alone, and the other demonstrating an application of PCR diagnosing sickle-cell anemia and authored by all the scientists who worked on the development of the test, including Erlich and Mullis. The applications article, however, did not list Mullis as the lead author. For a variety of reasons, the journal *Science* readily accepted the applications paper, and it was published on December 20, 1985,[10] but Mullis's theoretical paper was rejected by *Nature* in the same month and later by *Science*. Since the *Science* applications article on PCR, which illustrated PCR's use in conjunction with another Cetus technology, Oligomer Restriction, went to press almost immediately, Mullis's paper was rejected as merely technical in nature and unoriginal![11] Mullis was understandably stung by the rejections, and he reacted by accusing his colleagues of in effect pirating his invention.

Matters worsened when his solo article, resubmitted to the journal *Methods in Enzymology,* was unaccountably delayed for a year before it was finally published in 1987.[12] While this wound festered, however, James Watson, who then headed the famous Cold Spring Harbor laboratory, offered balm for Mullis by inviting him to present PCR at a May 1986 symposium on "The Molecular Biology of Homo Sapiens," a presentation that was published later in the year in the center's prestigious journal naming Mullis as the lead author. It was a high honor for Mullis, and his presentation was extremely well received.

Significantly, the PCR process was greatly enhanced during all of the publication chaos when scientists at Cetus began using Taq polymerase. In early experiments, additional polymerase had to be added after each cycle because the enzyme was destroyed by the heat. In addition to relieving scientists of this chore, Taq proved to be more specific and produced far more of the genetic target than the enzymes previously used.

Mullis and a team of researchers at Cetus under the direction of Henry Erlich went on to develop the reagents for PCR and perfect the technique. Frustrated by corporate life, however, Mullis resigned from Cetus in September 1986, taking with him from the company a $10,000 bonus, which would, in retrospect, seem a paltry sum for such a historic invention.

The invention of PCR would not go unchallenged, however. DuPont Corporation sued Cetus over the patents for PCR (which were approved in 1987), claiming that one of its scientists, Nobel Prize-winner Har Gobind Khorana of the Massachusetts Institute of Technology, had already invented PCR in the early 1970s. Citing two papers written by Khorana, published in 1971 and 1974, DuPont argued that PCR was nothing new at all, that all of the ingredients for PCR had been available in Khorana's lab since the late 1960s.[13] Under patent law, this was "prior art" that invalidated Cetus's patents, DuPont argued, confidently launching the sale of its own PCR diagnostic tests while the suit was pending. In fact, Khorana's experiments explored the possibility of making copies of a sequence of DNA that had been assembled in the laboratory with the use of the cellular enzyme DNA polymerase (not Taq), discovered in 1955 by Arthur Kornberg of Stanford University, who testified on behalf of DuPont in the patent case, but Khorana's experiments were abandoned before it was discovered how to keep the two separated strands of DNA from reconfiguring themselves into double-stranded DNA once the mixture was cooled. In the early 1970s, too, other scientists announced the invention of recombinant DNA cloning (which involves inserting foreign DNA segments into bacteria that reproduce the segments as they multiply), a technology that solved DuPont's immediate research problems, so Khorana's experiments were shelved. On February 28, 1991, a jury unanimously upheld Cetus's patents.[14]

Despite the legal victory over DuPont and accolades from the scientific community, Cetus as a corporation was in decline, rocked by years of corporate infighting and defections of key scientists. Cetus had embraced the war on cancer early on and pursued the "magic bullets" of cancer research. In particular, Cetus determined to become a principal supplier of a version of the drug interleukin-2 (IL-2), even though the exclusive patent rights on recombinant IL-2 were held by Hoffman-LaRoche, a unit of Switzerland's Roche Holdings Ltd., through a sub-licensing agreement with the primary patent holders. Cetus's version of the drug was originally devised (that is, chemically modified) as an end run around patent restrictions, but it soon became clear that Hoffman-LaRoche was in an excellent position to squash Cetus's plans for IL-2. By 1988, however, Hoffman-LaRoche became very interested in PCR, and a deal was struck: Cetus would be free to pursue its IL-2 plans and receive a cash infusion and, in return, Hoffman-LaRoche would jointly develop and market PCR diagnostic products and services. By 1990, Cetus was continuing to lose money, as well as disaffected scientists, and everything was riding on IL-2. Then in July 1990 disaster struck: An advisory panel of the Food and Drug Administration rebuffed Cetus's drug application for IL-2, sending it back for more data, the acquisition of which would unquestionably cost the company lots of money. Cetus's stock plummeted, and within a month the company's board of directors forced the company's president, Robert Fildes, to resign.[15] A year after the FDA decision, Chiron Corporation, another start-up biotechnology company that had begun by renting laboratory space from Cetus in the early 1980s,[16] bought its former landlord and soon turned Cetus's PCR into a financial bonanza by selling PCR outright to Hoffman-LaRoche for $300 million, plus additional royalties. The FDA subsequently approved IL-2, now Chiron's, for the treatment of kidney cancer.[17] Henry Erlich later presided over PCR as head of Roche Molecular Systems, a subsidiary of Hoffman-LaRoche.

For his part, Mullis came forward to defend in court Cetus's PCR patents in the fight with DuPont, despite the persistence of bad feelings with some of his former co-workers. After leaving Cetus, Mullis turned to consulting work, writing, the lecture circuit, and a variety of fringe scientific pursuits. He also co-founded the California-based company Stargene that endeavored for a time to hawk jewelry encasing the amplified DNA of various Hollywood stars and famous historical figures. He is also an avid surfer and (reportedly) a notorious womanizer. His lectures have become flashpoints because he has been known to intercut his slide presentations with revealing photographs of past and present girlfriends.[18] He also attracted criticism for aligning himself with a fringe group of scientists that continue (in the face of overwhelming evidence, much of which was developed using PCR) to dispute that HIV is really responsible for causing the symptoms associated with AIDS.[19] Mullis, moreover, has been pathologically open about his own recreational drug history, dating to his college days in the 1960s, and he admits to the continuing use of LSD on occasion "to be reminded of the complexity of things," as he explained to a reporter for The New York Times.[20]

Despite his curious scientific allegiances and questionable personal conduct, Mullis has earned the highest honors science has to offer. In 1993, Mullis was awarded the Nobel Prize in Chemistry for inventing PCR. According to news accounts, Mullis bragged about being drunk when he learned of receiving the award[21] and went surfing immediately afterward.[22] Aside from the sheer honor of the award, Mullis split $825,000 in prize money with a co-recipient of the Nobel (who won for an unrelated discovery). "I figured they had to give it to me eventually," Mullis told *Science* in 1993.[23] A few months later, Mullis received another valuable international science award, the Japan Prize, bestowed by the Science and Technology Foundation in Tokyo, a prize that included a $385,000 cash award.[24]

Mullis gained perhaps his greatest visibility when he was named as an expert witness by the legal team representing O. J. Simpson in his criminal trial. Referred to by name, in fact, by Defense Attorney Johnnie Cochran in his opening statement in that case, Mullis was billed as a DNA expert prepared to take the stand and proclaim that PCR, his own invention, was (at least in the Simpson case) unsuitable for forensic DNA analysis. PCR-based tests constituted the bulk of the DNA tests that weighed in against Simpson. According to Cochran, the Nobel Prize-winning inventor of PCR was the ideal expert witness to undermine the majority of the DNA evidence against his client. But not everybody was convinced, given Mullis's checkered past and pyrotechnic present.

Any doubts about Mullis's fitness for the role of scientific foil became not unreasonable doubts when Deputy District Attorney Rockne Harmon, one of the state's principal DNA experts in the Simpson case, vowed during the trial to savage Mullis on the stand and cross-examine him "on every aspect of his life which reflects on his credibility, competency and sobriety."[25] In arguments to the court outside the presence of the jury, Harmon suggested that perhaps Mullis could prevent questions about his use of LSD by entering a detoxification center before his appearance in court.[26] The defense attempted to head off Harmon's inquiries by asking the judge to preclude questions by Harmon about Mullis's past or current lifestyle. A defense motion filed in court in August 1995 explained: "Since the time of Plato, scholars have recognized that one cannot judge the merits of an argument based on the character and lifestyle of the person advancing the argument."[27] Before the judge had an opportunity to rule on this philosophical point, however, defense attorneys withdrew Mullis's name from their witness list and gave up on the idea of calling him as a witness in the case.

Henry Erlich's laboratory at Cetus developed the PCR-based systems for analyzing the HLA DQA1 gene and in 1990 rolled out a DQA1 test for the forensic science community.[28] PCR ultimately became a cornerstone technology in the Human Genome Project, and PCR-based tests are widely used in the diagnosis of hereditary and infectious diseases.

Acknowledging in 1989 that PCR had changed the scientific landscape forever, the journal *Science,* which is published by the National Academy of Sciences, honored the PCR technique, along with the enzyme Taq polymerase, by awarding both the journal's first annual designation of "Molecule of the Year." Reporting that PCR "applications are burgeoning," *Science* hailed PCR as "one of the most powerful tools of modern biology," a discovery "that is likely to have the greatest influence on history" (the necessary criterion for meriting the award).[29]

Since PCR has been widely available, scientists have studied several regions of the human genome for their suitability for forensic DNA typing using PCR. The first coding region to be exploited, and widely used in forensic casework, was the human leukocyte antigen (HLA) DQA1 gene system, which is itself a part of a larger genetic region referred to as the major histocompatibility complex (or MHC), which governs organ and tissue compatibility.

Human leukocyte (or lymphocyte) antigens are protein-sugar structures found on the surface of most cells, excluding, however, red blood cells. These are the immune system proteins that identify foreign cells and will trigger a battle to rid the body of invaders when they sense their presence. Once these structures detect outsiders, they signal the body's T cells to destroy them. These antigens first came under scrutiny in the 1960s during pioneering efforts to transplant organs and were found to be related to the body's acceptance or rejection of the organ or tissue grafts. It was subsequently discovered that the HLA genes are genetically variable (polymorphic) and that it is of critical importance to match the HLA genetic characteristics of recipients with donors. Organ transplantation is now a commonplace medical procedure and short-term survival rates are extremely high, but recipients of organs that have perfect HLA matches (of six loci) will survive far longer than recipients of organs with less than perfect matches.[30] Given the importance of this gene system to medicine, the HLA DQA1 gene complex is among the most thoroughly understood regions of the genome. (DQA1, however, is just a small part of a greater complex of genes in the HLA system located along the short arm of chromosome 6.)[31]

Although the DQA1 region is itself comprised of thousands of base pairs, the specific allelic variations that interest forensic scientists occur within a particular region of the gene consisting of only 242 base pairs. This is the segment of the gene targeted for amplification. The remainder of the HLA-DQA1 gene is the same in everyone (that is, it is nonpolymorphic), although a number of other variable loci occur elsewhere within the greater HLA complex.

There are a variety of alleles at the DQA1 locus, producing a large number of genotypes. The identified alleles fall into four major types, denominated types 1, 2, 3, and 4. Types 1 and 4 are further subtyped as 1.1, 1.2, and 1.3 and 4.1, 4.2, and 4.3. (Only the subtypes of type 1 were investigated in early forensic casework using this system, but the subtyping of 4 eventually became standard.) HLA DQA1 tests are designed to distinguish between the variations of the DQA1 gene. Since humans can inherit at most two alleles at a particular site, one from each parent, there are a total of 28 different pairings that can occur within a population. Whereas RFLP or STR analysis measures the size of DNA fragments as a means of differentiating between individuals, DQA1 is designed to detect whether a specific order of nucleotides exists at one particular allelic site. To determine which of the alleles are present, scientists employ specifically designed, short oligonucleotides (15 to 30 nucleotides long) referred to as allele-specific (or sequence-specific) oligonucleotide probes. Scientists mix the probes with the dissociated DNA strands generated in the amplification process to permit hybridization to take place. The probes and the DNA product will not hybridize unless the sequences complement each other perfectly.

To permit visualization of the results, DQA1 probes, like RFLP or STR probes, are labeled. When a DQA1 probe finds its complement in the PCR product, the transparent mixture—with the addition of chemicals—will turn blue. In the most common typing method, the DQA1 probes are immobilized horizontally on a nylon membrane and flooded with PCR product that has been denatured. The DQA1 alleles in the amplified DNA will be captured on the membrane by their matching probes. When the probes find perfect complementarity in the PCR product, hybridization occurs. After a series of washings and heat treatments, the membranes are placed in a solution to which chemicals are added, triggering an enzymatic reaction that causes the color change to occur.

Population geneticists consider the figures for frequencies of the alleles occurring at DQA1 to be extremely reliable. DQA1 typing is not meant to be a means of positively establishing identity, however, since all of the 28 different phenotypes that might randomly occur at the DQA1 locus are found among humans in relatively high numbers. According to FBI figures, the most frequent of the 28 phenotypes occurs in one of every five people, while the least frequent phenotype occurs in one in every 800 people. Even with these numbers, however, DQA1 typing is a powerful way to eliminate suspects.

To enhance the discriminating power of such PCR-based testing, DQA1 was eventually combined with another allele-specific DNA system, called Polymarker, and packaged in the same kit (the AmpliType PM+DQA1 system), which took advantage of the ability of scientists to analyze several genetic markers simultaneously (in the same test tube) in a process known as multiplexing. Polymarker targets five separate loci (LDLR, G4PA, HBGG, D7S8, and GC) with either two or three alleles possible at each site. With Polymarker, primer sets are used that seek all variations of the five loci, and 13 probes are used to indicate which alleles are present. Polymarker alleles are visualized on typing strips in the same way that DQA1 alleles are, by generating colored dots. Standing alone, the discrimination power of Polymarker is only one in 200; combined with DQA1, however, the power of discrimination is enhanced to up to one in 2,000. Although DQA1 and Polymarker analyses have been involved in many significant cases, including the O. J. Simpson case, the use of these systems has been essentially phased out in recent years in favor of the latest generation of forensic DNA tests, PCR-based STR systems, which will be described later in the book. The Amplitype PM+DQA1 kit was in common use by forensic laboratories for nearly a decade.[32]

The invention of PCR electrified scientists from a wide range of disciplines, many far afield of forensic science, and has even spawned a few specialized scientific disciplines previously unimaginable. These new disciplines include, for example, molecular archeology (which boasts, among other things, the genetic analysis of Egyptian mummies) and molecular paleontology—the new field that has given the world, aside from its actual accomplishments, the idea behind *Jurassic Park*. The pioneers in these fields may even have inspired forensic scientists to explore the use of PCR in a variety of criminal contexts once it had been demonstrated that PCR was helping solve the mysteries of archeology and paleontology. In addition, out of these efforts came some analytic techniques and DNA targets that subsequently found use in forensic casework, including Alu-repeat typing and, far more significantly, mitochondrial DNA testing.

Writing for *The New York Times* in the summer of 1991, following the release of Michael Crichton's novel *Jurassic Park*, Malcolm W. Browne asked, "Will it one day become possible to breed a living dinosaur from genes preserved in fossils?" A few scientists, according to Browne, believe the idea "can no longer be dismissed out of hand." Browne goes on to explain why anybody would ever even dream of such a thing: "A major reason for their surprising optimism is the recent development of a laboratory technique called the polymerase chain reaction, or P.C.R."[33] PCR caused a stampede by scientists the world over to hunt down and analyze every shred of biological material that might harbor interesting and intact genetic information. The ranks of the scientists participating in this gene rush included scientists from virtually every biological discipline. Beginning in the mid-1980s, headline-grabbing accomplishments began occurring with regularity: "Intact Genetic Material Extracted from an Ancient Egyptian Mummy,"[34] "40-Million-Year-Old Extinct Bee Yields Oldest Genetic Material,"[35] "DNA from the Age of Dinosaurs Found,"[36] "A Scientist Says He Has Isolated Dinosaur DNA,"[37] "30-Million-Year Sleep: Germ Is Declared Alive,"[38] and "Neanderthal DNA Sheds New Light on Human Origins,"[39] to name a few.

The question is: Were all of these reports true? If DNA could survive for millions of years, then it certainly could survive long enough on contemporary evidence samples to allow DNA analysts to determine the origin of a bloodstain. Right? On the other hand, if lawyers and forensic experts in criminal cases were prepared, practically, to beat each others' brains out over the reliability of DNA tests, how could scientists working with ancient materials so readily accept the results of DNA tests? The answer, of course, is that they didn't. Although there were no fierce courtroom battles, the debates were pretty lively.

The technologies used to pursue ancient DNA are virtually the same as those available to forensic scientists. The procedures and techniques are the same. And the problems encountered are, to a certain degree, identical. A review of the scientific literature demonstrates that forensic scientists maintained a watchful eye on the developments concerning ancient DNA, and vice versa.

Genetic material can, in fact, sometimes reveal valuable information and answer some intriguing, even pressing, questions: How is one species related to another? How long can the DNA molecule survive without vanishing altogether? Did such and such a historical figure suffer from this or that genetic disorder? Or father illegitimate children? Are these remains who we think they are?

In 1962, Dr. George O. Poinar, an insect pathologist at the University of California, Berkeley, stumbled upon a piece of amber while walking along a beach on the western coast of Denmark. So intrigued by the translucent gem was Poinar that he began a collection and, by the 1980s, was devoting his life to cataloguing and analyzing amber, and, more particularly, the things that were found inside it.

In 1982 Dr. Poinar, along with his wife, Dr. Roberta Hess Poinar, a microscopist in Berkeley's entomology department, made a startling discovery. Examining a fly that had become imbedded in Baltic amber some 40 million years before, they found their specimen to be much more than just an imprint of the original creature.[40] "Instead," Dr. Poinar wrote in *The Sciences* magazine, which is published by the New York Academy of Sciences, "we observed dark areas within the outline, indicating that the body of the organism itself, not an impression or fossil, was inside the amber."[41]

By the summer of 1993, the year the first *Jurassic Park* movie was released, Dr. Poinar had in his collection preserved creatures in amber that were alive at the dawn of the dinosaur age, 230 million years ago. "We're not talking about an imprint in stone," Dr. Poinar emphasized in an interview with *Omni* magazine in 1993. "This is the entire organism that is preserved to the point that we can actually make out cellular structures in exquisite detail, including the nuclei where the genes reside."[42]

Encouraged by these observations, Dr. Poinar got together with other amber enthusiasts and formed the Extinct DNA Study Group in the early 1980s, something of a fringe paleontology think tank. It was this group that came to the attention of Michael Crichton in 1983 and fired his imagination. Crichton acknowledges the Poinars' contribution in his novel *Jurassic Park*.[43] In a journal article about the Baltic amber, the Poinars fired the imaginations of many scientists when they explained that the ancient insect in their 40-million-year-old amber "corresponded to what one would expect to find in a routine examination of present-day insects. . . . In essence, what we are describing is an extreme case of mummification. . . ."[44]

The early 1980s was a time when the mental flywheels of a few dedicated scientists were in high gear. Their goal was to find some vestige of "ancient DNA," a term that loosely defines the objective of retrieving strings of nucleotides from something long dead. At first, for researchers determined to see if DNA could be recovered from relics of the past, the technology at their disposal was limited. But it was enough to get things started. Beginning in the early 1970s with the invention of recombinant DNA technology, or gene splicing, scientists began in earnest to investigate the mysteries and possibilities of DNA. This technology enabled scientists to insert, or recombine, the hereditary information of one living organism into another, either to study gene function or as a way of multiplying genetic material.

Before PCR came along, there were only two ways to multiply DNA. The first was simply to grow the cells of the organism that contained the desired DNA. But since this was sometimes impossible, the second method, cloning, became the preferred method for generating DNA. With cloning (used here in its broadest sense, unrelated to reproductive cloning), however, there were significant snags. For one thing, cloning is laborious and time-consuming. For another, the process is inefficient; few cloning attempts actually succeed. An additional genetic breakthrough occurred in 1977 when Frederick Sanger described a rapid and reliable method for sequencing (or spelling out) genetic information. Once the genetic material had been cloned, researchers could then easily translate the information into an—at least, partly—understandable language.

Cloning is a process for copying specific segments of DNA. To clone a particular piece of DNA, scientists first cut the longer DNA molecule into fragments with the help of restriction enzymes, which act as molecular scissors, and introduce all the fragments into bacteria. Each bacterium obligingly takes up one of the fragments. The whole batch is allowed to multiply, replicating the fragments along with the rest of the bacteria in every successive generation of daughter cells. The resulting clones are identical offspring of the starting bacterium, but with an important difference: They have been customized with a new snippet of DNA. Techniques, such as hybridization (using a complementary probe that has been specifically synthesized for the purpose), are used to distinguish the clones containing the DNA sequence of interest.

With these techniques in hand, a quiet race got underway in the early 1980s by scientists eager to be the first to extract DNA from something old. Some researchers prowled quietly around museums exhibiting extinct animal species. Others tried to sell the cloning idea to curators of museums with ancient human exhibits. In either case, the proposal to "borrow" a slice of a prized museum exhibit, perhaps from the big toe of an Egyptian mummy, was no doubt a tough sell. Nevertheless, the idea proved too tantalizing to refuse.

The watershed event occurred in 1984 when a team at the University of California, Berkeley, led by Allan Wilson, succeeded in cloning DNA from an extinct creature called the quagga, a member of the horse family whose remains had been carefully stored in a German museum for 140 years. The team was provided with a piece of dried muscle and connective tissue from the skin of the animal, the remains of which had been preserved at the San Diego Zoo in its frozen repository of extinct and endangered species. Originally of Southern Africa, the quagga had gone extinct in the late 1800s after being hunted down by farmers trying to preserve their grazing land for livestock; in 1883 the last known quagga died in the Amsterdam Zoo. Its physical appearance remained a curiosity, however, and inspired erudite debate. Because the quagga seemed to resemble a peculiar cross between a zebra and a horse, there had been long-standing disagreements over the animal's exact lineage. Remarkably, the Wilson team, which included Russell Higuchi, not only managed to extract some DNA from the quagga carcass, they extracted enough to demonstrate that the quagga was in fact more closely related to the zebra than it was to the domestic horse.[45]

To authenticate their find, Wilson's group multiplied their genetic quarry through bacterial cloning and then sequenced the results. The genetic sequences obtained were then compared to the corresponding genetic sequences of the quagga's nearest living relatives, the zebra and the horse. The results were not unequivocal, but they were definitive enough not to dampen the exuberant mood of the scientists. The problems that did arise were attributed, quite rightly as it turned out, to the shortcomings of cloning.

The quagga study was the first retrieval of sufficiently informative DNA sequences from a preserved specimen to perform a phylogenetic comparison, meaning the comparison of genetic family trees. The DNA extracted from the quagga included both nuclear DNA and mitochondrial DNA (mtDNA), which exists in hundreds, even thousands, of discrete, compact subunits outside the nucleus of the cell in the cytoplasm. The molecular weight of the DNA recovered by Wilson's team was low, fewer than 500 base pairs in length, indicating the DNA was significantly degraded. In reporting the results of the quagga study, the Berkeley group specifically tipped their hats to George and Roberta Poinar for their 1982 observations. They also recognized that their quagga study, far from being the culmination of a scientific quest, was just the beginning. "If the long-term survival of DNA proves to be a general phenomenon," they wrote, "several fields including palaeontology, evolutional biology, archaeology and forensic science may benefit."[46]

Using the traditional extraction and cloning methods, the ancient DNA researchers barreled ahead. With only these tools, Svante Paabo achieved a milestone by excavating DNA (Alu repeats) from an Egyptian mummy more than 2,000 years old. The particular Alu-repeat sequences Paabo targeted are part of human, and only human, nuclear DNA.[47] (Other versions of Alu repeats do, however, appear in other species.) In design, they are similar to the VNTRs that are sought in standard VNTR analysis, but are less informative. Moreover, the repeated sequences apparently do not code for proteins.[48] Alu-repeats have since been adapted for forensic use because of their ability to differentiate between population groups based on their geographic locations.[49] Though plagued with problems, Paabo's experiments demonstrated that human DNA can survive, at least to some degree, for thousands of years.

As spectacular as the quagga and mummy studies were, the study of ancient DNA was about to hit a brick wall. The problem was not that DNA could not be retrieved from ancient artifacts; subsequent experiments on other species and ancient remains had verified that the earliest studies were not flukes. DNA was proving to be a resilient molecule, but it was also showing itself to be far from indestructible. It would change, sometimes radically, over time. More significantly, there were big problems with the technology—cloning—being used to replicate the ancient DNA, problems that were inherent to the technology and, therefore, insurmountable. Ancient, even old, DNA can be heavily modified by the ravages of time, reducing most segments to a few hundred base pairs. Furthermore, these short segments are often pockmarked by lesions—that is, lost bases, oxidized pyrimidines, and cross-linkages—in other words, a tangled mess.[50]

The practical side of working with heavily degraded DNA is that efforts to clone it, although successful, are barely so. The bacteria into which the damaged DNA is introduced will struggle to reproduce it, but will fail more often than not.[51] Consequently, very few clones will be produced. Further bedeviling the process is the fact that when the bacteria do succeed in copying segments of DNA, the well-meaning bacteria will often introduce errors. Bacteria naturally repair broken or incomplete strands of DNA during replication, but this natural process has a certain error rate. The result, therefore, is mis-repaired DNA. The errors produced in this way are known as cloning artifacts.

The picture that emerges from ancient DNA that has been conventionally cloned can be a distorted and unreliable one. Worse, since so little DNA is produced by cloning in the first place, there is no way to go back and verify one's findings by repeating the analysis. At least not by cloning.

"Molecular evolutionists who were keen on time travel therefore found themselves in a depressing situation," Paabo explained in *Scientific American* in 1993 about the realization that cloning was a dead end. "Because they could not verify their results by repeating an experiment, the study of ancient DNA could not qualify as a fully respectable science."[52]

The shortcomings of cloning—in particular, inefficiency and the mis-repair of DNA—effectively sabotaged any hope of reproducing or independently verifying the trail-blazing studies of ancient DNA.[53] Wilson's team, for example, had analyzed two mtDNA sequences in the quagga study, one in the gene for cytochrome oxidase I and one in the gene for NADH dehydrogenase I.[54] Both segments were "identical" to the plains zebra, the quagga's nearest living relative, except for one problem. Each of the analyzed sequences also contained a single nucleotide substitution, a letter out of place, so, in reality, the sequences were not in absolute agreement. Out of a total of 229 bases compared, the quagga appeared to have two that differed from the plains zebra. These differences, however, which would have resulted in the production of different amino acids at the substitution sites, made no sense at all. The particular sequences analyzed should not have been different, because if they were, quagga mtDNA at these genetic locations would be different from all other vertebrates—including everything from frogs to humans.[55] Since this is extremely unlikely, the Wilson team concluded that the substitutions were in fact cloning artifacts and that, when it was alive, the quagga shared all 229 nucleotide positions in each gene with the plains zebra.[56] The substitutions, therefore, at least theoretically, must have arisen from postmortem damage (that is, lost bases being incorrectly repaired by the bacteria in the cloning process, thereby producing cloning artifacts). But the theory could not, of course, be proven by additional cloning.

All of this changed when PCR became available to scientists eager for an alternative to molecular cloning and all of its attendant headaches. With PCR the process of multiplying a desired section of DNA was dramatically streamlined. Scientists found PCR to be, as Dr. Harley Rotbart, a microbiologist at the University of Colorado School of Medicine, put it in 1991, "even more efficient than nature"[57] at reproducing genetic material. "Now there is a genetic time machine for looking back into the past," remarked Alec Jeffreys when PCR had begun to show its mettle.[58] One of the first research groups to put PCR to the test was Allan Wilson's team at Berkeley. Shortly after receiving his doctorate in 1986, Paabo joined the Berkeley team. Now, new challenges could be tackled.

The very first time that PCR was used to analyze ancient DNA, the experiment was performed by molecular geneticist Russell Higuchi. The subject was a 40,000-year-old baby woolly mammoth named Dima, which had been chiseled out of a frozen Siberian tundra in 1977. The extinction of the woolly mammoth has been attributed to any number of causes, including a changing global ecosystem as the glaciers of the last Ice Age melted and receded or over-hunting by early humans. Whatever the reason, woolly mammoths were completely gone by about 9,500 years ago, according to the fossil record, with the exception of a population of dwarf mammoths that lingered on in isolation on Wrangel Island in the Arctic Ocean until perhaps 4,000 years ago, well beyond the end of the Ice Age.[59] While resembling modern-day elephants, woolly mammoths had long hair (measured in feet) and lengthy, curved tusks. They stood approximately 10.5 feet tall and weighed about six tons. Dwarf mammoths reached a more modest average height of six feet and weighed only about a third of their larger cousins.[60]

Attempts by Higuchi to clone Dima's DNA while performing postdoctoral work at Wilson's laboratory in the early 1980s had failed due to all of the problems associated with cloning. When PCR came along (although it was conceived in 1983, PCR did not become available to other scientists for a few years), Higuchi was motivated to take another crack at Dima's DNA.[61] PCR enabled him to isolate and sequence 350 nucleotides of Dima's mitochondrial DNA. This information has helped establish that woolly mammoths are equally related to two living species, the Indian elephant and the African elephant, substantiating the theory of a three-way evolutionary split from a remote ancestor.[62]

Armed with PCR, Paabo and Wilson were able to revisit the original quagga study and amplify the genes from the original DNA extracts, publishing their findings in the summer of 1988.[63] The quagga gene sequences they amplified were the same two genes—cytochrome oxidase I and NADH dehydrogenase I—that had been cloned and sequenced earlier. They found them to be exactly as predicted, identical to the cloned quagga sequences at all positions *except* the two wayward nucleotides that had seemingly appeared out of nowhere the first time around. PCR confirmed that the two extraneous and puzzling nucleotides were indeed cloning artifacts, and not some freakish genetic anomaly confined to the doomed quagga. The PCR products generated two bases at these positions consistent with the plains zebra—and all vertebrates, for that matter. Using PCR, Paabo also confirmed that the DNA retrieved from an 8,000-year-old human brain recovered in Windover, Florida, was indeed human DNA, using a probe specific to an Alu-repeat sequence.[64] (This archeological site arose out of a Windover peat bog, originally stumbled upon by construction workers, where scientists eventually recovered 177 skeletons and 91 brains.[65] The remains were radiocarbon-dated to between 6,9990 and 8,130 years old.[66]) Paabo and Wilson subsequently extracted mtDNA sequences from a 7,000-year-old brain unearthed in North Port, Florida, at a place called Little Salt Spring. Little Salt Spring was a large, flooded sinkhole believed to be a source of fresh drinking water for prehistoric humans.[67] (What they found was mtDNA that distanced the individual dredged out of Little Salt Spring from the maternal lineages of Asian people, Native Americans, and Amerindians from the North American Southwest. The finding suggested three, rather than two, distinct maternal lineages among the earliest American colonists, a novel suggestion.[68])

"We and other laboratories could thus reproduce and verify the results," Paabo later said in *Scientific American*. "Molecular archaeology could for the first time claim to be a respectable branch of science."[69]

Two of PCR's principal strengths were demonstrated in the second quagga analysis: First, PCR can amplify pre-selected DNA segments to sufficient levels for direct sequencing, starting with infinitesimal traces of DNA. In 1988 researchers using PCR at Cetus Corporation first reported success detecting DNA polymorphisms in single human hairs, with as little starting material as a single molecule.[70] That year, too, another group of researchers, including some from Cetus, reported using PCR to analyze single sperm.[71] Second, since genetic material can be amplified from such tiny amounts, an analysis can be done again and again. The ability to reproduce and verify one's results—as Paabo stresses—is crucial to credible scientific analysis. In forensic science, the ability to revisit a genetic analysis and confirm its findings is also desirable. PCR alleviates the fear that a particular specimen will be completely used up by a single test. Or worse, by a botched test. With PCR now in the arsenal, the study of ancient DNA took off, as did forensic DNA analysis.

Throughout the 1990s, a relatively small band of scientific free spirits, George Poinar and Svante Paabo among them, continued to chase the DNA molecule back in time. Using the new tools of molecular biology and PCR, these scientists sought to determine the outer limits of DNA survival. How long can DNA last? Under what conditions will it survive the longest? What will it look like when we find it? And how do we verify that what we're actually looking at is, in fact, ancient DNA? These are some of the questions that inspired and tormented the pioneers of ancient DNA analysis.

The "successes" of these scientists have been widely publicized, while their failures, or the failures of other scientists to duplicate the results, have received much less attention. Have scientists really extracted genuine stretches of DNA from extinct insects that were trapped in amber millions of years ago? George Poinar and his colleagues have reported doing so, as have other groups of scientists. Was dinosaur DNA actually chiseled out of a Cretaceous-era fossil bone? One group of scientists led by Scott Woodward, in fact, reported extracting DNA from the leg bone of an 80-million-year-old *Tyrannosaurus rex* bone.[72] Was any of this true? The answers to these questions are hotly debated, but few scientists believe such reports anymore.

The history of science is replete with examples of claims that turned out to be either fraudulent or the result of seriously flawed scientific work. One of the most famous examples of scientific fraud is the anthropological "find" known as Piltdown Man, a humanlike skull that surfaced in 1912 in a gravel pit on Piltdown Common in southern England. Because of its unique attributes, the skull was embraced by many scholars for over 40 years as the "missing link" in human evolution. It wasn't until 1953 that Piltdown Man was formally revealed to be a hoax.[73] Although it has not drawn allegations of fraud, the claims of scientists that they had "resurrected" bits of ancient DNA from fossils, mostly amber-enclosed insects, that dated back as far as the age of the dinosaurs, has failed to withstand scientific scrutiny. All of the scientists were using PCR-based DNA techniques that sought to amplify whatever DNA might happen to have persisted in the fossil over the eons. Some critics raised the specter of DNA contamination, meaning, basically, that stray, modern-day DNA might well account for the results the scientists were getting. After the release of *Jurassic Park*, however, the premise of which was that dinosaurs could be cloned from the DNA retrieved from the bellies of ancient, dinosaur-era blood-sucking mosquitoes that had been entombed in amber, as far as the public was concerned the whole thing seemed perfectly reasonable.

The problem was, the ancient DNA results that had been reported could not be replicated in other laboratories, and reproducibility is an essential step in sustaining scientific credibility. Years of studies followed, along with a heated, but low-profile debate in the scientific literature. Ultimately, a consensus emerged among scientists that the ancient DNA results arose from laboratory contaminants, not hundred-million-year-old DNA. Further studies indicated that DNA could not last that long, not even close, even under the most fortuitous circumstances, such as being encased in amber. It might well last 100,000 years, however, many agreed. Because of the nature of the controversy, there were never any formal retractions in scientific publications (most of the ancient DNA scientists continue to maintain their results were authentic), and certainly not on the front pages of anything. The whole thing just slouched toward oblivion.[74]

DNA, in any event, can last a long, long time. In fact, a consensus has emerged that DNA can survive under favorable conditions for tens of thousands of years at least, certainly longer than the average unsolved cold case.

One of the ancient DNA feats that has survived scrutiny—at least insofar as the DNA retrieved was deemed actual, ancient DNA—was the landmark study of a Neanderthal fossil conducted by Svante Paabo. Paabo's research was significant both as a demonstration of the resilience of DNA and because the results of the study purported to shed light on one of the hottest debates in paleoanthropology, namely, to what degree are Neanderthals related to modern humans, if at all. The dispute is related to the "Out-of-Africa" or "mitochondrial Eve" hypothesis, which posits that modern humans arose from a distinct group of early humans who originated in Africa and spread out across the world beginning about 100,000 years ago. Significantly, the theory holds that early modern humans (known as Cro-Magnon) completely replaced any other early humans, referred to as "archaic" human species, including Neanderthals, that had already settled in various places outside Africa. They did not interbreed with Neanderthals, according to this more widely accepted view, even though the fossil record demonstrates that both early moderns and Neanderthals occupied the same geographic locales simultaneously. The reasons advanced for the complete obliteration of the Neanderthals include that they were simply outcompeted for resources by the rival Cro-Magnons, that it was biologically impossible for early modern humans and Neanderthals to produce offspring because they were separate hominid species, or that they were systematically killed off by early modern humans in perhaps the first known instance of organized genocide. Significantly, too, mtDNA analyses of modern-day human populations tend to support a single African origin for modern humans, based on statistical models of how similar the DNA is between the groups and estimates of how long it would take for variations to arise across the eons. An alternate view to this "replacement" theory is known as the "multiregional" theory, which, based primarily on the fossil record as it compares to the various physical attributes of modern humans, holds that early human ancestors arising out of Africa met, intermingled with, and ultimately interbred with other early humans already in residence around the world. The Neanderthals, according to this view, were never really extinguished at all. They simply merged with Cro-Magnon until their distinct physical characteristics faded into the upholstery of humanity.

The Neanderthals, whose brains were bigger than modern humans, are known to have inhabited Europe and western Asia. They had survived for several hundred thousand years, over the worst of the Earth's ice ages, only to go extinct, based on the fossil record, roughly 28,000 years ago. Neanderthals differed physically from modern humans and Cro-Magnons, however, in many significant ways. They were shorter, stockier, thicker boned, had broad noses and heavy eyebrow ridges. On average they were somewhat larger than their Cro-Magnon counterparts and a great deal stronger.

Paabo, who earned his Ph.D. in molecular biology from Uppsala University in Sweden in 1986 and now teaches at the University of Munich, analyzed the mtDNA of the very first Neanderthal fossil bones ever discovered, found in a cave in 1856 in the Neander Valley near Duesseldorf, Germany, which gives the Neanderthals their name. The bones are estimated to be between 30,000 and 100,000 years old. Paabo's group extracted mtDNA from the arm bone of the 1856 fossil and generated sufficient amounts of DNA for testing with the help of PCR. After sequencing the Neanderthal mtDNA, they compared the sequences with analogous sections of the mtDNA of about 2,000 modern humans from around the world, concluding that, on average, the genetic differences between Neanderthal mtDNA and modern human mtDNA were four times greater than those seen between different ethnic groups of modern humans, and, in any event, the tests didn't reveal any special similarities with modern European mtDNA. The cover of *Cell,* the journal that published the results, blared, "NEANDERTHALS WERE NOT OUR ANCESTORS."[75]

Even the most vocal critics of ancient DNA claims were convinced by Paabo's findings after scrutinizing his data. Tomas Lindahl of the Imperial Cancer Research Fund in Hertfordshire, England, a noted DNA expert, for example, who has repeatedly sneered at claims that DNA can survive for millions of years, called Paabo's Neanderthal paper "compelling and convincing" and "arguably the greatest achievement so far in the field of ancient DNA research."[76] Mark Stoneking of Pennsylvania State University independently confirmed Paabo's Neanderthal DNA results.

In late 1998, however, the discovery of 24,500-year-old remains of a four-year-old boy in Portugal gave new ammunition to those favoring the notion of Neanderthal/human assimilation. The boy's skeleton had features that seemed a mixture of modern humans and Neanderthals. On the one hand, he had the stocky build and short legs of a Neanderthal; on the other, his chin, jaw, and arms were those of a modern human. Since the boy dated to several thousand years after the Neanderthals presumably went extinct, the boy's physiology seemed to make the perfect case for the interbreeding hypothesis. Replacement theorists, of course, dismissed the physical traits of the apparent hybrid child as well within the normal range of modern human anatomical variation.[77] In addition to the questions this find posed, some critics of Paabo's Neanderthal mtDNA analysis raised the specter of contamination as somehow accounting for the unusual DNA sequences he had obtained.

A follow-up mtDNA study published in 2000, however, seems to have tipped the scales back in favor of the replacement theorists. That study, conducted on the 29,000-year-old rib bone of a Neanderthal infant discovered in southern Russia by a Russian archeological team, served to confirm Paabo's results in every way. The mtDNA analysis in that study was performed by Dr. William Goodwin, working with Swedish and Russian colleagues, of the Human Identification Center at the University of Glasgow. Goodwin's analysis found a 3.5 percent difference from the mtDNA sequences produced in Paabo's study, indicating that a diverse population existed, and further showed that the Neanderthal mtDNA sequences of the infant were starkly different from those of modern humans, which, like Paabo's study, suggested an evolutionary split from modern humans about 600,000 years ago. Scientists who cling to the assimilation theory point out that no Cro-Magnon mtDNA studies have yet to be performed from fossils that were contemporaries of the Neanderthals; such a study might demonstrate, they note, that Cro-Magnon mtDNA is closer to that of Neanderthals than the mtDNA of modern humans.[78]

Science is often in a state of flux, however, and a new discovery can—and often does—change everything. Initially, a new discovery may simply add an annoying new wrinkle that has to be puzzled over for awhile until someone figures out whether it fits into conventional theories (by possibly being wrong) or does, in fact, change everything. One of the more recent mtDNA studies of an early human fossil appears to have added just such an annoying new wrinkle to the genetic studies concerning the origins of modern humans. Unearthed in a desolate Australian lake bed known as Lake Mungo in 1974 and dated to be about 62,000 years old by several modern dating methods, Mungo Man, as the fossilized remains became known, has all the physical characteristics of *Homo sapiens*. An mtDNA study published in January 2001, however, concluded that Mungo Man's mtDNA was completely distinct. It didn't match the mtDNA of modern humans nor that of any of the Neanderthal fossils examined so far. Nor did the mtDNA sequences obtained by the Australian team seem to relate back to ancient African mtDNA sequences. Published by a team of researchers that included retired anthropologist Alan Thorne of Australian National University and Gregory Adcock, a doctoral candidate in genetics at the Australian National University's Research School of Pacific and Asian Studies, who actually performed the DNA analyses, the Mungo Man mtDNA data were a shock to many people in the field.[79]

"People just fell over when they read this new stuff," Princeton University anthropologist Alan Mann told *Discover* magazine in 2002. "The people at Mungo were totally modern looking and were expected to carry the DNA we have, but they didn't. I think that makes for an incredibly complicated story. It's a stunning development."[80]

Whatever else it all means, the study had apparently produced the oldest known sample of human DNA; this assumes, of course, that the study can withstand very close scrutiny and can be replicated in other laboratories. (Three rival laboratories were attempting to duplicate the study with the Mungo Man extracts at the time that the *Discover* article appeared.) In line with the multiregionalist view, Alan Thorne champions the theory that the direct descendants of *Homo sapiens* migrated out of Africa around two million years ago and settled Europe and Asia. The various human groups that evolved in these far-flung locations eventually moved around, intermingled, interbred, and finally produced the variegated assortment of modern humans that exist today. Needless to say, the proponents of the out-of-Africa/replacement model are discounting the Mungo Man data as either flawed or otherwise explainable within the scheme of the more popular theory they espouse.

The publication of the Mungo Man mtDNA study coincided with the publication of another paper seeking to resurrect the multiregionalist view of man's origins. Co-authored most conspicuously by anthropologist Milford H. Wolpoff of the University of Michigan, who has long been the most vocal defender of the multiregionalist view, the paper sought to challenge the replacement theory by borrowing a statistical technique commonly used in genetics, pairwise comparison, and applying it to the fossil record of human evolution. As with its genetics counterpart, Wolpoff and colleagues were looking for matching attributes between different fossil groupings. Wolpoff's team identified a list of 30 or more distinct physical features and then analyzed 25 ancient skulls from three groups: early modern humans (more than 80,000 years old) discovered in Israel and Africa, more recent modern humans (between 14,000 and 30,000 years old) from Australia and the Czech Republic, and archaic humans (more than 45,000 years old), including Neanderthals, from Australia and Central Europe. Based on the skull features identified and matched, the researchers concluded that some of the fossils distant from Africa were a blend of features from the various groups. The skulls, in other words, indicated mixed ancestry, suggesting that local archaic peoples interbred with migrants from other parts of the world, including Africa, in a process that ensured continual gene flow throughout the world as well as a degree of continuity, rather than replacement and extinction, for archaic humans that had long been in residence around the world.[81]

While proponents of the "mitochondrial Eve," or replacement, theory generally welcomed the effort by Wolpoff and colleagues to objectively quantify the assertions of multiregionalism using the fossil record, they also generally remained deeply skeptical about the group's conclusions. Deep skepticism also describes the reception of the Mungo Man mtDNA study by the majority of scientists who favor the replacement model. Replacement-theory adherents challenged the sexing of Mungo Man, the dating of the fossil, and virtually every aspect of the mtDNA analysis, suggesting that some of the mtDNA data may reflect artifacts from very old and damaged DNA or that all of the DNA extracted arose from modern contaminants. DNA is unlikely to survive that long, it was argued, particularly in the harsh, parched environment of Lake Mungo. Svante Paabo, for example, now of the Max Planck Institute for Evolutionary Anthropology in Munich, explained to *Science:* "We know from many failures in our laboratory that contamination can rear its head in many forms."[82] Paabo also pointed to the lack of independent verification. Others noted that even if the mtDNA is genuine, the oddball DNA sequence may simply mean that the maternal line it represents died out, as would be expected of many human maternal lines across prehistory.[83]

For his part, Gregory Adcock believes the Mungo Man mtDNA data are not a total refutation of the Out-of-Africa theory. Rather, the Mungo Man results suggests that the Out-of-Africa model embraced by most scientists is too simplistic. Human evolution, he says, occurred with far more complexity than assumed by the conventional wisdom.[84] Wolpoff, on the other hand, believes his skull-comparison evidence and the Mungo Man study have dealt a crushing blow to the Eve theorists. Undaunted by the criticisms leveled at the studies, Wolpoff predicts the two studies are just the beginning of what promises to be, in his words, "the multiregional millennium."[85]

One thing is certain. MtDNA lasts a long time.

Mitochondria are the little powerhouses of cellular activity found as organelles outside the nucleus. With the help of various enzymes, the mitochondria transform oxygen and nutrients into energy to power the cell. In a process called respiration, mitochondria produce the vital cellular fuel adenosine triphosphate (ATP), without which everything stops and the cell dies. MtDNA has several advantages over nuclear DNA for certain types of genetic analyses. Although it contains far less genetic information than even a single chromosome, mtDNA is far more plentiful in each cell; each cell carries hundreds, perhaps thousands, of (usually) identical copies of the power packs each with their own DNA. MtDNA also generates mutations at an accelerated and regular rate, and thus facilitates phylogenetic comparisons. MtDNA is maternally inherited, bypassing the mechanism by which the genetic information of parents is coupled (although there is some controversy about this), the process that results in the nuclear DNA of offspring. Mitochondria sneak past the normal channels of genetic inheritance by hitching a ride in the mother's egg. Although present in the spermatozoa as well, the father's mtDNA is jettisoned at conception. Even though they contain an infinitesimal amount of genetic information (16,596 base pairs[86]) when compared to nuclear DNA, mtDNA have nevertheless been linked to a few genetic diseases and may even play a role in aging and the chronic degenerative neurological and muscle diseases associated with old age.[87]

Whereas nuclear chromosomes come in pairs, and thus provide two (potentially different) copies of each gene sequence to the cell, mitochondria generally come in a single genetic version—a state of affairs referred to as homoplasmy. Thus, the hundreds of mitochondria in each cell of the body normally start out as identical copies of each other. These homoplasmic cells may be normal or bear a mutation. Occasionally, however, the mitochondria may come in two versions, one normal and one carrying a mutation; having two versions of mitochondrial DNA is referred to as heteroplasmy.

To further complicate matters, cells containing both normal and mutated versions of the mitochondrial DNA will spawn daughter cells that contain varying percentages of the mutated and normal versions. One daughter cell, for example, may end up with 25 percent of the mutated versions, and the other daughter cell may get the other 75 percent. Eventually, later generations of cells will have either a cell full of the mutated variety or a cell full of the normal variety of mtDNA. Thus, some tissues in the body are affected by the mutation while others are not.

Distinctive features of mtDNA make it useful in the context of comparing individuals with one another and groups of people with one another. Since mitochondria are maternally inherited, familial relationships can be readily established. In other words, an individual can be identified as a member of a family by comparing the mtDNA of maternal relatives, such as a sibling, mother, maternal grandmother, or maternal aunts and uncles.[88] The U.S. military, in fact, has established a DNA testing program, the Armed Forces DNA Identification Laboratory (AFDIL), to, among other things, identify the remains of soldiers killed during the wars in Korea and Vietnam, and even World War II; AFDIL relies heavily on the more plentiful mtDNA.[89] Because the remains of these soldiers are decades old and often fragmentary, the process of identification is typically dependent upon PCR's ability to retrieve tiny bits of DNA from such things as bones or teeth. The mtDNA of living relatives can be compared with the mtDNA extracted from the remains of the fallen soldiers to establish a possible linkage. Since mtDNA transmitted from mother to child does not (most believe) undergo genetic recombination, as nuclear DNA does during meiosis (the formation of eggs and sperm), mtDNA is faithfully inherited.[90] Except for occasional mutations, then, immediate relatives will have the same mtDNA type. Although mtDNA can never definitively establish identity, it can eliminate suspected relationships.

The military's DNA testing program for unidentified remains has been very successful, and many American families whose lost loved ones had long been listed as MIAs have been notified that the remains of their father or brother have finally been found, allowing, at least, a degree of closure. This certainly was the case for the family of Air Force First Lieutenant Michael Blassie, whose fighter jet was shot down over Vietnam in 1972. The circumstances surrounding Blassie's identification were highly unusual, however. Circumstantial evidence concerning Blassie's fate pointed to a surprising conclusion, first publicized in 1998. Blassie's remains, which were limited to six bones, might be among those housed in the Tomb of the Unknowns at Arlington National Cemetery in Virginia as a proxy for the more than 2,000 Vietnam veterans listed as missing in action. Earlier forensic examinations conducted prior to the interment of the remains in the Tomb in 1984 had indicated that the bones might belong to Blassie, but the examinations also pointed to Army Captain Rodney Strobridge, a helicopter pilot who was shot down on the same day as Blassie in a separate incident. Given the uncertainty, the remains were listed as unidentifiable. With the emergence of DNA testing, however, the Pentagon, reluctant to disturb the hallowed memorial, came under increasing pressure from the Blassie family and several MIA/POW veterans groups to exhume the remains and conduct DNA tests. Relenting in 1998 after a recommendation from a Defense Department research group, the Pentagon opened the Tomb in May 1998 and sent the remains to the Armed Forces Institute of Pathology. DNA tests were subsequently performed at the Armed Forces DNA Identification Laboratory in Rockville, Maryland, with assistance from other army facilities, and comparisons were made with the living family members of Blassie and Strobridge. Following the tests, the Army determined in June 1998 that the remains were those of Michael Blassie, and his remains were sent home to Missouri for burial. Emboldened by the mtDNA results in the Blassie case, the Pentagon subsequently redoubled its efforts to identify the remains of unknown soldiers buried around the country in various military cemeteries.[91]

Mitochondrial DNA is known to accumulate mutations at a relatively constant rate, a mutation rate that is faster than nuclear DNA. Based on its predictability, comparative analyses of mtDNA have been used to trace the migrations of modern humans (Homo sapiens) around the world, originating in Africa and moving through the Middle East and into Europe, Asia, and the Americas. Global migrations can be traced because of the distinctive mutations that arose as the populations fanned out around the world (and, to some extent, circled back). The populations of each continent, it has been found, tend to differ in the genetic sequences of their mtDNA.

In forensic mtDNA analyses, scientists focus on two hypervariable sections of a control region of the mtDNA known as the D-loop, which consists of 1,100 base pairs. This region is known to mutate at an especially high rate, at least five to ten times faster than nuclear DNA. The two regions of the D-loop (or displacement loop) targeted in forensic analysis are referred to as hypervariable region I (HV I) and hypervariable region II (HV II), each consisting of about 300 base pairs. MtDNA is expected to differ from 1 to 2 percent between any two unrelated individuals. Often, PCR is used first to generate sufficient DNA for analysis. MtDNA analysis may involve either direct sequencing of the DNA or sequence-specific hybridization, just like HLA DQA1 typing. Direct sequencing is possible in forensic work with mtDNA because, typically, there is one genetic variant. By contrast, most people carry two versions (are heterozygous) at the loci used in forensic nuclear DNA analysis, and this seriously complicates sorting out which version is being sequenced.[92] Since mitochondria are so numerous in the cell, mtDNA has been retrieved when cellular degradation is so severe that no usable amount of nuclear DNA remains. mtDNA has been successfully recovered from hair shafts (without the hair root), as well as bones, fingernails, and teeth. Nevertheless, mtDNA analysis remains a typing system of last resort because it is less informative than nuclear DNA.[93]

The sequencing step in mtDNA analysis is performed using a variation of the Sanger method that has been highly automated and requires a DNA Sequencer. Denatured mtDNA is exposed to fluorescently tagged bases that are designed to terminate the extension of the template once they have joined the growing strand of DNA. The process involves heating and cooling stages in a thermal cycler, just as in PCR, but the objective is different. The synthetic bases are color coded to indicate which of the four bases they represent, A, T, G, or C. Once the annealing process is done, the targeted sequence has been produced in every conceivable variation of its sequence, from the smallest to the largest, covering one end of the sequence to the other. The exact sequence of the analyzed sample is determined by running all of the DNA fragments through a gel (electrophoresis) where they naturally separate out according to length; a fluorescent detector in the DNA sequencing machine reads, and a computer records, the bases in the gel from bottom to top. The machine produces a colored graph, or electropherogram, that shows the order of the fluorescently tagged nucleotide bases. The result is a base-by-base read-out of the mtDNA sequence. MtDNA analysis is not a definitive method of identification. Frequencies of specific mtDNA sequences vary in the population from very rare to fairly common. The most common mtDNA type, according to the FBI's database, occurs in 4 percent of the population.[94]

MtDNA was used in the determination that remains buried in 1882 under the name Jesse James were indeed those of the famed American frontier outlaw (although this finding was challenged). It was also critical in confirming in the mid-1990s that remains discovered in Yekaterinburg, Russia, in the late 1970s were members of the Russian royal family, and not simply a group of related individuals in a mass grave.

The Romanov family, the last royal family of Russia, had been executed by firing squad by the Bolsheviks in 1918 as the communists seized power, but the bodies of the royal family had long been unaccounted for. The chance finding of a mass grave outside Yekaterinburg in the late 1970s seemed to end the mystery (which was not publicized until the fall of the Soviet Union in 1989), but there were many who doubted the find was authentic. The remains were not excavated until 1991. Extensive DNA tests conducted between 1993 and 1998 finally demonstrated to everyone's satisfaction that the Yekaterinburg remains were indeed the Romanovs. The testing was extremely complicated, in part due to a genetic anomaly associated with the bones believed to be those of Czar Nicholas. The existence of an mtDNA heteroplasmy in those remains at first perplexed genetic researchers of the British Forensic Service Laboratory, where much of the testing was conducted; because two types of the sequenced mtDNA were found in these bones, different only at one nucleotide position, they were not a definitive match to maternal relatives of Nicholas. The two versions of mitochodria seemed to invalidate the results. Later, however, the heteroplasmy added an unusual degree of certainty to the identification. The same heteroplasmy was subsequently confirmed in separate analyses by Erika Hagelberg, Mary-Claire King of the University of California, Berkeley, who had accumulated a sizable mtDNA datatbase, and Victor Weedn at AFDIL. Finally, Russian analyst Pavel Ivanov found the identical heteroplasmy in the exhumed remains of Czar Nicholas's brother, the Grand Duke of Russia Georgij Romanov, who died of tuberculosis in 1899. The mtDNA oddity simply ran in the family.

This was the first time that an mtDNA heteroplasmy played a role in a human identification.[95] (Partly because heteroplasmy is more readily seen in human hair than other body tissues, the FBI Laboratory has found that between 5 and 10 percent of their criminal cases involving mtDNA have exhibited heteroplasmy.[96]) MtDNA testing (along with STRs) was also used in 1994 to determine that Anna Anderson Manahan—who had long claimed to be the missing Grand Duchess Anastasia Romanov, who, according to many legends, somehow survived the night her family was massacred—was, in fact, not related to the Romanovs at all. Anna Anderson's claim had, to this point, survived a considerable amount of expert scrutiny over the decades, including handwriting analysis, photographic comparisons, and even a "scientific" ear-comparison examination.[97]

The first use of mtDNA analysis in an American criminal courtroom, which involved direct sequencing of the mtDNA, occurred in 1996 in Chattanooga, Tennessee. The defendant, 27-year-old Paul Ware, was accused of raping and murdering a four-year-old girl, Lindsey Green, in 1994. Ware showed up drunk at the house of a friend where another man was babysitting several children, including Lindsey, who was sleeping in a back bedroom. In the morning the girl's body was discovered by her frantic parents alongside Ware in a utility room. He was the obvious suspect. Although a visual hair comparison had further linked Ware to the crime, based on hair samples found in the girl's bed, bloodstains recovered in the case were found by FBI DNA technicians to be too degraded for PCR-based nuclear DNA tests. After consulting with Tennessee Assistant Attorney General Lee Davis, the FBI went ahead with an mtDNA analysis of a single pubic hair recovered from the victim's throat. The DNA results confirmed a match to Ware and confirmed the conventional hair-comparison evidence. Ware was convicted on September 4, 1996, and later sentenced to life in prison without the possibility of parole. The FBI had only approved its mtDNA laboratory protocol a few months earlier after six years of study and development.[98]

Over the next year, the FBI laboratory performed mitochondrial DNA analyses in five additional cases that led to convictions in Texas, North Carolina, South Carolina, Michigan, and a second case in Tennessee. In the Texas case, a human limb hair found stuck to some tape that was used to make a bomb matched the mtDNA profile of the suspect, Loran B. Pierson, and he was convicted in January 1997. In the North Carolina case hair found in the trunk of a car belonging to an ex-police officer, LaMont Underwood, was found to match the victim of a homicide both visually and after mtDNA analysis, resulting in Underwood's conviction in July 1997. The victim in that case was apparently having an affair with Underwood's wife. In the Michigan case, mtDNA showed its staying power by allowing investigators to solve a murder case dating back to 1983. On the night of February 6, 1983, a member of the Satan's Sidekick Motorcycle Club, Otis Charles Avery, disappeared after a party held with a rival motorcycle gang. Although Keith Noland Williams was seen with Avery shortly before Avery's disappearance, investigators felt there was insufficient evidence and closed the case in 1986. After a tip in 1996, however, remains believed to be those of Avery were found in the possession of Williams. Proof that it was Avery was established when the remains yielded mtDNA that matched a maternal relative of Avery's. Williams was convicted.[99]

MtDNA soon began to tackle some of the most significant criminal investigations in the country. Certainly one of the most elusive criminals in the nation was the so-called Green River killer, believed to be responsible for the deaths of 49 women in the Pacific Northwest between 1982 and 1984, 42 of whose remains were discovered either near the Green River in Washington State, where the first victims turned up, or scattered in the woods from Seattle to Portland, Oregon. (Seven apparent victims are still missing.) The victims, all young women, were believed to have been abducted from the streets of Seattle as runaways or while working as prostitutes. Since many of the remains were badly decomposed when they were found, by 1999, not all had been positively identified. Among those believed to be a victim of the Green River serial killer, but still unaccounted for, was Tracy Ann Winston. Winston had suddenly vanished in 1983 during the time the killer was active. In 1999, investigators in Seattle sent samples from five sets of remains accumulated in the case, along with blood samples from the mothers or daughters of twelve missing women. Among the remains sampled were those discovered in 1986 in a park near Seattle. These remains, the FBI laboratory's mtDNA analysis determined, were those of Tracy Ann Winston. The other four sets of remains did not come back as a match to any of the other families tested.[100] In December 2001, based on additional STR tests, authorities in Seattle formally charged Washington resident Gary Leon Ridgeway, 52, with some of the slayings in the Green River case. The evidence against Ridgeway, who had long been a suspect, included DNA matches from a saliva sample Ridgeway had provided to police back in 1987 that now linked him to semen found in three of the Green River victims. The DNA tests of the saliva sample, which was preserved on a piece of gauze, were conducted in March 2001. Other evidence in the case linked Ridgeway to a fourth victim. He has pleaded not guilty.[101]

In another high-profile case, mtDNA tests conducted by the FBI laboratory on a single strand of hair found near the home of slain Amherst, New York, doctor Barnett Slepian, who performed abortions at a local clinic in Buffalo, connected anti-abortion activist James Charles Kopp to the killing. Slepian had been gunned down as he stood in his kitchen by a sniper outside the house using a high-powered rifle. A hair retrieved from a baseball cap found near where investigators believe the shot originated was genetically matched by mtDNA analysis to hair recovered from Kopp's home. Kopp remained on the FBI's "10 Most Wanted" list for two years before being apprehended in France on March 29, 2001, pursuant to an international warrant. Following a protracted extradition battle, Kopp was finally returned to the United States in June 2002 to face both state and federal charges in Buffalo, arising from the killing of Dr. Slepian. In part, the extradition process was slowed by France's unwillingness to extradite prisoners who may face execution in the requesting jurisdiction. In June 2001, a French court authorized the extradition of Kopp after being satisfied with assurances from U.S. officials that Kopp would not be facing the death penalty upon his return to New York. Kopp fought the French court's order for another year before finally giving up. Shortly after returning to New York, Kopp was ordered by a New York judge to submit hair, blood, and saliva samples for DNA analysis.[102] During a jailhouse interview in November 2002, Kopp admitted to a reporter for *The Buffalo News* that he did indeed fire the shot that killed Dr. Slepian, but Kopp insisted that he only wanted to wound the doctor, not kill him. Through his lawyer, Kopp explained that he was trying to incapacitate the doctor as a way of preventing him from performing abortions scheduled for the next day.[103]

5 Exonerations:
Take No One's Word

IN THE SUMMER OF 1996, THE NATIONAL INSTITUTE OF JUSTICE (NIJ) ISSUED a report featuring the cases of 28 men who had been convicted of rape crimes (six of which involved the murder of the victims) and incarcerated for lengthy prison terms or even sentenced to death, but were later freed on the strength of DNA testing.[1] Entitled *Convicted by Juries, Exonerated by Science,* the report drew attention to two significant aspects of the American legal landscape: the burgeoning use of DNA analysis and the long-standing problem of innocent people being jailed for crimes they did not commit. Among the cases in the NIJ report was that of Ricky Hammond. By the end of 1996, more than a dozen other men were also released from jail after genetic tests produced results contrary to their convictions.

As DNA technology has improved, the exonerations have accelerated. In some cases, evidence used to convict has been turned on its head by new DNA tests. In October 2001, for instance, Albert Brown was cleared of a crime in Oklahoma, where he had been serving a life sentence since 1981, after cutting-edge DNA tests showed that the hair evidence originally used by prosecutors to implicate him in the murder of a firefighter wasn't a match to the head hair of Brown after all.[2] In about a dozen of the DNA exoneration cases, the actual perpetrators of the crimes have been apprehended, tried, and convicted.[3] In January 2002, Larry Mayes became the 100th convict to be exonerated by DNA evidence, and the exonerations continue. Mayes had been convicted of rape in 1982 in Indiana based on early serology tests and the eyewitness-identification testimony of the victim.[4] In recent years, the phenomenon of DNA exonerations has sparked legislative changes around the country and even became a presidential campaign issue in 2000, at least to the extent that DNA evidence bore on the issue of the death penalty.

Many of the issues that surfaced in the Ricky Hammond case have been similarly prominent in the other DNA-exoneration cases, including strong eyewitness identification testimony, arguments by defense counsel that photo-arrays or line-ups were suggestive, weak alibis, incriminating hair-comparison evidence, con-sciousness-of-guilt evidence, and allegations of police or prosecutorial mis-conduct. Hammond's case was unique in that he was convicted by a jury *in spite of* the DNA evidence. In the other cases, DNA analysis was not performed until after the convictions.

Guided by the idiosyncratic appellate rules of the various jurisdictions (until it fi-nally changed the law in 2001, Virginia only allowed new evidence to be introduced for 21 days after a conviction; after that, it wouldn't be considered), most of the ex-onerated were forced to explore the outer reaches of judicial relief in order to obtain freedom. Sometimes freedom came quickly, but more often not. Most eventually had their cases overturned and the charges dropped. In some cases, however, because the legal rules foreclosed any other form of relief, the defendants were forced to seek pardons in order to be released from jail, an avenue often subject to the uncertainties of the political environment. Furthermore, sometimes prosecutors embraced the DNA results and advocated on behalf of the defendant. On other occasions, the pros-ecutors were wary of the results and resisted strenuously.

Spurred by the rising tide of DNA exonerations, each state has had to reexamine its own legal framework to determine if its particular system is fairly considering pe-titions by convicted inmates seeking DNA tests or asking the courts to consider the results of DNA tests that indicate, or flatly prove, they are innocent. These reexami-nations have brought change. Besieged by criticism over its 21-day rule, for example, Virginia finally passed in 2001, and the governor signed, a law giving death-row in-mates and other felons the right to pursue exoneration through ever more sophisti-cated methods of DNA analysis well beyond 21 days and requires the state to retain and secure biological evidence that may be used for that purpose in the future. (In other jurisdictions, biological evidence is routinely destroyed once appeals have been exhausted.) Although a milestone, Virginia's law does not encompass other forms of newly discovered evidence, which are still subject to the 21-day rule, and requires that a post-conviction DNA test be certified by the state Supreme Court as demon-strating innocence. Further, a state constitutional amendment has yet to be passed fully authorizing the DNA exception to the 21-day rule, so even beneficiaries of the law—the first was Marvin Anderson, cleared in December 2001 of a 1982 rape—must still formally seek a pardon from the governor.[5]

By the time of Anderson's exoneration, the regularity of the DNA exonerations prompted about 20 states, including three of the biggest, New York, California, and Illinois, to pass laws allowing inmates greater access (to some degree, at least) to DNA testing. There was social change, as well. Overall, public support for the death penalty appears to have eroded due to the high-profile release of inmates from death row fol-lowing DNA tests.[6] In September 1999, the National Commission on the Future of

DNA Evidence, established by Attorney General Janet Reno in response to the 1996 NIJ report, published *Postconviction DNA Testing: Recommendations for Handling Requests,* which urged prosecutors and judges (despite the acknowledged legal obstacles) to more fully cooperate with efforts by convicts to demonstrate their innocence through DNA testing. The recommendations were being made, the attorney general wrote in a preface to the report, "because forensic DNA technology can strengthen our confidence in the judicial process."[7] (The recommendations were not universally cheered, of course.) Subsequently, the Innocence Protection Act, introduced by Senator Patrick Leahy (D–VT), which would mandate that prisoners, including those facing execution, have greater access to DNA testing at government expense and would prevent the premature destruction of biological evidence, was proposed in Congress (and is still being considered).

Eight of the cases outlined by the NIJ study involved RFLP, 17 depended on PCR (DQA1 and sometimes Polymarker), and two cases relied on a combination of RFLP and PCR (including the Hammond case). (Information on which method was used in one case is unavailable.) As DNA technology has matured, the newer, more sophisticated methods have come into play in the DNA appeals, enhancing the ability of scientists to elicit informative genetic material from the samples available, some of which are very old or were beyond the reach of the earlier DNA testing methods. These techniques include mtDNA and STRs. In all of the DNA-exoneration cases, the convicts were excluded as the source of the evidence sample based on the DNA results. Some of the freed men received compensation from the government for their wrongful convictions, others did not. One of the men, Ronald Cotten, went to work for a time at the DNA laboratory responsible for his exoneration.

The release of the 28 prisoners described in the NIJ report was generally hailed as proof that DNA evidence is a boon to the search for truth in the criminal justice system. Nevertheless, these cases, and those like them that followed, are not without considerable controversy. This is clear from the commentaries in the report written by individuals from both the prosecutorial and criminal-defense perspectives.

Defense attorneys Barry C. Scheck and Peter J. Neufeld, for instance, both of whom are best known as the expert DNA counsel for O. J. Simpson, wrote in the preface to the 1996 report that "there is a strong scientific basis for believing these [28] matters represent just the tip of a very deep and disturbing iceberg of cases."[8] As evidence, they note that of the 10,000 sexual assault cases sent to the FBI for DNA analysis between 1989 and 1996, the principal suspects were excluded about 25 percent of the time; moreover, a survey of private laboratories revealed a similar 26 percent exclusion rate. Given these numbers, and the fact that about 62 percent of individuals tried in state court on felony rape charges (and, presumably, the principal suspects were headed for trial before the DNA came back) are convicted, Scheck and Neufeld conclude that "DNA exonerations are tied to some strong, underlying systemic problems that generate erroneous accusations and convictions."[9] Acknowledging the possibility of false exclusions, they nonetheless maintain that there are multiple reasons for wrongful convictions and accusations that need to be addressed, including the following: "Mistaken eyewitness identification, coerced confessions, unreliable forensic laboratory work, law enforcement misconduct, and ineffective [lawyering]."[10]

Scheck and Neufeld were at the center of controversy over DNA evidence long before being drafted to represent O. J. Simpson in California. As defense attorneys, of course, they routinely opposed the introduction of DNA evidence. However, their approach was singular in that they founded and co-chaired the DNA Task Force of the National Association of Criminal Defense Lawyers. It was the unabashed mission of the DNA Task Force to debunk DNA typing in court wherever and whenever possible. The task force was, in fact, initially established as a response to a series of prosecutorial victories with DNA-typing evidence in the mid- to late 1980s. The string of convictions based on DNA evidence, many observers believed, was due in no small part to an unprepared, unschooled defense bar. The Task Force responded by coordinating an attack on DNA typing by conducting regular seminars on DNA typing, compiling scientific reports on DNA evidence and transcripts of DNA cases, and providing materials, witnesses, and support to defense attorneys around the country confronting DNA evidence. As part of this strategy, the DNA Task Force aggressively attacked the laboratories around the country performing DNA analysis, including the FBI laboratory. As a countermeasure to these efforts, the FBI established its own DNA task force to assist prosecutors around the country.

Scheck and Neufeld gained prominence when they persuaded a trial court judge in 1989, Judge Gerald Sheindlin, after a 12-week hearing, to limit the introduction of DNA evidence in a gruesome double-murder case. As a result of Sheindlin's ruling, the Bronx, New York, case of *People v. Jose Castro* became one of the most famous DNA cases ever. The case did not, however, foretell the doom of DNA typing in court.[11] (Some of the DNA evidence survived the ruling, and the defendant eventually pleaded guilty.) Rather, Sheindlin simply demanded a higher level of scrutiny for DNA evidence in court than had previously been required under the *Frye* test. Since the *Castro* decision, Judge Sheindlin has consistently referred to DNA typing as the greatest advance in forensic science since fingerprinting. The *Castro* case did, however, signal a new willingness on the part of judges to carefully examine the DNA evidence before them.

Under the auspices of the DNA Task Force, Scheck and Neufeld also established the Innocence Project, a pro-bono advocacy group for prisoners modeled after the pre-existing Committee to Free the Innocent of the National Association of Criminal Defense Lawyers. Maintained at the Benjamin N. Cardoza School of Law at Yeshiva

University in New York City, where Scheck is a professor of law, the Innocence Project was actively involved in many of the DNA cases catalogued in the 1996 NIJ report and the cases that followed. By 2002, the Innocence Project had set up shop at 18 more law schools around the country, relying heavily on student volunteers.[12] Based on periodic reviews of the cases they and others have handled, the Innocence Project has found that as many as 84 percent of the wrongful convictions studied were based, at least in part, on errors in eyewitness identification and that roughly a third involved questionable hair-comparison evidence.[13]

Further attention was drawn to the relentless DNA exonerations with the publication in 2000 of *Actual Innocence: Five Days to Execution and Other Dispatches from the Wrongly Convicted,* by Scheck, Neufeld, and Jim Dwyer, particularly as a growing number of those exonerated by post-conviction DNA evidence were on death row (eight at the time of publication). Recounting the nightmarish stories of ten of the men freed by DNA, *Actual Innocence* enumerates what the authors see as the principal weaknesses of a criminal justice system that managed to put so many innocent people in jail, including forensic incompetence and fraud, police and prosecutorial overzealousness and misconduct, the dubious reliance by investigators (and, ultimately, juries) on jailhouse informants, coerced or otherwise false confessions, and the infirmities of various types of evidence, including eyewitness identification and conventional hair-comparison evidence. The book ends with a list of proposals for reforming the system. Throughout 2000, a presidential election year, *Actual Innocence* gained in influence as the political contenders squared off on issues such as the death penalty, DNA exonerations generally, and the relative health of the nation's criminal justice system.[14] The book even drew praise from death-penalty proponents like syndicated columnist George F. Will. "Capital punishment," Will noted, "like the rest of the criminal justice system, is a government program, so skepticism is in order."[15]

Among those freed with the help of the Innocence Project was Walter Snyder, convicted in 1986 in Alexandria, Virginia, of raping, sodomizing, and burglarizing a neighborhood woman, who identified Snyder in a police station "show-up." With only his mother to back up his alibi that he was home sleeping at the time of the crime, and confronted with matching blood-type evidence (type A, which Snyder, the assailant, and roughly 40 percent of the population share), Snyder was convicted and sentenced to 45 years in jail. His appeals failed. After the Innocence Project took up the case in 1992, prosecutors agreed to release the evidence for DNA testing by a private laboratory in Boston, The Center for Blood Research. Twice the laboratory's DQA1 tests excluded Snyder as a contributor to the semen found on the vaginal swab and, after some deliberation, Virginia Commonwealth Attorney John E. Kloch, the prosecutor overseeing the case, joined the defense attorneys in a petition to Governor L. Douglas Wilder of Virginia requesting a pardon for Snyder, which was granted on April 23, 1993. Although the victim of the crime continues to maintain that Snyder was her attacker, three of the jurors in the original trial told *The New York Times* upon Snyder's release that the governor's action was warranted.[16]

Eighteen months later, on October 21, 1994, Wilder's successor, George F. Allen, signed a pardon for another client of the Innocence Project, Edward W. Honaker. Honaker's case had long been a source of consternation for advocates of the wrong-fully convicted because of some hard-to-digest facts surrounding his conviction. Despite these facts, Honaker's bid for freedom was a tortuous one.

Like Snyder, Honaker was convicted in a trial that predated DNA testing. On June 23, 1984, a woman asleep in a car with her boyfriend was abducted at gunpoint by someone carrying a flashlight and posing as a policeman. The man chased the boyfriend away with the weapon, held—the couple recalled—in the assailant's left hand. After taking the woman to a remote spot, he repeatedly raped and sodomized her, all the while complaining of his experiences in Vietnam. Following the rape of another woman 100 miles away near the home of Honaker—and the victim in that case telling police that her assailant resembled Honaker—Honaker's picture was displayed to the first victim and her boyfriend in a photo-array, and they agreed it was him. They both went on to identify Honaker in court. Honaker was never charged in the second case. In addition to the eyewitness identifications, the prosecution put on circumstantial evidence that included hair-comparison testimony from a state labora-tory forensic expert who told the jury that a head hair found on the woman's shorts "was unlikely to match anyone" but Honaker. Furthermore, sperm was detected by the state laboratory on the woman's clothes and on the vaginal swab.

Honaker countered that he was right-handed, had never been to Vietnam, was asleep in his house hours from the crime scene the night of the rape (an alibi corrob-orated by several relatives and friends), and had undergone in 1977 a vasectomy that made it impossible for him to have been the source of the sperm-rich semen stains. Evidently believing the prosecution's version of events, which included the sugges-tion that Honaker's vasectomy had been a failure, the jury convicted Honaker in two hours on February 7, 1985, and he received three life terms.

A Princeton-based advocacy group for the wrongfully convicted, Centurion Min-istries, eventually took up the case. After their investigation determined that the identification testimony of the victim and her boyfriend had in fact been induced by hypnosis, that the victim's initial description of her assailant was inconsistent in some respects to Honaker's appearance (both facts were kept secret from Honaker's trial attorney), and that Honaker's vasectomy had been downplayed at trial, Centurion Ministries sought help from the Innocence Project, which succeeded in gaining ac-cess to the physical evidence in the case for DNA testing. Prosecutors hinted that they would contest an exclusion by the defense laboratory when they argued that the sperm identified by the state laboratory could have originated from the victim's boyfriend.

The evidence, which included the vaginal swab, an oral swab, and a semen stain on the victim's shorts, along with Honaker's blood sample, was sent by Honaker's lawyers to California for PCR testing by Forensic Science Associates, whose director, Edward Blake, was one of the early pioneers of DNA evidence in court. Blake was also a consultant to O. J. Simpson's defense team. After conducting DQA1 tests on the evidentiary materials it was sent, Forensic Science Associates reported back in January 1994 that the semen stain on the swab and the one on the shorts did not match genetically; moreover, even assuming Honaker could produce sperm, he was excluded as the contributor of either stain. A second report by Forensic Science Associates in March 1994 concluded that the boyfriend could not be eliminated as the source of the sperm on the shorts, although both he and Honaker were eliminated as sources of the sperm on the vaginal swab.

The task of disentangling the genetic evidence grew more complicated, however, when the victim revealed in June 1994 that she had had a secret lover apart from her principal boyfriend at the time of the 1984 assault. Now, suddenly, the existence of an unapprehended rapist could only be established by excluding Honaker, the boyfriend in the car, and the secret lover. To this end, Virginia tracked down the second boyfriend and sent his blood sample to the Virginia State Laboratory for analysis; the state laboratory reported that this man, although his DQA1 type was not an exact match, could not be excluded as the source of the sperm on the vaginal swab. Despite the lingering ambiguities, prosecutors joined in a clemency petition for Honaker submitted June 29, 1994. Finally, Forensic Science Associates enhanced the discriminatory ability of the testing by repeating the DQA1 tests on all the evidence and blood samples, this time adding the Polymarker test. The final report in September 1994 concluded that neither Honaker nor the two boyfriends were responsible for the sperm on the vaginal swab. Within a month, Governor Allen signed the pardon. Honaker had been in jail for ten years.[17]

"The truth is that DNA is accurate,"Westchester County District Attorney Jeanine Pirro told *The New York Times* in 1995. "It's like the finger of God pointing down saying 'you did it.'"[18] In any case involving DNA analysis, however, only after careful review of the facts of the case and the circumstances of the analysis performed can a determination be made as to where God is pointing the finger.

Writing jointly in the 1996 NIJ report, Deputy District Attorneys George W. Clarke (a prosecutor in the O. J. Simpson case because of his expertise with DNA evidence) and Catherine Stephenson, both of San Diego County, California, emphasized that "[c]areful and timely collection and preservation of evidentiary material [are] critical." Moreover, "[c]areful examination" of exclusions are necessary.

"Typing results that exclude a suspected assailant," they explained, "may not demonstrate innocence. Not uncommonly, evidence collected and subjected to DNA profiling may reveal results from biological material left by other consensual sexual partners unrelated to the offense investigated or from other individuals having contact with the victim. Consideration of those results in the context of all other evidence in a specific case is essential to the determination of what took place."

And, they note, there is the other evidence to consider. "DNA profiling evidence can speak, but not with the passion of a victim's voice. DNA typing results can shed light on 'who'; it cannot explain precisely when, or how, or even why. The victim who survives the sexual assault must always be the primary and most important source of information."[19]

DNA evidence has been responsible for several astonishing exonerations. Another of the cases featured in the NIJ report was eventually traced through DNA to notorious Virginia serial killer Timothy Wilson Spencer. After making an apparent confession, David Vasquez, who has a diminished mental capacity, pleaded guilty to the sexual assault and murder of a Virginia woman and was sentenced to 35 years in prison. His plea bargain allowed him to avoid the death penalty. Ultimately, DNA tests conducted by the Virginia State Laboratory, Cellmark Diagnostics, and Lifecodes demonstrated that Spencer was responsible for the killing. Although never charged in that killing, Spencer was tried and convicted four times for the rapes and strangulation deaths of four other women (one 15 years old) in Virginia, all of whom were attacked in their homes. In the first three trials, Spencer was convicted in part based on RFLP tests performed on the evidence; in the fourth trial, DQA1 typing linked him to the crime. Spencer became the first person in the country sentenced to death on the strength of DNA evidence, a sentence that was eventually imposed four times. The Virginia Supreme Court upheld all four convictions and death sentences and the United States Supreme Court declined his appeals. On April 27, 1994, Spencer became the first person in the United States put to death on the basis of DNA evidence.

In another case, on the very day that the publication of the NIJ report was announced, June 14, 1996, three men were released from a Chicago courtroom to home monitoring by a judge persuaded by mounting evidence, including DNA tests, that they had most likely spent the last 18 years in prison for a crime they did not commit. One of the men had spent much of the time dodging a date with the executioner. A fourth man involved in the case, also sentenced to death, had been released on bond earlier in the year after the Illinois Supreme Court ruled that witness perjury in his trial necessitated a new trial.[20]

All four had been arrested on an anonymous tip following the widely publicized 1978 killing of a young Illinois couple engaged to be married. Abducted at gunpoint from a gas station in a suburb of Chicago, where one of them worked, they were taken to an abandoned townhouse where the 23-year-old woman, Carol Schmal, was repeatedly raped. Both she and her fiancé, Lawrence Lionberg, were found dead of gunshots to the head. The case quickly took on racial overtones when the four men arrested turned out to be black; the murder victims were white. Based on circumstantial evidence and on testimony from a woman named Paula Gray, who testified that she accompanied the defendants during the killings, but who was later jailed for perjury after changing her story several times, Dennis Williams and Verneal Jimerson were convicted in separate trials and sentenced to death, Willie Rainge was convicted and received a life term, and Kenneth Adams was convicted and sentenced to 75 years. Jimerson was originally let go after Gray, who was a sometime girlfriend of Adams at the time of the slayings, revised her story to exclude him, although she continued to insist the other three were involved and said so at their trials. Years later, Gray revised her story yet again while in prison, this time fingering Jimerson. So in 1984 he was tried, convicted, and sent to death row.

Although Jimerson was fortunate in that a prominent Chicago attorney, R. Ter Molen, took on his case for free and eventually won him another trial based on the perjured testimony, the case against the other three men only began to unravel in 1996 when journalism professor David Protess of Northwestern University invited his class to investigate the case. Three of his students began a six-month probe, which has since been chronicled in the book *A Promise of Justice* by Protess and Rob Warden.[21] With the assistance of a private investigator who had been working on the case intermittently since 1980, the students tracked down Paula Gray, Marvin Simpson (who claims he told the police within days of the killing who the real killers were, although no police report on this information has been found), and, based on Simpson's information, Ira Johnson. Johnson, an Illinois inmate who was serving a 74-year sentence for the 1990 strangulation murder of another woman, became a key figure when he admitted in writing to the students that he and three other men were really responsible for the 1978 double murder. Subsequently, one of the men Johnson identified as an accomplice, Arthur Robinson, also signed a statement admitting his role in the crime. A third man, Dennis Johnson, Ira's older brother, died of a drug overdose in 1993, and a fourth man, Juan Rodriguez, was still at large. Paula Gray now told a story to Protess and his students of being threatened by police and pressured to testify falsely in the earlier trial. To the extent that she knew any details about the crime, she said, the information was fed to her by police.

Based on these facts, prosecutors agreed to submit the evidence in the case for DNA testing. Despite the considerable age of the evidence specimens, DNA tests were successfully performed and excluded all four of the convicted men as donors of the sperm fraction retrieved upon autopsy from Carol Schmal's body. It was this evidence that convinced the judge to let Williams, Adams, and Rainge go home in June 1996, pending further review of the case. The same month, all charges were dropped against Jimerson, and on July 2, 1996, all charges were finally dropped against

Williams, Adams, and Rainge, ending their 18-year ordeal. They were subsequently pardoned by Governor Jim Edgar. Commenting on the case after their release, Cook County State's Attorney Jack O'Malley praised the American legal system but characterized the case as "a glaring example of its fallibility."[22] The next day charges were filed against Ira Johnson, Arthur Robinson, and Juan Rodriguez, who was located by authorities earlier in the year.[23] They were subsequently convicted. In 1999, Jimerson, Williams, Rainge, and Adams won a $36 million settlement against Cook County for their wrongful imprisonments, which their attorneys characterized as the largest such award to that time in a U.S. police-misconduct case.[24]

Ronald Cotten's case illustrates the power of DNA testing to force the criminal justice system to make a mid-course correction. Featured on PBS Television's *Frontline* in 1997, Cotten's case contains all of the elements that engender angst in legal observers.[25] In 1984, a young white woman banged on her neighbor's door in North Carolina in the early morning hours screaming that she'd been raped by a black man in her home. The victim, Jennifer Thompson, had just managed to escape out the back door of her apartment after being held captive for several hours, during which time she caught fleeting glances of her assailant in the mostly darkened apartment. Thompson was taken to the hospital for a forensic medical examination, which involved the storage of evidence in a rape kit and the preservation of Thompson's clothes, while a second woman was being raped in Thompson's neighborhood, apparently by the same man.

In the days that followed, Thompson helped produce a composite sketch of the rapist, the publication of which brought a tip that Ronald Cotten looked like the sketch. Cotten had been convicted at age 16 of attempting to rape a 14-year-old white girl, so police had a file on him. Presented with a photo-array that included Cotten, Thompson picked him out and later fingered him in a lineup, but not without first deliberating whether it was Cotten or another man in the lineup, Bobby Poole. The second victim failed to pick out Cotten and instead picked out Poole, so Cotten was put on trial only for the rape of Jennifer Thompson. Poole was not charged. The judge excluded evidence the defense attorney wanted to present on Cotten's behalf regarding the second rape and that victim's failure to identify Cotten, as well as the testimony of an expert witness that would have called into question Thompson's memory of the rape. The case thus turned entirely on Thompson's eyewitness identification as there was virtually no physical evidence tying Cotten to the crime, other than a piece of foam rubber seized at Thompson's house that seemed similar to the type of rubber used to make a pair of aging sneakers owned by Cotten. Blood-typing tests on semen stains proved inconclusive.

Thompson's testimony won over the jury. Cotten did not testify to prevent his criminal record from being introduced and to play down the fact that he had changed his alibi a number of times. Family members testified that he was in fact home sleeping on the couch at the time of the rape. Cotten was convicted and sentenced to life in prison.

From jail, Cotten wrote to his attorney claiming that Bobby Poole, who was also in jail on another matter, was in fact the rapist in his case. Another inmate backed this up by saying Poole had admitted his involvement to him. Poole, in fact, was bragging about it. Granted a second trial by the North Carolina Supreme Court on the grounds that the trial judge erred when he excluded evidence of the second rape and the victim's failure to identify Cotten, Cotten and his attorney must certainly have felt the scales had tipped in their favor. Unfortunately for Cotten, the second rape victim now identified him (that is, Cotten) as her attacker, and he was tried for both rapes and identified in court by both women. Furthermore, Bobby Poole took the stand in a hearing outside the presence of the jury to deny any involvement whatsoever, and neither of the rape victims were moved to change their identifications from Cotten to Poole after seeing them both in court. Thompson, in fact, was convinced that she had never seen Poole before in her life.[26] The judge, therefore, excluded Poole's testimony from the trial, and Cotten was convicted of both crimes and sentenced to two life terms, a conviction that was affirmed on appeal.

Cotten languished in jail for eight more years until University of North Carolina law professor Rich Rosen was struck by the lack of physical evidence in the case. By this time, 1994, DNA testing was routine in rape cases and, miraculously, the rape kit and clothing evidence had been preserved at the request of the principal detective in the case. Under normal circumstances, such evidence would have long been destroyed as evidence lockers are cleared to make room for new evidence. Although the samples from one of the victims' rape kits were too degraded for DNA testing of any kind, PCR-based DNA testing was successful with the rape kit and underwear of the other victim and the analysis excluded Cotten as the rapist. (Rape kit evidence is considered particularly useful because it is secured during a medical examination and is carefully preserved to avoid contamination; moreover, a chain of custody is maintained.) At the same time, the DNA results were compared with DNA samples of Bobby Poole that had been stored in North Carolina's DNA databank for convicted violent felons. Confronted by the DNA evidence that showed a match to him, Poole confessed to both rapes. (He died of cancer in prison in 2000.) On June 30, 1995, charges were dismissed against Ronald Cotten, and he was released after serving almost 11 years in jail. Pardoned by the governor a month later, Cotten received the statutory $5,000 compensation from the state for his wrongful incarceration.[27] (After persistent lobbying by Jennifer Thompson, the state eventually gave Cotten an additional $110,000 for his wrongful conviction.[28]) Soon after learning of the DNA exoneration, Jennifer Thompson begged Cotten to forgive her for her mistaken identification, but Cotten made clear that he harbored no grudge against her. The two have since become friends.[29]

"DNA testing is showing us with a great deal of scientific certainty," Barry Scheck commented on *Frontline*, "that [mistaken eyewitness identification is] an even greater problem than we suspected."

Jennifer Thompson told *Frontline* that she accepts the scientific results but still sees Ronald Cotten's face in her memory of the attack. She has become an outspoken opponent of the death penalty and a critic of eyewitness-identification evidence.[30] In 2000, Thompson wrote an op-ed piece for *The New York Times* entitled "I Was Certain, but I Was Wrong," in which she recounted her experience in the Cotten case.

"If anything good can come out of what Ronald Cotten suffered because of my limitations as a human being," she wrote, "let it be an awareness of the fact that eyewitnesses can and do make mistakes."[31]

The reasons that prosecutors have misgivings about some of the DNA exclusions are embodied in another case featured in the 1996 NIJ report, but which turned out to be a story only half told. On September 25, 1981, Kerry Kotler was arrested and charged with raping—on two separate occasions—a Long Island, New York, housewife in her East Farmingdale home. The first rape was in 1978 while the rapist was wearing a ski mask; the second rape occurred at knifepoint in 1981, but this time without the mask. During the second attack the victim quoted her assailant as saying he was "back for another visit." This time the victim saw his face. The first time the rapist stole jewelry, the second time he stole jewelry and $343 in cash. According to authorities, the victim picked Kotler out of a collection of 500 photographs and then later positively identified him in a police lineup, by both his appearance and by the sound of his voice.

At trial Kotler claimed it was simply a case of mistaken identification, but the prosecution bolstered the victim-identification testimony with serological blood evidence that showed a match between Kotler and a semen stain on the victim's underpants for three genetic markers, ABO blood type, PGM (phosphoglucomutase), and GLO (glyoxalase). Only 3 percent of the population exhibit this combination of markers, including Kotler, so he could not be excluded as the rapist.[32] In addition, his alibi was weak. After deliberating for two days, a Suffolk County jury convicted Kotler of both rapes as well as various burglary and robbery charges. He was sentenced to 25 to 50 years in prison. On December 2, 1983, the court denied a motion brought by Kotler to set aside the jury verdict on the grounds that, among other things, the prosecution had engaged in misconduct by falsifying and withholding evidence. Kotler's conviction was later affirmed by the Appellate Division on March 3, 1986. A year later Kotler filed a second motion to set aside the verdict, citing as justification alleged perjury by police detectives and concealment of evidence from Kotler's defense attorney prior to trial, which was denied.

In the meantime, however, Kotler had begun to pursue other remedies. In September 1988, after learning about DNA typing, Kotler asked for help from the Legal Aid Society to have the evidence in his case retested using the newer DNA methods.

With financial help from his father, Kotler succeeded in arranging for DNA tests on the rape kit and the victim's underwear. On February 15, 1989, these items, along with a sample of Kotler's blood, were sent to Lifecodes for analysis, but Lifecodes, one of the first private laboratories in the United States to conduct "DNA finger-printing" tests—based on their own patent—in criminal cases, only performed RFLP tests at the time and concluded there was insufficient genetic material on the evidence for RFLP testing. After some investigation by a Legal Aid attorney, the evidence was sent to Forensic Science Associates in California for PCR tests in February 1990. Although the California laboratory reported back that Kotler, based on DQA1 testing, was excluded as the source of the semen, the New York District Attorney's office proposed a different theory for the results.

Specifically, Kotler's DQA1 type was homozygous for the allele designated 4; his type was thus 4,4. (The test was done with the early version of DQA1, which did not subtype the 4 loci.) The victim had the exact same type (4,4), but her cellular contribution would presumably have been eliminated during the lysing (separation) step. The sperm fraction extracted from the victim's underwear contained two alleles, allele 1.1 and allele 4. Consequently, prosecutors argued that the DNA results on the semen stain could be explained as originating from both Kotler and a consenting sexual partner contributing the 1.1 allele. The prosecution's faith in the correctness of the jury's verdict, it seemed, would not be shaken easily. The hardest sell was Suffolk County, New York, District Attorney James Catterson.

The Innocence Project now embraced Kotler's cause and began a two-year battle with Catterson to have Kotler freed.[33] When efforts failed at persuading the prosecutor of Kotler's innocence, attorneys Scheck and Neufeld filed a legal memorandum in support of a motion to vacate the judgment in Kotler's case in March 1992. The memorandum stressed two main points: the PCR exclusion by Forensic Science Associates; and the claim that the original prosecutors had withheld evidence from the defense, including police reports indicating that the victims' original identification of Kotler had, in fact, not been a definitive one. In addition to showing discrepancies in age, height, and weight between Kotler and the victim's description of her assailant, the police reports noted that the victim had originally characterized Kotler as merely a "look-alike." The district attorney's office responded by filing its own memorandum opposing the motion to vacate. The matter seemed destined for a bitter court fight.

The evidence by this time had been sent for additional DNA tests to The Center for Blood Research in Boston, whereupon Forensic Science Associates' tests were confirmed. Finally, at the request of defense counsel, a sample of blood from the victim's husband, presumably the victim's only consenting sexual partner at the time of the rapes, was sent to both laboratories.

The husband's DQA1 type—according to both laboratories—contained alleles 2 and 3, so he was clearly excluded as being responsible for any part of the evidentiary semen stain. Ultimately, too, the vaginal swabs were tested and once again excluded Kotler. On November 24, 1992, the two laboratories issued a joint statement to the Suffolk County Court declaring that the semen stain excluded both Kotler and the victim's husband.

At this point, District Attorney Catterson hesitantly gave in and joined with the defense in a request that the court vacate Kotler's conviction. On December 1, 1992, the court vacated the conviction and released Kotler on his own recognizance. He had been in jail for 11 years. Two weeks later, on the prosecution's motion, the court dismissed all indictments. Subsequently, the serologist who conducted the original blood and enzyme tests in Kotler's case, Ira Dubey, pleaded guilty to perjury stemming from falsification of his qualifications and training in another case.

Overall, according to Scheck, Kotler's case involved "unbelievable prosecutorial misconduct."[34] Catterson, however, denied that evidence was withheld and told reporters after Kotler's release that he was still not convinced of Kotler's innocence, suggesting that contamination somehow accounted for the exclusionary results.[35]

The victim in the case immediately went out and bought a gun and soon fled the State of New York, reportedly fearing for her own safety and that of her daughter. Following Kotler's release, she told a reporter for The New York Post, she found a note on her car that read, cryptically, "you have a very Qute [sic] daughter. Do not pursue this."[36] No link was ever established between Kotler and the note.

"I don't know what went wrong in the lab," she told Newsweek shortly after Kotler's release. "Now that he's free, I look over my shoulder and listen for his voice."[37]

For his part, Kotler became something of a celebrity for being cleared by DNA evidence. In January 1993 he was featured in an article in Newsweek, and in May 1993 he appeared on NBC-TV's Donahue along with Scheck, Neufeld, and other men recently released after DNA testing. He also was not content to go quietly back home to resume his life after 11 years behind bars. Represented by Scheck and Neufeld, he sued the State of New York for wrongful imprisonment under New York's Unjust Conviction and Imprisonment Act and in January 1996 won the liability phase of his suit before Court of Claims Judge Leonard Silverman, who found, according to his ruling, "overwhelming evidence" of Kotler's innocence.[38] The damage phase was scheduled to begin some months hence, and Kotler and his counsel were set to ask for an award of $12 million.[39]

Suddenly, the trajectory of Kotler's new life abruptly ricocheted. On April 8, 1996, Kotler, now 37, was arrested for another rape, stemming from an August 12, 1995, incident involving the kidnapping and rape of a 20-year-old college student on Long Island. The victim managed to recall four digits of the license plate on her assailant's car, as well as a general description of the car, and police traced the information to a 1988 white Pontiac Grand Am, a car that belonged to Kerry Kotler's girl-friend. It was a car Kotler had been seen driving on occasion and one that he was in fact driving a few days after the rape when police originally stopped and questioned him in Montauk Point, Long Island, about the rape.[40] Inside the car, investigators found dog hair from Kotler's German shepherd, hairs investigators concluded were similar to those retrieved from the dress the victim was wearing when attacked. Kotler's girlfriend, Kelly Norman, posted a $25,000 cash bond for Kotler three days after his arraignment.[41]

According to prosecutors, the assailant flashed a badge while driving alongside the victim and induced her to pull over. When she asked to see the badge again, the assailant pulled a knife and abducted the woman. Her car was found, its engine still running, hours later. She was taken to a secluded area and raped. The rapist then used a water bottle to douche semen from inside of the victim and explained to her his rationale for doing so: He was removing evidence. This suggested to authorities, of course, that the rapist was savvy about DNA. These efforts on the part of the assailant failed, however, because, during the course of the eight-month investigation, forensic scientists were successful in extracting and analyzing DNA (using RFLP) from a semen stain recovered from the victim's dress. The DNA evidence in this new case, according to prosecutors, came back as a solid match for Kerry Kotler (six VNTR regions were probed), and the odds of any random person matching the DNA in the case were calculated to be one in 7.5 million.[42]

Kotler's criminal defense lawyer, Jack T. Litman, however, was quick to turn everything on its head, suggesting the case against his client was part of a vendetta against Kotler for the chagrin of losing the civil case. Detectives investigating the case, Litman told *The New York Post,* "mentioned" Kotler's civil suit to people they interviewed.[43] Litman vowed to vigorously challenge the allegations, even if that meant contesting the reliability of the DNA tests conducted in the case, a position, he said, that in no way undermined the earlier DNA tests that were pivotal in freeing Kotler in 1992.[44] Finally, Litman stressed that neither the rape victim nor another potential witness was able to identify Kotler.

In the civil case against the State of New York being handled by Scheck and Neufeld, State Attorney General Dennis Vacco rejected a $2 million settlement offer in 1996, calling the fight over the first case "far from over."

"We stand by the story of the woman who was raped twice," he told the press in 1996. "To find that this guy was wrongfully convicted you have to disbelieve the assertion of a woman who looked into his eyes during the second attack."[45]

Despite the strong sentiments of the state's attorney general, Kotler, following the damages phase of the civil suit, was awarded $1.5 million by Judge Silverman for his wrongful incarceration in the first case on the eve of this second trial.[46] On June 25, 1997, Kerry Kotler went on trial in Long Island for the 1995 rape. In his opening statement to the jury, Defense Attorney Litman told the jury that the biological evidence was planted by detectives in an effort to frame Kotler. The semen was probably procured by police, Litman explained, in an argument reminiscent of Scott Turow's best-selling novel *Presumed Innocent,* by stealing Kotler's used condoms from his garbage. To substantiate this claim, Litman offered the testimony of Edward Blake of Forensic Science Associates. Dr. Blake testified that extraneous bands on the autorads were due to bacterial and/or nonhuman contamination, which Blake contended was excessive in the evidentiary sample. Litman suggested that just such contamination would occur in a discarded condom. He also argued that Kotler was so knowledgeable about DNA evidence that if he had committed the rape, he would have used a condom to avoid leaving biological evidence behind; he would not have relied on a water bottle.[47]

After a three-week trial, Kotler was convicted on July 18, 1997, of first-degree rape. According to published accounts, the jurors found the DNA evidence against Kotler compelling and his version of events unconvincing. In September 1997, Kotler was sentenced to 7 to 21 years in prison.

"This is an interesting case, because it chips away at the clay feet of the Neufelds of this world," Catterson later commented to *The New Yorker.* "Isn't that ironic? Hoist by their own petard. It's really—it's justice."[48]

Despite the 1999 recommendations of the NIJ to be more amenable to DNA appeals, some prosecutors continue to fight efforts by prisoners to clear themselves through DNA, especially when the DNA results might be open to interpretation.

Defense attorneys have horror stories to tell as well, stories whose endings are likewise punctuated by a round of emphatic DNA tests. On February 25, 1983, ten-year-old Jeanine Nicarico was home alone from school with a cold when someone kicked in the front door of her home in Naperville, Illinois, and kidnapped her. Two days later a group hiking through a nearby nature preserve found her body, still clad in her nightgown, crumpled in a heap in a patch of weeds. She had been raped and brutally beaten to death, as evidenced by a grotesquely misshapen nose.

The crime caused horror in the affluent community of Naperville, located 35 miles west of Chicago, and a $10,000 reward was immediately offered for information leading to the killer's apprehension. Following an anonymous tip two weeks later, the sheriff's department sent detectives to question Alejandro Hernandez. Although he denied involvement, he allegedly implicated other men in the crime. Despite his denials, Hernandez was arrested on March 6, 1984.

Detectives next questioned Rolando Cruz, an acquaintance of Hernandez's, and, according to investigators, Cruz admitted that he'd had a "dream vision" about the crime that included details about the case that were not publicly known. On March 9, 1984, Cruz was indicted for the crime. The indictments coincided with a bitterly contested primary battle for the office of DuPage County State's Attorney. Ultimately, a third man was also indicted for the girl's killing, and the three were tried together in 1985.

Cruz's "dream vision" became a centerpiece of the prosecution's case against him. Several sheriff's detectives testified that they were present when Cruz recounted the dream, and a lieutenant in the sheriff's department corroborated their account when he testified that the others had reported the incident to him immediately afterward. Other incriminating statements were also introduced against the defendants through law enforcement personnel and civilian witnesses. Defense attorneys did not aggressively present an alibi defense, but the lawyer for Hernandez attempted to convince jurors that any statements he made against the other men were made merely in the hope of collecting the sizable reward. The efforts on behalf of Hernandez and Cruz failed, however, and they were both convicted and sentenced to death. The jury could not reach a verdict in the case against the third man, and in 1987 charges against him were dropped.

A vigorous and sustained appellate battle ensued on behalf of Cruz and Hernandez to prevent their executions. While these appeals met with repeated success, Cruz and Hernandez remained locked in a Sisyphean cycle of winning an appeal only to lose upon retrial. By 1988 both men had won reversals because they should have been tried separately. The original trial was deemed unfair by the Illinois Supreme Court because of the introduction of the statements of the codefendants against one another. Even with separate trials, however, both men were once again convicted. Cruz again received the death penalty, but Hernandez drew only 80 years this time. After some hesitancy, the Illinois Supreme Court reversed Cruz's second conviction on July 14, 1994, based primarily on the statements made by another man, Brian Dugan, that he was in reality the lone killer of Jeanine Nicarico. The statements, elicited during a hypothetical question and answer session in pursuit of a plea bargain in an unrelated case, were made to police in 1985. They were not, however, revealed to defense counsel for Cruz or Hernandez, which is a serious constitutional-rights violation. Just as distressing, DNA tests conducted in 1989 found that Dugan could potentially be linked to the crime after all, although the tests were not definitive.[49] A new trial was scheduled for Cruz in 1995, and he elected to be tried before a judge this time. Hernandez, too, won a reversal for his second conviction, but he was convicted a third time by a jury, only to have this conviction reversed, too. By this time, the cases had created a fissure within the law enforcement community itself. Several officials expressed their belief that the men were innocent. In 1992, a DuPage County prosecutor, Mary Brigid Kenney, resigned in protest over the continued pursuit of the case against the men.[50]

The cycle of conviction and reversal experienced arrhythmia when DNA tests were performed in September 1995 on the evidence, including semen, that had been secured from the body of the dead girl. The tests excluded both Cruz and Hernandez as contributors of the semen deposit, but did not exclude Brian Dugan, currently serving life for the rape/murders of a 7-year-old girl and a 27-year-old woman. Nevertheless, prosecutors persisted, arguing the tests did not definitely prove that the men were innocent of the girl's murder. Prosecutors elected, therefore, to retry Cruz a third time and Hernandez a fourth. It was a decision that would take the case onto uncharted legal terrain.

As Cruz's retrial was underway in the fall of 1995, the sheriff's department lieutenant who had corroborated the "dream vision" testimony of the other sheriffs unexpectedly recanted the testimony he'd given in the earlier trials. Whereas he previously told juries that the sheriff's detectives who'd been witness to the "dream vision" confession of Cruz had called him soon afterward about it, he now conceded to the judge in this case that he'd been in error about this. In fact, according to Sheriff's Lieutenant James T. Montesano, he had been in Florida at the time he was supposed to have learned of the "dream vision" from the other sheriffs.[51]

Montesano's testimony prompted the judge to immediately order an acquittal for Cruz and release him from jail, triggering a year-long grand jury investigation. By this time, Cruz had been on death row for 11 years. In the meantime, Hernandez was also freed after having spent 11 years in jail, three of those on death row. Upon freeing Cruz, Judge Ronald Mehling characterized the case against Cruz as one riddled with mistakes, sloppy police work, and outright lies.[52]

On December 12, 1996, the results of the grand jury probe sent shockwaves through the legal community and shivers down the spines of not a few people in law enforcement. In a damning assessment of the Cruz and Hernandez cases, the grand jury indicted three former DuPage County prosecutors who handled the case and four sheriff's deputies involved in the investigation, including Montesano. According to the indictment, the "vision statement" was a complete fabrication on the part of investigators. Moreover, the indictment alleged that prosecutors, in league with members of the sheriff's department, concealed from defense attorneys records of Brian Dugan's statement to police that suggested he alone killed Nicarico, a violation of the duty to turn over any potentially exculpatory information to the defense.

The 47-count indictment charged the former prosecutors with conspiracy to commit obstruction of justice and conspiracy to commit official misconduct; the sheriffs were all charged with perjury and official misconduct, and three of them were also charged with obstruction of justice. One of the original prosecutors, Robert K. Kilander, was by this time a county judge. Another, Patrick J. King, Jr., had gone on to be a United States Attorney in Chicago, and the third, Thomas L. Knight, was in private practice. In addition to Montesano, the sheriffs included Thomas E. Vesburgh, Dennis Kurzawa, and Robert L. Winkler. The lawyers for all the men indicted emphatically denied that their clients were guilty of any wrongdoing in the case.[53] By the time their trial began in early 1999, the men had become known as the DuPage Seven.

Such cases are extremely rare. Research by *The Chicago Tribune* turned up only six prior instances of prosecutors being tried for similar official misconduct in the 20th century. Only two of these resulted in convictions, for misdemeanors, and involved small fines. Cases against the four others were either dropped or resulted in acquittals.[54] At the same time, a *Tribune* analysis revealed that since 1963, 380 defendants in the United States had won reversals because prosecutors had either hidden evidence or lied.[55] About midway through the trial of the DuPage Seven, Judge William Kelly dismissed all charges against Judge Kilander and U.S. Attorney King, citing insufficient evidence to prove their complicity in the conspiracy. On June 5, 1999, following a two-month trial, a jury acquitted four of the remaining DuPage Seven, and Judge Kelly dismissed the charges against the fifth.[56]

Allegations of fabrication have been made against forensic scientists, too. Over time, the Innocence Project found that as many as a third of the DNA exoneration cases involved tainted evidence or forensic fraud.[57] Three of the 28 cases featured in the 1996 NIJ report apparently involved just such a scenario, all traceable to former forensic serologist Fred Zain, who worked for 13 years, until 1989, in the West Virginia State Police crime laboratory (a good part of that time as director of the laboratory's serology department). Later, until 1993, Zain worked as a forensic chemist for Bexar County, Texas. Significantly, during the time that Zain worked in West Virginia, his laboratory was not yet performing DNA typing. At least six men who had served a total of 40 years in prison on the strength of Zain's testimony were set free after an unprecedented investigation into his forensic science laboratory and a review of his testimony and forensic test results in dozens of criminal cases. DNA tests performed on evidence retained in the cases of these six men flatly contradicted the forensic tests offered as evidence against them by Zain.[58] Although acquitted of perjury in one trial in West Virginia in 1995 (a second count was dropped), Zain faces an additional fraud trial there arising out of the same general allegations of falsifying forensic test results. In the pending case, Zain is charged with four counts of obtaining money from the state of West Virginia (for example, his salary) under false pretenses. On September 18, 2001, a mistrial was declared after a jury found itself hopelessly deadlocked on the fraud allegations, but prosecutors have said they will retry the case.[59]

Forensic DNA analysis now allows a more definitive second look at evidence from serious criminal cases that have faded from memory. Such second looks frequently permit scientists to conduct DNA tests on evidence previously deemed untestable, which has led to many of the DNA exonerations. On other occasions, however, such as in the Zain cases, the new DNA tests reveal that the evidence used against a defendant in an earlier trial not only demonstrates innocence, it completely contradicts the original characterization of the physical evidence by the forensic scientist whose testimony was pivotal to the conviction. As things were turning out, the Zain case was not an isolated one. Just as DNA analysis has opened a window into the past for ancient DNA enthusiasts, it is shedding a new, sometimes unflattering, light on the annals of the criminal justice system.

When Jeffrey Pierce was convicted of rape in 1986, Oklahoma City police chemist Joyce Gilchrist testified that hairs recovered from the crime scene tied Pierce to the crime. Pierce was sentenced to 65 years in prison. When Pierce was freed from jail in May 2001 after DNA tests on the hairs and on semen preserved from the case exonerated him, Gilchrist suddenly found herself under the microscope. A previous DNA analysis in a Gilchrist case had freed Robert Miller from death row and led to the arrest of another man whom Gilchrist had said the hair evidence excluded as a suspect. An FBI review of eight of her cases revealed, according to the Bureau's report, that Gilchrist had either misidentified evidence or testified improperly in at least five of them. The hairs in the Pierce case were not, according to FBI analysts, a match to Pierce, as Gilchrist had testified they were.[60]

Given the stakes, Oklahoma Governor Frank Keating ordered a complete review of Gilchrist's forensic casework (covering roughly 3,000 cases from 1980 through 1993, after which time she was promoted to supervisor) and stayed the executions of 12 death row inmates in Oklahoma whose convictions were based, at least in part, on Gilchrist's forensic evidence. Unfortunately, by this time there had been 11 executions in Oklahoma of inmates whose trials featured the testimony of Joyce Gilchrist. The U.S. Department of Justice decided to launch its own investigation.

"The juries have been lied to for the past 20 years," Pierce told reporters upon his release. "There are going to be a lot more victims."

"Nobody has listened to us for 15 years," Pierce's attorney, David Autry, said, echoing his client, "and now science has advanced enough to prove what we've known all along: that Jeff Pierce is innocent."[61]

A few months after Pierce's release from jail, the U.S. Court of Appeals for the 10th Circuit overturned the death sentence of Oklahoma inmate Alfred Mitchell (although the court simultaneously upheld his conviction) because, the court said, Gilchrist gave "false and misleading" testimony at Mitchell's trial.[62]

By the time of the verdict in the DuPage Seven case, Rolando Cruz was one of 13 men from death row in Illinois who had been found, for a variety of reasons, including new DNA evidence, to have been wrongly convicted in that state since the death penalty had been reinstated in 1977. One of the men, Anthony Porter, had come within two days of being executed before students in the journalism class of David Protess turned up information that cast doubt on his guilt. He was subsequently released from jail in 1999.[63]

In January 2000, just one week after Democratic Senator Patrick Leahy of Vermont introduced the Innocence Protection Act in the Senate, Illinois Governor George Ryan, a pro-death penalty Republican, became the first governor in any of the 38 states with the death penalty to impose a moratorium on executions. "I now favor a moratorium," Governor Ryan explained, "because I have grave concerns about our state's shameful record of convicting innocent people and putting them on death row."[64]

Calling the system "fraught with error," Governor Ryan later established a 12-member commission in the state to scrutinize the erroneous verdicts and the administration of the death penalty overall. The commission subsequently recommended 85 changes to the Illinois death-penalty system, which the Illinois legislature is expected to take up in 2003. In October 2002, the Illinois Prisoner Review Board conducted clemency hearings for most of the 159 prisoners on death row in Illinois. While 142 of the death row inmates actively sought to have their death sentences commuted to life in prison, the remainder inexplicably declined to seek commutations (so no hearings were held concerning their cases). Governor Ryan—who decided in 2001 not to seek re-election in 2002 amid a growing bribery scandal in his administration—was poised to act on the board's recommendations before leaving office in January 2003, when he was replaced by a Democrat, Rod R. Blagojevich. Governor Ryan had said he would consider commuting the sentences of all the inmates facing execution, even those who did not request hearings before the review board. During the televised clemency hearings in October 2002, however, which included the heart-rending stories of many family members of murder victims—who often directed embittered recriminations at Governor Ryan—and a retelling of the often gruesome crimes by prosecutors, Governor Ryan back-pedaled from an earlier statement (which prompted the flood of clemency petitions in the first place and triggered the hearings) suggesting that he might grant blanket clemency to the 159 death-row inmates.[65] Despite the fevered emotions that characterized the hearings, the editors of the *New York Times* urged Governor Ryan to depart office on a "high note" by commuting the death sentences of all the inmates on Illinois's death row.[66] That surely seemed a difficult path to take in light of some of the testimony from victims' family members. Sam Evans, for example, spoke of his burning hatred for death-row inmate Fedell Caffey, who was convicted along with Jacqueline Williams for the 1995 killing of Evans's daughter, Debra, whose full-term fetus was cut out of her womb during the attack. "If you are having trouble finding someone to pull the switch, to make the injection, to shoot him, I volunteer," Evans told the review board. "I'll kill him myself if it's a problem for this state, this governor."[67] Nevertheless, on January 11, 2003, Governor Ryan announced during a televised speech that he was commuting to life in prison the death sentences of all Illinois inmates currently under sentence of death, a decision likely to be one of the most celebrated and reviled decisions in legal history. The day before the speech, Governor Ryan had granted full pardons to four death row inmates whose convictions, the governor believed, resulted from confessions obtained through torture. In his commutation speech, Governor Ryan characterized the Illinois death penalty system as "haunted by the demon of error."

Ryan's initial moratoriam action was perhaps the most visible example of a groundswell of concern over the fairness of the death-penalty system that had been building in recent years, precipitated in large part by the nationwide experience of seeing death-row inmates exonerated by DNA tests. Moratorium bills had been floated in 12 states already, and Nebraska had nearly passed a moratorium legislatively. There was fierce debate over the death penalty in Florida, too, where 18 death-penalty convictions had been reversed since 1977.[68] The Illinois moratorium propelled the anti-death penalty movement forward. Within seven months, according to a *Los Angeles Times* report, roughly 500 new groups had sprung up around the country promoting moratoriums on the death penalty, effectively doubling the number of such groups nationwide. Additionally, several states had launched studies to determine the impact of racial bias, inadequate representation, and a host of other problems on death-penalty verdicts.[69] Many of those people calling for reforms were proponents of the death penalty, like Governor Ryan, concerned about perceptions of fairness in the death-penalty system.[70]

Governor Ryan's action did not occur in a vacuum, of course. It followed years of consternation over death-penalty verdicts being upended and the harsh light such reversals cast on the criminal justice system. Since the mid-1970s, 87 people had been freed from death row throughout the country due to various problems with their convictions. Of late, the bad press was international. In 1999, while visiting the United States, Pope John Paul II urged an end to the death penalty. The same year, the United Nations Human Rights Commission passed a resolution calling for a worldwide moratorium on the practice.[71] Back at home, *The Chicago Tribune* published a five-part series on the death-penalty system in Illinois in the months leading up to Ryan's decision. The *Tribune* concluded that fully half of the state's 285 death-penalty convictions since the 1970s were problematic.[72] Despite these influences, Ryan certainly could not have foreseen the degree to which his decision would shake up the political landscape nationally.

Within six months of Ryan's declaring a moratorium, the cover of *Newsweek* read: "Rethinking the Death Penalty: DNA Tests and New Evidence Have Saved 87 Prisoners from Execution. The Growing Debate Over Who Should Die." Ryan's moratorium decision, wrote *Newsweek*'s Jonathan Alter, "has resonated as one of the most important national stories of the year."[73] Just as the *Newsweek* article hit newsstands, a comprehensive study of death-penalty cases covering 1973 through 1995 was published by Columbia University School of Law professor James S. Liebman and his colleagues. The study found that two out of every three death-penalty convictions in the United States had been overturned on appeal, which meant new trials.[74] The study analyzed the 4,578 death sentences that were appealed (of a total of 5,760) during the period. According to the study's authors, the rate of serious error in death-penalty cases nationally was 68 percent, with the errors traceable to incompetent lawyers, the suppression of evidence by police or prosecutors, bias on the part of judges or juries, faulty jury instructions, coerced confessions, reliance on snitches, and other defects. (Illinois actually had an error rate slightly lower than the national average, 66 percent, according to the study.) Further, 75 percent of the reversals led to lesser sentences and an additional 7 percent resulted in not guilty verdicts.[75] While the Liebman report was not embraced uncritically—writing for *The Wall Street Journal*, for example, University of Utah law professor Paul G. Cassell lambasted the report as statistical "legerdemain" and noted that death-penalty cases were accorded, by U.S. Supreme Court mandate, "super due process," which would explain the high reversal rate[76]—the report nevertheless raised the political stakes over capital punishment. *Actual Innocence,* the book by Barry Scheck, Peter Neufeld, and Jim Dwyer, moreover, was making the rounds.

The year 2000 was a presidential election year, and the death penalty, for years a safe issue for serious presidential contenders, so long as they were for it, had suddenly become a lot more complicated. Both of the leading candidates, Vice President Al Gore, a Democrat, and Texas Governor George W. Bush, a Republican, had declared their unqualified support for the death penalty, positions that were, if not heartfelt, at least pragmatic, given the consistently high level of support among American voters for capital punishment. For candidate Bush, however, the increasingly complex issue of the death penalty had to be confronted in a nuanced way because he hailed from Texas, a state that had become, somewhat inconveniently, the nation's leader in carrying out executions.

By late May 2000, things were looking bleak for Texas death-row inmate Ricky Nolen McGinn. His death-penalty appeals had all failed, and, in recent days, both the Texas Court of Criminal Appeals (the state's highest such court) and the Texas Board of Pardons and Paroles had declined to stay his June 1st execution to allow time for DNA tests that McGinn and his lawyers maintained could demonstrate that he was not guilty of a capital offense (although the tests were unlikely to show that he was completely innocent). The final blows seemed to come when both the U.S. Court of Appeals for the 5th Circuit and the U.S. Supreme Court also refused to grant a stay. McGinn's execution, if it went off on schedule, would be the 132nd execution in Texas since 1977. McGinn had one thing going for him, however. Timing.

By this time, McGinn had already had an earlier execution date postponed because his Texas attorney's office had been ransacked by a tornado and records were lost. Subsequently, Barry Scheck and Peter Neufeld put the weight of the Innocence Project behind McGinn's cause, even offering to pay for his DNA tests, and were successful in persuading a lower court judge to allow the tests (although the Texas Court of Criminal Appeals nixed the idea). The reason for Scheck and Neufeld's enthusiasm had to do with timing, too. Advances in DNA testing since McGinn's 1995 trial now held out the hope that evidence secured during the investigation—evidence that was deemed beyond the reach of DNA analysis at the time—could be genetically examined after all. If true, the lawyers argued, the DNA tests might well be dispositive of McGinn's innocence or guilt. During the month of May 2000, too, the Republican-controlled New Hampshire Legislature had become the first state in the nation to vote a bill to abolish the death penalty since its 1976 revival (although the effort fizzled when the state's Democratic governor, Jeanne Shaheen, vetoed the bill). Just a few months earlier, a Gallup poll had revealed that public support for the death penalty had slipped from its high of 80 percent in 1994 to 66 percent in 2000. Finally, presidential candidate Bush had actively sought to soften his image as staunchly pro-death penalty in recent weeks, publicly stating that he fully supported the use of DNA tests to "erase any doubts" in capital cases. Bush had, in fact, been keeping a watchful eye on developments in the McGinn case.

After the 5th Circuit and the Supreme Court declined to grant McGinn a stay, Bush stepped in at the 11th hour to grant McGinn a 30-day stay of execution so that the newer DNA tests could be performed. It was the first such stay he had ever granted in his five-and-a-half years as Texas governor, although he had pardoned three prisoners in noncapital cases after DNA had exonerated them.[77]

If there was a victor in all of this, it was DNA evidence. "Mr. Bush's unusual action in the case of Mr. McGinn," wrote The New York Times, "underscored the way in which modern DNA testing and technology have begun to affect the debate over capital punishment by casting doubt on convictions once considered just."[78] Bush's decision to stay McGinn's execution, The Boston Globe said, "is striking evidence of how advances in DNA technology are shaking a long-standing national consensus on crime, punishment, and the death penalty."[79]

McGinn had been convicted in 1995 of the 1993 rape and murder of his 12-year-old stepdaughter, Stephanie Rae Flanary, a capital crime. On May 22, 1993, McGinn was left at home to watch Stephanie while her mother was out of town. That night, McGinn called the sheriff's department to report that Stephanie had been missing for several hours after she had gone for an afternoon walk. When an initial search failed to find Stephanie over the course of that night, tracking dogs were brought in the following morning. When one of the dogs, trained to alert at the scent of cadavers, alerted near the back of one of McGinn's cars, the police searched it and found bloodstains throughout the interior. McGinn told investigators at the time that the blood was fish blood from the day before, but forensic tests conducted later revealed it to be human blood. Even before the tests were completed, however, police arrested McGinn. On May 25, 1993, police found Stephanie's body inside a metal culvert several miles from the house. An autopsy revealed that she died from a number of blows to the head from a blunt instrument, which fractured her skull, and that there had been sexual activity. Semen was discovered inside her body and on Stephanie's shorts. Among the hairs recovered during the investigation was a single pubic hair retrieved from inside Stephanie's vagina at autopsy. The pubic hair was found to have microscopic characteristics consistent with McGinn's pubic hair.[80]

The blood found in McGinn's car was subjected to ABO blood tests and found to be type A blood, as was Stephanie's. McGinn's was type O. Although it proved impossible to perform DNA typing on the blood in the car, conventional serology tests indicated that the blood was consistent with Stephanie's, to the exclusion of between 96.4 and 99.99 percent of the general population. Additionally, a head hair found adhering to one of the car bloodstains was found to have microscopic characteristics consistent with Stephanie's hair.[81]

The day after Stephanie's body was found, the police found an ax underneath the seat of a broken-down truck on McGinn's property, an item that had been missed in previous searches. The blunt end of the ax was stained with blood. The blood from the ax proved amenable to both enzyme tests and PCR-based DNA tests. Although the DNA tests indicated a strong probability that the ax blood was Stephanie's, the DNA tests were not the most discriminating tests available to forensic scientists. A hair mixed in with the ax blood was also determined to be microscopically similar to Stephanie's. Efforts by forensic analysts to conduct DNA tests on the pubic hair recovered at autopsy and the semen stain on the girl's shorts proved inconclusive. Additional blood spots on McGinn's clothing and shoes also linked him to the murder by conventional serology. Based on this evidence, the jury quickly returned guilty verdicts at his 1995 trial and recommended he be sentenced to die.[82]

Although McGinn continued to maintain he was completely innocent of the crime, McGinn's attorneys advanced a far more circumspect argument on his behalf in these last-ditch efforts to prevent his execution. By this time, McGinn had more than exhausted his appeals. Moreover, since his 1995 trial McGinn had been linked by DNA evidence to at least one other murder, that of 19-year-old Christy Jo Egger in 1992. No charges had ever been filed in the case because McGinn was already on death row and had previously dated Egger.[83] He could, therefore, arguably account for the semen sample recovered from Egger's body as the result of consensual sex. McGinn's name had also come up in the murder of 12-year-old Sherri Newman in the same county. McGinn was not a fresh face around the courthouse, even at the time of his 1993 arrest, having been tried and acquitted of murder during the 1980s. Although the 1995 jury was unaware of these facts, they had heard the testimony of two women who claimed that McGinn had raped them and his own biological daughter who testified that McGinn had molested her as a child. (Charges were never brought from any of these allegations.)

Among the few arguments that were still available to McGinn was that something new had arisen since his last (habeas corpus) petition, something that wasn't available to him before. In a May 2000 hearing, consequently, McGinn's lawyers asserted that "new" DNA tests had become available—namely, mitochondrial DNA and STR analysis—that could be used on two key pieces of evidence, the pubic hair from the autopsy and the semen stain on the shorts, both of which had yielded inconclusive DNA results prior to McGinn's 1995 trial. The newer DNA tests, McGinn's attorneys wrote in court papers, "would yield more positive results and remove any doubt as to the correctness of the guilt and death penalty verdicts of the jury."[84] The attorneys were not, however, suggesting—in so many words, anyway—that McGinn was entirely innocent. They acknowledged, in fact, the other compelling blood evidence in the case that pointed to McGinn as the murderer of Stephanie Rae Flanary. Their argument was confined to the fact that in order to receive a death sentence for murder in Texas (as in other jurisdictions), the murder had to be accompanied by an "aggravating factor" that transformed the murder into a capital offense. At the 1995 trial, the rape was the aggravating factor that led the jury to recommend a sentence of death for McGinn. Whatever McGinn's personal position on all of this, his attorneys were merely hoping that the new DNA tests would clear him of the aggravating factor so he could avoid execution.

The Texas Court of Criminal Appeals found this a bizarre argument. McGinn's formal petition, wrote the presiding justice, "essentially claims that the 'new' DNA technology could clear him of the sexual assault but not clear him of the murder. So what it all comes down to is that at best the 'new' DNA technology could establish that someone else sexually assaulted the victim and [McGinn] then murdered her."[85] That belief aside, the court ruled against McGinn's petition on the grounds that the DNA tests being sought, while not available for his 1995 trial, had been available to McGinn in the past few years while he was filing his earlier appeals. Too late, basically.

Governor Bush's stay of execution, however, allowed the DNA testing to go forward, anyway, and the physical evidence was distributed among three laboratories, the FBI's, the Texas Department of Public Safety's and Ed Blake's Forensic Science Associates in California, which tested some of the evidence on behalf of the Innocence Project. Ultimately, 18 pieces of physical evidence from the 1995 trial were submitted for retesting, not just the two items McGinn's attorneys had focused on.

Given the factual background of the case, it was perhaps not a shock to insiders that the newest round of DNA tests served only to confirm McGinn's guilt. The FBI Laboratory, in fact, was able to extract mtDNA from the pubic hair recovered during Stephanie Rae Flanary's autopsy and determine that the hair originated either from McGinn or a maternal relative (although none of McGinn's maternal relatives were ever even suspected of being involved in the crime). Similarly, the DNA tests conducted by the Texas Department of Public Safety's laboratory on the semen stain were successful in identifying a sufficient number of STR markers to determine that the probability that it belonged to someone other than McGinn was one in 65 quadrillion. Forensic Science Associates reached a similar conclusion on the DNA from the semen stain, although Blake's laboratory put the odds of a random match at a more modest one in 53 billion. At the time of the formal announcement of the results in mid-August 2000, mtDNA tests were ongoing in the death of Sherri Newman. McGinn's execution date was quickly reset, and he was executed by lethal injection on September 27, 2000.[86]

A month after the McGinn DNA test results were first made public, another Texas inmate, Roy Criner, was pardoned by Governor Bush and released from custody after additional DNA tests demonstrated that he was not the man who raped and murdered 16-year-old Deanna Ogg in 1986. Criner had served 10 years of a 99-year sentence for aggravated rape (the murder count was dropped), including three years following an initial DNA test on semen found in the dead girl that excluded Criner as the source. What clinched it for Criner were follow-up DNA tests on a cigarette butt found near the victim's body that matched the semen DNA profile, but not Criner's.[87]

By the end of the year, a number of states, including Arizona, California, Delaware, Oklahoma, Tennessee, Washington, and Illinois, had enacted legislation authorizing, under varying laws, post-conviction DNA testing.[88] A lower court federal judge in Virginia was even moved to rule that inmates claiming they had been wrongfully convicted have a constitutional right to DNA tests that might prove their innocence.[89] Another level of intrigue was added to the DNA testing debate, however, when *The Wall Street Journal* revealed that the principal laboratories so far involved in post-conviction DNA testing— Forensic Science Associates and Cellmark Diagnostics—acknowledged that the post-conviction DNA tests only result in exonerations about 40 percent of the time. "While exonerations have made headlines across the country, and are indeed highly newsworthy," *The Wall Street Journal* reported, "scant attention is paid to the 60% of cases where DNA confirms a prisoner's guilt."[90] In fact, the article notes, McGinn was the fourth death-row inmate in Texas whose guilt was confirmed by post-conviction DNA tests in the past three years.[91] He would not be the last. The question posed by the *Journal* was, why would a guilty man ask for DNA tests? Such a request is particularly confounding when a DNA confirmation of guilt might scuttle all hope of future appeals on other grounds. The answer, according to forensic psychologists surveyed by the paper, is that they "hope the test will create confusion about their guilt."[92] And then what? An O.J. defense? For those prosecutors who vigorously fight requests for post-conviction DNA testing, this is one of their concerns. Illinois prosecutor Kevin Lyons, for example, who fought such testing in an Illinois case that eventually resulted in a confirmation of guilt, explained to the *Journal* that prosecutors fear "any result other than absolute certainty will be spun by the defense and viewed as vindicating" the convict.[93]

A national debate about the death penalty in the United States continues to roil. Now, however, that debate is very often framed by the exonerations made possible through advances in DNA testing, even though DNA tests have been responsible for only a small percentage of death row exonerations (12 of 101 since 1976, as of mid-2002). The issue has even inspired some high-profile artistic endeavors. In 2002, author Scott Turow published the novel *Reversible Errors*,[94] in which he explores capital punishment against an Illinois backdrop, and a New York play entitled *The Exonerated,* which examines the stories of six freed death row inmates, opened in Manhattan featuring guest appearances by several famous Hollywood actors, including Richard Dreyfuss, Jill Clayburgh, Gabriel Byrne, and Mia Farrow.[95] In May 2002, Maryland Governor Parris Glendening, a Democrat who generally supports capital punishment, followed Governor Ryan's lead and declared a moratorium on executions in Maryland, at least until an official study of the state's death-penalty system is completed in late 2002, a study focused on alleged racial disparities within the system. (Nine of 13 men currently on death row in Maryland are black, even though blacks are only 28 percent of the state's population.) Governor Glendening noted the "reasonable questions" about capital punishment that have been raised around the country due to, among other things, the exoneration of death row inmates through DNA testing. (Glendening's successor, Governor-elect Robert Ehrlich, vowed to end Maryland's moratorium on executions, however, when he takes office in January 2003.) On July 1, 2002, United States District Court Judge Jed S. Rakoff declared the federal Death Penalty Act unconstitutional because, he ruled, it violated the Fifth Amendment's Due Process Clause. Pointing to the death-penalty study published in 2000 by James Liebman and the "groundbreaking DNA testing" conducted in recent years that has led to the exoneration of 12 people on death row, the judge concluded that there existed an "undue risk of executing innocent people" for the federal law to stand. (The ruling means the two defendants before Judge Rakoff, Alan Quinones and Diego Rodriguez, who stand charged with operating a heroin smuggling ring and torturing and killing a government informant, will not—if the ruling survives on appeal—face the death penalty in their upcoming trial. Only if the decision is upheld on appeal will the ruling have any legal force outside Judge Rakoff's courtroom.) "What DNA testing has proved, beyond cavil," the judge wrote, "is the remarkable degree of fallibility in the basic fact-finding processes on which we rely in criminal cases." Given these facts, the judge explained, "it is therefore fully foreseeable that in enforcing the death penalty, a meaningful number of innocent people will be executed who otherwise would eventually be able to prove their innocence." This made the federal Death Penalty Act "tantamount to foreseeable, state-sponsored murder of innocent human beings," Judge Rakoff concluded.[96] (On December 10, 2002, a three-judge panel of the U.S. Court of Appeals for the 2nd Circuit overturned Judge Rakoff's decision and upheld the constitutionality of the federal Death Penalty Act. Further appeals are possible.)

The impetus to refine DNA-typing technologies was not, of course, any generalized concern about freeing the wrongfully convicted (although the techniques were immediately pounced on for that purpose). It was the pursuit of shards of ancient DNA and the desire to solve old criminal cases or identify missing people that drove the science forward. Both ancient DNA sleuths and forensic scientists had hit practical barriers by the late 1980s that were, to a considerable degree, removed by the invention of PCR. In the late 1980s and early 1990s, therefore, scientists were carefully exploring their new analytic horizons. For a time ancient DNA seekers were left with the sense that, PCR or not, the relics that they could exploit were very rare indeed. The more celebrated DNA excavations thus far seemed to be uniformly originating from specimens that had been preserved in what can best be characterized as fortuitous circumstances. It was time for another breakthrough.

What turned out to be extraordinary was the extent to which bone harbors reasonably intact DNA, even over long periods of time. The idea that this would be true was not an obvious one; it was, in fact, counterintuitive since the number of cells found in bone is very small when compared to soft bodily tissues, where cells are highly concentrated. Luckily, though, bones are a plentiful commodity by any measure, for anthropologists, archeologists, paleontologists, and forensic scientists alike.

The breakthrough came in late 1989 when Erika Hagelberg, Bryan Sykes, and Robert Hedges reported in *Nature* that bone was, as far as they could tell, the best source of ancient or old DNA so far. Undertaking a novel series of experiments, this group succeeded in extracting and amplifying DNA from a prodigious pile of bones and teeth—including several recovered from an English Civil War cemetery and an Anglo-Saxon medieval cemetery in Abington, England—which ranged in age from about 300 years old to about 750 years old, as well as bones that were far older, including a Stone Age leg bone unearthed in the far-off Judean Desert.[97]

The implications of this study were that the far more plentiful cache of archeological bone specimens available to scholars could be a bountiful source of genetic information and that the bones themselves could originate from less than extraordinary locations. In other words, the bones need not be encased in amber or freeze-dried relics from a Siberian tundra to harbor substantial DNA. One of the most dramatic examples of this, of course, is Svante Paabo's extraction of DNA from a Neanderthal bone, described earlier.

But in addition to just being successful in recovering DNA, the Oxford team, based upon prevailing standards of the day, were wildly successful. "In a recent review," they wrote, "it was stated that it is generally not possible to amplify ancient DNA fragments that are longer than about 150 bp, although well-preserved specimens may yield fragments of up to 500 bp. We amplified a 600-bp mtDNA fragment from the 750-year-old bone. . . ."[98]

Originally, Robert Hedges, an archeologist at Oxford University's Research Laboratory for Archeology who specializes in new methods of radiocarbon dating, and

Bryan Sykes, an Oxford biochemist and expert in collagen who worked at Oxford's Institute of Molecular Medicine (John Radcliffe Hospital), joined forces to study collagen in the bones of ancient human remains. Soon afterward, Erika Hagelberg, an American expert in bones who trained at Cambridge University, was persuaded to join the team to conduct the actual experiments. The group had no intention of studying ancient DNA—until, that is, Hedges happened to visit Allan Wilson's laboratory and was given a crash course in PCR.

The reasons for the success of this study, however, are not readily apparent. Perhaps the obvious natural toughness of bones shields DNA from the rigors of the world outside and the body's own digestive enzymes. In order to get at the material to begin with, scientists must start by boring into the bone and in effect sandblasting away the outer surface. The reason for success might also be attributable to the morphological characteristics of internal structures of bones known as osteocytes.[99] Osteocytes are specialized cells located in the inner recesses of bone. Two types of these cells, known as osteoblasts and osteoclasts, operate to continuously remodel bone throughout life.[100] Whatever the reason, it worked.

In the end, Erika Hagelberg's cutting-edge work with bone earned her the nickname "the bone crusher," and she went on to participate in some groundbreaking criminal casework in the United Kingdom, as well as a variety of population studies involving the analysis of bones.[101]

In the summer of 1981, a young teenage girl ran away from an English children's home and simply vanished. In December 1989, construction workers stumbled upon a badly decomposed body that had been wrapped in a carpet and buried in the garden next to a home in Cardiff, Wales. Dental evidence established an approximate age and physical characteristics established that the young victim of this apparent homicide was a girl. Subsequently, a facial reconstruction by medical artist Richard Neave of Manchester University seemed to restore a semblance of life to the victim. She had been quite a pretty girl.

The model had been constructed in a single day and pictures of it were immediately published. Two days later, a social worker in Cardiff phoned the police and tentatively identified the girl in the photo as Karen Price, the teenage runaway who had disappeared in 1981. Armed with the name, investigators re-examined the physical evidence. Taken together, the dental records of Karen Price and the facial features of Neave's model, observed closely by witnesses, strongly suggested that the body of Karen Price had been found in the garden, but the identification was nevertheless equivocal.

In 1990, in an effort to shore up the investigation, and aware of the work Erika Hagelberg had done on the Abington bones, police sent Hagelberg a sample of the young victim's femur for genetic analysis. The fragment of femur that Hagelberg had to work with weighed only five grams. As a precaution against potential contamination, she sandblasted away one to two millimeters of the outer surface of the fragment. Dividing the results into two equal portions, she performed DNA extraction on each sample precisely as had been done in her landmark 1989 experiments with the Abington bone collection. Hagelberg managed to produce about 5 micrograms of DNA from each of the two samples, but only about 10 percent of it turned out to be human DNA, according to the ultimate report, as substantiated by hybridization with a human Alu-repeat probe. The remainder was presumably genetic contamination from bacteria and fungi. At this point, Hagelberg needed help, so she sent her extracts to Alec Jeffreys, who found the DNA to be severely degraded and only about 1 percent human nuclear DNA. Since the strands of DNA Jeffreys detected were almost all smaller than 300 nucleotides long, RFLP analysis, which requires continuous segments running into the thousands of base pairs, was out of the question. Nor was the newer technique of AFLP (Amplification Fragment Length Polymorphism) possible, a procedure where PCR is used to replicate common genetic repetitions such as minisatellites.[102]

Hagelberg and Jeffreys decided to use PCR to target six different microsatellite loci, variable genetic markers that repeat only two to seven base pairs over and over. (In this case the repeated segment was CA, bound to its complement, GT.) Microsatellites fell comfortably within the length constraints of the degraded DNA sample that Hagelberg and Jeffreys were working with from the putative remains of Karen Price. They had the disadvantage, however, of being less variable than the VNTRs of RFLP analysis and were thus less discriminating. Hagelberg and Jeffreys, therefore, decided to use no fewer than six such genetic segments because of the low allelic variability of the regions.[103] Fortunately, all six loci were successfully (and reproducibly) amplified from both of Hagelberg's extracts, with five of the six proving to be heterozygous. Their size ranged from 72 to 153 base pairs in length.

Now they needed something with which to compare the results, specifically a known reference sample that might provide a definitive link to Karen Price. Since the corpse had already been tentatively identified as that of Karen Price, blood was obtained from her parents, and the same microsatellite regions were amplified as referents. According to their report: "All six microsatellite loci gave results fully consistent with the femur DNA being derived from an offspring of [Karen Price's mother] and [father], with no instances of a femur microsatellite allele which could not be attributed to [mother] and/or [father]."[104] "[T]he DNA data," they reported, "establish with a high degree of confidence that the murder victim was indeed the daughter of [the presumptive parents]."[105]

Karen Price had been found. The chances that the body was someone other than the missing girl were 200,000 to 1, according to Hagelberg and Jeffreys.[106] Although in its infancy here, the method used to identify her came to be known as short tandem repeats, or STRs, and it would be a DNA methodology that would soon revolutionize forensic DNA analysis and DNA paternity testing.

"This analysis," they explained, "is the first example of the successful identification of skeletal remains by bone DNA analysis and provides very strong evidence that the human DNA extracted from the bone was from the victim and did not arise through contamination."[107]

Ultimately, these results became the first PCR-derived forensic tests to be accepted as proof of identity, or proof of anything for that matter, in a British Court.[108] Courts in the United States had, as mentioned earlier, and as might be expected, more of a pioneering spirit when it came to PCR. Based on their solid credentials in the field of genetic testing and bone analysis, the Jeffreys and Hagelberg team would be drawn into other important forensic investigations, one of which would personally affect millions of people from around the world, people whose lives had been permanently scarred by the subject of the investigation. Their contribution, it would turn out, would constitute another significant breakthrough in forensic DNA analysis.

6 The World's Most-Wanted Man (Part 1)

THE NEXT SIGNIFICANT CASE THAT ALEC JEFFREYS AND ERIKA HAGELBERG worked on together concerned, like the Karen Price case, the identification of human remains. Moreover, like the Price case, a great deal of forensic work had gone into the investigation before their participation was sought, and many people were satisfied that a positive identification had been made. But the identification was not a definitive one, and so their talents were enlisted to help resolve the issue once and for all.

In most particulars, however, this next inquiry could not have been more different. To begin with, the scientific conclusions would have international ramifications, well beyond providing another demonstration of the utility of DNA typing. In addition, the remains to be identified were not those of a victim of a crime but those of the perpetrator of, not a single crime, nor even a string of crimes, but a life's work of crime. And the world wanted to know if he was truly dead. To complicate matters, proof that he was in fact dead—for many people—would not necessarily be a welcome thing.

The forensic investigation began on June 6, 1985, quite inauspiciously, when Dr. Jose Antonio de Mello, the assistant director of the federal police department of São Paulo, Brazil, held up for the benefit of a gathered crowd and television cameras a skull that had just been hurriedly unearthed from a village cemetery in the town of Embu. Even though the grave was marked with the name Wolfgang Gerhard, the skull was believed to be that of Dr. Josef Mengele, the notorious fugitive war criminal who had earned the nickname "the Angel of Death" for his role as an SS camp doctor in the Nazis' most notorious death camp, Auschwitz, in Poland. After decades on the run, Mengele had become a "symbol of the Holocaust."[1]

The Holocaust is the 20th century's meteor strike. The wound the Nazis inflicted on humanity continues even today to cause agony and consternation, and will almost certainly define "man's inhumanity to man" for centuries to come. Survivors and their families continue to struggle, both physically and psychologically, in the Holocaust's aftermath. Nations have been reconfigured, and entire cultures, most significantly that of the Jewish people, have sought to reinvigorate themselves following the carnage. The fallout from the Third Reich has easily stretched into the 21st century, having triggered multi-billion-dollar lawsuits (and some settlements) stemming from stolen artwork and real estate, unpaid insurance claims, the use by thousands of companies in various industries of millions of prisoners as slave laborers during the Nazis' rule and the retention by Swiss banks of hundreds of millions of dollars that belonged to the victims of the Nazis, thus depriving their surviving heirs of their inheritances.[2] On almost every level, the Holocaust engenders a degree of controversy. On everything from the exact scale of the catastrophe to how the Holocaust should be taught to future generations, there is sometimes bitter scholarly disagreement. Following the collapse of the Soviet Union in 1989, a new round of claims was pressed against Germany, the international banking and insurance industries, and the myriad companies that profited from the Nazi regime. While these efforts raised the hope that many injustices would finally be redressed, it also sparked a disconcerting wave of anti-Semitism throughout Europe and the United States. At the same time, a number of well-respected intellectuals have been critical of the exploitative appearance of pursuing many of these claims so long after the fact, particularly as the claims are championed by politicians and lawyers with dubious motivations.[3]

Whatever the truth of these concerns, the Holocaust remains a singular trauma and continues to frame the debates surrounding any number of medical and ethical issues that have arisen since World War II, including human stem-cell research, cloning, designer children, medically assisted suicide, genetic engineering, and even mainstream genomic research. Certain viewpoints are automatically suspect because of the Nazi experience. When Australian philosopher and bioethicist Peter Singer was hired by Princeton University in 1998, for example, his appointment to the faculty as DeCamp Professor of Bioethics caused a public uproar because of his extreme utilitarian views that elevate the quality of human life over the sanctity of such life, which has led him to argue that, in certain circumstances, it would be morally justified to kill a disabled child. Singer is also an extreme advocate of animal rights (which are on par with human rights, he suggests), and although he has many followers, his views have been dismissed by many observers as akin to Nazism and he has been likened to Josef Mengele and Martin Bormann. His appearances are often accompanied by protests. Such modern-day disputes are inescapably guided by the Holocaust.[4]

The skull de Mello held up to the world in 1985 had been shattered during the exhumation after São Paulo Federal Police Chief Romeo Tuma, impatient with the rusted latches on the coffin, ordered the grave diggers to smash the coffin.[5] This was the same Romeu Tuma who, in 1967, had made something of a name for himself as a Nazi hunter when, acting on a tip from the Simon Wiesenthal Center, he had succeeded in capturing the notoriously cruel Franz Stangl, the commandant of two other infamous concentration camps also located in Poland, Sobibor and Treblinka.[6] In 1978, Tuma enhanced his Nazi-hunting reputation when he captured another notorious war criminal, Gustav Franz Wagner, the "Hangman of Sobibor."[7]

The images of the exhumation were seen on television all over the world and riveted the attention of millions of people. Telephones rang around the world, and politicians in Washington, D.C., immediately started to mobilize a forensic inquiry.

The scene at Embu was something of an anticlimax to the months of mounting effort in the United States to search the globe for Josef Mengele and bring him to justice. After a mock trial of Mengele was held in Israel in January 1985, conducted by the Los Angeles-based Simon Wiesenthal Center and televised over four nights around the world—and during which actual Auschwitz survivors took the stand to testify against an absent Mengele—the U.S. Senate convened hearings on Mengele in February. Simultaneously, the Simon Wiesenthal Center upped the ante by offering a $1 million reward for information on Mengele's whereabouts leading to his arrest and extradition. Together with the $1 million offered by the government of Israel, and another $1 million put up by *The Washington Times* newspaper, this brought the total bounty on Mengele's head, including additional private and governmental rewards, to $3,458,000.[8] These events, combined with embarrassment over international headlines that two U.S. Army veterans, Walter Kempthorne and Richard A. Schwartz, had personally seen Mengele *in custody* in 1945 in U.S. POW camps, prompted U.S. Attorney General William French Smith to order the Department of Justice Office of Special Investigations (OSI) and the U.S. Marshals Service to initiate a worldwide manhunt for Mengele. (Created in 1979, the OSI established a special division to pursue suspected Nazi war criminals living in the United States. By 1999, it had persuaded courts to revoke the U.S. citizenship of 61 individuals who, it was alleged, had misrepresented their wartime activities when they entered the country after World War II; 48 people had by this time been deported.[9]) At the same time, the military established the Department of the Army Mengele Task Force. The Pentagon and the Central Intelligence Agency were also to cooperate in the effort.[10] As it turned out, though, it was the West German federal police in cooperation with Brazilian authorities that located the grave site at Embu.

Political considerations aside, the question remained: Had the remains of Josef Mengele been found? This was strictly a medical question.

In 1985, after a tip from a German university professor, the West German government raided the house of Hans Sedlmeier in Guenzburg, Germany, the town where Josef Mengele was born. Sedlmeier, a lawyer, had long worked as one of the directors of the prosperous family business started in the early 1900s by Josef Mengele's father, Karl Mengele and Sons, a company that manufactures agricultural equipment and remains Guenzburg's largest employer. He had long been suspected of being a financial conduit between fugitive son Josef and his family. As a trusted Mengele family insider, Sedlmeier was even put in charge of the business when family members were either temporarily under arrest or forced aside during the postwar de-Nazification period.[11]

Among the items seized in the Sedlmeier raid was a letter cryptically announcing to Sedlmeier the death by drowning in Brazil of "our mutual friend," also referred to as "Uncle."[12] The letter was dated 1979. Signed by Wolfram Bossert, the recovered letter describes the afternoon drowning of "Uncle" during a picnic on the beach at Bertioga Bay, a futile rescue attempt and resuscitation efforts by a doctor who happened to be on the beach that day.[13]

From the outset, the Israeli government was extremely suspicious about the remains found in Embu. Menachem Russek, Israel's chief Nazi-hunter, denounced the claim that Mengele had been found as yet another hoax orchestrated by the Mengele family. Such suspicions were well-founded. Beginning immediately after the war, members of Mengele's family had engaged in a savvy sleight-of-hand that misled authorities and allowed the family to intervene at key moments in Josef Mengele's flight when he might otherwise have been apprehended. At the end of the war, for example, the Mengele family fanned rumors that Josef had been killed in the waning days of the war. His first wife, Irene, even went so far as to dress in black and arrange a memorial service. There is also one report that Irene Mengele took steps to have her husband officially declared dead.[14]

Sadly, such deceptions proved effective.

The problem was compounded by a singular law in the former West Germany (Germany has since been reunified), Article 52 of the German Penal Code, which provided that relatives of a fugitive (even distant ones, as well as fiancées) could refuse to testify and were under no obligation to assist authorities in their search for a wanted criminal.[15] To the extent that Sedlmeier was ever questioned (and he would not fall under the protections of Article 52), he lied, according to investigators from the U.S. Department of Justice.[16]

The death announcement to Sedlmeier, assuming it was genuine, was further demonstration of a pattern of deception. Wrote Bossert: "We also believe to be in agreement with your thinking if we plan to continue secrecy now as before. This is intended not only to avoid personal unpleasantness but also to compel the opposition to continue wasting money and effort on something that has already been superseded by events."[17]

By 1985, Josef Mengele was the world's most wanted Nazi war criminal. Yet, assuming he had lived up until that time, he had managed to elude capture for 40 years, despite the fact that his crimes during the war had become legendary.

The actual crimes Mengele committed, although they do indeed represent the very worst in the Nazi scheme, are easily confused with embellished and imagined ones, the artifice of playwrights and popular fiction writers. As a result, the true Mengele was difficult to distinguish from the mythological Mengele. In the words of psychiatrist and author Robert Jay Lifton, Mengele was surrounded by a "cult of demonic personality." [18]

As forensic teams were being assembled for a trip to South America, little could be said with certainty about what became of Josef Mengele after the end of World War II. In the course of the investigation spawned by the Embu grave, however, hard evidence began to surface that would—aside from assisting in the identification of the remains—provide a detailed map of Mengele's life underground. At the same time, these facts would help to demystify Mengele, although nothing diminished the nature of his crimes.

Mengele's significance only took hold gradually over the years. Demands for his capture reached fever pitch only sporadically until 1985. This was true despite the countless testimonies given against him by Auschwitz survivors beginning immediately after the war. According to Lifton, Mengele's "status of public infamy" only began with the publication in 1958 by Ernst Schnabel of *Anne Frank: A Portrait in Courage*. [19] Mengele is believed to have encountered the Frank family as they disembarked from the trains at Auschwitz. Only Otto Frank, Anne's father, survived captivity. After being transferred to another concentration camp, Bergen-Belsen, Anne contracted typhus and died within weeks of liberation.

Among the earliest fictional portrayals of Mengele was a character known simply as "the DOCTOR" in the play *The Deputy*, written by Rolf Hochhuth and translated into English in 1964. "The DOCTOR," wrote Hochhuth, "has the stature of Absolute Evil" and is an "uncanny visitant from another world," adding that "this 'doctor' stands in such sharp contrast not only to his fellows of the SS but to all human beings" and "is in reality comparable to no human being."[20] "The DOCTOR" was "only playing the part of a human being," wrote Hochhuth.[21]

Another fictionalized Mengele appeared in William Goldman's 1973 novel *Marathon Man* and the very popular 1976 movie that Goldman wrote for the screen. The Mengele character, a Nazi dentist named Christian Szell with the nickname "the White Angel," was played in the movie by Sir Laurence Olivier. Szell is pitted against a character played by Dustin Hoffman, Thomas Levy, who discovers that his brother is a spy mixed up with Szell. Szell has been forced to come to the United States, the plot goes, to retrieve a fortune in diamonds plundered during the Holocaust. Szell, the dentist (which the real Mengele was not), believes Hoffman has information necessary to his regaining possession of the diamonds. In one of the most horrific scenes in American cinema, Szell tortures Hoffman by strapping him into a chair and drilling into his *healthy* teeth, which maximizes the level of pain because the nerves are vibrant.[22]

The dentist scene is not only effective as theater, it was instrumental in bringing American political pressure to bear on Paraguay, where many believed (wrongly, as it turned out) that Mengele was hiding during the 1960s and 1970s. Former Israeli Ambassador to Paraguay Benno Weiser Varon suggested that it was *Marathon Man,* and the idea that Mengele would torture and attempt to kill Dustin Hoffman, that would, more than anything else, sufficiently outrage the American public to clamor for Mengele's capture.[23]

Ira Levin's *The Boys from Brazil,*[24] also made into a popular movie, followed shortly after and perhaps came closest to capturing the essential Mengele: a geneticist given a free hand, virtually unlimited resources, in a morally bankrupt world. Played by Gregory Peck, Mengele is locked in mortal combat with Yakov Lieberman, a world-renowned Nazi hunter, played (in a role-reversal) by Sir Laurence Olivier. After a tip from a fledgling Nazi hunter, Lieberman gradually uncovers Mengele's diabolical scheme: He has cloned Adolf Hitler 94 times and scattered the offspring around the world in homes meant to duplicate the unusual childhood experiences of the Führer.

As the investigation progressed into the background of the real-life Mengele, authorities determined that a fugitive Josef Mengele had at one time operated a pharmaceutical business in Argentina by the name of Fadrofarm.[25] He had also published an article on genetics in an Argentine journal under the alias "G. Helmut."[26] *The Boys from Brazil,* then, was simply expanding on the theme of Mengele's actual intellectual pursuits. In fact, a prisoner-doctor who worked with Mengele at Auschwitz (as many such doctors were forced to do), Dr. Lottie M., later recalled what interested Dr. Mengele the most: "That was his question . . . genetics . . . genetics and environment."[27] This, in fact, was a fixation of the Third Reich. Even the inventiveness of gifted fiction writers, however, cannot do justice to Josef Mengele or his Nazi colleagues, as a closer inspection would reveal.

At the Embu grave, the very best of science was called in to examine what—by all accounts—were the remains of the worst of science. Four independent teams of experts were ultimately sent to Brazil for the initial forensic investigation, two of which were assembled by U.S. authorities. The Marshals Service and the United States Department of Justice retained Dr. Ali Hameli, a forensic pathologist and the chief medical examiner for Delaware, Dr. Ellis Kerley, a forensic anthropologist from the University of Maryland, and Dr. Lowell Levine, a noted forensic odontologist. At the same time, the Justice Department enlisted handwriting and document specialists to compare purportedly recent writings with writing samples obtained from Mengele's original SS file, which held a treasure trove of information. Also at issue was the exact age of the more recent writings. The analysts for this aspect of the investigation included handwriting specialist Gideon Epstein of the Immigration and Naturalization Service, Dr. Antonio Cantu, a highly respected expert in ink analysis, who was at the time with the FBI, and independent expert Dr. David Crown, whose specialty was document examination.[28] The SS file, dating back to the 1930s and 1940s, was provided by the Berlin Documentation Center, which serves as a central repository of Nazi war documents.

Under the auspices of the Simon Wiesenthal Center, Dr. Leslie Lukash, Dr. Clyde Snow, and Dr. John Fitzpatrick were also dispatched to Embu in their respective capacities as medical examiner, forensic anthropologist, and radiologist. The fourth team assembled for this initial phase was sent by West Germany and included Dr. Rolf Edris, a forensic odontologist, and Dr. Richard Helmer, a forensic anthropologist from the University of Kiel. Dr. Helmer, in particular, would bring some unique expertise to the case.

As the Mengele investigation got under way, the forensic tools available, for all practical purposes, did not include DNA typing. After all, 1985 was the year Alec Jeffreys had first performed RFLP DNA typing in a forensic setting. PCR, too, was in its infancy, so it would be a long time before the technology would mature to the point where a suitable test could be devised that would have any meaning to the Embu inquiry, assuming such a test could. At every conceivable opportunity, however, attempts were made to exploit the evident power of DNA technology. Before Jeffreys and Hagelberg were brought into the case, the best DNA-testing laboratories in the world, including those of the FBI[29] and Cellmark,[30] had tried and failed to obtain any DNA results on the Embu remains. Clearly, some innovative procedures were needed in this case to successfully extract DNA from the bones at Embu. In the meantime, other forensic methods would have to suffice.

The first order of business was to gather in all the pieces of the body found at Embu for comparison with the records of the person whose name was on the grave marker, Wolfgang Gerhard, as well as those being compiled about Mengele. Because the exhumation had been so badly handled, shattering some of the recovered bones as well as the skull, a second exhumation was undertaken that turned up additional bones and a few teeth.[31] It fell to West Germany's Dr. Richard Helmer to painstakingly reassemble the broken skull. Precision was imperative in this since Helmer planned to use a novel scientific method of his own devising that purported to compare the dimensions and characteristics of the reassembled skull with photographs of Mengele to determine whether there was a visual match. Extrapolating skin depth and facial contour based on the proportions of the skull, Helmer would superimpose images of Mengele and the recovered skull captured on video and draw his conclusions. Needless to say, Helmer's proposed technique inspired a certain angst among members of the Justice Department, so the United States initiated an independent assessment of the technique in an effort to lend credence to any conclusions Helmer reached.[32]

At the same time, all of the teeth and bones were closely examined for irregularities that matched known idiosyncracies of Josef Mengele. (Note that many of the individual's original teeth had been replaced by removable partial dentures, which were also recovered from the grave.)[33] Among the more significant facts known about Mengele was that he had suffered from some sort of a bone disease during his life and, a fact even more definitively known based on photographs contained in his SS file, he had a wide gap between his upper front teeth known as a diastema.

Clyde Snow, Lowell Levine, and Ellis Kerley looked to the SS documents provided by the Berlin Document Center for anything that could be used for comparative purposes, such as the records of a 1938 physical examination performed on Mengele. Taken together, as authors Christopher Joyce and Eric Stover have remarked, "These early SS files were the forensic team's Rosetta stone."[34] The records being pored over, however, provided a great deal more than documentation of physical characteristics or anomalies. The records began to paint a more complete picture of who Josef Mengele was and what kind of medicine he practiced.

One particularly telling set of documents contained in the Berlin records was a series of questionnaires that Mengele submitted to Nazi Germany's Central Office for Race and Resettlement (a division of the SS) in the late 1930s. These included Mengele's application for membership in the SS and his application to marry Irene Schoenbein. In addition to providing valuable biographical and medical information on Mengele, these documents very succinctly capture the essence of Nazi racial theory.

The information was required by the SS principally to establish the applicant's human pedigree. Each applicant was required to file forms tracing their maternal and paternal lineage back to the year 1750 to insure racial "purity" (that is, the applicant had no Jewish blood or congenital defects).[35]

To qualify for membership in the SS (or *Schutzstaffel*, meaning defense squadron or shock troops), the applicant had to meet certain criteria measured against an ideal Aryan model, which was rigidly defined by the overall head of the organization, Heinrich Himmler, appointed by Hitler in 1929. The applicant had to be in good health, the right race, have certain facial and other physical characteristics, needed to have good teeth, and so on. To keep this racial ideal on track, Himmler established a program in 1936 known as *Lebensborn*, or "Spring of Life," which was essentially a state-controlled baby factory. The purpose of *Lebensborn*, according to Himmler, was "to breed the SS into a biological elite, . . . [a] racial nucleus from which Germany could replenish an Aryan inheritance now dangerously diluted through generations of race-mixing."[36] It was considered an Aryan soldier's duty to contribute offspring to the state by sleeping with the Aryan women chosen for the program. To augment the program, "biologically valuable" children were kidnapped from occupied areas.[37]

Among the Nazis' early efforts to wipe out the Jewish population were the notorious *Einsatzgruppen*, or Special Groups, formed largely from the ranks of the Waffen SS to act as mobile hit squads. It soon became apparent, however, that something more efficient would have to be devised. Technology would provide the solution.

Mengele's SS application did not pose any problems, and he was accepted into the SS in 1938. His family, after all, was well established in Guenzburg, and he was nominally Roman Catholic. Moreover, Mengele himself had long been active in political activities. From 1924 until 1930, he had been a member of the Greater German Youth Alliance; from 1931 until 1933 he was a member of the "Steelhelmets," a paramilitary organization; and from 1933 until October 1934, he was a member of the ill-fated paramilitary SA (or *Sturmabteilung*, meaning "Storm Troops"), the one-time parent group to the SS that fell into disfavor with Hitler as it grew in strength and began asserting its political muscle. The SA's leader, Ernst Röhm, in fact, had taken to publicly denouncing Hitler in a vain attempt to get concessions from him. Hitler responded by having the entire SA leadership, including Röhm, killed on June 30, 1934, in what became known as the "Night of the Long Knives." After this, the SS took the lead role in Hitler's Germany. It was at this point that Mengele managed an opportune and reasonably honorable departure from the SA due to his recurrent bone disease.

The bone disease was believed to have been caused by an infection during Mengele's adolescence. Medical records described the ailment as "sepsis, osteomyelitis, nephritis."[38] In any event, it was not hereditary. Given the extremely rigid physical requirements for SS membership demanded by Himmler, another individual might well have been disqualified. In fact, in the early years of SS recruitment efforts men who had even a single tooth filled were summarily rejected by the SS. (As the war progressed, however, and more and more of the original SS members were killed in battle, these SS standards were relaxed.) But Mengele brought academic credentials to the table, a Ph.D. in Anthropology and a medical degree. On the documents he submitted to the SS, he described his chosen field as "Genetics and Eugenics." By this time, too, Mengele was already well connected professionally and a rising star in the German medical establishment. Since doctors were at the heart of the Nazi racial program, Mengele was on the fast track.

Mengele's prospective wife, on the other hand, had serious problems with her pedigree, a fact that stalled their marriage plans. Questions were raised by a senior SS official about Irene's maternal grandfather, who, possibly, but not definitely, was the illegitimate son of an American, Harry Lyons Dumler, who was also, inconveniently, a Jew. Dumler had been in Nice, Germany, while serving in the U.S. consulate. Court records from 1886 surfaced showing Dumler making provisions for the support of Irene's maternal grandfather, while expressly disclaiming paternity and insisting he be left "totally in peace from now on."[39] The issue simply hovered there for awhile for Irene and Josef. Ultimately, the snag was resolved after Mengele filed additional documents, including character references, that attested to Irene's "very Nordic ways."[40] The marriage was approved in March 1939, and they were wed that summer.

"As of the date of the forensic examination" in the summer of 1985, according to the OSI's final report on Mengele, "almost all of the reliable, medical data on Mengele could be found in his SS file."[41] The salient medical facts culled from those records boiled down to the following:

1. Mengele was born on March 16, 1911, and was therefore almost 68 on the date of this purported death.
2. Mengele was male and Caucasian.
3. Mengele's height was 174 centimeters.
4. Mengele had a distinctive high brow as revealed in the photographs in the SS file.
5. Mengele had a wide gap between his top front teeth as displayed in the photographs.
6. Mengele received medical treatment in "1926/27" (age 15/16) for "sepsis, osteomyelitis, nephritis," according to a medical history dated February 16, 1938, in the SS file.[42]

Osteomyelitis is an infection of the bone marrow, whereas sepsis is a systemic blood infection and nephritis is an inflammatory disease of the kidneys. Of the three ailments, only osteomyelitis would leave a telltale trace discernible in a skeleton.

The forensic scientists on hand at Embu set about the task of comparing the documentary data against the raw data that could be gleaned from the recovered remains. Where they were confronted with anomalies or incomplete skeletal features, they would be forced to develop working theories and draw conclusions. Next, they would compare notes and determine areas of agreement and, inevitably, lack of agreement. Lastly, they would defend their conclusions.

Mengele had examined countless corpses in his heyday, many of whom died by his own hand. As the OSI acknowledges, the fact that Mengele was himself an expert in such examinations was extremely troubling. Even total forensic agreement would be viewed by some with deep skepticism. "[S]ome raised the possibility that Mengele, himself a physician and anthropologist," the OSI report notes, "could have somehow secured a body with characteristics similar to his known [characteristics] and buried that body instead. In essence, there was speculation that the discovery in Embu was a hoax."[43]

Eugenics, one of Mengele's specialties, was something of an old notion that the Nazis had inherited, not one they invented. But they seized upon it with relish and very early on made it a cornerstone of national, legal, and public policy. By 1934, a year after Hitler assumed power in Germany, Deputy Party Leader Rudolf Hess declared at a mass meeting that "National Socialism is nothing but applied biology."[44]

In the early part of 1985, the Department of Justice realized that there could be an enormous problem if the worldwide manhunt managed to locate someone believed to be Mengele: Proof of identity was essential but lacking. When Israeli agents kidnapped Adolf Eichmann off the streets of Buenos Aires in 1961, the first order of business was establishing his identity. After all, the man they abducted was not living openly in Buenos Aires under the name of one of the chief architects of the Final Solution. (Eichmann was one of the key bureaucratic masterminds behind the Nazi effort to exterminate the Jews; among other things, he coordinated the railroad system that fed the burgeoning network of concentration camps.) Moreover, many years had passed and available information was sketchy. So before leaving the country with the man they believed was Eichmann, the agents took him to a safe house in Buenos Aires for an interrogation and identification. Very fortunately, for the government of Israel at least, which was acting without the permission, let alone the cooperation, of Argentine authorities, Eichmann broke down and admitted his identity almost immediately. Eichmann was smuggled out of Argentina, put on trial for war crimes in Israel and executed after his conviction.

Over the years there had been countless questionable Mengele sightings and even one book that claimed the author—a Brazilian police agent—had gunned down Mengele in a 1968 shootout. While some of these claims were flatly ludicrous, others had the ring of truth; of these, some proved embarrassing. A 1981 *Life* magazine story, for example, made the claim that Mengele was secretly living in Bedford Hills, New York, 30 miles outside New York City.[45] But perhaps the worst episode of misidentification occurred in 1985 just as resolve for Mengele's capture was stiffening. In late 1984, based on a tip to a West German journalist and a retired Argentine policeman from a Uruguayan prostitute, who claimed to have Mengele as a customer, a reporter from *The New York Post* was dispatched to Uruguay to check out the story. His target was a man living under the name Walter Branaa who owned several estates. Once in Uruguay, the reporter enlisted the help of a freelance writer who was also hot on the trail of Branaa. After flying back to New York with a series of hastily taken photographs of the suspect that were not clear enough to be of any use, the two returned to Uruguay for a second try, now accompanied by a professional photographer. This time managing to capture better pictures of Branaa in a number of poses at his ocean villa, the *Post* farmed out three sets of the photographs to three independent forensic analysts, including Dr. Ellis Kerley. All three analysts compared the 1985 photographs of Branaa with a 1938 SS photograph of Josef Mengele, without knowing the identity of the person they were evaluating, and all three arrived at the same conclusion: There was between a 90 and 95 percent probability that they were the same man—thus, Mengele had been found alive, it was concluded. After some investigation by the Mossad—Israel's foreign intelligence service—Israel concluded that these conclusions were simply wrong. And when confronted in Uruguay, a dumbfounded Walter Branaa provided more than adequate proof of his identity. It was a fortunate thing, too, since efforts were underway to arrange for his kidnapping and trial as a Nazi war criminal.[46]

Certainly the most definitive proof of identity available at the time was fingerprint evidence, but as the Mengele search got underway the U.S. authorities had no verified fingerprints of Mengele on file that they could authenticate, despite the fact that two sets had been provided by the government of Israel and West Germany. In May 1985, however, OSI received unprecedented access to an original International Red Cross document dating from 1949 entitled "Application for Travel Document" signed by "Helmut Gregor." It was an application for permission to travel from Italy to Argentina. A handwriting analysis performed by the U.S. Immigration and Naturalization Service (INS) Forensic Document Laboratory established that the signatory of that document was the same person who filled out the documents contained in Josef Mengele's SS file. With a crucial link established, the FBI was able to lift a print of the right index finger from the 1949 travel paper and compare it against the fingerprints provided by Israel and West Germany. Concluding that there was a match between all three sources, U.S. authorities were now prepared to establish the identity of a live—or, at least, recently alive—suspect.[47]

The body found at Embu, however, had long since lost its fingerprints to the forces of decomposition. Nevertheless, the effort was not wasted since valuable proof now existed that began to chart the course of Mengele's life.

Fingerprinting and theories about genetics—or, more accurately, eugenics—have long been linked, but chiefly because the same scientist, Sir Francis Galton, is rightly credited with advancing both of them. Although inheritance plays a role in the development of general fingerprint patterns—the familiar loops, arches, and whorls—even identical twins do not have identical fingerprint ridge characteristics. Nor are the irises or freckle patterns of identical twins exactly alike.[48] The unique development of fingerprints, among other things, is thus largely determined by environmental factors during gestation. The cousin of Charles Darwin (whose *On the Origin of Species by Means of Natural Selection,* published in 1859, introduced the theory of evolution), Galton went on to leave his own mark in science after earning degrees in both medicine and mathematics. Seizing upon two letters about fingerprints published in the journal *Nature* in 1880 by Henry Faulds and William Herschel, two rivals working independently on the subject, Galton began his own intensive investigation of the field. With data provided generously by Herschel, Galton set out to establish the uniqueness of fingerprints, something never before systematically demonstrated. Although the classification system he devised would later be greatly improved and standardized by Edward Richard Henry, the publication of Galton's book *Fingerprints* in 1892 was a watershed. For the first time, science provided a reliable system of identification in criminal investigations. To this day, Galton's legacy on fingerprinting is unchallenged.

Galton leaves quite a different legacy on the subject of eugenics, however, although his influence around the world on the subject was enormous. Genetics is solid science, of course. Eugenics, on the other hand, is not science. Eugenics is ideology. Inspired by the evolutionary theories spelled out by his cousin in *Origin of Species*, Galton became consumed with the idea that human talents were hereditary. His own illustrious family, in any event, seemed to prove the rule. He first proposed the notion that society could be improved "through better breeding" in an article entitled "Hereditary Talent and Character"[49] that he published, no doubt coincidentally, the same year that Gregor Mendel's famous paper on inheritance first appeared, 1865. Galton coined the word "eugenics" in 1883, from the Greek root meaning "noble in heredity" or "good in birth."[50] Over time, eugenics theories were routinely featured in biology textbooks and were considered mainstream around the world. As part of his research, Galton studied identical twins, something that remains a staple of genetic research, and his observations only bolstered his view that "nature prevails enormously over nurture."[51] In 1875, Galton published a study entitled "History of Twins as a Criterion of the Relative Powers of Nature and Nurture."[52]

The list of notables who endorsed eugenics in subsequent decades includes playwright George Bernard Shaw, aviator Charles Lindbergh, novelist H. G. Wells, U.S. Supreme Court Justice Oliver Wendell Holmes, British Prime Minister Winston Churchill, birth-control advocate Margaret Sanger, and Charles B. Davenport, the founder of both the world-famous Cold Spring Harbor Laboratory (which received funding for eugenics from the Carnegie Institution) and the Eugenics Record Office. The American Eugenics Society began in the 1920s to sponsor "Fitter Families Contests" at state fairs around the United States. Such things as "pauperism" and criminality were said to be genetically determined.

The first compulsory eugenic sterilization act passed in the United States was adopted on March 9, 1907, by the state of Indiana.[53] Over the years, as many as 30 states passed similar eugenic sterilization laws and in the movement's heyday such laws existed on the books in 27 states simultaneously. In 1927, the U.S. Supreme Court gave its blessing in the case of *Buck v. Bell* to a Virginia sterilization law, which authorized the sterilization of people held in state institutions who were found, after a perfunctory hearing, to be afflicted with hereditary insanity or imbecility.[54] Writing for the court, Justice Oliver Wendell Holmes explained: "It is better for all the world, if, instead of waiting to execute degenerate offspring for crime, or to let them starve for their imbecility, society can prevent those who are manifestly unfit from continuing their kind. The principle that sustains compulsory vaccination is broad enough to cover the cutting of the fallopian tubes—Three generations of imbeciles are enough." Although the practice would later be significantly curtailed in the United States, lawful sterilizations nonetheless continued until 1972, by which time a total of 70,000 Americans had been sterilized.

The American eugenics movement also spawned miscegenation laws, which pro-scribed racial intermarriage, and laws forbidding marriage by people suffering from such things as epilepsy and mental retardation.[55] Such laws, in retrospect, had all the earmarks of the Nazis' Nuremberg Laws. Under the Nazis, Germany embarked on an extremely aggressive sterilization campaign, one that American eugenists would later come to envy. American Dr. Joseph S. De Jarnette of Virginia, for example, declared (in 1934): "The Germans are beating us at our own game!"[56]

At the end of the 19th century, German Social Darwinists (who were part of a worldwide school of thought that addressed social problems in Darwinian terms and with Darwinian analogues) began the movement of racial hygiene, which advocated eugenics and the purification of the German race. In 1904, Alfred Ploetz founded the German Society for Racial Hygiene. One branch of the society established in 1910 in Stuttgart, Germany, was presided over by Wilhelm Weinberg,[57] a scientist now famous to all DNA aficionados as one of the authors of the Hardy-Weinberg Law, which undergirds the calculations of statistics in all DNA identification matters. As with its American counterpart, the German movement was intended to forestall the "degeneration of the human race." Ultimately, 20 or so university institutes of racial hygiene were established in Germany, all *before* the Nazis rose to power.[58] Moreover, two separate institutes were devoted exclusively to the principal of racial hygiene, the Kaiser Wilhelm Institute for Anthropology, Human Genetics, and Eugenics in Berlin (1927–45) and the Kaiser Wilhelm Institute for Genealogy in Munich (1919-45).[59] After the Nazis came to power, both of these institutes were involved in the compilation of "genetic registries" that were used to round up the Jews and Gypsies in the late 1930s and early 40s.[60] Efforts to identify people based on flimsy genetic criteria were a problem that bedeviled the Nazis, however, since surnames did not al-ways give people away and looks can be deceiving. Some German racial hygienists were convinced that a simple blood test could be developed to differentiate between the races.[61] Serious scientific investigations were undertaken (and lavishly funded) by the Nazis to identify such tests, efforts that continued in such places as Josef Men-gele's laboratory at Auschwitz. These efforts did not succeed.

At the racial hygiene institutes, the study of identical twins, especially those reared apart, was a major focus of study. The objective was to disentangle the influences of nature versus nurture and identify human behaviors that were controlled by genetics.[62] The focus was generally on identical twins, referred to as monozygotic (MZ) twins because they originate from a single egg (zygote) that splits early in devel-opment. Fraternal twins are dizygotic (DZ), so-called because two eggs were fertilized simultaneously; such children are genetically alike only to the same extent as other siblings.

Adolf Hitler became Chancellor of Germany on January 30, 1933, entirely by legal and constitutional means. On March 20, 1933, Heinrich Himmler set up the first concentration camp at Dachau, which quickly became a dumping ground for all political dissidents being consigned to slave labor. A sign was erected at the entrance, one that was later also installed at the entrance to Auschwitz, that read "Arbeit macht frei"—"work makes you free." Before the end of World War II, there would be 5,000 concentration camps throughout Europe, mostly for slave labor. On March 23, 1933, the Reichstag passed the Enabling Act by a wide margin, 441 to 84, legally establishing Hitler as the dictator of Germany. The vestigial powers reserved in President Hindenburg—aside from his role as rubber stamp, Hindenburg retained nominal control of Germany's military—died with him on August 2, 1934, when he expired of natural causes. Immediately thereafter, Hitler's cabinet consolidated the offices of President and Chancellor. With this, Hitler had absolute power.

Very quickly the racial hygiene movement acquired teeth. The three principal phases of the Nazi racial hygiene program (which led inexorably to the Holocaust) included aggressive sterilization laws, the Nuremberg Laws, and a sophisticated euthanasia operation. These first three legs provided the legal and technological groundwork for the Holocaust.

Although there were sterilization laws and marriage prohibitions in the United States, the third item on the Nazi agenda—euthanasia—was a new wrinkle in the eugenics movement not attempted in America. (The idea did, however, have advocates in the United States.)

In the 1930s a series of sophisticated propaganda films were produced and shown throughout Germany meant to acclimate people to the notion that "life unworthy of life" (that is, the insane, the deformed) is best dispatched in a humane fashion, lest the rest of Germany be overrun by hordes of "useless eaters" and be driven into economic ruin.[63]

This debate, of course, continues today in the United States over the activities of Dr. Jack Kevorkian and the 1997 assisted suicide law in Oregon, known as the Death with Dignity law, that permits doctors to help terminally ill patients end their lives.[64]

On July 14, 1933, several months after the passage of the Enabling Act, the Nazi government passed the Law for the Prevention of Genetically Diseased Offspring, or the "Sterilization Law."[65] The law sanctioned forced sterilization for anyone diagnosed with certain genetically determined illnesses.[66]

The German Medical Association founded a journal called *The Genetic Doctor* (edited by a mentor of Josef Mengele's, Dr. Otmar von Verschuer) to help doctors determine who should be sterilized.[67] Under the Sterilization Law, between 350,000 and

500,000 people were in fact sterilized by the end of the war. But it was only seen as a first step in a much grander sterilization campaign that was envisioned by the most ardent proponents of the eugenics movement in Germany. It was an essential first step, however, because by design the sterilization drive co-opted the legal and, more significantly, medical communities. Doctors were made the logistical linchpin of these bold efforts, and were hard-wired into the procedural trappings that gave the program legitimacy. Some 1,700 Hereditary Health Courts and 27 Hereditary Health Supreme Courts were established under the law, whose principal officers included medical doctors.[68]

Robert Jay Lifton notes that Professor Rudolf Ramm of the medical faculty of the University of Berlin wrote an influential pamphlet during this era advocating the view that, beyond taking care of the sick, doctors should become "cultivator[s] of the genes" and "alert biological soldier[s]," and further spoke of "breakthroughs in biological thinking" that would reverse racial decay.[69] By the early years of World War II, German citizens were grumbling about the genetic police.[70]

So-called criminobiologic research institutes were also established (such as the Institute for Racial Protection, Criminal Psychology, and Criminal Biology[71]) and studies were undertaken to determine eligibility for castration or sterilization of habitual criminals, or presumptively habitual criminals (based on their personal histories) who had only been apprehended once.[72]

Nazi researchers theorized that specific physical characteristics were emblematic of criminality, just as they were for racial types or feeblemindedness. One article published in Germany at the time argued that you could tell if someone was a criminal by, among other things, examining his ears. Physical traits were believed to be a window into one's genetic predisposition, so long as you could read them right. Such notions had been around at least since Galton's time, of course. A hugely influential late-nineteenth-century book entitled *Criminal Man*, written by Italian criminologist Cesare Lombroso, argued, for example, that tell-tale physical characteristics, so-called stigmata, were a dead giveaway to criminal behavior. Such traits included harelips, prominent jaws, dark skin, and low brows, and so well-accepted was this view that Lombroso's theories influenced the writings of Sir Arthur Conan Doyle and Bram Stoker. Lombroso was often called upon to testify as an expert witness for the prosecution and God help the defendant if he looked guilty of anything![73]

With the passage in 1935 of the Physicians Law by the Nazis,[74] the medical community was tightly drawn into what Lifton has called the "Nazi biomedical vision."[75] A series of subsequent laws intended to clarify this vision defined doctors as "biological state officers."[76] The participation of the medical profession was, of course, integral to advancing Nazi racial theories and provided a gloss of respectability to what became, in essence, an industrial killing machine. Ultimately, 45 percent of the German medical community joined the Nazi Party. The reasons are complex, but may boil down to simple economics. Although in 1933, one percent of Germany's people were Jewish, 13 percent of the doctors in Germany were.[77] As Jewish doctors were systematically driven out of the profession (7,000 of them, altogether[78]), non-Jewish doctors found work.

Autumn 1935 saw the passage of still more laws to accelerate the winnowing away of undesirable genetic influences in Germany, but now they veered hard into racial territory and gave full expression to an ever present anti-Semitism. The so-called Nuremberg Laws of September 15, 1935, are the most notorious of these laws and were subsequently amended a number of times over the ensuing years until it was clear that their aim was the extinction of the Jewish people. In this original version, the Nuremberg Laws prohibited marriage or sexual relations between Jews and non-Jews. In administering these laws, Nazi medical theorists would turn for guidance to America's miscegenation laws—which specified degrees of racial-relatedness and forbade intermarriage in some states—to determine who was Jewish and who was not.

Formal measures against the Jews began as early as 1933. Among these were measures that forbade Jewish doctors to hold memberships in important medical organizations and restricting their practices. The rights and privileges of citizenship were gradually eroded for all Jews, most dramatically with the passage of the Reich Law of Citizenship, which drew a distinction between "citizens" (Aryans) and "inhabitants." This draconian law shunted German Jews into a legal twilight zone with all legal recourse against persecution swept away.[79] With the passage of the Nuremberg Laws, the Jewish pogrom was unleashed, symbolically culminating in 1938 in Kristallnacht, or the Night of the Broken Glass, when the windows of thousands of Jewish-owned businesses were smashed and hundreds of people were killed throughout Germany. On August 3, 1939, in the "fourth amendment" to the Nuremberg Laws, the medical licenses of all Jewish doctors were nullified.[80] A law passed on September 1, 1941, required all Jews to prominently wear a Star of David in public.

The final leg of the Nazi racial program was inaugurated in early October 1939 when Hitler decreed in a secret memo that his underlings were authorized to perform "mercy killings."[81] Thus began an aggressive euthanasia program focused, at first, on genetically "unfit" German children. To the extent there were half measures in the Nazi euthanasia program, they were confined to cloaking the program in protocol, secrecy, and deception in an effort to avoid a public outcry. No actual euthanasia law was ever passed.[82] Committees were established to "select" those who were to die. They were taken to designated medical institutions and killed. Soon the program was expanded.[83] Bodies were immediately cremated to destroy evidence.

The first gassings of mental patients took place on October 15, 1939, at Posen in Poland, shortly after Germany had invaded that country and officially triggered the start of World War II.[84] In places like Grafeneck, one of six principal German psychiatric institutions participating, all of which were eventually equipped with gas chambers,[85] the euthanasia program began with children three years old or younger, "nature's mistakes," as they were called, but soon expanded to older children and adults. In these institutions, drug overdoses and poisons were first administered, but there were simply too many waiting in line for this approach to remain practical. Thus began gassings and the use of such things as sham shower rooms into which doctors pumped carbon monoxide.[86] Bodies were immediately burned in crematoria down the hall. By August 1941, the program broadened, and euthanasia was routine at hospitals throughout Germany for the physically and mentally handicapped, as well as psychiatric patients. The program was so widespread, in fact, that prominent psychiatrists began to worry aloud that their profession no longer had any patients left.[87] As the first phase of this program was winding down in August 1941 (gassings were discontinued), more than 70,000 people had been killed in this manner in over 100 German hospitals.[88] Before the end of the war, the total who died under the banner of euthanasia would reach 100,000. Eventually, a public outcry in Germany stalled the euthanasia program. But the machinery and expertise of the euthanasia program did not lie idle for long; everything was dismantled and shipped eastward for the final phase of Nazi racial theory, the elimination of what the Nazis characterized as "a diseased race."[89]

Just as the tools of euthanasia were heading eastward into Nazi Germany's expanding territory, the instruments of sterilization were also being refined by the Nazi's to tackle their larger problem, the "Jewish problem." For years, Nazi propaganda had assailed the Jews as "vermin," a "pestilence," or "insects"[90] jeopardizing the quest to restore Germany to a race of pure-bred Aryans.

It was, in fact, according to Nazi propaganda of the time—and backed up by "scientific" reports—a matter of medical urgency that Jews be avoided and segregated. The Nazis pointed to studies by racial hygienists that claimed Jews suffered from more genetic ailments, both mental and physical, than other races. Dr. Gerhard Wagner, Führer of the Nazi Physicians League, condemned the Jews as a "diseased race,"[91] but the sentiment was not limited to concerns over genetic diseases. One of the initial measures taken against Jews in Poland after the Nazi invasion was to ban them from the railways because they were "germ-carriers."[92] After the passage of the November 23, 1939, law in Poland requiring all Jews to wear a Yellow Star, a German medical journal hailed the law as creating "an externally visible separation between the Jewish and Aryan population."[93] Once the Jewish population of Poland had been herded into ghettos in cities like Lodz and Warsaw, as part of a population-control strategy called for in the same medical journal, the inevitable outbreaks of contagious diseases in the ghettos provided a rationale for the Nazis to institute quarantines, shoot escapees on sight, and, later, exterminate the inhabitants.[94] Technology, it seemed, was now offering some more permanent solutions to the "Jewish problem." Since the Jews, Russians and other "undesirable" peoples numbered in the millions, the techniques of surgical sterilization were simply not practical as a method of curbing their reproduction. Methods would have to be developed that were much quicker and, ideally, could be performed on people without their noticing. Based on this need, experiments were launched at a number of concentration camps, including Ravensbrück, Buchenwald, and Auschwitz. The experiments were performed on inmates and included attempts at chemical sterilization and sterilization by means of irradiation, a method that had been lawfully used since 1936 as part of the sterilization campaign, although the law required the consent of the patient. No consent was sought from concentration camp inmates, of course.[95]

Ultimately, the sterilization program was rejected because Hitler disapproved, perhaps deeming it a mere half-measure.

Into this maelstrom strode SS Captain Josef Mengele, at age 32, who, after a brief tour of duty on the Russian front where he was wounded, volunteered to work in Auschwitz in May 1943. Although Auschwitz is remembered as the Nazis' most infamous death camp, the Nazis viewed it, to a considerable extent, as a medical research installation. Mengele arrived on May 30 and would stay 21 months, fleeing only as Russian troops closed in from the east. Among his other duties, Mengele worked with a medical team attempting to develop a chemical means of mass sterilization.[96] The principal reason for his being assigned there, however, was his perfect suitability for the kind of work being done at Auschwitz for the Reich. In fact, on March 9, 1943, with the Final Solution well underway, Himmler had ordered that only doctors trained in anthropology be permitted to perform "selections,"[97] that being the term—now with a decidedly medical air about it—for deciding who was fit to live and who was not. The Nazis had borrowed the term from the lexicons of the Social Darwinists and the racial hygienists (that is, artificial and natural selection).

Mengele's personal and professional objective, however, was to design a new race of human beings, while another "race" was being exterminated.

7 The World's Most-Wanted Man (Part 2)

AUSCHWITZ IS LOCATED ABOUT 38 MILES WEST OF CRACOW NEAR THE VIL-lage of Oswiecim (the Polish word for Auschwitz). Originally a collection of army barracks, Auschwitz came into being in June 1940 with the arrival of 728 Polish political prisoners, the first of about 150,000 who were to come later, half of whom eventually died there. After Hitler invaded the Soviet Union in 1941, Auschwitz became a dumping ground for Soviet POWs, too. As the numbers of POWs swelled, Himmler ordered that a second camp be built two miles away in a forest of birch trees. Given the sea of manpower there, Auschwitz soon became a source of slave labor for the German company I. G. Farben, a manufacturer of synthetic oil and rubber, located nearby to exploit area coal reserves. Once the POW arrivals subsided, however, Himmler altered his building plans to accommodate larger numbers of people as part of the Final Solution.

Although Auschwitz evolved from a POW camp to an extermination facility, it always retained the character of a business enterprise due to its relationship to German industries. The huge numbers of incoming people were inventoried; the expected life span of slave laborers was calculated, even regulated; clothing, property, and valuables were seized and warehoused in an area of the camp that came to be known as "Canada"; luggage inscribed with the owners' names was carefully stacked; gold fillings from the teeth of inmates were extracted for the ore; the shaved hair of inmates was baled, earmarked for the clothing and carpet industries of the Reich; and eyeglasses and shoes were stockpiled.[1] Everything was done with assembly-line efficiency, especially the killing. The camp, in fact, was run by the SS Economic and Administrative Department (WVHA).

Auschwitz was, in fact, not a single camp, but a cluster of perhaps 140 interconnected camps and sub-camps with the main facilities designated Auschwitz I (Auschwitz proper), Auschwitz II (known as Birkenau, so-called because of the species of birch trees that surrounded the camp), and Auschwitz III (known as Monowitz and located immediately adjacent to I. G. Farben's huge plant, referred to as "Buna"). It was in Birkenau that Mengele was posted as camp doctor and where he established an "experimental" station that became his passion. "In comparison [to Auschwitz]," wrote Auschwitz Commandant Rudolf Hoess in his autobiography, "Dante's inferno is almost a comedy."[2]

The manner in which people were herded into Auschwitz and executed is well known. What is perhaps less well known is the degree to which medical personnel were responsible for camp operations. They were, in fact, in charge of day-to-day operations. Auschwitz, after all, was perceived (or, at least, rationalized) by the Nazis as a center for scientific advancement. Concentration camp medical commander SS Colonel Enno Lolling explained to an audience in April 1943, a month before Mengele's arrival, that Auschwitz was to become "an experimental physiological, pathological station,"[3] one directly linked to the internationally respected German medical centers of learning.

At Auschwitz, doctors did little traditional medical work. They were there principally to supervise and carry out the Final Solution, the mechanized genocide of the Jewish people. That was the main purpose of Auschwitz. First, doctors "selected" arriving Jews as they disembarked from the trains at the station in the Birkenau camp, Jews arriving from every corner of Europe. This was typically a formulaic procedure where the old, the very young, the infirm, and mothers with children were immediately directed to the gas chambers, with able-bodied young people temporarily preserved for work details; perhaps 20 to 25 percent were spared at the ramp. For those arriving on the train, this procedure was frantic and disorienting. "The door opened with a crash," wrote Primo Levi in *Survival in Auschwitz,* "and the dark echoed with outlandish orders in that curt, barbaric barking of Germans in command which seems to give vent to a millennial anger."[4]

There were, of course, periods of time when the influx of people overwhelmed the staff and protocol broke down, such as during the spring and summer of 1944 when the Nazis purged 400,000 Jews from their Hungarian homeland and sent them to Auschwitz, an event precipitated by an order from Himmler to "cleanse" conquered territories[5] and prepare them for German settlers. According to Primo Levi: "We also know that not even this tenuous principle of discrimination between fit and unfit was always followed, and that later the simpler method of merely opening both doors of the wagon without warning or instructions to the new arrivals [prevailed]. Those who by chance climbed down on one side of the convoy entered the camp; the others went to the gas chamber."[6]

Everything told to the new arrivals was a deception designed to avoid causing a panic and keep everybody moving along. If they inquired about their luggage, they were told not to worry, they would get it all back later. If they seemed anxious about family members going off in a different direction, they were told that they would all be reunited very soon.

Once the selection was completed, a presiding doctor—along with a technician and the necessary gas pellets—was driven to the building that housed the gas chamber and adjoining crematoria. Often the doctor made the trip in an SS vehicle bearing a red cross.[7] The doctor was there to supervise the entire gassing procedure.

The buildings the prisoners entered were innocent looking to the new arrivals. As one survivor recalled, "the houses that the crematoria had—you know, brick houses, windows, curtains, white picket fences around the front. And people never thought of anything—regardless of chimneys smoking. They could not believe it. . . . There was a touch of diabolic genius."[8]

Once at the death chamber, the presiding doctor was responsible for determining how many gas pellets were needed to kill the number of people being presented; he would then assign the task of dropping the pellets into the holes in the ceiling.

Although the Nazis had used carbon monoxide in the euthanasia program, they had since substituted hydrogen cyanide (prussic acid) as the gassing agent of choice. Available under the trade name Zyklon-B, and readily on hand since it was also used as a pesticide and delousing agent, the gas first proved its effectiveness in a series of tests on several hundred Russian POWs in August and September of 1941. The use of Zyklon-B in the mass killing of Jews began shortly thereafter or in early 1942. The decision to use Zyklon-B was made jointly by Commandant Hoess and Adolf Eichmann.[9] Supplied by a company controlled by I. G. Farben, Zyklon-B was distributed within the SS under the direction of the SS Hygienic Institute in Berlin, which had an office at Auschwitz. Once its use broadened beyond rodent and insect control, Zyklon-B was put under stricter and stricter medical control. Because of its original purpose, German law had required Zyklon-B to be produced with an ingredient that acted as a chemical irritant to warn human beings of its lingering presence in fumigated areas. Sometime in 1943, this ingredient was removed and the canisters bore the warning: "Attention! No irritant!" At this point, the chemical was put under total medical authority and was to be handled solely by medical personnel.[10] In August 1998, a lawsuit was filed in United States federal court in Newark by Holocaust survivors against the conglomerate Degussa AG of Frankfurt, Germany, and its American subsidiary, Degussa Corporation of Ridgefield Park, New Jersey, charging that Degussa profited during the war by producing Zyklon-B for the Nazis and smelting gold confiscated from Jewish Holocaust victims. The plaintiffs are seeking damages equivalent to the entire current value of the company, $5.56 billion.[11]

Once the prisoners were herded into the gas chambers, with beatings whenever necessary, the doctors on hand observed the executions through portals and declared when the prisoners were dead. Cyanide gas impairs the ability of red blood cells to carry oxygen, so victims of cyanide poisoning suffocate, often while experiencing loss of bowel control, hemorrhage, and paralysis of the respiratory tract. Once everyone had apparently expired, the bodies were removed by *Sonderkommandos* (inmate workers) and conveyed to the crematoria for burning. No registry was kept of prisoners disposed of in this manner. They simply vanished from the earth.

Other duties of the camp doctors included carrying out periodic selections from camp inmates to thin out the prison population. One form of this involved lining up the inmates outside and weeding out the weak and undesirable. At other times, the selection occurred on the medical blocks and took the form of what Lifton has called "medical triage-murder," reminiscent of the earlier euthanasia program conducted at German hospitals.[12] Those selected in this way went to the gas chambers or were administered phenol injections by the doctors or their subordinates right on the medical block. If, as was common, contagions swept the barracks, doctors ordered the sick or exposed to the gas chambers. Lastly, doctors were called upon to execute certain prisoners for political reasons.[13]

Related medical duties included signing false death certificates (unnecessary for those immediately sent to the gas chambers), performing abortions, observing physical punishments of prisoners, and consulting on the best method for outside burning of bodies when the crematoria were overloaded, an acute problem during the summer of 1944 with the arrival of the Hungarian Jews. "They were all doctors," recalled one Auschwitz survivor.[14]

The number of people killed at Auschwitz in the gas chambers or by other means was long believed to be four million. This was the number that the Soviet Union had initially publicized and, since many of the documents seized at Auschwitz remained in Soviet archives, inaccessible to the West, there was no opportunity to study them. Moreover, the retreating Nazis had made every effort to destroy the buildings housing the crematoria and gas chambers, so the principal evidence of this machinery is documentary. With the political changes in Russia following the collapse of the Berlin Wall, scholars in the West gained access to the documentary evidence left behind by the Nazis, including the architectural plans of the crematoria and gas chambers. Based on these plans (and architectural plans used to prosecute Commandant Hoess in 1947, but since forgotten), as well as other logistical considerations, it is now believed that the death toll at Auschwitz was closer to 1,100,000 people, still a staggering figure.

Josef Mengele is remembered by many survivors as exhibiting an unusual zeal for selections. He is remembered in this capacity variously as good natured (for example, whistling all the time), perfunctory, solicitous, and extremely cruel, but always cutting a handsome and well-groomed figure as he performed his assigned duties on the ramp. By all accounts he was a fastidious dresser and quite vain. It is believed that it was, at least partly, this vanity that prompted Mengele to forgo having his blood type tattooed under his left arm, as required of SS members. For U.S. authorities looking for possible war criminals at the end of the war, these tattoos were deemed a "litmus test" of wrongdoing.[15]

Mengele had a particularly keen interest in the selection process due to the research he was conducting at Auschwitz, research that reportedly brought him to the ramp on many occasions when his name did not appear on the duty roster, according to witnesses. Some survivors insist he was there all the time.

Mengele's passion, of course, was genetics, and he was forever scanning the sea of faces at Auschwitz for evidence of fraternal and identical twins, young or old, it didn't matter. As noted earlier, racial hygiene was long the preoccupation of two preeminent medical research institutes in Germany, the Kaiser Wilhelm Institute for Anthropology, Human Genetics, and Eugenics in Berlin and the Kaiser Wilhelm Institute for Genealogy in Munich. Altogether there were more than 20 research institutes devoted to racial hygiene at various German universities. The study of twins, and particularly twins reared apart, was (and is) central to disentangling the influences of nature and nurture in the development of human character and abilities. Since Nazi racial hygienists believed that individual human behaviors (as well as physical characteristics) were genetically determined, they believed problematic behaviors could be rooted out by controlled breeding. Twin studies would confirm this theory and therefore "scientifically" establish the prevailing racial theories.

The Hungarian infusion in the summer of 1944 was a boom time for Mengele, according to one prisoner who worked with the doctor. She estimated that about 250 twins were added to his collection that season.[16] Twins were immediately segregated and directed to special blocks or one of Mengele's three special facilities for twin research scattered throughout the camp. Whereas those people who survived the selection were given standard tattoos on their left forearms, twins were tattooed with special serial numbers. One prisoner-doctor, Lottie M., recalled Mengele's reaction upon finding a pair of twins: "Mengele beamed—he was happy, . . . in a kind of trance."[17] For some twins this discovery meant special privileges and survival; others faced the prospect of gruesome surgical experimentation and death. Altogether, 3,000 twins passed through Auschwitz, but only 157 survived.[18]

The earliest document mentioning Mengele produced by the U.S. investigation into Auschwitz after the war is dated April 30, 1945; entitled "Report on War Crimes," it was written by Lieutenant Gerard Meillet. Meillet stated that "Mengele . . . had a hobby about twins,"[19] one that involved surgery. (The United States, however, did not have jurisdiction over Auschwitz since it fell to the Soviet army.) Less than a year before, an SS captain charged with writing an evaluation of Mengele's work performance said of Mengele: "He has filled all of the tasks assigned to him with circumspection, perseverance, and energy to the complete satisfaction of his superiors and proven himself in command of every situation. More than that, he has as an anthropologist used the little free time at his disposal in order to extend his studies and has made a valuable contribution in the field of anthropology, using the scientific materials at his disposal."[20]

As author Gerald Astor has noted, "Mengele's behavior at the [concentration camp], his research, and the concepts and philosophies of race superiority behind it were *not* unique but reflected, though in extreme degree, the mainstream science of Germany at the time."[21]

Josef Mengele did his doctoral research at the largest of the racial hygiene facilities, the University of Frankfurt's Institute for Racial Hygiene, working under Otmar von Verschuer. In 1942, von Verschuer became the director of the Kaiser Wilhelm Institute of Human Genetics and Eugenics in Dahlem, near Berlin. Throughout his career, Mengele maintained a close professional relationship with von Verschuer, who in fact is believed to have directed the twin research Mengele conducted at Auschwitz. Von Verschuer called twin research the "sovereign method for genetic research in humans."[22] Mengele forwarded "experimental materials" back to the institute as part of a study on the racial specificity of blood types,[23] among other studies. These materials included eyes, blood, and extraneous body parts, even heads. When the German Society for Blood Group Research was established in 1926, its founder, Otto Reche, cited as one of the society's principal missions the identification of a scientifically reliable in vitro test to differentiate between Aryans and non-Aryans.[24] The German Research Council funded the studies into the racial specificity of blood groups.[25]

The legacy of such Nazi racial thinking continues today in the form of alleged attempts by various scientists or governments to create genetically engineered biological weapons that can harm only specific ethnic groups. According to some experts, it is theoretically possible to create or mass produce a biological agent with a previously known, or specifically designed, ability to attack specific racial groups, either because the agent has an affinity for cell receptor sites prevalent in a given population or because the people in a given region lack immunity to a particular pathogen. United Nations weapons inspector David Kelly, for instance, leveled accusations at the Iraqi government over an alleged bioweapons program to mass produce "camel pox," for which Arabs have acquired an immunity but which could cause a widespread, and potentially fatal, outbreak in Europe or North America. Additionally, apartheid-era South African scientist Wouter Basson—called "Dr. Death" by the media—was accused of attempting to develop biological and chemical weapons that attack only blacks, in addition to researching methods to reduce the fertility of black women. (A lengthy trial in South Africa ended in 2002 with Basson's acquittal, however.) Although many scientists downplay the risks of such a scenario because the genetic similarities between the various human ethnic groups far outweigh the differences, the British Medical Association warned in a 1999 report entitled "Biotechnology, Weapons and Humanity" that "weapons could theoretically be developed which affect particular versions of genes clustered in specific ethnic or family groups."[26]

There were other "medical" studies under the Nazis, too, that at Mengele's laboratory resulted in the shipment of body parts back to von Verschuer, such as the study of eye color. One of the prisoner-doctors that worked alongside Mengele, forensic pathologist Miklos Nyiszli, recalled that the rare occurrence of heterochromes in identical twins, making one eye blue and the other brown, warranted such special treatment. Such twins were routinely killed and dissected because of the anomaly, according to Nyiszli, and their organs shipped to the institute marked "War Material— Urgent,"[27] as were all organ shipments. One of the few documents pertaining to Mengele's service at Auschwitz that survives is a document he signed that accompanied the head of a twelve-year-old Gypsy boy back to the institute.[28]

During the time Mengele was in residence at the institute in Frankfurt in 1937, von Verschuer published an article praising the Führer's leadership role in the racial hygiene movement, saying "Hitler is the first statesman who has come to recognize hereditary biological and race hygiene and make it a leading principle of statesmanship."[29] In 1942, von Verschuer said: "We have laws to protect German blood and German hereditary health but not only these special laws but the entire leadership and the achievement of the present state are fully conscious of the value of the concepts of heredity and race. This concept of race has now become the underlying principle of the solution of the Jewish problem."[30] After the war, investigators turned up von Verschuer's name in connection with Mengele and the twin studies conducted at Auschwitz. Von Verschuer was only interrogated, however, never indicted for war crimes. He even attempted to mitigate accusations his interrogators leveled at Mengele, suggesting that Mengele was posted to Auschwitz against his will and had sought a transfer, also informing them that all correspondence, papers, and specimens relative to Mengele's research (innocently enough, he assured them) had been destroyed.[31] War crimes investigators concluded that von Verschuer himself had destroyed the evidence, but they took no further action.[32] Immediately after the war, von Verschuer appealed to the Bureau of Human Heredity in London for permission to resume his research. After a period of years, during which colleagues came to his defense, von Verschuer won an appointment as professor of medicine at the University of Münster, retaining the position until his death in 1969. Historians have noted with alarm that von Verschuer's twin studies continue to be cited even today in scholarly publications by leading geneticists.[33]

Mengele is remembered by colleagues and survivors of Auschwitz as very passionate about his research. His research objectives were evidently twofold: First, he was seeking to control inherited traits and substantiate the Nazi racial theories concerning genetic determinism. Second, he was seeking a method for producing multiple births, such as twins, triplets, and so on. Together these goals, of course, were aimed at rapidly populating the world with Aryan look-alikes.

Whatever his research focus, Mengele evidently believed he was on the brink of extraordinary scientific breakthroughs in genetics. He wasn't alone in this view. The Reich Research Council apparently thought so, too, because they were his source for research funding at Auschwitz,[34] funds approved by a review committee made up of leading German physicians. Mengele also impressed some medical colleagues and inmate doctors at Auschwitz as atypically intelligent, cultured, and driven. Dr. Hans Munch, a camp doctor at Auschwitz who was later acquitted of war crimes at Nuremberg because of his refusal to participate in selections and other duties (for which he was *not* punished by the Nazis, incidentally), told author Gerald Astor in the 1980s that Mengele "was not only intelligent but generally and scientifically a very interesting person."[35] Inmate-doctor Ella Lingens, a non-Jew, said: "He could have had a first-rate academic career."[36] Such views of Mengele, however, are very much in the minority since he is remembered, above all, for exhibiting an unflinching cruelty in the pursuit of his research objectives.

Mengele's research grants were approved on August 18, 1943. Von Verschuer subsequently wrote that Mengele's research had been authorized by Himmler himself and concerned "anthropological examinations" and noted that "blood tests are sent to my laboratories."[37] By the terms of the grants, Mengele's work involved "specific albuminous matter," whatever that is, and a study of "eye color."[38]

Mengele was not a creature committed to his work on twins solely for the sake of science, however. He had practical career objectives in mind at the same time. According to one friend at Auschwitz, Dr. Ernst B., as reported by Lifton, Mengele coupled his research at the camp with the preparation of his *Habilitation,* an academic presentation required for gaining a post as a university lecturer and, eventually, a professorship.[39] Even prisoners of the camp who observed Mengele at his work concede that, but for Auschwitz, he seemed well-suited for academia.[40] One prisoner-doctor put it this way: "In ordinary times he could have been slightly sadistic German professor."[41]

Science, for Mengele, was a means to an end, not the end in itself. It was personal ambition that animated him. According to one of his contemporaries, Julius Diesbach, "He always wanted to do something outstanding to be a great scientist."[42] The ambition that possessed him either blinded him to the suffering and torment of his human subjects at Auschwitz or his heart was made of steel. Certainly, the twins and the other people he studied were merely lab animals to him, not sentient beings. As one prisoner-doctor put it, "he had no problems—not with his conscience, not with anybody, not with anything."[43] Through it all, said another, "he was absolutely convinced he was doing the right thing."[44]

The exact nature of the crimes Mengele committed are based on countless eyewitness accounts given by survivors of Auschwitz over the years. Dr. Mengele, of course, was never captured and put on trial for his crimes, so their enumeration necessarily remains in the form of allegations. None of the allegations has been proven beyond a reasonable doubt to the satisfaction of a jury or a panel of judges. No doubt, too, many survivors mistakenly believe they encountered Mengele at Auschwitz because he has become so closely identified with the place. Nevertheless, the sheer number of witnesses and the gravity of the charges leveled against Mengele are compelling.

One such damning testimonial is provided by Eva Mozes-Kor, an Auschwitz survivor, who has recounted the ordeals of herself and her twin sister Miriam. Both aged nine when they arrived in Auschwitz in the spring of 1944, they were immediately segregated by Mengele and sent to a twin laboratory:

> . . . Our interactions with other prisoners [were] extremely limited, as the twin experiments were top secret. . . .
>
> . . . In fact, we were there for one reason: to be used as experimental objects and then to be killed. Mengele had two types of research programs. One set of experiments dealt with germ warfare. In the germ experiments, Mengele would inject one twin with the germ. Then, if and when that twin died, he would kill the other twin in order to compare the organs at autopsy.
>
> . . . One of the twins, who was 19 years old, told of experiments involving a set of teenage boys and a set of teenage girls. Cross-transfusions were carried out in an attempt to "make boys into girls and girls into boys." Some of the boys were castrated. Transfusion reactions were similarly studied in the adolescent twins.
>
> In the area of genetics, Mengele collected dwarfs, giants, hunchbacks, and people with abnormalities and defects. He studied genetic traits in the hope of "purifying" the "Aryan superrace." He closely monitored eye and hair color.
>
> A set of Gypsy twins was brought back from Mengele's lab after they were sewn back to back. Mengele had attempted to create a Siamese twin by connecting blood vessels and organs. The twins screamed day and night until gangrene set in, and after 3 days they died. Mengele also attempted to connect the urinary tract of a 7-year-old girl to her own colon. Many experiments were performed on the male and female genitals.[45]

Despite the immediate accumulation of testimonial evidence implicating Mengele in war crimes, West German authorities were extremely slow to search for Mengele or even issue warrants for his arrest. During this time, Auschwitz survivor Hermann Langbein (a political prisoner), serving as the general secretary of the International Auschwitz Committee, uncovered Mengele's divorce records indicating he was living in Buenos Aires. Together with Simon Wiesenthal, Langbein lobbied the West German government to arrest Mengele.[46] As a result, the first warrants were issued on February 25 and June 5, 1959; up until that time, West Germany had made no effort to find him. These warrants, although they did result in Mengele's extradition being sought in Buenos Aires, languished until January 19, 1981, when a comprehensive new arrest warrant was issued superseding the older, unexecuted ones. The 1981 warrant is 33 pages long and catalogues a series of crimes that recall the Hoess commentary that Auschwitz makes Dante's Inferno seem a comedy by comparison.

Dr. Mengele was, of course, not the only physician experimenting on live patients, and such experiments were taking place not only at Auschwitz but at a variety of other Nazi installations and concentration camps, not to mention Japanese concentration camps. Many German doctors participated in these experiments. To represent these crimes, 20 high-ranking Nazi doctors were captured and put on trial along with three non-doctors, in the so-called Doctors' Trial at Nuremberg, which ran from December 9, 1946, to August 19, 1947. Of this group, however, only one had worked in a concentration camp, and none had worked in a death camp.[47] The Doctors' Trial was, then, partly symbolic. It was part of the judgment of this trial that became known as the Nuremberg Code, a ten-point enumeration of ethical standards in medicine for human experimentation that continues to guide the medical profession today. Of the defendants, seven were ultimately hanged for their crimes, including four doctors, among them the personal physician to Adolf Hitler and the former head of the German Red Cross. Seven were acquitted. "The 20 physicians in the dock," said chief counsel and Brigadier General Telford Taylor in his opening statement to the tribunal, "range from leaders of German scientific medicine, with excellent international reputations, down to the dregs of the German medical profession."[48] The medical experiments in question involved live involuntary subjects and included high-altitude experiments (carried out in low-pressure chambers and meant to benefit the German Air Force), freezing experiments (where prisoners in tanks of ice water were frozen to death—meant to benefit the German Air Force), malaria experiments (infections and drug treatments), mustard gas experiments (experimental treatments of wounds caused by the gas), sulfanilamide experiments (application of this substance to wounds infected with streptococcus, gangrene, and tetanus), bone, muscle, and nerve regeneration and bone transplantation experiments (which involved the mutilation of prisoners), seawater experiments (efforts to make seawater drinkable, which involved forcing prisoners to ingest it), epidemic jaundice experiments, sterilization experiments, spotted fever experiments (vaccine experiments), experiments with poison, and incendiary bomb experiments. In addition, it was charged that 112 Jews had been selected for a skeleton collection—they were photographed and measured while alive, then killed, examined and de-fleshed. Finally, four of the defendants were tried for, among other things, their principal roles in the Nazi euthanasia program.[49]

As Taylor noted, the "policy of mass extermination could not have been so effectively carried out without the active participation of German medical scientists."[50] Also, "these crimes were the logical and inevitable outcome of the prostitution of German medicine under the Nazis."[51] Taylor reiterated the remarks of Justice Robert H. Jackson before the court from the previous year: "The wrongs which we seek to condemn and punish have been so calculated, so malignant, and so devastating, that civilization cannot tolerate their being ignored because it cannot survive their being repeated."[52]

By all accounts, Josef Mengele continued working on his research until the last possible moment. When the thunder of Russian guns made the liberation of the camp imminent, Mengele gathered up some of his research materials and fled. This was on or about January 17, 1945. The camp was liberated on or about January 27, 1945. Mengele at first slipped back behind German lines, and with the fall of Nazi Germany in May Mengele was absorbed into the chaos of the initial postwar period. He would live for decades in freedom to comment to friends about the world's first heart transplant and the first walk on the moon, calling these achievements "fantastic."[53]

Exactly how Mengele eluded capture in the first days and years following the war is a question that has provoked recriminations and rancorous debate. The Department of Justice has attempted to put to rest many of the claims that U.S. authorities either seriously bungled the search for Mengele or in fact aided and abetted his escape, in the same way that U.S. intelligence did in fact aid the flight of Klaus Barbie, the former Gestapo chief of Lyons, France, who brokered for his freedom based on intelligence deemed valuable to the United States. In exchange, the United States took Barbie out of Europe via the so-called Rat Line.[54] There is, the Office of Special Investigations maintains, not "even a scintilla" of evidence that Mengele received similar treatment.[55]

The OSI discounts the claims of the two U.S. Army veterans, Walter Kempthorne and Richard A. Schwarz, that a "sterilization doctor" namely Mengele, was interned on their respective watches at a U.S. POW camp in the summer of 1945. Vague details and lack of corroborating evidence lead to this conclusion. A reconstruction of his immediate postwar movements by the OSI includes the following: Based on German Federal Archives, Mengele was assigned to Gross Rosen Concentration Camp after fleeing Auschwitz. Unconfirmed reports put him at Mauthausen Concentration Camp after that. On or near May 2, 1945, Mengele arrived at a German military field hospital in the Sudetenland wearing a German Army (Wehrmacht) uniform, instead of his own SS uniform. There he contacted a medical colleague, Dr. Otto-Hans Kahler, with whom he'd worked at Dr. von Verschuer's Institute in Frankfurt before the war. Dr. Kahler reported to the OSI that Mengele was severely depressed at this meeting and spoke openly about selection duties at Auschwitz. With Dr. Kahler's help, Mengele joined a German unit that traveled to an unoccupied area, or so-called no man's land. Neurologist Dr. Fritz Ulmann, attached to this unit, recounted to the OSI Mengele's presence during this period. Dreading they would fall into the hands of advancing Soviet troops, members of this unit, including Mengele, willingly moved into the nearby American zone where they were taken into U.S. custody and interned at a POW camp in Schauenstein sometime in June. From the outset, according to Dr. Kahler—since reunited with Mengele—Mengele used the alias Joseph Memling.[56] This was not the first alias he used; many others would follow.

After staying in Schauenstein for about six weeks, Mengele, Kahler, and Ulmann were transferred to a second POW camp to the south, probably at Helmbrechts. Because there were an estimated 3 million-plus German POWs in custody at the end of the war, there was a great deal of pressure to release people, in large part due to the daunting task of feeding them all.[57] To complicate matters, German personnel were used as translators and administrators. (Coincidentally, and ironically, during the time Mengele was interned, U.S. officials interrogated Irene Mengele at her home outside Guenzburg about the whereabouts of her wanted husband.[58]) SS members were not eligible for release until they had undergone a rigorous interrogation, and in the vast majority of cases, they were identified by the standard blood-type tattoo under their left arms and/or identification papers that all prisoners were required to retain. The tattoo, in particular, was considered a litmus test, although the military later learned it was not universally done.[59] As noted earlier, Mengele did not have the blood-type tattoo; further, he either presented no ID papers or false ones.[60] Lists were circulated of people identified as known war criminals or people subject to automatic arrest,[61] which included all high-ranking Nazi party or SS members.[62] Joseph Mengele was identified by the U.N. War Crimes Commission as wanted for war crimes on List No. 8 (no. 240) issued in May 1945, but this list evidently had not reached Helmbrechts.[63] At least one other list identified him as wanted as well.

Mengele may have been more rigorously interrogated at Helmbrechts because of a lack of papers, but either way he was lucky to be there—"Helmbrechts was dedicated to discharging POWs and had a large turnover of prisoners. Mengele was discharged sometime during the first week of August 1945, the OSI concluded.[64] According to both Kahler and Ulmann, interviewed by the OSI in 1985, Mengele was released under his own name, something the OSI found impossible to conclusively determine.[65] (One additional witness insists he had two discharge papers, one in his own name.[66]) Whether this is true or not, one thing is certain: Mengele obtained from Ulmann a duplicate discharge certificate which he later doctored for identification purposes,[67] adopting, at least for a time, the confirmed alias Fritz Holman.[68] The OSI believes it is possible Mengele used a duplicate of Ulmann's discharge paper from Schauenstein to expedite his clearance out of Helmbrechts.[69] He thus somehow managed an in-custody name change under the nose of his American captors. This would not have been unprecedented, because authorities now know that Adolf Eichmann managed a similar escape from U.S. captivity.[70] Mengele was reportedly transported by truck to Ingolstadt, a town north of Munich and east of Guenzburg.[71] Mengele then made his way to the town of Donauwoerth, near Guenzburg, where he lunched with a former schoolmate and his wife. He had arrived there on a loaned bicycle, having stashed the set of discharge papers bearing his real name in the handlebars. These he forgot to take with him when he left.[72] Mengele next appears to have gone into the Soviet zone to visit a nurse he'd met at the German field hospital, a great risk given the general fear of apprehension by the Soviets.[73] Abandoning this location, he returned to the American zone by mid-October 1945 and worked on a farm in Mangolding until August 1, 1948, under the alias Fritz Holmann.[74]

During this period meetings occurred between Mengele and his family members, Sedlmeier and Irene Mengele. It soon became clear that Irene did not want a life underground and divorce was inevitable. It is not known exactly where Mengele hid out between August 1948 and mid-April 1949 when he finally left Germany. It is clear, however, that nobody seemed to be actively searching for Josef Mengele during his residence in Germany. Indeed, the OSI concluded that "there appears to have been no specific high-level [U.S.] instruction to find him"[75] and "no prolonged manhunt was undertaken."[76] In part this is due to the fact that the United States had no direct jurisdiction over war crimes at Auschwitz; possibly for this reason, Mengele's name did not appear on any U.S. "specialized" list of wanted war criminals, such as the so-called Rogues Gallery.[77] Furthermore, as the years passed, a blossoming Cold War eclipsed any imperative to apprehend Nazi war criminals.[78] Complicating everything, false reports that Mengele had been arrested by the Americans had been circulating since December 15, 1946, when he was first reported arrested by a Viennese newspaper.[79] So convincing were these reports that in the first few months after the initial report, testimonies against Mengele poured in from all over Europe to assist potential war crimes prosecutors.[80] The confusion snowballed until the misinformation had circumnavigated the globe, extending even to officials in the Pentagon.[81] To make matters worse, on January 19, 1948, U.S. Chief of Counsel for War Crimes in Nuremberg, Brigadier General Telford Taylor, responded to an inquiry by letter stating Mengele had died in October 1946.[82] Although the exact origin of the rumor of the death of Josef Mengele has never been traced, it is certainly well known that the Mengele family and Irene Mengele actively fanned this rumor.[83] It was during this time that Irene reportedly arranged for a memorial mass in her husband's name, wore black and took steps to have him officially declared dead.[84]

But Mengele was very much alive and with the financial assistance of his family he purchased forged documents and made his way south to Austria, crossing into Italy at Brenner, and ultimately to Genoa, led by a series of well-paid guides.[85] In Italy, Mengele obtained an identity card issued under the name Helmut Gregor[86] and later obtained an Argentinean *permisso de libero desembargo,* or Permit of Free Passage.[87] Once in Genoa, Mengele's last guide booked passage on the *North King,* scheduled to leave in five days. In the interim, Mengele went to the Swiss Consulate and, based on the identity card, received an International Red Cross passport. Next, he secured some technical releases from Italian obligations enabling him to get a visa from the Argentinean Consulate, had a perfunctory physical, and had only to get an Italian exit visa. When he attempted this, the Italians arrested him for reasons that remain unclear, but certainly don't appear related to his wartime atrocities. After a brief detention, Mengele was released.[88] On May 26, 1949, at 2:45 P.M., the *North King* set sail for South America with Josef Mengele on board.

When Josef Mengele disembarked from the *North King* in Buenos Aires, he could not have found a more hospitable shore. Argentina had long been a country with pro-Hitler inclinations and had lately become a haven for escaping Nazis. Finally, and haltingly, released in the 1990s under President Carlos Saul Menem, Argentina's "Nazi files" show an extraordinary complicity with the Nazi underground to harbor Nazi war criminals and collaborators. Whereas most outside investigators had assumed that perhaps several dozen Nazi war criminals had, at one time or another, found refuge in Argentina, the newly released "Nazi files" reveal that the true number is well over a thousand.[89] Among them were, of course, Josef Mengele, who had an extensive and revealing Argentine file, and SS Lieutenant Colonel Adolf Eichmann, the immediate subordinate to Himmler's chief deputy, Reinhard Heydrich. Among the new names on the list of fugitives was Ante Pavelic, the leader of Croatia's pro-Nazi government that ruled from 1941 to 1945, who settled in Argentina with eight of his deputies and has been linked to oppressive police activity under former Argentine President Juan Perón.[90]

By the time Mengele arrived in Argentina, an influential network of ex-Nazis were comfortably situated in Buenos Aires. Welcomed into this community, Mengele was able to exploit the business and government contacts made available by the Nazi cabal. This underground network was reportedly presided over by Hans Rudel, the Nazi's most decorated pilot (someone not wanted as a war criminal). Rudel's influence stemmed, in part, from his willingness to help Juan Perón launch an air force. For Mengele, the next decade was, in many respects, far from the typical life of a notorious fugitive, his halcyon days. During this period he even regained an interest in science; this is when he published an article on genetics in an Argentine science journal under the name "G. Helmut."

Perhaps the worst thing that happened in the early years was when Josef's father visited in 1954 and informed his son that Irene was at last pressing for divorce. In granting her wish in March of that year, Mengele executed a power of attorney—using his true name—at the West German Embassy in Buenos Aires. While the divorce documents would haunt him later when unearthed by Hermann Langbein, they did not seem to cause a stir with West German authorities at the time. By 1956, Mengele dispensed with the alias of Helmut Gregor for most purposes and obtained an Argentine foreign resident's permit under his own name. Through the West German embassy in Buenos Aires he requested a copy of his birth certificate from Guenzburg, accompanied to the embassy on this occasion by Hans Sedlmeier. Later that same year, Mengele traveled to Switzerland, where he visited with his then-12-year-old son, Rolf, who was told he was meeting an uncle from South America. On this trip, as well, he met with his brother Karl Jr.'s widow, Martha, accompanied by her son Karl-Heinz. Karl Jr. had died in 1949 at age 37, and it was Karl Sr.'s hope, at least, that two problems might be dealt with at once. In other words, he wished to keep family members secure and the family business safe from the influence of outsiders. Josef, although he was a fugitive and had agreed to relinquish his share of the family business as a condition of the family's continued support during his exile, had the benefit, at least, of being an insider. Martha and Karl-Heinz would join Josef in Buenos Aires later that year. In September 1956, Josef Mengele began living openly in Buenos Aires under his real name. He applied for and received an Argentine identity card as Josef Mengele after receiving a certification from the West German embassy that this is his true name. His name and number were published that year in a Buenos Aires telephone directory. Although Mengele had reportedly been involved in a variety of businesses up until that time, his last known business, Fadrofarm, was started in 1957. Registered in Argentina to sell chemicals and pharmaceuticals, the business was well capitalized and listed Josef Mengele as a partner. In 1958, Martha and Josef were married in Uruguay. Also that year, Mengele—whose medical credentials had been stripped after the war—petitioned Goethe University in Frankfurt to have his medical degree restored.

The following year Hermann Langbein and Simon Wiesenthal turned up the heat and West Germany issued an arrest warrant for Mengele and sought his extradition from Argentina. Although Perón was in exile by this time, the political climate in Argentina continued to be favorable to fugitive war criminals, and, claiming Mengele could not be located, government officials delayed an entire year, until July 1960, before an Argentine court finally issued an order for Mengele's arrest and extradition. By this time, Mengele was long gone. Simon Wiesenthal has lamented the "widespread antipathy toward extradition"[91] that prevails in South America. Historically, extradition in South America has been next to impossible due to the ever-changing political fortunes of South American leaders, many of whom came to rely on the sanctuary of other countries.

Aware from media accounts that his whereabouts were known, Mengele fled to Paraguay in 1959. By virtue of his ties to Hans Rudel, Mengele found himself with influence all the way to the top of this small South American country. President Alfredo Stroessner, a general and the son of a Bavarian immigrant, was a close ally of Rudel and had been dictator since 1956. After various perjured affidavits were filed on Mengele's behalf attesting to his five-year residency in Paraguay, he received a passport and on November 27, 1959 was granted full citizenship. For many years afterwards all efforts to have Mengele extradited from Paraguay were resisted by Stroessner and his successors on the grounds that as a citizen of Paraguay, Mengele was off limits. In fact, however, this was all part of a clever ruse, because Mengele very soon departed from Paraguay to go permanently underground in Brazil. By the time Paraguay revoked Mengele's citizenship in 1979 after years of international pressure, it mattered not at all.[92]

By this time, Martha and Karl-Heinz had returned to Germany and another divorce was in the works. Although abandoned by his wife, Mengele continued to receive financial support from his family and fell under the protection of Wolfgang Gerhard, a pro-Nazi Austrian who had settled in Brazil in 1948, having been introduced to Gerhard once again by Hans Rudel. Gerhard, in turn, introduced his charge to two families, both deemed politically suitable, who would take Mengele in and provide him sanctuary. The first couple, Geta and Gitta Stammer, were introduced to Mengele under the name Peter Hochbichler, said to be Swiss, and agreed to take him on to manage their farm in Nova Europa in Araraura, northwest of São Paulo. In 1962, they all moved to another farm at Serra Negra, closer to São Paulo.[93] Although their suspicions had been aroused, it was only there that the Stammers learned Peter's true identity after Gitta confronted him with a published photo of Mengele that displayed his distinctive smile and the gap between his front teeth. After the Stammers learned the truth, Mengele nevertheless stayed with the couple for the next 12 years, largely due, it is believed, to the intervention of Gerhard and Sedlmeier and money paid by the Mengele family. As an indication of his state of mind during this period, Josef Mengele built a watchtower on the farmhouse at Serra Negra.[94]

Although Mengele had met Liselotte and Wolfram Bossert sometime in 1970 while living with the Stammers, he did not fall under this couple's protection until 1975, when he moved into a building they owned in Eldorado, a suburb of São Paulo. By this time, a rift had brought the Stammer era to an end. He lived in Eldorado alone but for the occasional company of housemaids who picked up after him. By this time, with the assistance of Wolfram Bossert, Mengele had assumed, for some purposes, the identity of Wolfgang Gerhard, who returned to Austria in 1971 but left behind his Brazilian identity card for Mengele's use. As an amateur photographer, Bossert substituted Mengele's photo for that of Gerhard. Before leaving for Austria, moreover, Gerhard secured a burial plot under his own name at Our Lady of the Rosary Cemetery, just outside São Paulo.[95] This was the grave site at Embu. He would also return in 1976 to obtain a new identification card for Mengele with the expiration of the first.

Because of their self-declared Nazi sympathies, the information provided to authorities by the Stammers and the Bosserts was necessarily viewed with deep suspicion. Might they not be part of a well-orchestrated plot to fabricate the death of Josef

Mengele? Meanwhile, Mengele could be living virtually anywhere, no? Certainly, this was a guiding sentiment as investigators pored over the accumulating evidence.

The forensic scientists at Embu were, thus, confronted by a case that demanded certainty. While the Embu experts were able to agree that the remains buried in the plot of Wolfgang Gerhard were in fact Josef Mengele's "within a reasonable scientific certainty," there were nevertheless substantial lingering doubts about the identification. There was nothing definitive about the evidence. At the urging of Israel and West Germany, therefore, the case remained open.

Of significance to those people with doubts was not so much the evidence they had before them, it was the lack of certain items of evidence. Also, much of the circumstantial evidence that had been pulled together by the investigation thus far originated from the Bosserts and the Stammers, ardent Nazis. Chief among the items that were lacking were X-rays of any kind that might serve to identify the bones and teeth. Further, the investigators were unable to detect any tangible evidence that the skeleton had suffered from osteomyelitis, even with X-rays. This, in particular, would remain a contentious issue for some time and invite a review of the evidence by some of the best scientists in the world. Another area of dispute centered around a hole exhibited in the left cheekbone of the Embu remains, which Clyde Snow believed was a postmortem artifact and Lowell Levine insisted was a fissure that formed during life from a tooth infection.[96]

In addition, the shattered skull fragments recovered at Embu caused a split among the experts[97] concerning the use of evidence of a wide incisor canal as necessarily meaning the corpse had a diastema, such as Mengele was known to have. Moreover, the Embu skeleton showed signs of a hip fracture, a subject that Mengele's SS file is at best equivocal about, and most likely does not account for. A potentially related problem was that the skeleton's left leg was 1.5 centimeters longer than the right, so some believed this person should have limped, which Mengele did not. Regrettably, too, the fingerprints that investigators had only recently managed to verify were of no use to the scientists, given the state of decomposition of the corpse. Finally, there was respectful concern over the validity of Dr. Richard Helmer's method of skull-photograph comparison, a method that produced compelling exhibits but nevertheless inspired certain reservations. A consensus of opinion emerged at the Department of Justice that Dr. Helmer's method needed further testing under rigorous laboratory conditions and with a variety of skulls and photographs. Thus, this procedure was deemed merely probative, not definitive.[98]

At a news conference on the morning of June 21, 1985, the scientists endeavored to explain their preliminary finding that the skeletal remains they examined belonged to Josef Mengele "within a reasonable scientific certainty."[99] Nevertheless, the Israeli government announced two days later that the case would remain open until all the evidence had been re-examined by Israeli forensic experts.[100] On the state of the evidence, the reluctance of Israel to close the case was, of course, not entirely unreasonable.

A number of experts were later brought in to examine the remains to determine if the bones showed signs of disease, but the Department of Justice ultimately concluded that the osteomyelitis issue didn't really matter all that much anyway, in light of other evidence that surfaced.[101] At the same time that the skeletal evidence was being reviewed, the OSI launched an aggressive search for X-rays that might have been taken of Mengele sometime during his life. X-rays, of course, can definitively establish identification of a corpse. Contained within Mengele's diaries were entries concerning the medical and dental treatment he'd undergone during the 1970s, including three references to X-rays. One occasion was a root canal performed in 1978.[102]

In March 1986, a dentist was approached who had treated a "Pedro Hochbichler" in 1978 for a root canal; his patient had been referred to him by a Dr. Kasumasa Tutiya, a general dentist. The investigation hit pay dirt when Dr. Tutiya turned over dental records for "Hochbichler" that included eight dental X-rays taken in 1976,[103] each meticulously inscribed in pencil with the name "Pedro H." by the careful dentist. With the X-rays in hand, Dr. Lowell Levine, the original team's forensic odontologist, flew back to Brazil to re-examine the recovered teeth and compare them to the X-rays taken of the Embu skeleton and a 1938 SS dental chart of Mengele. In his report, Dr. Levine concluded: "Radiographically the skeleton is absolutely the dental patient who identified himself as Pedro Hochbichler and is actually Josef Mengele based upon the scientific comparisons by all the forensic disciplines and modalities."[104] The X-rays confirmed that the deceased had a large gap between his front teeth, something that could not be independently verified by the remains, given their condition.

On the basis of the newly gathered evidence, the U.S. forensic team submitted a report dated November 6, 1986, concluding that the "remains exhumed at Embu Cemetery, near São Paulo, Brazil, were those of Doctor Josef Mengele."[105] They further concluded: "The probability of any two people having this many specific points of agreement is virtually nil."[106]

Despite this confidence level, representatives of the Israeli government prevailed upon U.S. investigators at a meeting in June 1987 to withhold publication of their findings until four additional aspects of the investigation were completed. These included an additional forensic examination, a wider search by West German authorities for medical records of Mengele, an effort to persuade Liselotte Bossert to submit to a polygraph examination (given her key role in the burial), and an attempt to conduct DNA testing on the remains, something contingent on the cooperation of Rolf and Mengele's first wife, Irene.

Before the year was out, the forensic examination had been completed, and the West Germans had uncovered Mengele's school records. Moreover, Liselotte Bossert passed a polygraph test. All that remained to be done was the DNA.

The FBI, however, was not yet up to speed on DNA typing. Aggressive efforts were ongoing to validate techniques for forensic DNA analysis, and these would continue throughout 1988. By December of that year, the FBI began to accept criminal casework involving RFLP for the first time. Validation studies continued, however, on the PCR dot blot test, and it was PCR that everyone was relying upon to close the Mengele case. In January 1989, the Embu remains were made available to the FBI laboratory for DNA testing by the government of Brazil; but before the FBI could begin analyzing these important remains, further validation studies were conducted on bones of similar age and condition provided by the Smithsonian Institution. Unfortunately, the FBI researchers concluded that they would be unable to extract high molecular weight DNA from bones that were as old as the Embu bones using PCR. This finding, of course, ruled out conducting standard RFLP tests. Next the FBI attempted to salvage even low molecular weight (highly degraded) DNA from sample bones of similar age. Once again, they came up empty. Finally, the OSI provided the FBI laboratory with hair samples unearthed with the Embu skeleton. Although approximately 500 hair roots were analyzed, this PCR test also was a failure. On June 22, 1989, the FBI notified OSI Director Neal M. Sher that they were unsuccessful in extracting DNA in the Mengele case.[107] In October, West German scientists also admitted defeat after attempting to extract DNA from the Embu bones.

This led the state prosecutor of Frankfurt, Germany, to call in the top gun, Professor Alec J. Jeffreys of the Department of Genetics, University of Leicester in England. Jeffreys, in turn, recruited Erika Hagelberg, and they soon set to work. By summertime, Jeffreys and Hagelberg reported success, at least to the extent that they had coaxed minute traces of degraded (low molecular weight) DNA from the femur of the Embu remains.

As might be expected from old remains, Jeffreys and Hagelberg initially extracted a lot of DNA from the bones, the vast majority of which turned out to be nonhuman DNA, probably microbial. With the humerus bone extracts, virtually none of the DNA turned out to be human. With the femur extracts, on the other hand, a paltry amount of highly degraded human DNA, amounting to ~0.14 percent of total DNA, was available for testing. Given the battered condition of the DNA (it was clear from the analysis that the DNA molecule had been reduced to very short fragments), there were very few testing options available. Jeffreys and Hagelberg, therefore, proceeded with the same methodology that they had used in the Karen Price case and targeted polymorphic microsatellite loci. This time, however, they increased the number of microsatellites analyzed to enhance the discrimination power of the overall test. In particular, they used five of the six loci they had used in the Price case and five new ones, all with relatively high heterozygote genotypes and with alleles that are rare in Caucasians.

Microsatellites were another form of repeating segment that DNA scientists were then finding scattered throughout the human genome. While longer VNTRs (variable number tandem repeats) remained the leading method for distinguishing one person's DNA from somebody else's at the time, there were already instances when these shorter genetic repeats had been utilized by forensic investigators endeavoring to make an identification with only small quantities of degraded DNA available. In fact, the remains of some American soldiers killed during Operation Desert Storm were identified using similar types of these short repeating segments in 1991.

The repeating segments targeted by Jeffreys and Hagelberg are extremely short segments of the genome and far less discriminating, on an individual basis, than the much larger variable number tandem repeats targeted in RFLP testing. Whereas some VNTRs contain repeat segments of 30 or more nucleotides, these shorter tandem repeats range from between two to seven nucleotides in each repeating segment and occur in limited allelic variations. Their ultimate lengths, then, will only run to hundreds of nucleotides, as opposed to the many thousands of nucleotides that comprise RFLP VNTR alleles. But since their ultimate lengths are shorter, they are ideal for cases like Mengele's where the DNA has been subjected to environmental insults over time. Whether long or short, the distinguishing characteristic of a genetic repeat is how long the string of repeating segments ultimately gets, or its allelic size. Therefore, these shorter repeat segments (short tandem repeats, or STRs, as they later became known) are ultimately distinguishable by reference to the same type of sizing ladder used to differentiate between the longer VNTRs. On the other hand, since they do not vary as much in length as their longer counterparts, STRs are not as discriminating.

It was for this reason that Jeffreys and Hagelberg ultimately decided to analyze a total of ten of these short repeating, or microsatellite, loci, thereby increasing the statistical likelihood that the overall DNA pattern would be unique. Just as in the Karen Price case, where six such loci were analyzed, the sites targeted here were based on polymorphic dinucleotide repeats (that is, the two bases CA repeating over and over). While all were premised on the same CA repeat, each loci was known to inhabit a different locale across six chromosomes of the human genome. Chromosomes 1 and 9 each harbored two of the loci targeted, but these loci were far enough apart so as to avoid the problem of linkage disequilibrium. In other words, the individual loci, despite being on the same chromosome, were randomly inherited, not joined at the hip, as it were, genetically. In fact, all of the loci are known to follow the same inheritance rules that Mendel laid down for genes that express themselves, even though the dinucleotide repeats do not code for anything. These short tandem repeat units are stable and maintain a constant length within an individual.

As with the Price case, Jeffries intended to run a paternity case in reverse. The plan was to compare the polymorphic characteristics of the extracted DNA against the individual's putative offspring, Rolf (who now goes by the name Rolf Jenke), cross-referenced with the DNA characteristics of Rolf's mother, Irene. Unfortunately, the analysis was stalled when Rolf and Irene refused to provide blood samples for comparison. Eventually, in the winter of 1991/92, Rolf and Irene had a change of heart and agreed to cooperate. In February 1992, Jeffreys received their blood samples. All amplified alleles were compared by agarose gel electrophoresis and then stained with ethidium bromide for visualization. Based on published Caucasian allele frequencies for all the microsatellite loci, they estimated that only one male Caucasian in every 36,000 would by chance have a genotype consistent with paternity for Rolf. On April 8, 1992, Jeffreys and Hagelberg reported that "the skeletal DNA has a consistent genotype compatible with the father of Rolf, and that [more than] 99.9% of Caucasians unrelated to Rolf would be excluded from paternity by this analysis." None of the paternal alleles in Rolf's DNA were absent from the Embu DNA. In sum, they said, "the skeletal remains are beyond reasonable doubt those of Josef Mengele."

Based on these conclusions, the state prosecutor of Frankfurt announced the same day that he would ask the Hesse state court in Frankfurt to close the Mengele case. Also that day the Israeli Justice Ministry announced that "all reasonable doubt was [now] removed, and it is possible to determine that Josef Mengele . . . died in 1979."[108]

Even the most vocal skeptics were convinced by the DNA tests. What remained was only a sense of bewilderment, as Anti-Defamation League Director Elliot Welles said at the time, that "so well-known and notorious a Nazi could have lived so long in freedom."[109]

Another great mystery of the Nazi era was also finally put to rest through DNA testing in 1998. At the conclusion of World War II in 1945, most of the higher-echelon Nazi figures were accounted for as either dead or captured. Conspicuously unaccounted for was Martin Bormann, the second most powerful man in the Third Reich, who served as Hitler's private secretary, close confidant, and principal aide. Even though Bormann could not be located following the war, his complicity in the Nazi horrors had been so extensive that he was put on trial in absentia for war crimes in 1946 by an international military tribunal in Nuremberg, convicted and sentenced to death. In the decades immediately following the war, rumors abounded about the fate of Bormann, and there were numerous reported sightings of him living well in various parts of the world. Most believable, perhaps, were the rumors that he had fled to South America and was living in either Argentina, Brazil, Chile, or Paraguay. In 1972, human remains were unearthed at a Berlin construction site that were established, primarily through dental records and skeletal comparisons to Bormann's known medical history, to be those of Bormann, according to forensic investigators at the time. According to these forensic experts, Bormann apparently committed suicide by taking poison on May 2, 1945, just as Soviet troops were advancing toward Berlin. Despite the identification, the rumors that Bormann was still alive persisted well into the 1990s. In 1993, for example, a Paraguayan newspaper reported that Bormann had lived in that country after the war and was buried in a small town outside Asunción following his death in 1959. In 1996, a British book asserted that the British secret service had smuggled Bormann out of Germany to England immediately after the war to help locate the stockpiles of Nazi gold. In an effort to put such rumors to rest once and for all, German authorities decided to conduct forensic DNA tests on the remains found in 1972. Living relatives of Bormann readily agreed to give a needed blood sample because they, too, were anxious to end the wild speculation about Bormann's whereabouts. The crucial blood sample was supplied by an 83-year-old living niece of Bormann's. Finally, in 1998, Professor Wolfgang Eisenmenger, director of Munich University's Institute for Forensic Medicine, concluded—based on mitochondrial DNA tests—that the DNA results showed "beyond any doubt" that the skeletal remains found in 1972 were those of Martin Bormann. In 1999, the remains were cremated and scattered in the Baltic Sea.[110]

Demonstrations such as that performed by Alec Jeffreys and Erika Hagelberg in the Karen Price and Josef Mengele investigations, the successful targeting of the lowest common denominator of genetic repeat segments and using them to confirm an identification, was compelling to forensic investigators for a number of reasons. While other PCR-based tests had demonstrated that they could break the forensic equivalent of the sound barrier, this innovation appeared to break the light barrier. Now investigators could extract highly discriminating genetic information from biological evidence in cases where the DNA was either minimal or highly of degraded. Furthermore, now investigators could re-examine older, unsolved cases (known in the trade as "cold cases") that time and the earlier limits of technology had forced them to write off as unsolvable. Now there was hope.

8   Less Is More

IN 1991, THE SAME YEAR THAT JEFFREYS AND HAGELBERG WERE COMPLETING their Mengele analysis, Dr. Thomas Caskey, a highly regarded professor at the Baylor College of Medicine, University of Texas, and several colleagues published an article suggesting the use of short tandem repeats (STRs)—although they trained their sights on slightly longer short repeating segments—for forensic DNA analysis in criminal casework.[1] Subsequently, companies like Perkin-Elmer Corporation (which later spun off PE Biosystems), in collaboration with Roche Molecular Systems, began the independent development of various commercial kits for forensic work using STR typing. By the end of the decade, STRs sparked yet another revolution in forensic DNA analysis.

In the years leading up to and after the Mengele DNA analysis, scientists were discovering and investigating a bounty of these short repeating segments of DNA that were turning up throughout the human genome. With an eye toward determining the suitability of some of these segments for forensic work, forensic scientists from around the world began the exacting process of running these shorter pieces of DNA through the paces of forensic DNA analysis on an experimental basis. These genetic oddities had to demonstrate their reliability as forensic DNA markers through a series of rigorous validation studies that examined, among other things, how discriminating they were, what—if any—interpretive difficulties arose when mixtures of DNA were analyzed, the effect of environmental insults (from such things as the dye in denim, which had long been known to inhibit PCR), and whether the STR alleles under study were being inherited independently in accordance with the rules of Mendelian inheritance. Since STR typing is heavily reliant upon PCR, the scientists also looked at what level of differential or preferential amplification occurs during the PCR process—meaning, in the first instance, that one locus within a given set of loci amplifies differently, and, in the second, potentially more serious, instance, that one allele is shown to amplify preferentially over another at the same loci, which can lead to a mistyping—and the degree to which results can be obtained from highly degraded DNA.

In the end, forensically suitable STRs were found among the repeat segments ranging from four to seven bases in length. They come with a moderate to high degree of polymorphism (between 5 and 20 alleles per loci). Although the alleles of STRs vary by size, in the same way that the longer RFLP loci are differentiated, they are a great deal shorter and, very significantly, occur in only so many length variations (or discrete alleles), making DNA comparison by gel visualization easier and more predictable. Because STRs are relatively short compared to the much longer VNTRs, PCR primers could be designed that would produce segments of DNA between 100 and 500 base pairs long.[2]

Such brevity was the order of the day for the pioneers of ancient DNA such as Svante Paabo and Allan Wilson and something they had to live with rather than rejoice in. As things were turning out, however, forensic scientists were discovering that less really is more. STR validation studies were conducted in academic settings, in the private sector, by the FBI's Forensic Science Research Unit, in various state and local forensic science laboratories around the country, and under the auspices of the Technical Working Group on DNA Analysis Methods (TWGDAM), which had long established guidelines for evaluating novel forensic DNA techniques. In 1999, the name was changed from TWGDAM to SWGDAM, for Scientific Working Group on DNA Analysis Methods. SWGDAM is a federally sponsored, independent research group composed of both private and government forensic DNA specialists that is universally acknowledged as an authoritative voice in the field; its success ultimately spawned a number of similar groups around the country, known now as regional SWGDAMs. The FBI's validation studies, always the more deliberative in matters of novel DNA techniques, continued well into 1998. As has typified the use of forensic DNA techniques from the beginning, laboratories in the private sector took the lead in offering the newest technique as evidence in court. Cellmark Diagnostics, for example, augmented its DQA1 and polymarker tests in 1995 with a commercially available STR kit in a Massachusetts murder case.[3]

In that case, Adam Rosier was put on trial for the brutal 1994 murder of 16-year-old Kristal Hopkins. After attending a drinking party, Hopkins had been seen leaving the party in Rosier's borrowed car, a Ford Mercury Capri, with Rosier at the wheel. The following day, hunters found Hopkins in the Pittsfield State Forest, barely alive. She was only partly clothed, covered in blood, and had sustained severe injuries to her face, head, neck, torso, and legs, which included what were later described as "road burns." Two hours after being transported to a hospital, Kristal Hopkins died from what the medical examiner would later describe as large blunt trauma consistent with having been struck, run over, and dragged by an automobile from two different directions. The next day Rosier admitted to a friend, according to testimony at the trial, that he had had a violent argument with Hopkins after the two had sexual intercourse and "to shut her up," as Rosier was quoted as saying, he first "stomped" her and then drove his car over her, first backwards and then forwards. Blood was later recovered from a tire, from the undercarriage of the car, and from inside the car on the passenger's side. What appeared to be human tissue was also discovered adhering to the car's undercarriage. Cellmark Diagnostics was subsequently hired by the prosecution to conduct DNA tests on the evidence blood and tissue, along with blood samples from the victim and the defendant. (The company has since changed its name to Orchid Cellmark and is now a division of Orchid Biosciences, Inc., of Princeton, New Jersey.)

Preliminarily, Cellmark analysts found that there wasn't enough blood to conduct RFLP testing. In the first round of tests, therefore, Cellmark performed PCR-based DQA1 and polymarker tests, which didn't exclude the victim, but didn't exactly nail the case shut either. The population frequencies generated by this combination of tests indicated that, for Caucasians, the probability that the blood originated from someone other than Kristal Hopkins was 1 in 5,000. To augment the evidentiary value of the DNA tests, Cellmark turned to STR testing. By this time Cellmark had already performed STR analysis in about 50 cases and been permitted to testify in about five criminal cases. (The majority of DNA tests, of whatever type, either exonerate the principal suspects or tend to inspire guilty pleas, so the issue of the test's validity never gets to court.) Cellmark had even been involved, back in 1991, using some of the earliest-known STR markers, in the identification of the remains of soldiers who had died in Operation Desert Storm.

In the Massachusetts case, Cellmark used the GenePrint STR System developed by Promega Corporation, involving the co-amplification of three STR loci. Co-amplification (or multiplexing) means that the DNA regions of interest are all amplified together in the same test tube using three separate primer sets for each loci. Aside from the obvious benefit of reducing the clutter of test tubes in the laboratory, multiplexing has the added advantages of saving time, minimizing the risk of contamination, and conserving sample DNA.[4] (An STR kit subsequently made available amplifies an impressive 16 STR markers, all at the same time in one test tube.) The three loci targeted by Cellmark were known to have, respectively, nine alleles (giving rise to 45 possible genotypes), seven alleles (giving rise to 28 possible genotypes), and eight alleles (giving rise to 36 possible genotypes). Before the trial, an evidentiary hearing was held to determine the validity of Cellmark's DNA tests and the statistical calculations it would offer into evidence, after which (based on the expert testimony of Dr. Charlotte J. Word, of Cellmark, and Dr. Christopher J. Basten, a population geneticist, from North Carolina State University) the trial judge ruled that the DNA tests and calculations met both the federal and state admissibility standards for novel scientific evidence.[5] An important consideration was the 1996 publication of a validation study of the GenePrint STR System.[6] (Since then, the majority of courts that have considered the validity of STR typing have ruled it admissible, but a few trial courts have—for various reasons—declined to allow it.[7])

Based on Cellmark's STR DNA database derived from its own paternity studies (which had been cross-checked against three other databases by Dr. Basten), Cellmark reported at trial the combined statistical significance of the battery of DNA tests it had conducted, as enhanced by the STR typing. Now the chances of a random match were 1 in 770,000, a far cry from the earlier odds. (Note that 770,000 is roughly the population of Boston.) In the end, Rosier was convicted of murder and sentenced to life without parole, and the conviction was upheld by the Supreme Judicial Court of Massachusetts.

Whether STRs or any other DNA markers or combination of tests are used, the ultimate question in court becomes, as it did in the Rosier case: What's the bottom line (assuming the suspect hasn't conceded the DNA identification and offered an explanation for the presence of his DNA)? What do all of these tests add up to, numerically? What are the odds of a random match with the DNA profile produced? For years, this was one of the most contentious debates in forensic DNA analysis.

On one side of the issue were a few population geneticists who claimed that the statistical calculations accompanying DNA matches were suspect because of the phenomenon of substructuring, meaning that there are certain smaller, genetically isolated subpopulations (or subgroups) existing within the various larger populations of Caucasians, Hispanics, Blacks, Asians, or other ethnic groups, each of them having unique genetic characteristics because they tend to marry chiefly among themselves. According to some critics of forensic DNA analysis, substructuring alters the distribution of genetic markers for those subgroups, as compared to the general populations, and erodes the confidence that can be placed in the genetic frequencies that DNA laboratories use to make their statistical calculations.[8] On the other side of the controversy were the majority of population geneticists and DNA experts who were convinced the substructuring argument was overblown.[9] Dozens of research studies published in the 1990s have only reinforced the latter view that the statistical calculations used in forensic DNA analysis are valid.[10] In late 1994, two of the country's leading adversaries on the utility of the ceiling principle, Dr. Eric Lander of the Whitehead Institute for Biomedical Research and Dr. Bruce Budowle of the FBI's Forensic Science Research and Training Center, jointly published an article in *Nature* declaring that the "DNA fingerprinting wars are over."[11]

The statistical measure of a DNA match, of course, can be a crucial component of a criminal prosecution. In 1996, the National Research Council gave its blessing to the practice of expert witnesses declaring that, under certain circumstances, a DNA match truly was a match to a specific defendant or individual, assuming a sufficient number of VNTR loci—nine was recommended—had been analyzed. In other words, the odds that the DNA was anybody else's were determined to be as close to zero as a statistician can get under these circumstances. The issue, known as "uniqueness," was relatively new. With STR markers, the issue has surfaced again, because, especially when as many as 13 (or even 16) STRs are analyzed to create a DNA profile, the odds of a random match often reach one in several hundred millions of times the population of the Earth, let alone that of the United States. Given such staggering numbers, the FBI determined in July 2000 that their experts would be authorized to testify about making a "source determination" and declaring with 99 percent certainty that no other individual in the country could have contributed the DNA in the evidence sample. [12]

Few developments relating to forensic DNA analysis have engendered as much debate as the ever increasing use of DNA databases as a means of solving crimes. DNA databases exist on both the state and federal levels. As of the year 2000, all 50 states had passed laws authorizing the collection of DNA samples from various classes of criminal offenders. Between 1989—when the collection started—and 2000, state DNA repositories had amassed more than 750,000 DNA samples from convicted felons. [13] As appellate courts generally ruled that no abridgement of convicted felons' constitutional rights occurred by requiring (or even forcing) them to contribute a DNA sample to the databases, the reach of the various state laws continued to expand.

There were, of course, occasional setbacks for proponents of the DNA databases, even before the issue was heard on the appellate level. In Massachusetts, for example, which was the 48th state to pass a law—in 1997—creating a DNA database, a lower court judge ruled in 1998 that the law violated not only the Fourth Amendment's restrictions against unreasonable searches and seizures but also similar provisions of the Massachusetts Constitution. In 1999, however, the Massachusetts Supreme Judicial Court overturned the lower court and reinstated the DNA database law, declaring, "The high government interest in a particularly reliable form of identification outweighs the minimal intrusion of a pinprick." [14] By this time, all 50 states had such laws (although a few had yet to begin collecting anything), but the decision was nevertheless decried by civil libertarians. John Reinstein of the Civil Liberties Union of Massachusetts, for example, which originally brought the lawsuit on behalf of seven inmates, responded: "I think the court has trivialized the dangers that are inherent in DNA testing." [15]

The trend among the states has been to collect more and more samples and expand the types of crimes that fall under their respective database laws. (The term databank is broader than the term database, although they are often used interchangeably. Database refers specifically to the computer repository of the DNA-related information, whereas a databank is a more global collection of information including such things as offender names and criminal histories, DNA-related computer software, sample collection methods, and authorizing legislation. [16]) The largest repositories exist in Virginia, California, Alabama, and Florida, in that order. By the end of 1999, the nation's first DNA database in Virginia had already amassed more than 190,000 DNA samples on its own.[17] For law enforcement, the benefits of such databases are clear: When the DNA profile obtained from a crime scene is processed, and no suspect has been immediately identified, the crime-scene DNA data can be entered into the computer catalogue of the DNA samples already on file. With any luck, there will be a match, or so-called cold hit. This is just what occurred in early 1999 in Connecticut based on its 1,800 DNA database profiles. When a New Haven–area rape investigation seemed to lead nowhere, a DNA cold hit led investigators to 33-year-old Eric Jones, whose DNA was on file from a previous conviction, and he was arrested. This was the second such cold hit in Connecticut.[18] By this time, the cold hit scenario had become almost a matter of routine in Virginia. Around the country, according to the FBI, by mid-2000 there had been about 1,300 such cold hit scenarios derived from state DNA databases.[19] By October 2002, the national DNA database included almost 1.2 million DNA samples from convicted offenders around the country and another 39,000 from ongoing criminal investigations. By this time, 5,700 cases had been solved through hits on the national DNA database. Virginia, with its early lead in the DNA database business, accounted, all by itself, for a sizeable percentage of these. In November 2002, Virginia authorities announced that they had made their 1,000th DNA cold hit.[20]

Although the volume of DNA samples has been accumulating from earlier versions of the DNA database laws, a number of states nevertheless continued to expand the categories of crime covered by their laws. At the same time, some states began to balk at expanding their current laws to embrace additional crimes. By mid-2000, all of the 50 states required convicted sex offenders (except, perhaps, for the lowest level of such crimes) to deposit DNA samples for their databases. Forty states had added to the list people convicted of offenses against children, 37 states included murder, 28 assault and battery, 25 attempted felonies, 22 kidnapping, 19 robbery, 18 burglary, and 24 states included juvenile offenders as subject to the laws. Seven states, however, had gone so far as to require that all people convicted of a felony, of whatever category, deposit a DNA sample in their state's database. These states included Alabama, Georgia, New Mexico, Tennessee, Virginia, Wisconsin, and Wyoming. States that had recently declined to take this last step include California, Washington, and Rhode Island, where bills failed on the issue. Even less inclusive expansion bills failed in Hawaii and Pennsylvania.[21] Conservatives and civil libertarians alike have voiced their opposition to governmental efforts to expand DNA databases. Gun enthusiasts, for example, already under pressure from expanding gun registration laws, have voiced concerns that prospective gun buyers may soon be required to submit DNA samples as part of a background check. Civil libertarians are also becoming more vocal.[22] Nevertheless, the trend toward expanding the list of crimes falling under DNA database laws continues at a rapid pace. By August 2002, 23 states had passed legislation requiring that all felons submit a DNA sample for inclusion in the database, more than tripling the number of such laws in only eighteen months. It is predicted that by 2007, 45 states will have passed similar laws. In addition, two states, Arizona and Virginia, had passed laws by May 2002 authorizing the taking of DNA samples from arrestees.[23]

"We've moved quickly from [testing for] crimes that DNA is most useful in solving to folks like embezzlers, whose crimes have little to do with it," ACLU associate director and privacy law specialist Barry Steinhardt told *USA TODAY*. "People are beginning to ask: Do we really want government to have all this stuff?"[24]

These concerns are accentuated, according to critics of expanded databases, when some of the latest technological advances are considered. DNA microchip technology, for example, which is expected to be available soon, may allow near instantaneous analysis of samples, much of which is predicated on STRs. Single nucleotide polymorphisms (or SNPs, pronounced "snips"), which identify a single nucleotide difference between people and are identified through DNA sequencing techniques, are also being investigated for possible uses in a forensic context. One such SNP test currently being heralded by the law enforcement community can identify the genetic trait for red hair (with a 90 percent accuracy rating), and other genetic tests identifying physical characteristics (for hair, eye, skin color, and perhaps even baldness) appear around the corner. Forensic DNA tests for ethnicity have already been in use for some years (and ethnicity can, to some degree, be inferred from RFLP or STR tests), and biotechnology companies are now aggressively seeking to identify SNPs that correlate to ethnic background in the hopes that these minute genetic distinctions will assist in prescribing appropriate medical treatments for patients with genetic diseases or genetically related drug reactions; others fear that such information might lend itself to abuse.

Compounding these problems are instances of what some have denounced as overzealous police work. In several states, including Florida, New York, and Massachusetts, police investigators have obtained DNA samples from suspects without the benefit of a search warrant or court order. In these situations, the tactic allowed analysts to link the suspects to various crimes by virtue of the DNA profile. However, the police simply took DNA samples from biological specimens left behind by the unsuspecting suspects, such as spit from the sidewalk or saliva from a coffee cup or cigarette butt left behind after an interrogation. No search warrants were obtained before securing the DNA sample. While law enforcement spokesmen liken the situation to the lifting of fingerprints, others have challenged the tactics as far more intrusive.

Even for advocates of expanded DNA databases, there are practical problems. Perhaps the most acute is the enormous backlog of DNA samples that have been deposited under the database laws but, for lack of funding and due to limitations of personnel, have yet to be analyzed. Because of the increased workloads, some states, including California and Florida, have turned increasingly to robotics to more quickly process the glut of DNA samples.[25] By early 2000, there were estimated to be 350,000 such samples in cold storage around the country, with the problem growing worse because of the expansion of categories of offenders required to make a deposit. In New York, for example, just as the state laboratory was polishing off the last of its DNA backlog in late 1999, new DNA database legislation kicked in that expanded the crimes covered by the law from 21 to 107, automatically requiring the DNA analysis of an additional 100,000 samples from people convicted of those crimes.[26] More alarming still are the approximately 180,000 rape kits around the country that remained unanalyzed, providing no hope that the DNA evidence from those cases would be connected to registered offenders any time soon.[27] The solution to all of these problems has come in the form of increased state and federal funding (for example, a $1.25 billion grant program was established to assist states under the 1998 federal Crime Identification Technology Act), with much of the actual DNA work being subcontracted to private DNA laboratories.

To further complicate matters, the technology suddenly changed. Based on the continued success of the PCR-based STR system in forensic DNA analysis validation studies, and due to the speed with which DNA samples could be processed by the newer STR method, the federal government established the STR system as the new standard for its overarching, national DNA database system. In January 1999, the FBI mandated that state laboratories switch over to the STR system. Since the state laboratories are subject to federal mandates, guidelines, and standards established by the Department of Justice's DNA Advisory Board (DAB), SWGDAM, and the National Institute of Justice (NIJ), the decision to establish STRs as the new standard system for DNA typing effectively required the states to convert all of the earlier samples analyzed by the RFLP method to the newer STR system, in addition to adopting the new system for future DNA cases. The conversion process will take years. However, since the analysis of STRs is much faster and far cheaper than DNA analysis under the RFLP method, the conversion is far from impossible. Moreover, the 1999 federal Violent Offender DNA Registration Act will provide tens of millions of additional dollars to states to assist the conversion, in addition to requiring the collection of DNA samples from about 15,000 federal prisoners and the myriad more on probation, parole, or some other form of supervision. Nevertheless, the conversion problems remain a little confusing, particularly after DAB recommended in 2000 the inclusion of additional STR markers beyond the ones recently established as the STR standard.[28]

On October 13, 1998, the FBI launched the National DNA Index System (NDIS), which is the capstone of the national DNA database system begun in 1990 as a pilot program. In 1990, the FBI set up the Combined DNA Index System (CODIS), using a software package designed to digitalize the genetic profiles of entered DNA samples and facilitate computerized cross-checks. Originally available to 14 state and local laboratories, by September 2001 NDIS was being used in 116 laboratories in 37 states.[29] NDIS was designed for use by all 50 states.

In the final level of the CODIS system, NDIS employs a specific set of 13 STR loci for use in its national repository of DNA profiles. Participating forensic laboratories can submit the computerized data derived from the STR tests, as reduced to specific numbers, and search the database electronically on an interstate level. The combination of these 13 loci, according to the FBI, will permit analysts to statistically calculate matching odds on par with, or even better, than those producible with RFLP. A 2000 report by the DAB predicted the use of 20 STR loci in the years to come, which caught some off guard.

The director of Virginia's Division of Forensic Services, Dr. Paul Ferrara, commented at an April 2000 meeting of the National Commission on the Future of DNA Evidence that the decision to convert to the 13 STR loci as the standard was "an extremely expensive decision." In fact, he said, referring to his own state's efforts to type a huge backlog of database samples that had piled up in Virginia, "they dragged me kicking and screaming into the 13 loci because it doubled the cost." As for the new 13 STR loci, Dr. Ferrara added, "I sure as heck don't want to see those added to or changed in the next 10 to 20 years."[30]

Because the STR system is PCR based, the processing time for an STR DNA profile has been reduced to a matter of days from the weeks or months a complete RFLP testing can take. Other innovations currently being investigated, such as laser desorption mass spectrometry, may cut the time to minutes, and microchip technology may reduce further. The NIJ—the research arm of the DOJ—is currently funding efforts to develop credit-card size DNA analyzers for use at crime scenes based on microchip technology that seeks out STR sites. The DNA profile thus generated will enable investigators to instantaneously search the NDIS. Research continues at Nanogen, a San Diego company, and the Massachusetts Institute of Technology's Whitehead Institute in Cambridge, both recipients of NIJ grant money.

CODIS uses two indexes to generate investigative leads from biological evidence. The Convicted Offender Index contains the DNA profiles of convicted offenders, and the Forensic Index contains the DNA profiles derived from crime-scene evidence. The computer software scans the separate indexes for matches and flags matching profiles in the Forensic Index because they signal serial crimes. During a three-year experimental phase of the database program involving 17 states, the database linked 193 crime scenes to the DNA of convicted felons. While the conversion to STR technology is taking place, database searches will continue to be available based on pre-existing RFLP profiles.

The first NDIS cold hit occurred about ten months after the system went online and after 14 states had signed on to the new national database system. (By this time, however, NDIS had been instrumental in solving hundreds of other crimes.) In July 1999, with about 180,000 DNA entries, Florida investigators announced that a dead man lying in the medical examiner's office in Jacksonville was genetically linked through NDIS to a series of rape cases in both Washington, D.C., and Jacksonville. The man, 38-year-old Leon Dundas, who had been killed in an as yet unsolved crime, had been questioned the previous year by investigators about the Jacksonville rapes. At that time, however, Dundas declined to volunteer a blood sample for DNA testing, and he had never been required to contribute a DNA sample to a DNA database. Now he was in no position to argue. Following a round of DNA tests, it was determined that the DNA from Dundas matched the crime-scene DNA from eight rape cases in Washington, D.C., and three in Jacksonville, although only one of the Washington rape cases had been entered into NDIS at the time of the cold hit.[31]

A new twist to the DNA database debate was added in late 1999 when authorities from around the country, including those in Wisconsin, New York, and California, began the novel practice of swearing out warrants for unknown individuals based solely on genetic profiles obtained from crime-scene evidence. Pennsylvania has since issued similar warrants. The practice was begun in an effort to beat the ticking clock of the statute of limitations for the crimes charged, usually rape. In California, for example, the authorities filed a warrant just in time to avoid the lapse of the statute of limitations in a six-year-old rape case. Without much in the way of concrete knowledge about the suspects, the authorities simply identified the "John Does" by reference to the genetic profiles, including STR markers. The practice has provoked court challenges after arrests were made (usually because of "hits" on the states' DNA databases) due to concerns by civil libertarians and the defense bar that the tactic is a ruse to circumvent existing statutes of limitation. The first-ever conviction arising from a John Doe/DNA warrant occurred on February 6, 2002, in Milwaukee following the trial before a judge of Bobby Richard Dabney, Jr., for the 1994 kidnapping and rape of a high school sophomore. Dabney was later sentenced to 120 years in prison and plans to appeal.[32]

Originally, only the worst offenders (rapists initially and then murderers), after a felony conviction, were required to provide DNA samples for inclusion in a state or federal DNA database, on the theory that any new offense would be immediately detected by comparing a DNA profile obtained from a crime scene with all of the profiles in the DNA registry. More recently, some states have included even categories of misdemeanors under their DNA registry laws, and Louisiana law authorizes the inclusion of DNA profiles in its databank for all people merely *arrested* for a felony sex offense. In Britain, which has the most sweeping national DNA database, anyone arrested for any crime that could lead to a prison term if convicted must provide a DNA sample for inclusion in the national DNA database; if the person is not convicted, the DNA profile is removed from the database.[33] In January 2003, Virginia became the first state to implement a DNA-collection scheme modeled on the British law, augmenting Virginia's pre-existing, and prodigious, DNA databank for convicted felons; the new Virginia law authorizes the taking of a DNA sample from people accused of a violent felony *upon their arrest*. As in Britain, the DNA sample will be destroyed if there is no conviction.[34] A number of officials around the United States have proposed taking DNA from all people arrested for any crime serious enough to require processing at a station house. Former New York City Mayor Rudolph Giuliani seriously proposed that all babies born in New York State have their DNA automatically registered with the state. He is not alone.

In 2001, three highly influential legal scholars—all serving as members of the National Commission on the Future of DNA Evidence, University of Wisconsin law professor Michael E. Smith, Arizona State University law professor David H. Kaye, and University of California-Davis law professor Edward J. Imwinkelried, began publicly advocating "a single, population-wide" DNA database for the United States. Since cold hits from DNA registries are now a matter of routine, they argue, a universal DNA database would clearly be an effective weapon against crime. At the same time, they submit, making a DNA database universal would eliminate the specter of a racially skewed national DNA database. According to Smith, Kaye, and Imwinkelried: "The Bureau of Justice Statistics reports that a black man is six times more likely to be imprisoned during his life than a white man. Reasons range from prejudice to race-neutral deployment patterns that lead to more aggressive policing in minority communities. Whatever the reasons, it seems clear to many that racial minorities are unfairly over represented among arrestees and convicts. Given this, the construction of huge convict or arrestee DNA databases could exacerbate the racial divisions that plague us."[35]

Currently, a blood or saliva sample used as a source of DNA that is destined for a DNA database is retained and refrigerated by law enforcement agencies. DNA databases, whether state or federal, focus on the variable repeating sections of the genome. The repeating segments of DNA used in forensic DNA analysis are noncoding, meaning they do not translate into meaningful genetic information. At the same time, there are DNA tests currently in use, as well as others in the pipeline, that target actual genes (HLA DQA1, for example). Nothing currently prevents investigators from analyzing a retained blood sample for other genetic markers. Civil libertarians worry that, in the future, DNA might be probed—by police or perhaps as part of a broad population study—for such things as genetic predispositions to certain behaviors or even medical information. Already in study are population-specific DNA markers that can—with some reliability—identify a suspect's race or ethnicity, although the conclusions are based on probabilities. Moreover, some STR types are more prevalent in certain population groups, allowing investigators to conclude, based on probabilities, that a particular DNA profile came from one ethnic group rather than another. By 2010, according to a report published in November 2000 by the National Commission on the Future of DNA Evidence, "there should be a number of markers available that identify physical traits of the individual contributing the DNA," which is expected to help police narrow their search for a suspect.[36]

The universal DNA database envisioned by Smith, Kaye, and Imwinkelried would focus exclusively on the noncoding regions of the genome and take law enforcement out of the business of DNA sample storage. "Law enforcement authorities would not need to—and should not be permitted to—handle or retain the samples," they write.[37] The idea is that the samples would be genetically typed elsewhere, perhaps by health professionals, for the noncoding regions of the genome and that that data, and that data alone, would be submitted to the national DNA database.

At present, many issues arising out of the ever-expanding national DNA database remain unresolved. Thus, many people continue to worry about the sanctity of the DNA information in the hands of the government and the possibility that the information may become accessible to outside parties, such as insurance companies or employers, or be abused by unscrupulous law enforcement personnel. As laws have been stiffened to more broadly protect medical privacy, these fears seem to have diminished, but these laws do not apply specifically to law enforcement agencies.[38] In any event, laws concerning medical and genetic privacy are complicated and continually evolving.[39] Suffice it to say, Americans generally seem to value their genetic privacy, just as they do their privacy overall. In several instances, employers who performed unauthorized genetic testing on their employees have attracted both legal action and general opprobrium. Lawrence Berkeley Laboratory in San Francisco, for example, ultimately paid $2.2 million in 1999 to settle claims brought by thousands of workers who were unknowingly subjected to tests for sexually transmitted diseases, pregnancy, and various genetic conditions (such as sickle-cell anemia) as part of pre-employment screening.[40] When it was learned that Burlington Northern Sante Fe Railroad, based in Fort Worth, had performed unauthorized genetic tests on employees who had filed disability claims for carpal tunnel syndrome (injuries related to repetitive physical motions of the hands), the U.S. Equal Employment Opportunity Commission (EEOC) sued the company in federal court for violating the Americans With Disabilities Act (ADA). A court settlement was reached in April 2001, and the company admitted conducting (or planning to conduct) the tests on 35 employees to determine if the claimants had a genetic predisposition to the condition.[41]

Complicating DNA privacy concerns—beyond the national DNA database—is the evident boom in recent years of private and governmental biorepositories. Private laboratories, academic researchers, local law enforcement agencies, private detective bureaus, hospitals (which often retain blood samples drawn from newborns for genetic screening), and even funeral homes and autopsy companies, all are accumulating burgeoning DNA collections for their own purposes, often without the knowledge or consent of the DNA contributors. This is an extremely worrisome trend to many observers, including Alec Jeffreys, who told *U.S. News & World Report* in late 2002, "I feel pride that [DNA fingerprinting] has spread so far so fast—and twinges of alarm that it's starting to get out of hand."[42]

Unquestionably, the recent gains in knowledge about the human genome are a double-edged sword. While the knowledge may be used for benevolent purposes, there is also the risk of abuse. In the forensic context, at least, there is far less concern about the uses of DNA derived from lower organisms, such as plants and animals, in solving criminal cases. Fortunately for investigators, the breadth of knowledge about plant and animal genomics has been expanding right along with that of human genetics. And this has resulted in some unique DNA tests.

Even broad knowledge of plant genetics can translate into forensic DNA tests. Plant DNA evidence was first accepted as evidence in a criminal case in Arizona in 1993. An Arizona appellate court upheld the use of the evidence in 1995 even though the specific analytic technique employed had never before been used in court and the acceptability of DNA (that is, human DNA) evidence generally was something of an open question at the time in Arizona due to reservations by the state's highest court about the methodology used for computing the odds of a random match—reservations that have since been dispelled. By the time of the original trial, the study of plant genetics was a hot field as university and biotechnology laboratories were busily reinventing agriculture. Moreover, the technique involved, known as Randomly Amplified Polymorphic DNA, or RAPD, was not brand new. It was a technique developed in the late 1980s and came complete with a set of testing protocols published by DuPont Laboratories that was an industry standard. RAPD was a widely used technique in research laboratories. It had never been used in court before, however.

The case involved the murder of Denise Johnson on May 2, 1992. Her nude body was discovered early the next morning by a man riding a dirt bike through a remote part of Maricopa County known as the Caterpillar Proving Grounds, so named because the location is where the manufacturer of Caterpillar heavy equipment tested out its products. When investigators arrived, they found Johnson's body lying face down in the brush near a cluster of palo verde trees, one of which had a fresh gash along a lower branch. What appeared to be Johnson's clothing was found scattered around and a path of crushed grass showed that the body had been dragged. Blood on Johnson's body was still wet, some of her fingernails were either bent back or broken, and her body remained at least partially bound. There was a cloth tied around her neck and left wrist, a shoelace around her left ankle, braided wire tied around her right wrist and ankle, and two apparent ligatures draped over her neck, a vinyl strap and a second piece of braided wire attached to a metal ring. It was later confirmed that Johnson had been strangled to death. Near the body, police found a pager that was registered to Earl Bogan, one that turned out to be used almost exclusively by his son, Mark Alan Bogan.

While the crime scene was being scoured, Chad Gilliam approached police with information he thought would help. Gilliam said that at about 1:30 that morning, as he was heading home from a party, he saw a white "dually" pickup truck with amber clearance lights on top of the cab run a stop sign and race away from the proving grounds toward the highway. A "dually" is slang for a heavy-duty pickup truck with four wheels on the rear axle. Gilliam's description fit the pickup truck that Mark Bogan drove, so he was called in for questioning and the truck was seized for an evidence search. During the search, police found two seed pods from a palo verde tree in the bed of the truck. When police questioned Mark's live-in girlfriend, Rebecca Franklin, she explained that Mark had been drinking heavily on the evening of May 2 and had gone out sometime before 11:30 P.M. When he returned about 2:00 A.M., she told police, he had fresh scratch marks on his face, which he explained away as resulting from a bar fight. Franklin later said she saw braided wire attached to a ring in Mark's truck sometime before the murder, but no such wire was found in the truck during the search. After he was sufficiently cornered by the accumulating evidence, Mark Bogan conceded to police that he had picked up a hitchhiker fitting Johnson's description on the evening of May 2, but he adamantly denied killing anyone and said he hadn't been to the Caterpillar Proving Grounds in years. According to Bogan, he and the hitchhiker had sexual intercourse in the cab of his truck and then drove around. As they drove toward the highway, Bogan said an argument erupted, and he stopped to order the hitchhiker out of the truck. Upon leaving, according to Bogan, she stole his pager and wallet, among other things, so he chased her to get back his belongings. It was at this point that the hitchhiker scratched him, Bogan said, and he didn't realize he had failed to retrieve his pager until the next day.

Although the evidence against Bogan was already fairly compelling, Homicide Detective Charlie Norton, who had noticed the tree abrasion at the crime scene, hoped it could be even better. Norton contacted Dr. Timothy Helentjaris, a professor of molecular genetics at the University of Arizona, and asked if it were possible to link the seed pods found in the truck bed to one of the palo verde trees at the crime scene. Dr. Helentjaris took up the challenge and began to extract DNA from the evidence seed pods and the trees at the scene, including the tree with the scratch mark that became known as PV-30. Using DuPont's RAPD procedures, which rely on PCR, Helentjaris compared the DNA from the two seed pods, 12 palo verde trees at the scene of the crime, and 16 other palo verde trees of the same subspecies found in various places in Maricopa County. Significantly, RAPD is a process that can elicit individualizing genetic characteristics from an organism even where virtually nothing is known about the genome of the organism, as was the case with palo verde DNA.[43] The particular subspecies of palo verde tree found at the crime scene is *Cercidium floridum,* which is one of three varieties native to Arizona. With RAPD, short stretches of randomly generated DNA sequence, 10 base pairs in length, are used as primers to seek out complementary stretches of DNA that naturally exist in the tree's genome. Once the primers line up on the genome exactly (with the A's paired with the T's and the G's with the C's), primer pairs are created that essentially mark off sections of DNA that can be measured, like the various bookends of a home library. Using PCR, Dr. Helentjaris created testable batches of DNA that singled out 47 sections of DNA of varying lengths, or markers, that were used to compare the genetic profiles of the individual trees against each other and against the DNA profiles of the evidence seed pods. The DNA from the seed pods was extracted from the pods rather than the seeds in order to get the DNA of the tree of origin, rather than a mix of DNA in the seed from the tree of origin (mom) and the tree that fertilized it (dad). Dr. Helentjaris conducted his DNA tests on the trees "blind," meaning he didn't know until all the tests were completed which sample was which or whether the pod DNA would line up with PV-30 or not. As things turned out, all the trees tested showed distinct genetic profiles and the RAPD profiles of PV-30 and the evidence pods appeared to be an exact match, even to the point where the marker signals (band intensities) coincided. In an effort to bolster confidence in the DNA tests, Detective Norton submitted to Dr. Helentjaris a second sample of PV-30 (along with a batch of samples from area palo verde trees) intentionally mislabeled as originating from just another Arizona palo verde tree unrelated to the crime. When Dr. Helentjaris reported back that there appeared to be a problem—believing that a random match had occurred from an area tree—Norton explained the ruse, and both agreed that the DNA tests appeared to be working perfectly.[44]

At trial, Dr. Helentjaris was permitted to testify that the seed pods "matched" PV-30, but the court would not allow him to assign a numerical value to that assessment because Arizona's highest court had temporarily nixed the admissibility of all DNA statistical evidence regarding the probability of a random match. (Consistent with rulings in a few other states, the trial and appellate courts treated the idea of DNA "matches" just like any other match related to physical evidence, such as footprints or fingerprints, which don't require quantification.) The defense's contentions that the RAPD technique was unreliable and the use of the term "match" misleading under the circumstances were unavailing. A jury convicted Mark Bogan of first-degree murder and sentenced him to life in prison. While the majority of the appellate court, which affirmed the conviction, was not troubled by the use of the word "match," one of the judges was. He felt that the majority's reasoning that DNA matches are the same as other types of physical matches, such as shoe prints or fingerprints, which aren't accompanied by statistical calculations, was flawed. First, DNA profiles are produced by complex scientific processes, he argued, thus are not comparable to the kind of look-and-see similarities that a jury can easily understand when it is asked to compare, for example, one shoe print with another. Second, there was no palo verde DNA database available as a referent, beyond the small one Dr. Helentjaris generated himself for the Bogan case. This didn't seem sufficient to the judge. Given those concerns, the judge would have ruled the plant DNA evidence inadmissible, but, at the same time, he voted to uphold the conviction of Mark Bogan because all of the other evidence in the case persuaded him that Bogan was indeed guilty anyway.[45]

Both in and out of the courtroom, plant DNA has found itself at the center of some significant controversies. One DNA study, for example, shook the wine industry to its foundations when a variation of forensic DNA paternity testing revealed that some of the world's most prized and expensive wines are, in fact, not entirely of noble origins but are instead descended from the wine equivalent of peasant stock. In 1999, a team of researchers at the University of California at Davis, led by Dr. Carole Meredith of the Department of Viticulture (grape growing) and Enology (wine making), where most of California's winemakers are schooled, published the results of their DNA investigation into the backgrounds of hundreds of wine varieties. Sending chills through the hearts of many a wine connoisseur and rattling the French wine-making business, the UC Davis researchers showed that 16 highly prized French wine varieties were actually the descendants of a long-ago cross between one of the royals of winery, pinot, and one of its despised and outcast third cousins, gouais blanc, a variety that the French had long thought was disdained out of existence. Somehow the whole thing seemed vaguely criminal, the equivalent of mixing Moët champagne and Thunderbird, then charging more for it. Officially, gouais (pronounced GOO-ay and derived from a medieval French term of derision) was long banished from French soil, where some of the world's finest wines are grown.[46] Dr. Meredith's DNA-typing technique focused on the short repeating segments (microsatellites) that occur throughout the genomes of the various grape varieties and vary in length (as they do in other organisms). Apart from the species difference, the technique is essentially the same as that used by forensic science laboratories in criminal casework, except that the grape study analyzed more of the repeat segments (20 to 30). The DNA repeat segment data were derived from a database created by Dr. Meredith, along with nine other countries, known as the Vitus Microsatellite Consortium.[47]

In late 2001, Dr. Meredith, working with two researchers from the University of Zagreb in Croatia, Dr. Ivan Pejic and Dr. Edi Maletic, used DNA profiling to solve a long-standing viticulture mystery concerning the origins of one of America's favorite wines, zinfandel, a wine variety that had been in America for such a long time, at least since the 1820s, that it had been effectively naturalized and was deemed America's only native grape. Experts knew the wine had to originate from elsewhere, but where? And from which grape variety—of the more than 10,000 grape varieties in the world—did it derive? There were a number of leading candidate varieties, however, including primitivo, which is grown largely in southern Italy and Sicily but is apparently a transplant, and plavac mali from Croatia, but no one was certain. In 1995, DNA tests conducted by Dr. Meredith showed that primitivo and zinfandel were the same variety, but this only deepened the mystery. Primitivo labels only began to appear in the 1890s, well into zinfandel's adolescence in America. Focusing on the plavac mali theory, Dr. Meredith traveled to Croatia in 1998 to collect samples, which were then genetically analyzed back at UC Davis. Her DNA tests revealed that none of the plavac mali varieties she collected was a match to zinfandel. Over the course of 2000 and 2001, the Croatian researchers continued to collect vine samples from the region around Zagreb with an eye toward finding vines that appeared similar to zinfandel. Eventually, they found a rare variety of red grape known as crljenak kastelanski (the pronunciation of which is obscure because few people have ever heard of it, even in Croatia), which originated in Dalmatia along the Adriatic coast facing Italy, and they sent vine samples to Dr. Meredith. In December 2001, a battery of DNA tests conducted at UC Davis demonstrated that crljenak kastelanski was an exact DNA match to both zinfandel and primitivo, ending the pedigree mystery. Later it was determined that plavac mali is a cross between crljenak kastelanski and the dobricic grape.[48]

There is a dark side to the wine industry, of course, just as there is perfidy in other trades. As a consequence, the DNA profiling efforts at UC Davis and elsewhere around the world can have practical applications in crime solving. To combat widespread wine fraud, which typically involves the sale of inferior wines labeled as fine wines to the public and collectors, scientists at France's National Institute of Agronomic Research announced in August 2002 that they had compiled genetic profiles for 45 of France's grape varieties. The DNA profiles are intended for use by investigators in cases involving allegations that a putative fine wine is really something much less. About the same time, an Australian winemaker, BRL Hardy, undertook to protect and certify some of its finest labels by mixing the DNA of some of their 100-plus-year-old McLaren Vale vines into the printer's ink that appears on their vintage wine's capsules, an idea inspired by the use of DNA-laced security labels on merchandise and memorabilia sold at the 2000 Sydney Olympics.[49]

The study of animal DNA, of course, is also of keen interest to researchers and has had payoffs in resolving research mysteries as well as criminal cases. Among the DNA projects undertaken in recent years involving animals are the Mouse Sequencing Consortium,[50] the Horse Genome Project,[51] the fruit fly sequencing project, and even a rat genome project.[52] Also underway are efforts to map the genomes of most domesticated animals used in American agriculture, as well as to preserve the genomes of endangered species. The United States, in fact, has at least seven "frozen zoos" to store the genetic material of endangered species for research and revival efforts.[53] One DNA study led to further genetic subdivision of the right whale, which had long been divided into two separate species, those in the northern oceans and those in the southern. Right whales, so named because they were dubbed by early whalers as the "right" whale to hunt, possibly because they are one of the few whales that float after being killed, are clearly one of the world's most endangered animals.

Their total numbers collapsed from about 60,000 worldwide in 1900 to an estimated 7,000 at the end of the century, even after a notable rebound. The DNA tests confirmed that the northern and southern varieties of right whale are sufficiently genetically distinct to be considered separate species, but it also concluded that, among northern right whales, those living in the Atlantic and Pacific Oceans are also distinct enough genetically to be classified as separate species.[54]

Whales, of all kinds, are an endangered lot, of course, having been hunted to the brink of extinction because of their commercial value. In 1986 a worldwide ban on whaling was declared by the International Whaling Commission (IWC)—although that may well be lifted in the early years of the 21st century—but several countries, including Japan and Norway, defied the ban. It was in this environment that forensic DNA analysis began to play an increasing role in the trafficking and sale of illegally obtained whale products. Technically, Japan has permission to conduct whaling expeditions for scientific purposes, but many environmentalists have accused the country of using the scientific permits as a cover for ongoing commercial whale fishing. To demonstrate this, scientific sleuths have occasionally lugged portable DNA testing equipment into Japanese fish markets and eateries to obtain food samples suspected of originating from contraband whale meat. Over a period of years, scientists from New Zealand found genetic evidence that a brisk black market exists in Japan for banned whale meat, and that Japanese whalers were killing whale species that are protected from all hunting, even from scientific research.[55] On one occasion, two American scientists, Frank Cipriano and Stephen Palumbi of the Center for Conservation and Evolutionary Genetics at Harvard University, not only found evidence of illegal trafficking in whale meat in Japan, but identified—through DNA identification techniques—a specific protected blue whale that scientists had been tracking since its birth in the North Atlantic in 1965. DNA samples from the raw meat were compared with genetic information on protected whales stored in a genetic database and an exact match was found.[56]

There is, of course, a worldwide black market for rare, exotic, endangered, or simply off-season animals. Some are valued as pets, some as trophies, some for supposed medicinal properties, some as decorative items or clothing, and still others as food. Increasingly, forensic DNA typing is used to solve poaching and smuggling crimes. The first laboratory established in the United States to combat wildlife crimes was the National Fish and Wildlife Service's Forensics Laboratory in Ashland, Oregon, which now routinely uses DNA analysis, along with other forensic techniques, to catch poachers and identify illegally traded items such as bear gall bladders and rhinoceros horns. "We are actually running the equivalent of a homicide investigation," laboratory director Kenneth W. Goddard told The New York Times. "We're almost drawing a chalk line around where the deer was found." Similar laboratories have been established in Wyoming, California, and elsewhere. The techniques used in these investigations are just as sophisticated as those used at forensic laboratories investigating criminal cases, and as the technology improved and matured in criminal cases, so it did in the wildlife laboratories.

A case in Wyoming began to unfold after police received an anonymous tip on the state's "Stop Poaching" hot line reporting that there were six headless antelope carcasses near the city dump. An investigation turned up the name Kenneth R. Nelson of Casper, Wyoming, and police later seized the head of an antelope buck left at a taxidermy shop by Nelson. DNA tests later linked the head to one of the carcasses and a jury convicted Nelson of six counts of wanton destruction in 1998.[57] In another incident, hunter Michael Autry was accused of killing a 240-pound female black bear in her den, which is illegal because such animals are typically pregnant, and then dragging the animal to a checkpoint and falsely claiming to officials that he killed the bear five miles from the den. After another hunter reported seeing the killing at the den, officials took hair and blood samples from the checkpoint and the den and sent them to the National Fish and Wildlife Service's Forensics Laboratory in Ashland, Oregon, which found a DNA match between the two sets of samples. Autry was found guilty, fined $1,000 and faces loss of his hunting license, although he is appealing the conviction.[58]

Sometimes it is not human misbehavior, but animal misbehavior, that is the subject of forensic DNA analysis. "The suspect had already fled the scene by the time a ranger arrived at Yosemite's Curry Village parking lot in the wee hours of March 27," reported *USA TODAY* in May 2001, describing a rash of car break-ins at Yosemite National Park. "But one look at the looted car gave her a good idea as to the culprit's identity."[59] Judging from the telltale dark brown hairs recovered from the car's shattered glass, it was yet another miscreant bear in search of a late-night meal. As to exactly which bear it was, the ranger would have to await the results of the DNA tests done at the National Fish and Wildlife Service's Forensics Laboratory. Since the park had for years been collecting and analyzing the DNA of its resident black bears—in an effort to monitor the estimated 350 to 500 of them living there—scientists had amassed about 250 black bear DNA profiles. It was from this de facto black bear DNA database that park rangers were hopeful they would yield a "cold hit" on the latest "car clouting" case, as it is called by park rangers.

When Yosemite launched a campaign in the mid-1990s to curb bear appetites for the food of campers, park visitors were instructed to secure their food in designated "bear safes" installed throughout the park. Unfortunately, many campers falsely assume their personnel vehicles are as secure as the official storage containers, so they simply leave the food in their cars or vans at night, which is usually more convenient, anyway. Ironically, this makes the food a little more accessible to the bears, too, who have apparently honed their car-breaking skills into an art form and have even learned which cars pose the least resistance (Honda and Toyota sedans are easy pickings, evidently). Furthermore, mother bears are commonly observed teaching their "car clouting" skills to their cubs. In 1997, the Ma Barker of bears and her cubs went on such unrelenting nightly rampages that the whole family was eventually euthanized as a danger to campers. Since then, the standards for euthanizing bears have been stiffened and only the worst offenders are candidates for receiving fatal injections. Since there are so many bear incidents (1,590 in 1999, ranging from "car clouting" to dumpster diving), a harsher standard might well prove to be, as biological technician Kathryn McCurdy put it in 1999, "a good way to kill off all your bears."[60]

On the other hand, a primary concern of park rangers everywhere is the safety of the humans who visit their parks, since bears do occasionally attack people. This is a guiding principle even when the offending bears are on the Endangered Species List. Only five people were killed by bears in Yellowstone National Park between 1839 and 1998 and ten people at Glacier National Park between 1910 and 1998. Over all, about two people a year are killed by bears somewhere in North America. Grizzly bears, which can weigh up to 1,000 pounds, account for half of such deaths, even though their numbers are tiny when compared to the continent's ubiquitous black bear population. Most bear attacks are not fatal but can nevertheless result in severe injury. Attacks by bears are most commonly defensive in nature and typically result from hikers stumbling upon a mother bear with cubs or a bear protecting its food. On rare occasions, however, bears can become predatory around people and see them as a food source. They have been known to drag sleeping campers out of their tents to eat them. Experts who observe bear behavior have developed something of a criminal profile to predict when bears are likely to become aggressive around people sometime in the future. One of the warning signs is when the bear loses its innate fear of humans and becomes what the experts term "habituated" to a lifestyle of foraging for food around human homes, facilities, or campsites, which is why the epidemic of "car cloutings" is so worrisome. At the national parks, such bears even cause what are known as "bear jams," although they are really just traffic jams with a lot of rubbernecking on park roads caused by bears wandering close to or into the roads and looking for handouts from park visitors, who are often obliging. If caught in time, problem bears are most often relocated to remote regions of a park or wilderness area where they will be unlikely to encounter people. When bears do kill people, wildlife authorities are confronted with some tough decisions, so, among other things, park officials want to make certain they have identified the actual bear or bears that may have killed someone. With the bear's life at stake, park officials typically conduct a reasonably thorough investigation. Was the bear provoked? Injured? Scavenging an already dead body? Were there witnesses?

One such investigation followed the 1998 death of 26-year-old Craig Dahl at Glacier National Park in Montana. Dahl, who worked at one of the park's concessions, took a hike on May 17 along a steep trail above an area of the park known as Two Medicine Valley, but he never returned. On May 20, park rangers found his partially eaten body concealed in some vegetation off the trail on a downward slope. Investigators collected hair samples as well as samples of bear droppings found at the scene, which were analyzed at the National Fish and Wildlife Service's Forensics Laboratory. Within a few days, the DNA tests revealed human DNA was contained in the bear droppings. Equally significantly, bear DNA extracted from the hairs matched the DNA of two grizzly bears whose DNA had been collected a year earlier by wildlife biologist Daniel Carney, who worked for the Blackfeet Tribe. It was a mother and one of her cubs. It was the first time DNA evidence had been used in a bear crime. Since Carney had recently captured the bears (which included a mother and a male and female cub), installed tags in their ears, and even fitted the mother with an electronic collar, there was little doubt the bears could be located. The question was, then what?

Although there was some uncertainty, the mother grizzly bear appeared to be a bear known as Chocolate Legs (because of her coloring) who had something of a juvenile record at Glacier National Park dating back to 1983. As an 18-month-old grizzly bear, Chocolate Legs had gotten into trouble for, among other things, causing "bear jams." In 1983, consequently, park rangers moved her to a back region of the park, where she evidently stayed out of trouble until at least 1995, possibly 1997, when she and her cubs were observed venturing uncomfortably close to people again. (Her original collar and ear tags either rotted out or were pulled off, and no one thought to save a DNA sample in 1983.) On one occasion in 1997, Chocolate Legs and her cubs walked right into a campground on the Blackfeet Reservation, which is where Carney captured and tagged them. This kind of habituation was a huge concern for park officials, particularly in light of the Dahl killing. Some bear experts, however, argued that the evidence in the Dahl case was purely circumstantial and that the bears might have been acting defensively or possibly only scavenging, assuming Dahl was dead when they found him. One expert maintained that at least the female cub was salvageable. Her DNA, at least, was not identified at the crime scene. Park officials who reviewed the case, however, said they were convinced by the crime-scene evidence that the entire family of bears had chased Dahl down the slope and participated in his demise. It was very troubling, according to Christopher Servheen, coordinator of the Fish and Wildlife Service's grizzly bear recovery program, that the cubs were accomplices. "Offspring learn from their mother, and since they were involved in pursuing and eating a human, those offspring were likely to do it again." It was decided that all three were now a danger, so they were tracked down and killed.[61]

The U.S. Fish and Wildlife Service has also conducted wide-scale investigations involving forensic DNA analysis. Beginning in April 1998, the service was charged with implementing new provisions of the Convention on International Trade in Endangered Species of Wild Fauna and Flora (or CITES), first signed in 1973, which governed, among other things, the expensive delicacy caviar. Under the new provisions, which apply to the United States and 150-plus other countries, all 27 species of sturgeon—whose eggs, or roe, are the source of caviar—were deemed protected species, including the three species of sturgeon fished from the Caspian Sea region, which has historically accounted for most of the world's supply of high-end caviar. Pursuant to the treaty and regulations promulgated by the Fish and Wildlife Service, anyone importing or exporting more than a half pound of caviar must produce a CITES certificate that guarantees the caviar was obtained legally from sustainable stocks and did not originate from 4 of the 27 species that are officially listed as endangered. DNA tests became an integral part of these monitoring efforts. For a variety of reasons, the most highly prized caviar comes from three species of sturgeon harvested from the Caspian Sea, the world's largest inland sea, and the great rivers that feed it. These are the beluga sturgeon (the source of the world's beluga caviar), the stellate sturgeon (which produces sevruga caviar) and the Russian sturgeon (which produces osetra caviar). Five countries now border the Caspian Sea, Russia, Kazakhstan, Azerbaijan, Turkmenistan, and Iran, a situation that has complicated international efforts to manage the sturgeon population. Before the collapse of the Soviet Union in 1991, only the Soviet Union and Iran bordered the sea, and both carefully controlled the stocking and harvesting of the native sturgeon population. Iran still does, but economic hardship and political turmoil in the newly formed nations of the former Soviet Union have left the sturgeon industry there in chaos. Poaching was rampant throughout the 1990s, and the world market was flooded with black market caviar, which temporarily drove down prices and increased demand. Unfortunately, the caviar heyday was short-lived because the illegal fishing (estimated to be five to ten times the officially sanctioned catch) seriously depleted the wild stocks of sturgeon, especially that of beluga sturgeon. Over all, scientists estimate that the adult sturgeon population decreased fivefold between 1986 and 2000. The official catch in the spring of 2000 was 400 tons, the lowest level in a century and a 60 percent drop over the previous year. Because sturgeon are killed to harvest their eggs and many males are killed as fishermen and poachers alike search for females, the situation is especially grave for the beluga. In December 2000, conservation groups called for a boycott of beluga caviar by U.S. consumers in an effort to stave off extinction and raise awareness of the problem. (By late 2002, the U.S. Fish and Wildlife Service was holding hearings to consider the option of listing beluga sturgeon as an endangered species and imposing a ban in 2003 on the importation of beluga caviar.) In February 2001, a CITES committee recommended that the countries bordering the Caspian (excluding Iran, because CITES officials are satisfied with Iran's sturgeon management program) reduce their planned exports of caviar for the year by 80 percent until they develop plans to control poaching. Otherwise, CITES promised, a total ban would be imposed. Resistance to the proposal was palpable, but since CITES effectively controls the movement of caviar in world markets, the four border nations ultimately agreed to a program of managed harvest, restricted trade, and putting an end to poaching.

Aside from providing a bounty of caviar, the sturgeon is revered as a "living fossil," since the fish dates back 250 million years. Sturgeons can grow to more than 15 feet in length, weigh as much as a ton and live 100 years in the wild. Sturgeons take a long time to reach sexual maturation (between 10 and 20 years or more), which is a problem for fish farmers (although genetic engineers are on the case), and they are covered with spikes and bony plates instead of scales, making them a challenge to capture. When the Fish and Wildlife Service entered the caviar fray in 1998, its criminal investigators were on the lookout for several types of criminal activity, smuggling, importers with falsified CITES documentation, mislabeled caviar products, caviar products consisting of endangered species of sturgeon, or some combination of the above. Things can get complicated because some deceptive practices involve mixing different batches of caviar eggs together.[62]

The tools brought into service for the caviar surveillance effort had been available for several years. In 1996, two molecular biologists at the American Museum of Natural History, Rob DeSalle and Vadim J. Birstein, demonstrated that PCR-based DNA tests could differentiate between the various species of sturgeon when they used their tests to reveal that several endangered species of sturgeon were being sold under false labels in the United States. Their focus was the three principal sturgeon varieties from the Caspian Sea. Reporting their laboratory results in the scientific journal *Nature*, DeSalle and Birstein found that five of 25 caviar products purchased on the retail market in New York City were mislabeled by species, with three of the five substitutions being threatened or endangered species. Although the labels on these three stated that the caviar was either beluga or osetra, the genetic tests revealed that the cans contained ship sturgeon, Amur River sturgeon, and Siberian sturgeon, all critically depressed varieties (although only one was officially listed on an international list as threatened). The other two, labeled American sturgeon and beluga, actually contained osetra and sevruga, respectively. For these tests, Drs. DeSalle and Birstein targeted mitochondrial, ribosomal, and nuclear DNA segments, and also performed DNA sequencing, to make their species determinations. The genetic targets were originally identified at the University of Moscow, where Dr. Birstein formerly worked.[63] Dr. Birstein is regarded as the world's leading authority on sturgeon.

By the time the U.S. Fish and Wildlife Forensics Laboratory began receiving field samples of allegedly contraband caviar in 1998, the scientists there had developed similar PCR-based DNA tests for all 27 species of sturgeon, with all of the comparative DNA sequencing data carefully charted in the laboratory's computer database. Laboratory personnel had even developed a sense of mission working on DNA projects involving endangered species. A cardboard cutout of Clint Eastwood as Dirty Harry kept in the laboratory seems to tell the story. Its caption read: "Go ahead, make my DNA."[64] By the end of 1998, agents had confiscated about two tons of caviar in the New York area alone for DNA testing. When tests reveal the caviar is mislabeled, it is destroyed. Much of the work by special agents is conducted at airports, where cold storage hangars are searched or caviar cans brought in by passengers are examined. Samples of caviar eggs are sent to Oregon for analysis.

During one of the routine airport checks in October 1998, U.S. Customs and Wildlife agents uncovered a large-scale smuggling scheme by a caviar distributor, Gino International, based in Stamford, Connecticut. After an investigation, three men were indicted in federal court on seven counts related to smuggling endangered wildlife into the country. According to the indictment, the three conspired to smuggle large shipments of caviar by paying courier fees of $500 to airline passengers, often off-duty airline workers, who would conceal the caviar in their luggage and then pass it on to the defendants upon arrival in New York after a flight from Poland. When the scheme first came to light, 1,000 pounds of caviar was seized from passengers at Kennedy Airport and another 1,000 pounds was seized from one of the defendants' garages in Stamford. No permits accompanied the caviar confiscated. Altogether, the government charged, 19,000 pounds of caviar, worth about $8 million retail, was smuggled between April 1 and November 3, 1998, by the defendants, even though Gino International had official permission to import only 88 pounds.[65] This was the first prosecution under the CITES law. After a trial, the owner of Gino International, Eugeniusz Koczuk, was found guilty of smuggling 21,000 pounds of Russian caviar through Kennedy airport, sentenced to 20 months in federal prison and fined $25,000. An appellate court later increased Koczuk's prison term to 48 months.[66]

A more elaborate scheme began to unravel about the time Gino International was indicted when a wildlife service inspector at Baltimore-Washington International Airport noticed that the labels on some caviar tins were of the peel-off variety as opposed to the more typical label of Russian caviar that is securely glued on. Following a lengthy investigation, three people and the company they worked for, Caviar & Caviar of Rockville, Maryland, were indicted by a federal grand jury on several counts related to the illegal importation and fraudulent labeling of tons of caviar sold over a period of four years ending in late 1999. Because the company was such a large wholesaler of caviar, the indictments came as something of an embarrassment to some of the company's major customers, which included American Airlines and Fresh Fields, the natural foods chain. It was also something of a surprise since no one seemed to recall ever receiving a complaint from customers, even though, the government alleged, the scheme involved the sale of vastly inferior types of roe under the label of higher grades such as beluga and sevruga caviar. In the end, the government demonstrated that the company and its principals were illegally smuggling tons of roe each year from the Caspian Sea region using counterfeit labels, many bearing fake certification seals from the Russian health agency. Sometimes labels were sent to the Middle East for packaging of the roe. In 1998 alone, according to reports, the company smuggled in more caviar (18 tons) than CITES allowed the country of origin to export. Based on DNA tests, it was further established that Caviar & Caviar was passing off relatively cheap American roe varieties—such as shovelnose sturgeon and paddlefish (which is actually from an entirely different family of fish) as far more expensive Russian caviar. Given the evidence, three people pleaded guilty to fraud, smuggling, and wildlife endangerment charges, and the company agreed to pay a $10.4 million fine, described by prosecutors as the largest ever levied in a wildlife smuggling conviction.[67]

In late 2002, an intensive investigation by New York federal prosecutors and the U.S. Fish and Wildlife Service culminated in a guilty plea in federal court by perhaps the largest caviar distributor in the United States, Russian immigrant Arkady Panchernikov, a naturalized American citizen, who ran the caviar distribution company Caspian Star Caviar, based at Kennedy International Airport, and is a co-owner of Caviar Russe, a Madison Avenue caviar bar. Panchernikov, who was known to many through his appearances on the Martha Stewart television show and who reportedly distributed 60 percent or more of the caviar eaten in the United States over the previous four years, pleaded guilty to multiple counts of importing and exporting Russian and American caviar without the proper permits and false labeling. Authorities initially began investigating him due to his dealings with Gino International's Eugeniusz Koczuk. Facing up to 30 years in jail and a $1.5 million fine, as well as the possibility of additional criminal charges being lodged against him, Panchernikov entered his pleas on November 1, 2002, in exchange for a sentence of up to 21 months in prison and a $400,000 fine. Among other things, Panchernikov, who forfeited his import license in the plea arrangement, admitted in court to labeling medium-grade caviar as a premium brand.[68]

Animal DNA of a far more pedestrian variety has also found its way into the courtroom. In fact, the first use of animal DNA in a criminal case involved the DNA of a domestic cat. It was a groundbreaking use of DNA technology that augured a shift in the way investigators think about solving cases. It also made cat DNA something of a hot commodity for forensic science. Cats, of course, are presumably accustomed to being at the center of controversy. Cats can pit neighbor against neighbor and even divide households between the cat lovers and cat detractors. Some people irritate their neighbors when they love cats too much and appear to be hoarding the creatures. But even run-of-the-mill cat owners can run into trouble with neighbors. This is especially true when the cats are allowed to run free, and the neighbor is a dedicated bird watcher who occasionally finds the viewing not unlike one of the more graphic scenes of predation seen on the Animal Planet channel. Indeed, bird lovers launched a national campaign in 2001 to encourage cat owners to keep their cats indoors or, at least, leashed in an effort to prevent the cats from killing millions of birds across the country each year. The American Bird Conservancy declared May 12, 2001, as the first annual National Keep Your Cat Indoors Day as a way of drawing attention to the bird carnage and in the hopes that the practice eventually spreads to every day of the year.[69] There have even been proposals to regulate the comings and goings of cats as strictly as that of dogs, but, so far, the proposals have been met with a sort of cat-like indifference.

While they have their detractors, domestic cats are beloved by millions. And for criminal investigators and forensic scientists, the era of DNA analysis has provided yet another reason to look upon cats with fondness. They can help solve cases. The first case cracked with the help of feline DNA involved the October 3, 1994, disappearance of 32-year-old mother of five Shirley Duguay from her home in Richmond, Prince Edward Island, Canada. A few days later, her abandoned car was located and a search inside revealed bloodstains that were later shown to be hers. A few weeks after that, a man's leather jacket was discovered in the woods several miles from her home. The jacket, too, was covered with her blood. Even more significantly, as it turned out, investigators with the Royal Canadian Mounted Police discovered 27 white hairs clinging to the lining of the leather jacket. Eight months after her disappearance, on May 6, 1995, the woman's body was unearthed from a shallow grave. With this, her estranged common-law husband, Douglas Beamish, was arrested and charged with her murder. Initially, Canadian mounties were hoping that the hairs retrieved from the jacket would match those of the defendant, but the forensic laboratory quickly reported back that the hairs belonged to a cat, not a human. Normally, this would have been deemed a dead end, but investigators (who may well have been keeping up with the O. J. Simpson trial in California, which dominated the news intergalactically from the summer of 1994 through the fall of 1995) decided that in DNA there may be a solution. If people can be differentiated by their DNA, why not cats? More to the point, if people can be specifically identified through their DNA, shouldn't DNA analysis be able to link the cat hairs to a specific cat? If so, this could clinch the case because the defendant lived with his parents and a white American shorthair cat named Snowball.

Unfortunately, most of the laboratories that investigators called were unwilling to tackle a cat DNA case, given the anthro-centric approach to DNA analysis that typified the field at the time. Eventually, however, Canadian investigators contacted Dr. Stephen J. O'Brien of the U.S. National Cancer Institute and the Laboratory of Genomic Diversity, which specialized in the study of domestic cat DNA. Among other things, O'Brien and his colleagues had identified nearly 400 dinucleotide positions on the cat genome that varied according to size in the same way that the equivalent DNA sites on the human genome varied. These variable STR sites are in the same class of genetic variations that now are the backbone of forensic DNA analysis, both for humans and for cats. Using PCR O'Brien and his team managed to extract and amplify DNA from one of the 27 cat hairs found on the leather jacket. Specifically, they focused on ten of the dinucleotide repeat STR loci. The same ten repeat segments were isolated from a sample of Snowball's blood for comparison with the hair DNA.[70] (A dinucleotide repeat segment is the simplest, or shortest, of all repeating segments, because only two letters of the DNA code—di means two—repeat over and over again, such as CACACA.) Snowball's blood was obtained by subpoena on January 3, 1995. Based on the laboratory's match criteria, a match was declared. The odds that the cat DNA from the hair recovered from the leather jacket could have come from some other cat were calculated by reference to the DNA profiles of 26 other cats from two cat populations (those of Prince Edward Island and U.S. cats) used to create two small DNA databases. The evidence was admitted at the trial on Prince Edward Island, and Beamish was convicted of second-degree murder on July 19, 1996. He was sentenced to 18 years in prison. The conviction in the Beamish case, combined with a series of subsequent cases that involved the analysis of domestic cat DNA, convinced O'Brien and his colleagues that they were really on to something. Consequently, they sought and received federal funding of $265,000 in 2000 for a two-year project to establish a National Feline Database that includes profiling data on the approximately 33 breeds of cat in the United States. Cat owners and cat breeders across the country were asked to contribute the cat DNA at their disposal. O'Brien told a reporter at the time that the cat DNA project began that he expected a dog DNA database would logically be next.[71]

By this time, canine DNA, for a variety of reasons, had already become another favorite of researchers. Mitochondrial DNA studies, for instance, had seemed to demonstrate that dogs had split genetically from wolves 135,000 years ago, about the time they became, eternally it appears, man's best friend. As befits this status, a loose confederation of scientists from around the world undertook a Dog Genome Project to unlock the genetic mysteries of, among other things, extreme loyalty. Subsequently, in September 2002, the National Human Genome Research Institute announced its plans to entirely decode the dog genome. Dogs, in fact, share some of the genetic afflictions of their keepers (not to mention personality quirks), so many scientists are eager to study them. (There are over 400 breeds of dogs.) By the year 2000, scientists had identified 21 specific disease genes for dogs, including the narcolepsy gene. Some are predicting that DNA tests may well be used to eliminate many dog diseases over the coming decades. In addition to identifying genes for the dog genome research, researchers are mapping repetitive DNA sequences, such as microsatellites, that are used for locating genes and determining family relationships among dogs. These repetitive sequences, of course, are the principal focus of forensic DNA analysis and can be used to identify a particular dog out of the 55 to 59 million dogs in the United States.[72] The American Kennel Club, in fact, routinely uses DNA paternity tests based on these markers to certify pedigrees.

The first forensic use of this genetic knowledge came in the case of a double murder in Seattle, committed on December 9, 1996, although true dog lovers may well consider it a triple murder. When Jay Johnson, 22, and Racquel Rivera, 20, refused to sell marijuana to gang members Kenneth Leuluaialii and George Tuilefano, according to prosecutors, Leuluaialii and Tuilefano kicked down the couple's door and entered the house with guns drawn. The first thing the gunmen confronted was the couple's dog, a one-year-old pit bull–Labrador mix named Chief. To eliminate the threat from the dog, Leuluaialii shot Chief, who clung to life for 30 hours before expiring. Both intruders then proceeded to torture and ultimately shoot to death Johnson and Rivera. After Leuluaialii and Tuilefano were apprehended, investigators discovered blood on Leuluaialii's jacket. When the blood turned out to be canine blood, police—motivated by the case involving Snowball the cat—turned to veterinarian and animal geneticist Joy Halverson, a senior scientist at PE AgGen, a private California biotechnology laboratory, who agreed to conduct DNA tests on the dog blood. After analyzing the blood on the jacket for ten repeat STR segments and comparing the results for the same ten DNA sites in Chief's blood, Halverson concluded, based on available DNA database information for dogs, that the odds were only one in 300 billion that the blood on the jacket originated from a dog other than Chief. After a trial, where the DNA evidence was admitted into evidence, both men were found guilty of the murders. Convicted of two counts of aggravated first-degree murder and one count of animal cruelty, Leuluaialii was sentenced in October 1998 to life in prison without the possibility of release. Convicted of two first-degree murder counts, Tuilefano was sentenced in December 1998 to 26 years in prison.[73]

In the years since, pet DNA has played an increasingly significant role in solving crimes. Between 1998 and 2002, prosecutors managed to obtain convictions against 14 additional defendants charged with serious crimes around the country. In one of the high-profile child abduction cases of 2002, the trial of the prime suspect, David A. Westerfield, turned in part on dog DNA extracted from hairs clinging to one of the defendant's belongings. On the morning of February 2, 2002, Danielle van Dam, 7, was discovered missing from her bedroom by her father in their San Diego home. On February 27, her nude body was found east of El Cajon, California, too decomposed to allow the pathologist to determine the cause of death or whether she had been sexually assaulted. A neighbor of the van Dams, Westerfield told police he had been traveling alone in his mobile home in the days immediately following Danielle's disappearance, but police found the child's blood, fingerprints, and blond hairs in the mobile home, as well as fibers from the vehicle on a choker necklace Danielle was wearing. Police also found Danielle's blood on Westerfield's jacket and hairs from the van Dams' dog, a one-year-old Weimaraner named Layla, on a comforter owned by Westerfield, despite Westerfield's efforts to have the items dry cleaned. Danielle was known to enjoy curling up with her new dog in front of the television. The dog hairs were genetically analyzed once again by Joy Halverson of Davis, California, who had by this time been involved in a number of criminal cases. For this case, Halverson was only able to extract mtDNA from the dog hairs, which she compared with Layla's mtDNA type. Based on her own dog DNA database, Halverson estimated that the likelihood of a match was 91 percent. Danielle's DNA profile came back a definitive match to the jacket blood as well as the spots found elsewhere in the mobile home. Despite the claims by the defense that the DNA from Danielle found in the trailer and on Westerfield's jacket were explainable because the girl had visited the mobile home prior to her abduction, the jury found the DNA evidence compelling and convicted Westerfield of first-degree murder and kidnapping, among other charges, in August 2002, subsequently recommending he be put to death for the crime (a sentence that was later imposed). The dog DNA evidence was presented to the jury by San Diego prosecutor George "Woody" Clarke, the noted DNA litigation expert who had served on the prosecution team in the O. J. Simpson criminal trial.[74]

The ability of scientists to resurrect tiny traces of identifiable DNA from biological materials once thought untestable has taken scientists in many new directions. Nowadays, scientists are limited perhaps only by their imaginations, because they are far less limited by the scientific technology at their disposal. While this is unquestionably very liberating, some people have argued that scientists are occasionally abusing their newfound freedom when they explore the mysteries of history. Less is not necessarily more in this context.

The DNA of many great figures in history, heroes and scoundrels alike, has been targeted. At any given moment, scientific researchers around the world are undertaking dozens of studies using DNA in an effort to resolve historical uncertainties. It is a trend with little sign of slowing down. For researchers investigating historically important, unsolved criminal cases (or even long-forgotten, little-noticed ones), the motivation seems reasonably pure: to solve the crime. And indeed, scientists have taken a stab at several major unsolved criminal cases in recent years using DNA. The complete and utter disappearance in 1975 of former Teamsters Union President Jimmy Hoffa, then aged 62, has always baffled investigators, for example. Despite a massive investigation, Hoffa's body was never found and no one was ever charged in the case. In 2001, therefore, it was perfectly appropriate for the FBI to conduct DNA tests on a strand of hair recovered in the early stages of the investigation from a car thought to be connected to Hoffa's disappearance. Although the car's interior had been thoroughly cleaned before it was seized as evidence, investigators did recover the hair from the car and trained dogs alerted to Hoffa's scent while exploring the car. In an intriguing twist to the investigation, the hair was, in fact, determined to be a genetic match to Hoffa, based on a comparison with the genetic profile generated from hairs retrieved from his hairbrush. In the end, however, the DNA match was insufficient to bring an indictment against a long-standing suspect in the case, Charles

"Chuckie" O'Brien, who was seen driving the car in question, a 1975 maroon Mercury Marquis Brougham owned by the son of a Detroit mob boss, on the day Hoffa vanished. (O'Brien, a longtime confidant of Hoffa's—Hoffa in fact took O'Brien in as a child after O'Brien's father died—was still alive at the time of the DNA testing and has repeatedly denied any involvement in the crime.) O'Brien did admit driving the car on the day of Hoffa's disappearance, however, and even cleaning out the interior, but he said he had his reasons.[75] This story is almost certainly not over, though, because now authorities have a DNA profile of Jimmy Hoffa to compare with any remains they find in the future that are believed to be his.

Although equally tantalizing, the very expensive efforts by mystery novelist Patricia Cornwell tried to solve the notorious case of Jack the Ripper using the latest that forensic science has to offer, including DNA testing, are unlikely to be the last word on that infamous crime spree, despite the title of her book on the subject. A year of research and forensic analysis orchestrated by Cornwell, much of it conducted by Dr. Paul Ferrara and his colleagues at the Virginia Institute of Forensic Science and Medicine, and an expenditure of about $6 million by the popular novelist, culminated in the 2002 nonfiction book by Cornwell, *Portrait of a Killer: Jack the Ripper, Case Closed*.[76] The book, which became an instant best seller, identifies the Impressionist painter Walter Richard Sickert, a former actor and a reputed master of disguise, as the infamous serial killer Jack the Ripper, who terrorized London for four months in 1888 by killing and mutilating a string of prostitutes (at least five). Sickert's paintings of woman, including some prostitutes, according to Cornwell, mirror the themes of brutality that the Ripper expressed in real life. A 1908 painting by Sickert was, in fact, titled *Jack the Ripper's Bedroom*. Moreover, Sickert may have suffered from a severe genital malformation that was at the core of a rage against women. None of the Ripper's victims appeared to have been raped. In addition, Cornwell notes, Sickert was in and around London at the time of the murders. Jack the Ripper famously taunted the police and local press, as well as other officials, with crowing letters recounting his crimes, and it is the letters, in fact, that serve as a cornerstone of the case Cornwell builds against Sickert. Watermarks (which identify the paper manufacturer), linguistics analysis, and handwriting from the letters seem to point to Sickert (although not definitively), based on comparisons with known items of the painter's. There was even DNA retrieved from under a partial stamp on one of the Ripper letters, one sent to Dr. Thomas Openshaw of the London Hospital Museum, but it proved too battered by time to offer anything beyond an mtDNA profile. In fact, none of the 55 items swabbed for the DNA testing, items that included not only Ripper letters but letters and stamps associated with Walter Sickert or people associated with him, as well as other suspects in the Ripper murders, revealed any trace of nuclear DNA. Most of them, however, did produce adequate mtDNA sequences for comparisons. Consequently, the mtDNA analysis provides another significant building block in Cornwell's case against Sickert after it is found to match mtDNA obtained from Sickert's own stationery. (Importantly, at least, it does not appear to exclude him.) However, the company that performed the mtDNA analysis, Bode Technology Group of Springfield, Virginia, concluded that most of the items tested contained a mixture of mtDNA profiles, having been contaminated evidently by contributions from other people who handled the letters. Unfortunately, too, Walter Sickert's body was cremated upon his death, so Cornwell did not have a reliable reference sample for her prime suspect. From the mtDNA profiles generated from the Sickert items, therefore, it was necessary to take a leap of faith and assume that Sickert's mtDNA type was present somewhere in the mixture. As it turned out, though, the partial stamp on the Ripper letter to Dr. Openshaw produced only a single mtDNA profile (as sequenced by Bode), presumably because only one person's mtDNA was present. This may indeed be the mtDNA of the infamous Jack the Ripper. When this mtDNA profile was compared with the other items tested, the same sequence was found among the multiple profiles found on items associated with Walter Sickert, as well as two other items related to the Ripper (an envelope carrying another Ripper letter and another stamp adhering to the opposite side of the letter sent to Dr. Openshaw). The Sickert items included an envelope belonging to Sickert's then-wife and an envelope and stamp from separate items of correspondence belonging to Walter Sickert. Unfortunately, all of these items were a jumble of mtDNA profiles, and it was not possible to say which mtDNA profile belonged to Sickert. Even assuming there is a match to Sickert, that still leaves a fairly sizeable portion of the population in London at the time that would also have matched the Ripper mtDNA profile.[77] "It is possible that the DNA matching sequences might be a coincidence," Cornwell conceded to *The New York Times* on the day her book was released. "This is not conclusive evidence, but another layer in what I consider conclusive evidence."[78] Cornwell considers the most compelling evidence against Sickert the psychological profile she developed on the painter, which, she says, indicates he was a "psychopath."[79] However, many die-hard Ripperologists, as Ripper fans are known, remain, at first blush, less than entirely persuaded by Cornwell's circumstantial case against Sickert, preferring to leave the case open for now.

Somewhat more satisfying DNA results were obtained in another infamous criminal case that has tormented investigators for decades, that of the Zodiac Killer, who terrorized the San Francisco Bay area in 1968 and 1969, when he senselessly shot or stabbed to death at least five people, and who may, in fact, have been responsible for a total of 37 deaths. Like Jack the Ripper before him, the Zodiac Killer was a writer of letters, writing boastful and cryptic letters or cards, 21 in all, to *The San Francisco Chronicle* and other local newspapers, always signing them with a cross inside a circle. (Unfortunately, more than half of the original envelopes from the Zodiac case were lost over the years, apparently by investigators.) The Zodiac Killer occasionally included cryptograms in his letters, which authorities were eager to decipher in an effort to identify him. One of them, written with strange symbols, was ultimately decoded to read: "I like killing people because it is so much fun. It is more fun than killing wild game in the forest because man is the most dangerous animal of all." As in the Ripper case, the Zodiac letters became the focus of the latest, 21st-century–style, investigation into the notorious crime spree. Prompted by an ABC-TV *Primetime* investigation into the case, an anonymous source—a retired investigator—rediscovered three of the original Zodiac envelopes among his own files and the items were returned to San Francisco authorities. As a result, in 2002 scientists in the San Francisco Police Department's DNA laboratory, under the supervision of Dr. Cydne Holt, re-examined them and were able to retrieve four (out of a possible nine probed) STR markers from the stamp adhering to one of the newly found Zodiac Killer envelopes, as well as a genetic marker indicating that the individual who licked the stamp was male. While the partial DNA profile is not enough to make a definitive identification, it did prove sufficiently telling to clear three men who had long been considered suspects by amateur sleuths. Had a more complete DNA profile been generated, investigators were hopeful that a "cold hit" might have been obtained once the DNA data were fed into the national DNA database, perhaps solving the crime once and for all. Since tests are continuing on the Zodiac evidence, this may well be the eventual outcome, assuming additional DNA markers are detected.[80]

The rationale for aiming DNA analysis at other historical figures is sometimes less compelling, but not always. And one may wonder what the dead think of all the morbid curiosity (in the true sense of those words). Historians can only guess in most cases, of course, but they do, at least, know how "the DNA of English literature," William Shakespeare, felt about it. Famously inscribed on Shakespeare's tombstone is the following:

Good friend, for Jesus sake, forbear
To dig the dust enclosed here.
Blest be the man who spares these stones,
And curst be he who moves my bones.

Although Shakespeare has so far escaped exhumation for a peek at his DNA, a great many other notable characters of history have not. Nothing is sacred, apparently, when it comes to DNA and history. Even the legendary Shroud of Turin, the presumed burial cloth of Jesus that bears a striking image of a man many faithful Roman Catholics believe to be the face of Jesus Christ, has been tested for DNA. (Shakespeare might well appreciate the irony here, having invoked Jesus to spare him the indignity of being dug up, but he might also be a little alarmed at the direction this trend of DNA-mining expeditions is taking.) The authenticity of the Shroud of Turin has long been debated, and the subject remains highly controversial. The cloth has undergone countless forensic tests in recent decades in an effort to resolve the question of the shroud's origins, but the results have often seemed contradictory. Depending on whom you talk to, some of the evidence suggests the relic dates to the time of Jesus and can even be linked to the region around Jerusalem; other evidence suggests it dates to medieval times, so it must be a forgery. In 1998, scientists in Texas found minute traces of apparently human DNA on a portion of the shroud, just large enough to indicate that the contributor was male. There was insufficient DNA to make a determination of ethnicity using STRs, according to one of the scientists involved, Leoncio Garza-Valdes, a pediatrician and adjunct professor of microbiology at the Health Sciences Center of the University of Texas. Garza-Valdes went on to write a book about this finding with the intriguing title *The DNA of God?*[81] Of course, the DNA might well have been a contaminant from someone who handled the prized cloth over the centuries, but who knows? Dr. Garza-Valdes was already a major figure in the shroud debates because he developed a theory that a "bioplastic" layer on the sacred cloth produced by bacteria over the centuries had dramatically skewed radiocarbon dating done on the shroud in 1988, which suggested the relic was only 700 years old. (The DNA tests conducted in Texas were not authorized by the Catholic Church, and the Vatican recalled the samples of the shroud on loan to the Texas researchers after learning of the results.) The controversy will likely continue for some time.[82]

Generally, it is necessary to have been a pretty important person before the DNA detectives maneuver in the backhoe to your grave site. But it is sometimes enough to have been a blood relative of one to catch the eye of investigators. This is what happened to the brother of Christopher Columbus, Diego Colón, whose bones were dug out of a garden near Seville, Spain, in September 2002 for the purpose of extracting DNA from his bones in an effort to resolve a dispute over where the actual remains of Christopher Columbus are entombed. There are now two sets of remains that purport to be those of Christopher Columbus, both buried under the explorer's Spanish name, Cristobal Colón, one in a tomb at the cathedral in Seville and the other in a monument in Santo Domingo in the Dominican Republic. The Spanish researchers involved in the project also plan to dig up other relatives of Columbus in order to make a DNA comparison with the two disputed sets of remains and perform, in essence, a DNA paternity evaluation to see if Diego and Christopher had the same mother or whether, as one theory has it, Christopher was the illegitimate son of a prince. Their journey is likely to be a great deal more circuitous than was Columbus's trip to the Americas because, at the time of the exhumation of Diego, they had yet to obtain the necessary governmental and church permissions for disinterring Christopher Columbus, either one of them. Moreover, there is uncertainty over whether they in fact dug up Diego because his bones were moved around a lot over the years. This was also the case with the remains of Christopher Columbus, whose bones may have traveled as much dead as they did alive, which is what led to the confusion.[83]

There are perhaps even more daunting challenges facing researchers who hope to determine who has the real remains of American revolutionary pamphleteer Thomas Paine, whose bones not only traveled extensively in death but apparently went off in different directions, some perhaps into oblivion, others reportedly made into keepsake buttons. The trouble started in 1819 when Paine's remains were exhumed and later spirited off to England—without permission—where, reportedly, some of the button work was eventually done. It was William Corbett who took the bones to England, with plans to build a memorial to honor Paine, but the memorial plans fizzled and Corbett's descendants appear to have gradually sold off Paine's remains—which remained stored in a trunk—a few bones at a time. (It is also possible that Paine's bones were buried in the Corbett family plot in England.) At present, there is a rib in France that may be Paine's and a putative descendant of Paine's in Australia claims to have possession of his skull. (The man who has the skull claims that he was descended from an illegitimate son of Paine's, and the family hopes to have DNA paternity tests conducted one day to substantiate the claim.) It may well take more than a little common sense to pull this off, but the Citizen Paine Restoration Initiative of New Rochelle, New York, where Paine settled in 1784 and was initially buried after his death in 1809, began an earnest project in 2001 to round up all of Paine's bits and pieces for reburial in New Rochelle, with any luck by the 200th anniversary of his death in 2009. Such an effort could never even have been attempted before the advent of modern DNA-identification methods, at least not with any genuine expectation of complete success. (Before DNA typing, people might well have tried to sell the eager historians bones by the ton from places like Paraguay and Laos. Who'd know?) But now the researchers have (at least potentially) a genetic baseline, against which they can measure the artifacts that are alleged to be those of Thomas Paine, possibly even the buttons. The Thomas Paine National Historical Association of New Rochelle believes it has the mummified brain stem and a lock of hair that once animated the American Revolution in the person of Thomas Paine. The hair remains at the Thomas Paine Museum in New Rochelle, and the brain stem is buried in a secret location on the grounds of the museum. The museum hopes that DNA comparisons derived from the hair will ultimately authenticate the skull and other bones scattered around the world and believed to be those of Paine, so they can be brought home. (If necessary, the museum will exhume the brain stem.)[84] Even as things stand now, the president of the historical association, Gary Berton, told the Associated Press, it is not entirely inappropriate for the remains of Paine to be scattered around the globe: "This is the man who said, 'The world is my country.'"[85]

There is a certain point where these DNA investigations cross a fuzzy line into what *The New York Times Magazine* once dubbed "Tabloid History."[86] (Attempting to make a determination about the legitimacy of Christopher Columbus's birth seems to qualify.) According to the *Times*, "Using DNA testing and other high-tech tools, scholars are ransacking the past—looking for the historical equivalents of the blue Gap dress," referring to the dress that produced the DNA profile of President Bill Clinton and, eventually, triggered impeachment hearings against the president.[87] Among those whose DNA is, or has been, coveted by this new breed of DNA cowboy, as the *Times* article recounts, are Ludwig van Beethoven, Jesse James, and Meriwether Lewis. A genetic profile was sought from a well-pedigreed lock of Beethoven's hair by a doctor in Arizona, Alfred Guevara, who is also interested in determining whether the tormented composer suffered from syphilis. Beethoven's head had taken on the shorn look of a punk rock musician by the time he was buried in 1827 because so many people took locks of his hair as mementos after his death, so DNA from the "Guevara lock," as it is known, may eventually be used to confirm who else has the real thing. (There are also a couple of love child candidates, so DNA may prove useful here, too.) In 2000, *Beethoven's Hair*[88] was published, a book written by Russell Martin, which details a number of forensic tests ultimately conducted on Beethoven's hair, including mtDNA sequencing, that shed light on why the composer was in such poor health during much of his life and eventually became deaf (significantly elevated lead levels). The tests did not find evidence that Beethoven suffered from syphilis,[89] and no conclusions have reportedly been made about a possible love child. (Dr. Guevara also speculates that Abraham Lincoln may have acquired syphilis in his rambunctious youth, then passed it on to his wife, Mary Todd Lincoln, based on the evident mental instability that characterized her life after her husband's assassination.) Although a team of forensic scientists, led by professor of law and forensic science at George Washington University James E. Starrs, made a determination in 1995—based on DNA tests—that the famed outlaw Jesse James was buried in Missouri, confirming the historical account that James was shot to death by Robert Ford in St. Joseph, Missouri, in 1882, other historians subsequently challenged that finding and are seeking a court order to exhume another body in Granbury, Texas, buried under the name J. Frank Dalton, for DNA tests. Their hope is to demonstrate that James really faked his death in 1882, as many had long contended prior to the 1995 exhumation, and went on to live a long and low-profile life under a variety of aliases, including J. Frank Dalton.[90]

The tragic half of the famed Lewis and Clark team that explored the untamed West in the early years of the Republic, Meriwether Lewis, may also be exhumed for various analyses. James Starrs has pressed for an exhumation of Lewis's grave in Tennessee, an effort resisted by the U.S. Park Service, to determine—through DNA—whether the grave actually contains Lewis and—through a pathological examination—whether Lewis really did kill himself in 1809, as most historians believe, or was murdered, as Starrs and others conjecture. Starrs also wants to see if the bones indicate that Lewis had syphilis. According to the *Times* magazine piece, literary scholar Harold Bloom speculates that William Shakespeare suffered from syphilis, too, which may well number Shakespeare's days of undisturbed repose.

When excavations were occurring all around Madrid in 1999 in search of the bones of celebrated Spanish painter Diego Velázquez, coinciding with the 400th anniversary of his birth (and dubbed the Year of Velázquez), Spanish writer Francisco Ayala wrote in the Spanish newspaper *El País*, "Let the illustrious dead rest in peace."[91] (It does not appear that anyone was successful in finding the remains, but officials in Spain are prepared to conduct DNA tests on any good candidate bones that are discovered, which can be compared to the DNA of Velázquez descendants.) Ayala's criticism of the efforts to find the remains of Velázquez (and he was not the sole dissenter in Spain) are echoed throughout the world by those who see the pursuit of the dead as more ghastly than glorious.

A similar controversy has erupted in France over proposals by French historian Bruno Roy-Henry, among others, to exhume the remains currently inside the tomb of Napoleon Bonaparte in Paris to conduct a variety of scientific examinations, including DNA analysis. Napoleon rose from humble beginnings in Corsica to become emperor of France and ruler of most of Europe before being defeated, finally, at Waterloo by the British. His reign brought, on the one hand, a revolutionary civil law code to feudal Europe, but was, on the other hand, characterized by brutality and unbridled militarism. It is this mixed legacy that has made Napoleon one of the most studied men in history. Roy-Henry believes that the British may have pulled a switch when they returned Napoleon's coffin to France in 1840, substituting someone else's body for the great French emperor's to conceal evidence that Napoleon had been poisoned. It has been popularly theorized that Napoleon was killed by arsenic administered gradually over time, perhaps even cyanide at the end. Napoleon was sent into exile by the British in 1815 to the desolate island of St. Helena located 1,200 miles off southern Africa, where he died, some say under mysterious circumstances although the official autopsy says of cancer, in 1821 at age 51. He remained buried on the island for 19 years before being returned to France. The British are not the only suspects in an alleged murder conspiracy against the former emperor, however, since French people loyal to King Louis XVIII would have been equally happy to see Napoleon done in to prevent his making any further forays into French politics. In 1994, the FBI conducted tests on two hairs said to be Napoleon's and determined that the elevated levels of arsenic present in the hairs were insufficient to have killed him, assuming the hairs were genuine, but the tests did not finally resolve the matter. A number of French academics have pressed for an exhumation of the remains in Napoleon's Tomb for years. Roy-Henry believes that a DNA test is perhaps the best way to resolve the controversy over who is buried in the tomb. He therefore wrote to the Defense Ministry in France in mid-2002 to request that DNA tests be conducted on a strand of hair taken by French doctors in 1840 as the returning coffin was nearing Paris, a hair that, long in the hands of Napoleon's descendants, was eventually put on permanent display at the French Army Museum in Paris in 1936. The

Defense Ministry declined to allow the test, however, let alone an exhumation, but the ministry did suggest that Roy-Henry seek permission, and comparative DNA samples, from living descendants of Napoleon. DNA tests on Napoleon, therefore, appear to remain very much a possibility.[92]

It is sometimes difficult to choose sides in the "tabloid history" controversy because some of the famous dead are almost certainly not resting in peace. They want to be found, perhaps, or have some degree of justice surrounding the murky circumstances of their death. Presumably, for example, famed American aviator Amelia Earhart and her navigator Fred Noonan, who both disappeared over the Pacific Ocean on July 2, 1937, during an attempt by Earhart to become the first female flyer to circumnavigate the globe, would have no objection to being found. (But, then again, who knows?) Their trip would have been the first successful flight ever around the world at the equator. Their plane was lost as they reportedly neared Howland Island for a scheduled refueling stop en route to Hawaii. On this leg of the journey, they had traveled an arduous 2,500 miles from New Guinea. Their final destination was California, where the adventure had begun, so they were agonizingly close to completing the trip. They were last heard from complaining in a radio transmission to a U.S. Coast Guard cutter that they were nearly out of fuel as they searched for the tiny island. A massive 250,000-square-mile rescue search by the U.S. Navy over the next several weeks found no sign of either the flyers or Earhart's Lockheed Electra airplane. While most observers believe the plane simply ditched in the ocean and went down, over the years a number of other theories have been floated about what became of Earhart and Noonan. Some have suggested that they were actually captured (and later executed) by the Japanese because they were secretly spying on them for the United States; others that they managed to land on an alternative island and decided to live the simple life in the Philippines, and still others that they succeeded only in crashing on an alternative island and soon perished there or perhaps survived the ordeal and eventually found their way, under assumed names, back to New Jersey. Various research teams have begun in recent years to investigate reported wreckage sites and search for bones that may be those of the downed pilots. A team of forensic anthropologists is actively searching for remains reportedly located somewhere on the Polynesian island of Nikumaroro, remains identified in a 1941 autopsy report on a dozen bones and a skull found on the island by British soldiers in 1940. The remains were later lost (although they may be stored at a medical school in Fiji). A modern-day reexamination of the report concluded that the remains could be those of Amelia Earhart. Should the bones be located, the researchers plan to conduct, at a minimum, mtDNA analyses on the remains, which will be compared with the mtDNA of Earhart's niece. So the search continues. Meanwhile, a competing deep-sea expedition, supported by PBS-Television's *Nova* program, has gotten underway to search for the wreckage of Earhart's plane around Howland Island.[93]

Occasionally, a DNA investigation gives renewed credibility to the whole enterprise of using DNA to solve historical riddles and provides momentum to other investigations. Hoping to end 200 years of speculation about what became of the heir to the French throne, ten-year-old Louis XVII (born Louis-Charles)—who would have been king but for the French Revolution, the execution by guillotine of his parents, King Louis XVI and Queen Marie-Antoinette, in 1793, and his imprisonment after the beheadings—scientists conducted a series of mtDNA tests in 1999 and 2000 and offered up a solid answer. Almost immediately after the boy's reported death from tuberculosis—if not criminal neglect—in 1795 while being held in Temple prison in Paris, rumors began to circulate that the young prince, or Dauphin (the official title at the time for the eldest son of a king of France), had actually been smuggled out of the prison (in a bathtub, by one account) by royalists hoping to one day restore him to the throne. According to this version of events, it was either an imposter child at the prison who really died or, at least, a substitute dead boy introduced into his coffin. (The boy's actual grave has not been located.) Over the past two centuries, speculation about the true fate of Louis XVII has spawned hundreds of books and countless academic articles, as well as a spate of pretenders and putative descendants. One man who claimed to be the Dauphin, Charles-Guillaume Naundorff, was so convincing that the government of the Netherlands was persuaded to bury him after his death in 1845 under a tombstone identifying him as the real Louis XVII. "Here lies the king of France," reads the inscription. The legend even had currency in America, and Mark Twain, naturally, found the tale irresistible, incorporating it into his 1884 novel, *The Adventures of Huckleberry Finn,* at one point having Huck relate the story of "the little boy dolphin" to Tom Sawyer and later introducing a character Huck meets who claims to be the Dauphin himself, as Huck narrates, the son of "Looy the Sixteen" and "Marry Antonette."

As it happened, an autopsy was performed by several doctors on the boy who evidently died in Temple prison in 1795. Unbeknownst to his colleagues, one of the attending doctors, Philippe-Jean Pelletan, stole the boy's heart by wrapping the organ in his handkerchief and slipping it into his pocket when the others weren't looking. Taking the heart home, Dr. Pelletan attempted to preserve it in alcohol in a glass container, but eventually the alcohol evaporated and the heart petrified. From this point on, the rock-solid heart endures something of a mosh pit of handlers over the years, even spilling out onto the floor on one occasion, where it lay for days. An assistant to the doctor later stole the heart, but it was returned after the man's death by his widow. Dr. Pelletan subsequently tried to return the relic to Louis XVIII after the monarchy had regained power in France, but the offer was spurned because of the heart's questionable provenance, and the organ eventually ended up in the hands of the archbishop of Paris for safekeeping. Unfortunately, civil unrest in 1831 led to the archbishop's palace being sacked. A printer on the scene during the siege named

Lescroart tried to rescue the heart and some papers attesting to its authenticity, but he was foiled when he came to blows with a national guardsmen and the two knocked the crystal urn in which the heart had been placed to the ground, smashing it to bits. For days, the little heart remained on the floor among the shards of glass until Lescroart and Dr. Pelletan's son returned to the palace and retrieved it. It stayed in the hands of the doctor's son until it was transferred by bequest to one of the families (on the Spanish side) descended from the French royal line, who kept it stored in a chateau for more than 80 years. Finally, in 1975 the heart was placed in a crystal urn and stored in a royal crypt at the Memorial for France at St.-Denis outside Paris, where it stayed until the 1990s, when historians and scientists started to "Think DNA."

Needless to say, the history of the peripatetic heart suggested something of an iffy chain of custody, but it was nevertheless decided to allow some reputable scientists an opportunity to examine the heart in hopes of extracting DNA from it. After years of lobbying by historians and a decision by the Memorial of France at St.-Denis to fund the experiments, two experts in human genetics, Jean-Jacques Cassiman of the Center for Human Genetics at Louvain University in Belgium and Bernd Brinkmann of the University of Münster in Germany, came to Paris in December 1999 and, after some suitable ceremonial honors, sliced two small pieces from the heart with a razor blade. Working independently back at their own laboratories over the next several months, both scientists succeeded in extracting sufficient DNA from the heart slivers to perform mtDNA sequencing. In the end, three samples from the heart produced mtDNA sequences of sufficient quality to compare with mtDNA sequences obtained from hair samples reliably linked to Marie Antoinette and two of her sisters, as well as the mtDNA of two living maternal relatives of Marie Antoinette, Queen Anna of Romania and her brother André. (Some of the preliminary work had been done earlier by Dr. Cassiman, who published a scientific paper in 1998 that concluded that Charles-Guillaume Naundorff was not, his headstone notwithstanding, related to Marie Antoinette or her sisters, based on a similar comparative analysis using the sisters' hair and tissue from Naundorff's remains.) Relative to the heart investigation, the scientists concluded that all of the mtDNA sequences they had obtained were "identical," substantiating the claim that the heart belonged to the son of Marie Antoinette, Louis XVII. The official announcement was made at a Paris news conference on April 19, 2000, in an august hall at the René Descartes School of Medicine, an event attended by historians, scientists, descendants of royalty and a crush of media people. It was intended to be the last word on the 200-year-old mystery, but there were still doubters. Following the announcement of the scientific results, there was immediate speculation as to whether the heart might really belong to another son of Marie Antoinette, one that history had completely neglected to record, and not the fabled Dauphin after all. Surprisingly, perhaps, no one asked if there was any evidence that the boy had suffered from syphilis.[94]

9 Forensic DNA Analysis and Customized Life

ORENSIC DNA ANALYSIS HAS WADED INTO A NUMBER OF DISPUTES IN THE field of genetic engineering, and even cloning. After a 1980 decision by the United States Supreme Court that authorized the patenting of a genetically modified organism (or GMO), there was a stampede of researchers and biotech companies to register claims. Since then the marketplace has been flooded with "designer life forms," with little indication that the practice will slow down any time soon. The public response has ranged from gratitude to absolute indifference to domestic terrorism.

So far, the most manipulated creature on the planet appears to be the mouse, largely due to its utility in studying human diseases and genetics. For some of the mice, genetic information, such as a human gene, is added; for others, known as "knock out" mice, genes are deleted or disrupted (in other words, knocked out).[1] Mice have been bred to have Alzheimer's (and then apparently cured!),[2] to have good or faltering memories,[3] to glow in the dark (under ultraviolet light),[4] to have whale genes, to be smarter (the so-called Doogie mouse, named after a TV child prodigy),[5] to suffer from learning disabilities,[6] to produce human antibodies,[7] to see red (as humans do),[8] to resist certain cancers,[9] and to grow hair thicker and faster.[10]

Genetic engineering has swept through the animal kingdom; it has reconfigured mammals, fish, plants, insects, and microbes. And the natural boundaries are frequently breached. Logically, the end point will be human genetic engineering and designer babies, a prospect that engenders considerable public discussion but has yet to generate much in the way of public policy. Whatever the reservations, a significant step toward designer humans seems to have been taken with the birth announcement in January 2001 of a genetically altered rhesus monkey named ANDi (for inserted DNA, spelled backward).[11] Produced by a team led by Drs. Gerald Schatten and Anthony Chan at the Oregon Health Science University,[12] ANDi was born in October 2000 out of a desire by researchers to create a new, primate transgenic model to study human genetics and disease, since mice have some universally acknowledged limitations. This was the same group of scientists who announced the first successful cloning of a primate in the year 2000 (although the method did not involve the use of adult cells, as with Dolly the sheep). Born at the Oregon Regional Primate Research Center in Beaverton, Oregon, ANDi was retrofitted with a jellyfish gene, GFP (for green fluorescent protein), just before conception. The gene, when it works, produces a protein that causes the cells to glow green under a fluorescent light.

Given this ability, the jellyfish gene has become something of a favorite among biologists and has even been turned into some novelty items. A Pittsburgh company, Prolume, for example, began using jellyfish genes, as well as those of other bioluminescent sea creatures, to fill up water pistols that were sold to raise money for research expeditions. The items were so popular that the company expanded its arsenal to include a two-chambered water rifle.[13] However, a storm of protest greeted the revelation that artist Eduardo Kac, working with French government geneticists, had created a transgenic rabbit named Alba that glows in the dark (all over) because of its ubiquitous jellyfish genes. Kac maintained that he was simply a pioneer of "transgenic art," which, given the new powers of genetic engineering, society must confront.[14]

ANDi was not easy to produce, and he was only partially successful. Although the scientists produced 126 embryos, there were only five successful pregnancies, resulting in three live births. A pair of twins miscarried; interestingly, their toenails and hair follicles glowed green when exposed to fluorescent light. Unfortunately, none of the live monkeys glowed. When DNA tests were performed on the three newborns using PCR, however, ANDi's tissues consistently revealed the presence of the jellyfish gene. He appears in all other respects to be a normal, healthy rhesus monkey. The DNA tests demonstrated that the gene was throughout his body—in his cheek cells, hair, urine, and cord blood. The only difficulty was that the gene was not actively producing the glow protein, which puzzled the research team.[15]

The ANDi experiment confirmed the worst fears of many people, that the genetic enhancement of human beings will soon take root and begin to deflect the natural course of human evolution. It "is perhaps the gravest imaginable crisis," commented syndicated columnist George F. Will. While some observers see only the hope of eradicating certain human diseases and deformities, Will sees darker forces at work. "Enhancement is not therapy," he writes, "it is [positive] eugenics." Negative eugenics, the prevention of specific traits, Will noted, is already commonplace, such as when an amniocentesis or chorionic villus sampling turns up a severe genetic abnormality and an abortion takes place.[16] Growing increasingly popular, too, in the context of in vitro fertilization (IVF), is preimplantation genetic diagnosis (PGD). With PGD, a couple at high risk for a particular genetic disease can have embryos created at a laboratory. The embryos will then be subjected to a series of PCR-based genetic tests to determine whether they carry the genes for the deadly disease of concern to the parents. Embryos with the diseased genes are discarded and those without them are then implanted in the woman's uterus. By mid-2002, PGD had been used to produce hundreds of babies worldwide. Perfected in the 1990s, the technique has been used to screen out embryos with cystic fibrosis, Tay-Sachs disease, Marfan's syndrome, sickle-cell anemia, Huntington's, and various sex-linked diseases, such as hemophilia.[17] The use of PGD is not without controversy, however.

In a first, Adam Nash was born in August 2000 to a Colorado couple after PGD was used to select an embryo determined to be the best candidate for donating tissue to an older sibling with a rare genetic disease. Adam's older sister, six-year-old Molly, inherited Franconi anemia. Of the embryos created in the laboratory, Adam's tested free of the disease itself and was an exact tissue match for Molly. The six human leukocyte antigen, or HLA, genes—of which DQA1 is a part—were the genes tested. Lack of genetic agreement would run the risk of tissue rejection. A month after his birth, blood from Adam's umbilical cord and placenta, preserved from his birth and rich in stem cells, were transfused into Molly to boost her blood supply and immune system.[18] The case caused deep concern among some bioethicists who argue the Nash case crossed the line to positive eugenics.[19] Similar objections have been raised by the increasing use of the technology for sex selection.[20]

In the United States, the signs of a continental drift toward positive eugenics already abound. There are of course sperm banks, including one dedicated to preserving and disseminating the sperm of Nobel Prize winners. Also, an advertisement was run in 1999 at Ivy League colleges around the country offering up to $50,000 for an ovarian egg donation from a tall, high IQ, athletic college woman, a rate of payment at least ten times that for the average egg donor.[21] (The ad spawned an egg-donation industry, as things turned out.[22]) A web site launched in late 1999 began soliciting bids for the eggs of beautiful supermodels for use in IVF procedures.[23]

Parents the world over already go to extraordinary lengths to confer advantages on their children intended to improve their chances of success in life, from subjecting them to endless hours of Mozart to paying for expensive SAT-preparation courses. So what's so different about starting a little earlier by, for example, equipping them with the genes for beauty, athleticism, or intelligence? According to Case Western University professor of law and bioethics Maxwell Mehlman, "Enough people will want this for their kids that I think it will happen."[24] A promotional ad campaign for the science fiction movie *Gattaca* in the summer and fall of 1997 sought to tap into the public's angst over the looming issue of designer children. In television spots, full-page newspaper ads, and on rolling billboards, the movie-makers juxtaposed a picture of a bright-eyed baby and the marketing pitch, "Children made to order. 1-888-4-BEST-DNA."[25] The movie depicts a society that encourages designer children and is tightly controlled by routine genetic screenings (such as when entering the workplace) to ensure that only the genetic elites have access to positions of power. The *Gattaca* scenario, more or less, is a genuine possibility, according to Princeton University molecular biologist Lee M. Silver. In his 1997 book, *Remaking Eden: Cloning and Beyond in a Brave New World*, Silver predicts that over the next few hundred years society will evolve into two distinct classes, the powerful GenRich and the lowly Naturals. Gradually, Silver expects, the Naturals and the GenRich will lose their ability to interbreed and become separate species.[26] Some bioethicists, including Boston University's George Annas, have expressed their concern that such a division of humanity could lay the groundwork for a kind of "genetic genocide."[27]

According to Professor of Public Policy at George Mason University Francis Fukuyama, commenting on the announcement that the human genome had been roughly sequenced in June 2000, "it seems almost inevitable that we will eventually seek to use genetic knowledge to actively reshape human nature. This could take many forms, from wealthy parents creating 'designer children' with superior looks and intelligence, to an egalitarian state trying to remedy natural inequality through a new form of eugenics. Once we better understand the genetic sources of behavior, we will be able to develop powerful new tools to better control it. The way is then open to superseding the human race with something different."[28] According to Michael Rose, a geneticist at the University of California at Irvine, this is "one of the most important questions for the human species: the extent to which it will direct its own evolution."[29] Bioethicist Margaret Somerville of McGill University told a reporter that the human genome is "the patrimony of the entire species, held in trust for us by our ancestors and in trust by us for our descendants. It has taken millions of years to evolve; should we really be changing it in a generation or two?"[30]

There are many voices, however, to counter these skeptics, not to mention market forces. When a March of Dimes poll asked people if they would improve their child's appearance or intelligence through genetic engineering, assuming the technology existed, 42 percent responded yes.[31] Some of the most powerful voices in science are open to the possibility, as well, including Dr. James Watson.[32] "I strongly favour controlling our children's genetic destinies," Dr. Watson wrote in *The Independent* in April 2001.[33] World-famous physicist, mathematician and cosmologist Stephen W. Hawking also subscribes to this view. Hawking has suffered from Lou Gehrig's disease, or amytrophic lateral sclerosis (ALS), since 1963, and is now completely wheel-chair bound and unable to speak, except with the aid of a computer. Despite the disability, Hawking is widely viewed as a contemporary Einstein because of his contributions to science. "There has been no significant change in human DNA in the last 10,000 years, but it is likely that we will be able to completely redesign it in the next thousand," he predicts. "I think the human race, and its DNA, will increase its complexity quite rapidly. In a way, the human race needs to improve its mental and physical qualities if it is to deal with the increasingly complex world around it and meet new challenges like space travel."³⁴ Hawking offered a specific example during a television interview. "I think the biggest challenge we face is from our aggressive instincts," he said. "In caveman . . . days, these gave definite survival advantages and were imprinted in our genetic code by Darwinian natural selection. But with nuclear weapons, they threaten our destruction. We don't have time for Darwinian evolution to remove our aggression. We will have to use genetic engineering."³⁵ Ironically, of course, Stephen Hawking would very likely not have made it out of the petri dish using modern methods, assuming his genetic illness was identified.

Whatever the fate of humans, genetic engineering has produced an entirely new McDonald's Farm. Scientists have manipulated the genes of domesticated animals, crops, insects, and fish in an effort to produce commercial products. So-called pharm animals, for example, are a new breed of livestock that have been equipped with either human genes or the genes of other species to produce various pharmaceutical drugs.³⁶ A herd of genetically altered goats were fitted with spider genes that enable the goats to produce a super-strength silk protein in their milk, a substance called Biosteel by its developers (who envision its use in bulletproof vests and fishing line).³⁷ A number of fish breeds have been genetically rejiggered to grow faster than their natural counterparts. Atlantic salmon, for example, were equipped with an extra growth hormone gene from chinook salmon and a promoter gene from an ocean pout (which together function as an unrelenting "on" gene), resulting in a breed of salmon that grows six times faster than in its natural state and reaches market size in about 18 months as opposed to the 36 months it would normally have taken. If licensed by the United States Food and Drug Administration (FDA), the supersalmon would be the first genetically altered animal to be approved for human consumption in America.³⁸ Critics worry, however, that the salmon will escape into the wild and outcompete wild salmon, potentially wiping out the natural fish altogether.³⁹ Such concerns were echoed in a report released in August 2002 by the National Research Council, a division of the National Academy of Sciences, entitled *Animal Biotechnology: Science-based Concerns.*⁴⁰ A 12-member panel of NRC scientists drafted the report at the behest of the FDA, which was poised to rule on the safety of food products made from genetically manipulated animals, whether through cloning or, as with the fast-growing salmon, genetic engineering. By late 2002, the FDA had yet to approve any of them for human consumption and planned to issue approval guidelines in 2003. While the sale of products from cloned animals, such as dairy cows, appeared not to pose any significant risk to consumers, according to the report, the panel called for vigilance to ensure that transgenic animal products don't produce allergens that can harm consumers and to prevent the introduction of "pharm" products containing bio-introduced medicines (such as in milk) from entering the food supply. With these caveats, the panel deemed the products from transgenic animals otherwise safe to eat. A major concern of the NRC panel, however, was the risk to the environment should transgenic animals such as the new salmon escape into the natural environment. This could devastate the natural counterparts of the genetically modified animals if they have been conferred a reproductive edge. (The inventor of the supersalmon, Aqua Bounty Farms, Inc., of Waltham, Massachusetts, is attempting to address these concerns by rendering its transgenic salmon sterile so that any escaping fish will pose no danger to wild salmon.) The report also called for stepped-up regulatory efforts and greater coordination by government agencies charged with monitoring these emerging technologies.[41]

Genetically modified plants are already entrenched in American agriculture and a regular ingredient of the American diet. This quiet horticultural revolution has not been without controversy, however. In the same way that commercial viability is the goal with genetically altered animals, genetic tinkerers have sought to develop useful products with the plants they produce. The Scotts Company, for example, is developing a variety of genetically altered grasses with Monsanto and Rutgers University. Among the varieties envisioned for golf courses and suburban homes across America are grasses that are drought-resistant and impervious to certain weed killers, "low mow" grass, and grasses that come in different colors or even glow in the dark.[42] There are critics, of course. Among the concerns is the extent to which such grasses will spread into the wild and encroach upon, or supplant, natural grasses. There are also efforts underway to create genetically modified plants that produce medicines.

By the turn of the century, the majority of processed foods eaten by Americans contained, to some degree or another, genetically modified ingredients. Many unprocessed foods, too, are GMOs. Currently available GMOs include herbicide-tolerant soybeans, canola, and corn; insect-resistant corn; and virus-and-insect-resistant potatoes.[43] To make crops resistant to pests, for example, the Monsanto Company spliced a bacterial protein derived from *Bacillus thuringiensis* (Bt) into its seeds for corn. Monsanto's Bt corn, consequently, produces its own pesticide and eliminates the need for costly chemical spraying. (Marker genes are also inserted for identification purposes.) The Environmental Protection Agency (EPA) conducted safety tests on the genetically produced pesticide and determined they were not harmful to mammals. Moreover, the FDA has ruled that genetically altered foods require no special labeling[44] because there is "substantial equivalence" between GMOs and natural plants.[45] (An Oregon voter initiative that would have required such labels in Oregon—in defiance of the FDA's position—was defeated in November 2002 by a wide margin.)[46] Similarly, the EPA concluded that genetically engineered corn, cotton, and potatoes are neither a health risk nor an environmental threat.[47] In 2001, the American Medical Association published a study that reached the same conclusions.[48] In the United Nations' 11th annual Human Development Report released in 2001, the UN Development Program endorsed biotech crops and criticized those in the United States and Europe who foment fear of GM crops; the UN Development Program believes that biotechnology offers the hope of eradicating hunger in developing countries and may also lead to medicinal cures.[49] In the spring of 2001, U.S. farmers continued, as *The Wall Street Journal* put it, "their stampede into crop biotechnology" by planting 82.3 million acres of genetically modified seed, an 18 percent increase over the previous year.[50]

Despite industry and government reassurances, however, public resistance to genetically modified foods has been growing in the United States. By 2001, there were 54 nonprofit groups lobbying the FDA for stricter regulation of GMOs.[51] Among the concerns raised are the unknown environmental effects, the fear that insects resistant to the genetic toxins will develop, and the possibility that some of the new foods might produce allergens or toxins harmful to some consumers. By 2002, European consumers had largely turned their backs on GMOs and a de facto moratorium on modified foods remained in place throughout Europe due to a four-year-old moratorium on the approval of new GMOs imposed by the European Union, whose 15 member nations have also required that foods containing GMOs be clearly labeled to alert consumers.[52] "You started to see it, first of all, in Germany and Austria, where there was almost a paranoia about anything to do with genetic modification," European Parliament member David Bowe told PBS's *Frontline/Nova* for its April 23, 2001, broadcast of "Harvest of Fear." "Eugenics as an issue is a very, very sensitive one because of recent history in Germany, and I think it was there that you first started to see real public concern." In addition to peaceful protests, groups such as the underground Earth Liberation Front (ELF) have taken credit for dozens of acts of arson, vandalism, and sabotage at facilities around the country studying GM crops. ELF has admitted uprooting and destroying crops at the University of Minnesota and Cold Spring Harbor Laboratory, among others. The group also set a fire at Michigan State University's Agricultural Hall causing $400,000 damage.[53] The continued pressure from consumers and environmental groups has persuaded some companies—including McDonald's and Frito-Lay—to reduce or eliminate the use of genetically modified ingredients in their products.[54] In an odd twist, Novartis A.G., one of the largest agricultural biotechnology companies in the world, announced in 2000 that it had completely eliminated genetically modified ingredients from all of the company's food product lines, which includes Gerber baby foods and Ovaltine.[55] In April 2001, candy companies Hershey Foods Corp. and M&M/Mars announced that they were not interested in buying any genetically modified sugar beets to use in their products, the announcement coming just as American farmers were poised to begin planting the new herbicide-tolerant sugar crops.[56] About the same time, the legislatures of New York and Massachusetts were contemplating imposing moratoriums on genetically engineered crops in those states.[57]

A principal fear of GMO opponents is the phenomenon of genetic contamination. Broadly, the concern is that the genetic modifications will find their way into places they were not intended to go. For GMOs that have wild counterparts, opponents worry that the pollen of the GMO crops will travel (via birds, insects, or the wind) to the wild plants, cause interbreeding, and, perhaps, spawn "superweeds" that will be difficult to contain.[58] A second concern is that—even where the crops have no natural counterpart in the wild—there will be unintended cross-pollination of the GMOs with nearby crops that are unmodified. This kind of contamination is called genetic (or pollen) drift and is of great concern to farmers who are targeting the "organic," or "non-GMO," market. Although buffer zones are required by federal law around genetically modified crops in the field, critics of biotech maintain that pollen can drift for miles, well beyond the buffer zones. A third concern is that GM crops will become commingled with natural crops due to some kind of storage or transportation mixup. (Of course, GM crops have been treated as indistinguishable from the non-GMO crops, so a great deal of mixing was simply routine, which is how GMOs found their way into most processed foods.) The result can be approved GMO crops intermingling with non-GMO crops that are supposed to be segregated or, more alarmingly, GMO crops that have not been approved for human consumption getting into the food supply. A fourth area of concern is proprietary rights. Biotech companies are, or course, interested in protecting their patented products. Some of the contamination that occurs, in their view, is intentional.[59] Lastly, there is concern that resistant insects will emerge that aren't fazed by the GM crops. This will not only render the new crops obsolete, but more significantly, it will destroy a line of defense against crop failure that farmers have long relied on, namely, the pesticide that was inserted genetically. Since the EPA acknowledges this potential problem, the agency mandates that farmers plant unmodified crops alongside the modified ones to act as "refuges" for the bugs. Since bugs in these areas will not develop resistance—and they will, presumably, interbreed with resistant bugs that arise in the GM crop areas—the likelihood that resistant bugs will come to dominate is reduced.

Needless to say, the farming of genetically modified crops has produced a litigation brier patch. A great deal is at stake because the U.S. food business is a $600 billion-a-year industry.[60] Organic farmers who believe their crops have become contaminated by genetic drift from nearby GMO crops are beginning to file lawsuits in increasing numbers in state courts against neighboring farmers and the companies producing the GMOs. To further complicate matters, biotech companies are suing farmers when their GMOs are found growing in fields without the proper licenses. When Monsanto sued a Canadian farmer, Percy Schmeiser, over GM canola plants discovered in his field, the farmer countered that the plants were the result of genetic drift from area GMO crops and, thus, Monsanto's fault. In a March 2001 decision that worries organic farmers, a Canadian judge ruled in Monsanto's favor and ordered the farmer to pay for using Monsanto's intellectual property.[61] In October 1998, Pioneer Hi-Bred International, Inc., sued Cargill, Inc., one of the nation's biggest grain processors, for selling corn seeds that contained specific, proprietary genetic traits developed by

Pioneer's researchers. Although Cargill initially denied the claims, Pioneer offered up the results of sophisticated DNA-identification tests that substantiated the allegations. Within a few months, Cargill conceded publicly that Pioneer was right. Its own internal investigation determined that a former Pioneer researcher, hired by Cargill, had introduced the traits into Cargill's breeding lines.[62]

The biggest GMO calamity, however, involved a genetically modified corn seed called StarLink developed and sold by Aventis CropScience, at the time a division of Aventis, the French-German pharmaceutical company. (The division has since been purchased by Bayer.) Due to concerns the corn might produce human allergens, it was only approved by the EPA for use in animal feed, but it somehow found its way into the human food supply. When it first came to light, *The New York Times* called the fiasco "A Texas-Size Whodunit."[63] StarLink corn is a yellow corn bioengineered to contain, among other things, the protein Cry 9C (Cry is short for crystalline). Like Monsanto's Bt corn, StarLink is designed to be toxic to the European corn borer, but it also contains genes resistant to commonly used herbicides. Originally, Aventis developed StarLink with human appetites in mind, but StarLink got hung up by the EPA in the approval process. As part of that process, a company must demonstrate that a novel protein is not an allergen. To do this, tests that re-create the chemical conditions of the human digestive system must examine how long it takes the protein to break down. Proteins that readily break down are not considered a problem, but those that take longer are suspect. Cry 9C, it turned out, was on the longish side, so the EPA deemed it a potential allergen and approved StarLink only for industrial use (such as for ethanol) and animal feed, pending further study.[64] This is what is known as a split registration, which assumes that farmers will segregate animal-feed corn from corn destined for human consumption. (Farmers were required to sign agreements to this effect.) StarLink was first grown in 1998. The actual DNA that produces Cry 9C is not a problem, because DNA breaks down quickly in the stomach.

When the environmentalist group Friends of the Earth learned about StarLink from an EPA web site in July 2000, the group purchased a variety of corn-based products at a local store in Washington, D.C., and sent them to Genetic ID, an Iowa laboratory that specializes in testing foods for major food makers. Using PCR, and testing each food sample three times to avoid the possibility of a false result, Genetic ID reported back that they had detected the DNA of Cry 9C in taco shells sold in grocery stores by Kraft Foods (which, of course, indicated that the protein Cry 9C might well be present in the food, too). Upon learning of the laboratory results in September, Kraft, which sold the taco shells under a license from the fast-food chain Taco Bell, immediately began to recall millions of its taco shells from supermarkets across the country.[65] Unfortunately, this was only the beginning of a nationwide agricultural trauma.

Although Taco Bell's core customer base appeared largely unconcerned by the Star-Link crisis (and its parent company's stock actually went up in the weeks after the announcement of the recall),[66] other companies moved aggressively to rid their inventories of the tainted taco shells for fear of negative customer response. When the Safeway Supermarket chain learned in October that Genetic ID DNA tests had discovered StarLink in its house-brand taco shells, they recalled all their taco shells and offered refunds. As it turned out, Safeway's supplier, Mission Foods, is owned by a Mexican company that owns a Texas flour mill to which Kraft had first traced its StarLink problem, but it soon became clear that StarLink corn had spread much farther than that.[67] Shaw's Supermarkets followed with a recall of its store-brand taco shells because it had the same supplier as Safeway.[68] The supplier itself recalled its entire line of tortillas, taco shells, and snack chips, and the Texas flour mill recalled all of its flour made from yellow corn.[69] Many companies switched from using yellow corn in their products to using white corn, believing that this would sidestep the contamination problem. Before the recall wave was through, more than 300 food products originating from Mission Foods were recalled.[70] In December 2000, Starlink corn was found in corn shipments sent to Japan, even though Japan had earlier suspended U.S. corn shipments to avoid the problem.[71] It also turned up in other parts of the world. The first of many lawsuits filed against Aventis came barely a week after the initial news reports in September, this one alleging a serious allergic reaction to Cry 9C. Among the lawsuits that followed was a class-action suit brought by farmers alleging that Aventis was negligent in bringing StarLink to the marketplace. Among the problems outlined in the suit was that ill-informed farmers had allowed their StarLink corn crops to cross-pollinate with other corns, thus spreading the unapproved Cry 9C gene.[72] The U.S. company that licensed StarLink from Aventis for sale and distribution in the United States, Garst Seed Co., was also sued multiple times. By the end of 2002, dozens of lawsuits were still pending in federal court, cases set to go to trial over the course of 2003 and 2004. Even Taco Bell (about 100 outlets) complained in legal papers that their businesses were stigmatized by the StarLink disaster, even though Star-Link was never detected in any of the products sold in Taco Bell restaurants.[73]

Critics of Genetic ID initially charged that the laboratory was simply pursuing an anti-biotech agenda, suggesting its DNA results regarding StarLink couldn't be trusted. However, the DNA results were subsequently confirmed by Kraft and the FDA. The real question was: Is StarLink dangerous? Since dozens of GMOs had already entered the food supply without incident, perhaps this was all much ado about nothing.[74]

Things got more complicated when Garst Seed Company announced in November 2000 that field tests had revealed the presence of the Cry 9C protein in a variety of corn seed that is not genetically modified, apparently through cross-pollination or some kind of seed mixup.[75] As the StarLink problem spread, various test kits were developed to test for Cry 9C in the field. After grinding up corn samples and adding water, chemical strips were dipped in to see if there was a reaction. By December 2000, according to *The New York Times,* every major U.S. food and agricultural company was conducting field tests such as these on all shipments of corn. As a result of the tests, which indicated some level of contamination when the strips turned red, a large number of shipments were turned away, a number wildly disproportionate to the actual percentage of corn acreage planted using StarLink, 0.5 percent of the year 2000's total corn acreage. The reasons suggested by observers ranged from flawed tests to commingling during storage and transportation to cross-pollination. Whatever the reasons, the grain industry was in an uproar.[76]

In March 2001, the USDA announced that less than 1 percent of the country's corn seed supply would be contaminated with StarLink corn in 2001. In an effort to minimize even that level of contamination, the USDA announced a program to buy back from growers any corn seed suspected of being contaminated.[77] A day later, however, the environmentalist group Greenpeace announced that DNA tests conducted by Genetic ID had found StarLink contamination in vegetarian corn dogs made by Kellogg and recently purchased in Baltimore.[78] Simultaneously, the EPA admitted that its earlier policy of permitting split registrations would "no longer be considered a regulatory option."[79] A subsequent report by the USDA, which used stringent, and more accurate, field tests, found StarLink corn in 22 percent of grain samples tested around the country.[80]

In June 2001, the FDA conducted DNA tests on a product submitted by a Florida optometrist who claimed that he suffered an allergic reaction to a corn product. In a first, the FDA determined that traces of StarLink DNA were found in a white corn product, even though Aventis CropScience only modified yellow corn to create StarLink. The product was Kash 'n' Karry White Corn Tortilla Chips and was voluntarily pulled from shelves by the Kash 'n' Karry and Food Lion grocery store chains on the heels of the report. The finding sent another shiver through the food industry because many food companies had switched to white corn varieties in an effort to avoid StarLink altogether.[81]

The StarLink fiasco was a nightmare for Aventis. The company immediately forfeited its license to sell StarLink (under EPA pressure) and began a costly search-and-destroy program. Within months of the discovery, Aventis had forced out several executives and signed agreements with 17 states' attorneys general to reimburse farmers and grain elevators for their losses.[82] Although Aventis earned only $1 million in licensing fees for StarLink corn, the company spent hundreds of millions in the effort to eradicate it from the food supply. More than $1 billion was spent by other companies, including food makers and processors, farmers, and seed companies, in the first half of 2001 on the StarLink cleanup.[83] Aventis also put its CropSciences division up for sale. All the while, Aventis continued to argue that the scientific evidence suggests Cry 9C is unlikely to be an allergen and, besides, exposed consumers would only be eating trace amounts, if that.[84] Food processing destroys the Cry 9C protein, Aventis said. Furthermore, to the extent that DNA has been detected, DNA is not an allergen.[85]

A study by the Centers for Disease Control and Prevention (CDC) released in June 2001 found no evidence that 17 individuals (of dozens that came forward) who claimed allergic reactions to StarLink actually suffered as a result of exposure to the unapproved corn. The CDC failed to find any antibodies to the Cry 9C protein in the blood of any of those people, suggesting there was no immune response to the protein. (Failure to find such antibodies, however, is not considered definitive proof the individuals did not suffer allergic reactions to StarLink.)[86] Other claimants were ruled out for other reasons. An additional FDA study found that almost none of the food samples provided by those claiming to be ill from StarLink in fact contained StarLink at all. The sample that showed the DNA of Cry 9C did not reveal the protein. The product that did reveal the DNA of Cry 9C was the product submitted by the Florida optometrist, the white corn tortilla chips. It was not the original bag, however, but was purchased subsequent to the alleged allergic reaction.[87]

In the final analysis, the StarLink episode was edifying for consumer groups and industry insiders alike. The ongoing efforts to create genetically modified plants that produce pharmaceutical drugs (known as bio-pharming)—including, for example, an enzyme to treat cystic fibrosis and an antibody to treat herpes simplex virus— now inspires strident criticism from consumer advocates and environmentalists. Moreover, even though strict FDA guidelines on bio-pharming were announced in September 2002, a subsequent battle erupted between trade groups representing the food and biotechnology industries over concerns by food companies that the pharmaceutical crops will somehow worm their way into the food supply, either through mishandling or cross-contamination, just as with StarLink.[88] Tensions were only heightened after there were a couple of apparent near-misses in November 2002 when it was discovered that corn plants genetically engineered to produce a medical treatment for pigs, bioengineered by ProdiGene of College Station, Texas, came dangerously close to entering the food supply, if only on a minuscule scale. According to newspaper accounts, two small test sites of ProdiGene's new corn, one in Nebraska and one in Iowa, had been mishandled in various ways (although it was unclear exactly who did the mishandling), forcing the USDA to take prompt action to contain the new corn. In Nebraska, a soybean crop was grown in 2002 over a small test plot that ProdiGene leased from a farmer in 2001 to grow its bio-pharm corn. Unfortunately, a few corn plants sprang up from leftover ProdiGene seeds among the soybean crops and the two plants—despite a USDA directive to remove the corn plants—were harvested together. As a result, the corn stalks went with 500 bushels of soybeans to a grain elevator, where the shipment was intermixed with about 500,000 bushels of soybeans, an amount estimated to be worth about $2.7 million.

Upon learning of the possible contamination, the USDA quarantined the grain elevator and ordered the destruction of its contents. A day later, the USDA ordered an Iowa cornfield destroyed that surrounded another small test site used by ProdiGene over concerns that test corn had cross-pollinated with corn growing in nearby fields. ProdiGene later agreed to pay a civil fine of $250,000 under the Plant Protection Act of 2000 and reimburse the government for the costs of buying and incinerating the contents of the Nebraska grain elevator.[89]

DNA analysis made another contribution to the issue of genetically modified foods when *The Wall Street Journal* announced DNA test results from a laboratory it had commissioned to analyze foods touted as "non-GMO" or "GMO-free" to see if the products were in fact as advertised. The foods included products specifically labeled as "non-GMO" or "GMO-free," as well as the products of companies that publicly announced their aversion to GMOs, but did not include such labels. "Of the 20 products tested," according to the *Journal*, "11 contained evidence of genetic material used to modify plants and another five contained more substantial amounts."[90] Apparently through no fault of their own, the natural and health food companies that produced the labeled products appeared to be in violation of the federal Food, Drug and Cosmetic Act, which prohibits the use of misleading labels of food products. (The FDA subsequently sent warning letters to several companies.) The unlabeled products simply weren't living up to their companies' aspirations to be GMO-free. Beginning in October 2002, the USDA began enforcing a new national standard for all products labeled "organic." Depending on the ingredients, the products carry a USDA seal indicating organic contents based on a three-tiered approval system, from "100 Percent Organic" to "Organic" (at least 95 percent organic) to "Made with Organic Ingredients" (at least 70 percent organic). USDA guidelines identify organic ingredients as those produced without any synthetic pesticides, petroleum-based or sludge-based fertilizers, irradiation, antibiotics, growth hormones, or genetically modified organisms.[91]

The DNA tests were conducted for the *Journal* in 2001 at GeneScan USA in Belle Chasse, Louisiana, a laboratory that, like Genetic ID, performs DNA analyses for major American food companies. The samples sent to GeneScan bore only a code, so the laboratory did not know the brand names being tested. GeneScan used PCR to amplify the genetic material in the products, but since food processing often destroys most of the original plant DNA, the laboratory was unable to determine how much of a given product's original ingredients came from GM crops. Nevertheless, GeneScan did report averaged levels of GM DNA found in each product, with the highest amount, about 40 percent, found in a Canadian vegetarian product labeled as GM-free. A consensus seemed to be emerging that it was becoming impossible to certify food products as completely free of GM ingredients.[92] In January 2002, a group of about 1,000 Canadian farmers filed a class-action lawsuit against Monsanto and Aventis seeking damages for the cross-pollination of their organic fields with the spores of the companies' GM canola, which the farmers contend has significantly reduced the value of their "organic" canola crops. The farmers cite several DNA-based studies that have tracked the degree to which organic crops are contaminated by GMOs. The farmers' suit also seeks to prevent the introduction of genetically modified wheat, which seemed imminent. A similar instance of cross-pollination of GM wheat with natural varieties, the farmers fear, might be the death knell for their industry.[93] Monsanto had planned to bring its new Roundup Ready wheat to market in 2005, but a rising chorus of resistance to the new product appears to have stalled those plans. Monsanto announced through a spokesman in mid-2002 that it would not launch the new GM wheat until there was "industry acceptance across the board."[94]

Cloning remains one of the most controversial subjects of the modern age, even though the issue dates to the early 20th century. The prospect of human cloning, in fact, is partly responsible for the creation of a whole new field of endeavor, bioethics. Despite misgivings, there have been several high-profile declarations by various scientists that they intend to be the first to clone a human being. A week after Dolly was announced in February 1997, President Bill Clinton issued an executive order prohibiting federal funding for human cloning efforts for five years and implored the private sector to voluntarily observe a moratorium on human cloning research until the implications of the new technology could be fully studied. President Clinton also asked his National Bioethics Advisory Commission to evaluate the issue and the group promptly characterized human cloning as "morally unacceptable," given the risks involved with the current technology.[95] No federal law was immediately forthcoming, however, although four states (California, Rhode Island, Louisiana, and Michigan) moved quickly to impose their own bans on human cloning. By 2002, with Congress still debating the contours of a federal ban, two other states, Iowa and Virginia, had imposed their own idiosyncratic bans and 22 more states were considering legislative proposals to do the same thing. The state laws passed thus far vary. Iowa and Michigan banned all types of human cloning, while the other states outlawed reproductive cloning but, to different degrees, left open the possibility of cloning human embryos for research or therapeutic purposes. California's law bans only reproductive cloning and is set to expire at the end of 2002.[96]

Dr. Richard Seed, a physicist, was the first to gain national headlines for his open defiance of the voluntary ban by announcing at a scientific conference on human reproduction in December 1997 that he planned to open a human cloning clinic at the earliest opportunity. White House spokesman Mike McCurry was quick to describe Dr. Seed as "irresponsible, unethical and unprofessional," which turned out to be one of the nicer things people said about him. Within a month, 19 European countries, all members of the Council of Europe, signed the first international ban on human cloning.[97] Dr. Seed, it turned out, never succeeded in attracting the necessary financial backing for his clinic.[98]

More (apparently) tangible plans for human cloning were soon revealed by the Raelian movement, a bizarre UFO cult with 55,000 adherents that dates to 1973 when its French founder, Claude Vorilhon, a one-time race car driver and journalist who now goes by the name of Rael, claims to have been taken aboard an extraterrestrial spacecraft piloted by beings known as the ELOHIM. According to the Raelians, the ELOHIM created life on Earth some 25,000 years ago through genetic engineering. Cloning, as it happens, is an integral aspect of the Raelian movement, which Rael described to *The New York Times Magazine* writer Margaret Talbot as "the most fanatically pro-science of all religions."[99] Immediately after Dolly the sheep was announced in 1997, Rael and a group of investors under the corporate name Valiant Venture Ltd established CLONAID in the Bahamas. CLONAID is billed on its web site as "the first human cloning company in the world"[100] and has been actively pursuing the goal of cloning a human since 1997, although there have been some serious setbacks. Raelian zeal is not so much about being the first to clone a human, as it was with Dr. Seed, but about advancing to the next technological phase of their spiritual quest. Contrary to what most other religions teach, according to one of the Raelians, "DNA is the soul." Therefore, "reincarnation can only happen through science—through cloning."[101]

Unfortunately for the Raelians, pressure from the French press prompted the authorities in the Bahamas to close down the first CLONAID headquarters (from this point on, its corporate structure gets murky), and CLONAID has been trying to stay one step ahead of the law ever since. After moving the operation to a secret location in the United States, CLONAID attracted major U.S. newspaper coverage in the year 2000 when the company announced its intention to clone the deceased 10-month-old son of a wealthy American couple who had contributed $500,000 to CLONAID. The boy had died during what was alleged to have been a botched operation; upon learning about CLONAID on the Web, the couple decided to use the malpractice settlement to give their son's genome another chance at life. The Raelians set up a laboratory, enlisted 50 prospective surrogate mothers and assembled a team of scientists and doctors. The team leader was Dr. Brigitte Boisselier, a French-born Raelian biochemist with two doctorates. Most experts condemned the effort because, at the very least, any such attempt would pose serious risks of deformity or death to a developing fetus and produce an abnormal human. In fact, Dolly the sheep was the single success from 277 embryos created from adult sheep cells (most failed to grow, a few died *in utero,* and several were born dead), and her health appears to have suffered due to mysterious anomalies related to the cloning process. Problems were also encountered in the other species subsequently cloned (mice, cows, goats, pigs, and cats), including large-birth syndrome, high infant mortality, immunological problems, organ malfunction, and inexplicable developmental problems, such as obesity. "It is absolutely criminal to try this in a human," said Ian Wilmut, the leader of the group of scientists who cloned Dolly in 1996.[102]

To complicate matters, a third group of scientists jumped into the fray in early 2001 when they announced their intention to clone a human within the next couple of years. It was a serious bid. At a conference in Lexington, Kentucky, on January 26, 2001, Dr. Panayiotis M. Zavos, a retired professor of reproductive physiology at the University of Kentucky, and Dr. Severino Antinori, an Italian reproductive specialist who had gained notoriety previously for assisting women well past menopause to become pregnant (including, in 1994, a 62-year-old grandmother), announced that they had already lined up ten infertile people willing to be cloned.[103] The reaction was similar to the one that greeted CLONAID's plans, near universal condemnation. The Vatican accused Dr. Antinori of trying to "emulate Hitler" and pursuing "Nazi madness."[104] Both this team and CLONAID found themselves in deep water after the FDA asserted jurisdiction over the clinical research involved in human cloning and sent them warning letters.

To sort out the FDA's claim of jurisdiction, a hastily convened congressional hearing was held in March 2001 before the Subcommittee on Oversight and Investigations of the House Energy and Commerce Committee.[105] Both Zavos and Boisselier testified at the hearing, as did bioethicists and experts in biotechnology and assisted reproduction. Director of the FDA's Center for Biologics Evaluation and Research Kathryn Zoan told the committee in a prepared statement, "Because of unresolved safety questions on the use of cloning technology to clone a human being, FDA would not permit the use of cloning technology to clone a human being at this time."[106] Dr. Zavos attempted to reassure the committee members about his intentions by stating, "We have no intention of stepping over dead bodies or deformed babies in order to accomplish this," explaining that careful monitoring for genetic defects would occur at every stage of the process. Others who testified emphasized that the problems arising with cloned animals are not always traceable to genetic anomalies, so they would escape early detection. Proceeding with human cloning at this juncture would be nothing short of bald human experimentation, it was argued.[107] Rael testified, too, asserting, "Nothing should stop science."[108] The tension was raised when a biotech company, Advanced Cell Technology of Worcester, Massachusetts, announced in the summer of 2001 that it planned to clone human embryos for research purposes. Later in the year, the company made headlines again when it announced that it had, in fact, produced three human embryos through the cloning process, although the embryos divided only briefly before expiring (the longest-lived of the three died after only six cell divisions).[109] In May 2002, it was revealed by documents obtained under the California Public Records Act that researchers at the University of California in San Francisco, under the direction of embryologist Roger Pederson, had conducted similar cloning experiments with human embryos in 1998 and 2001. Reportedly, the experiments were unsuccessful. The UCSF experiments were undertaken in an effort to produce stem cells genetically identical to cell donors for possible use in medical therapeutics. The research objective was not reproductive cloning.[110]

Although questions still remained about the FDA's jurisdiction, the agency nevertheless tracked down CLONAID's laboratory in West Virginia after the 2000 hearing and pressed it to cease operations.[111] A second round of hearings was also held in Congress on the issue of human cloning, which led to the passage in 2001 of a bill in the House of Representatives (by a vote of 265 to 162) to ban all forms of human cloning, including therapeutic cloning. (Therapeutic cloning, as opposed to reproductive cloning, is focused on treating a patient's disease by creating an embryo from his own cells, or those of someone else, to generate replacement tissues designed to ameliorate the condition. As of January 2001, British law allows therapeutic cloning, as well as embryonic stem-cell production through cloning. Under British law, however, cloned human embryos must be destroyed after 14 days.[112]) An additional blow to CLONAID came when the parents of the ten-month-old deceased boy pulled out of their deal and shut the Raelians out of the West Virginia laboratory.[113] Despite the setbacks, and despite the criticisms leveled at the Raelians and Drs. Zavos and Antinori (who have since gone their separate ways in their cloning pursuits), these maverick scientists have vowed to press ahead with their cloning efforts.[114]

The furor over human cloning only escalated in 2002 with repeated claims by several scientists that women under their care had in fact been impregnated by cloned embryos. By this time, there were a reported five separate research teams in various parts of the world attempting to clone humans. In April 2002, Panayiotis Zavos, who holds a Ph.D. in physiology, not a medical degree, and runs two fertility clinics in Kentucky, confirmed that he and Dr. Antinori had severed their relationship in late 2001 over questions of methodology. Zavos and Antinori then began pursuing cloning efforts independently and with different research teams. The Raelians also continued their efforts. In addition to these three groups, it was reported that research teams in Russia and China were also trying to clone humans. Beginning in April 2002, rumors began to circulate that women under the care of Dr. Antinori, a gynecologist, were pregnant with human clones rumors he later confirmed. Many scientists were skeptical. MIT Professor of Biology Rudolf Jaenisch, who is also affiliated with the Whitehead Institute, said, "I do not trust these people to tell us the truth."[115] Later in the year, Dr. Jaenisch and some colleagues published a study in *The Proceedings of the National Academy of Sciences USA*[116] showing that the genomes of cloned mice—which they examined using gene-chip technology—contain hundreds of mutations across dozens of genes as a result of the cloning process. According to Dr. Jaenisch, the study explains why so many clones die in utero or shortly after birth and demonstrates, he told a reporter, that it is "very irresponsible to think this method could be used for the reproductive cloning of humans."[117]

In May 2002, the House reform subcommittee on criminal justice, drug policy and human resources convened another hearing on human cloning, during which Zavos and a number of law enforcement experts testified. Zavos told the subcommittee that his own research team intended to proceed over the summer with aggressive efforts to impregnate a number of women with clones of their husbands or partners. He predicted the birth of the first human clone by early 2003. In a statement delivered to the House subcommittee, the Department of Justice warned of the difficulty of policing the field of assisted reproduction in the event that reproductive cloning is outlawed and therapeutic cloning is lawful. Embryos created through either conventional IVF procedures or therapeutic cloning would certainly look identical to those created for clandestine efforts at reproductive cloning. Some fear reproductive cloning will simply become an underground activity. Forensic DNA analysis would provide perhaps the only way to establish that an illegal cloning procedure had taken place, assuming the "parent" can be identified. Criminal cloning investigations, in other words, could be a real headache for law enforcement and exceedingly invasive for research laboratories, fertility clinics, and patients. So far, however, the groups attempting human cloning are operating outside the United States.

Finally, on December 27, 2002, Dr. Brigitte Boisselier announced that CLONAID scientists had succeeded in producing the first human clone, who they nicknamed "Eve." The baby was born the previous day, according to Boisselier, at an undisclosed location outside the U.S. to an unnamed 31-year-old American woman, who was the subject of the cloning procedure. No proof of any kind was offered to substantiate the claim, but Boisselier promised that arrangements were being made through an independent freelance journalist, Michael Guillen, to have DNA samples obtained for a battery of DNA tests by scientists Guillen characterized as "independent, world-class experts." Despite the assurances for DNA tests, however, CLONAID soon backed away from submitting "Eve" and her "mother" for any tests after a Florida lawyer, unconnected with the group, sought the appointment of a guardian for the child. (The petition was later thrown out by a judge.) The credibility faux pas created by the lack of DNA tests for "Eve" did not, however, prevent CLONAID from announcing the births of two additional human clones over the course of January 2003 (with a fourth and fifth promised by early February), also without proof of any kind.[118]

How will we know when someone has actually cloned a human being? Will the mere claim that they have succeeded be enough? Would scientific tests, such as DNA tests, offered by the cloners be credible? Since everyone seems to agree that human cloning is inevitable, isn't the stage set for a truly colossal bogus claim? Certainly CLONAID's claims are worthless unless independent DNA tests are conducted by reputable scientists to authenticate them. In fact, though how do we know any of the cloning claims—of lower mammals—are really true?

To be sure, there were doubting Thomases aplenty when a team of obscure scientists in Scotland announced that they had cloned a mammal from adult cells. Indeed, the cloners of Dolly were formally challenged about their claim. For one thing, by the time Dolly was born, most scientists had long since written off cloning a mammal from adult, or even slightly differentiated, cells as impossible. For another, the history of cloning was littered with instances of suspect claims and outright fraud. There had been lawsuits, investigations, and wrecked careers over cloning claims. To a great extent, the sorry history of cloning explains why only obscure scientists were attempting to do it. The vast majority of scientists rely on institutional grants to fund their research, so very few of them wanted to risk professional suicide by pursuing what conventional wisdom held was science fiction.

Most of the cloning attempts conducted in the 20th century were not done with the primary goal of cloning for the sake of cloning. Most such experiments were performed to test a theory about cell biology or, as was the case with Dolly, merely as a stepping stone to a greater scientific objective. All of the attempts focused on the phenomenon of cell differentiation, the process by which early embryonic cells gracefully shift gears from totipotency (the ability to generate or regenerate a whole organism from a part—a genetic state with seemingly infinite possibilities) to a state of specialization when the dividing cells are designated as either muscle cells or liver cells or any of the myriad cell types that make up the body (in humans, there are about 260 types). In the late 19th century, August Weismann, a German professor of zoology and comparative anatomy at the University of Freiberg, proposed that each time a cell divided and differentiated, it lost a greater and greater share of the hereditary information that makes up the entire organism, ending up with only the specific information that it absolutely needed to function.[119] (Note that forensic DNA analysis is predicated on the knowledge that Weismann was wrong and that virtually all cells in the body contain a complete set of genetic instructions, although most of it is dormant.) For more than half a century, scientists fiddled around with animal embryos in an effort to prove or disprove Weismann's theory. The first experiments, conducted by German embryologist Wilhelm Roux, seemed to bear out Weismann's theory. Roux destroyed half of two-celled frog embryos with a hot needle and watched as the remaining cells produced only half embryos. Was this confirmation?[120] Many believed so. Another German, Han Dreisch, however, raised doubts when he experimented on sea urchin eggs. Dreisch shook the embryos apart (first two-celled, then four-celled) and produced identical twin and quadruplet sea urchins, although they were smaller than normal. Not only did this contradict Weismann, but it was probably the first time a human had successfully cloned an animal of any kind from a single cell.[121] The next significant experiment was conducted by Hans Spemann, who was later awarded the Nobel Prize in 1935 for his work in embryology. In 1902, when the dispute between the Dreisch and

Roux camps still raged, Spemann managed to separate the cells of two-celled salaman-
der embryos (which he found were too delicate to brook shaking) by fashioning a
noose from the hair of his newborn son and tightening it. Each of the cells went on to
become separate embryos and grew into normal salamanders, a result that seemed
to refute Weismann's theory. Subsequent experiments by Spemann and others demonstrated that it is the cytoplasm—the sophisticated cellular material that surrounds the
nucleus—that dictates whether embryonic cells, after several cell divisions, can be restored to totipotency. Although the technology did not yet exist for such a feat, Spemann famously proposed in 1938 that one day, somehow, a "fantastical experiment"
should be undertaken whereby the nucleus from a differentiated cell, even, perhaps,
one taken from an adult, would be placed inside the cytoplasm of an egg whose own
nucleus had been removed, just to see what happened. The crude instruments of the
time, however, would have caused far too much damage to the cellular material, so
Spemann never dared to try it.[122]

Throughout this period, and even into the early 1950s, Weismann's theory of
hereditary information loss was still a leading theory on cell differentiation. After all,
just because early embryonic cells appeared to retain all of the original genetic information didn't necessarily mean that more advanced cells had any reason to. It fell to
Robert Briggs of the Institute of Cancer Research in Philadelphia and Thomas J. King
of New York University to test the theory. Their experiments on the eggs and embryos of American frogs *(Rana pipien)* produced the first, at least partially, successful
clones by nuclear transfer, as envisioned by Spemann. Retrieving the nuclei from
blastula cells—embryos that contained 8,000 to 16,000 cells—and placing them in
eggs whose nuclei had been removed, Briggs and King succeeded in 1952 in creating
197 reconstructed frog embryos, from which 27 tadpoles resulted (none grew beyond the tadpole stage, however). Once again, Weismann's theory took a hit. Quite
coincidentally, Briggs and King had ushered in the modern age of cloning.[123] Subsequent experiments conducted in other laboratories, however, revealed that the more
advanced the cells (experiments were tried using cells right up to the tadpole stage),
the less likely it was that functional embryos would develop. The older cells produced developmental problems and early death. For decades, consequently, it was
widely held that cloning was impossible with adult cells.[124]

In the late 1950s, Oxford developmental biologist John Gurdon began experimenting with the nuclear transfer method hoping to disprove Weismann's theory
once and for all. In 1962, Gurdon reported success in producing embryos, tadpoles,
and even fully developed, sexually mature albino frogs from cell nuclei taken from
the intestinal cells of feeding tadpoles (and placed in the embryos of a different
species). While his success rate was low (1.5 percent for producing the full-grown
frogs), his feat nevertheless attracted worldwide attention and clearly seemed to refute Weismann. Unfortunately, the lore that arose from Gurdon's work exaggerated

his actual accomplishment. It was widely held, and even recounted in textbooks, that Gurdon had cloned frogs from the cells of adult frogs, when in fact that wasn't true. He had obtained his cells from the intestines of tadpoles.

Even so, not everyone was convinced. Dennis Smith, a student of Robert Briggs, conducted follow-up experiments and found that he could only clone frogs from undifferentiated embryo cells. Furthermore, Smith pointed out, 2 to 5 percent of the cells Gurdon had used from the tadpole intestines were actually primordial sex cells (cells in the process of becoming sperm or eggs that had, in some instances, yet to finish undergoing meiosis, so they still contained the full complement of the frog's genes), which typically migrate from the stomach walls through the intestines on the way to the gonads. Therefore, Smith suggested, Gurdon was not cloning from differentiated cells at all, but instead from immature cells little different than very early-stage embryonic cells. To drive the point home, Smith repeated Gurdon's experiments using a species of frog whose primordial sex cells do not move through the intestines and reported that the experiments failed utterly.[125]

Whatever the truth of Gurdon's results, the idea that cloning from adult cells was possible entered the public dialogue. It was true, it seemed, at least for amphibians. Eventually, of course, scientists would crack the mammal barrier and clone humans, many believed. Consequently, from the 1960s onward there were many bold pronouncements about what it all meant for mankind. Some felt it was great, others were horrified. Among the most notable early proponents of cloning was Nobel laureate Joshua Lederberg. In the mid- to late 1960s, Lederberg wrote a couple of high-profile articles, including one for *The Washington Post,* in which he explained that cloning offered a way to improve humans. By this time, two-time Nobel laureate Linus Pauling (1954 for chemistry and 1962 for peace) was already on board, having seriously proposed in a 1968 paper published in the *UCLA Law Review* that people with recessive, disease-causing genes have their foreheads tattooed to prevent their mating with another carrier of the trait. In 1971, James Watson wrote an article for *Atlantic Monthly* entitled "Moving Toward Clonal Man," in which he warned that the cloning of humans was not far off.[126]

Such views were very unsettling to many people, among them National Institutes of Health biochemist and medical doctor Leon Kass. Inspired to write a reply to Lederberg's *Washington Post* article, the tone of which Kass found "cavalier," Kass found himself driven away from biology and into philosophy and ethics because of the cloning debate. In 1969, he joined the Hastings Center, an ethics institute dedicated to examining the philosophical and theological questions posed by modern science and medicine.[127] Kass, the bioethicist, ultimately joined the faculty at the University of Chicago in 1976, where he remains. Over the last 35 years, Kass has been steadfast in his opposition to human cloning. In August 2001, President George W. Bush authorized the use of federal funds for the study of some 60 or so pre-existing

embryonic stem-cell lines (totipotent early embryo cells that can be endlessly reproduced in culture). Shortly afterward, President Bush tapped Dr. Kass to chair a presidential advisory committee, the Council on Bioethics, to examine, among other
things, stem-cell research and oversee the allocation of those funds to researchers.
The council, which replaces President Clinton's National Bioethics Advisory Commission, will also address assisted reproduction, cloning, genetic screening, PGD,
gene therapy, euthanasia, and other medical issues.[128] Just prior to his appointment
to the Council on Bioethics and before the House finally voted to ban human cloning,
Dr. Kass co-wrote an article for *The New Republic* on cloning with Daniel Callahan, a
co-founder of The Hastings Center. They argued: "[Human cloning] is only and emphatically about baby design and manufacture, the opening skirmish of a long battle
against eugenics and the post-human future. Once embryonic clones are produced in
laboratories, the eugenic revolution will have begun."[129] In July 2002, the President's
Council on Bioethics released a report entitled *Human Cloning and Human Dignity: An
Ethical Inquiry*[130] calling upon Congress to impose a four-year moratorium on human
cloning for biomedical research (therapeutic cloning) and an outright ban on reproductive cloning. The 18-member council was nearly unanimous in its denunciation of
reproductive cloning. The council voted ten to seven (one abstention) for a moratorium on biomedical-research cloning to allow time for additional public debate on
the issue. "Cloning touches many of the most fundamental aspects of our humanity
and our competing ideas of the good life," wrote council director Professor Kass in
an opinion piece published in *The Wall Street Journal* at the time the council's report
was released, "and is a harbinger of even more daunting biotechnologies."[131]

For all the exaltations and dire predictions over cloning, a series of developments in
the late 1970s and early 1980s almost killed off cloning research entirely. In the late
1970s, the cloning of humans seemed just within reach, not only to the general public
but even to some scientists. Consequently, when a popular science writer, David
Rorvik, published a book in 1978, on the nonfiction list of a respected publishing
house, J. B. Lippincott, entitled *In His Image: The Cloning of a Man,* purporting to recount the inside story of a multimillionaire's successful quest to have himself cloned,
many people were inclined to believe it. The actual arrival of the first test-tube baby,
Louise Brown, at a clinic in England that same year surely added credibility to Rorvik's
revelation. Although the book's claim prodded Congress into holding its first official
hearing on cloning in 1978, Rorvik's book was later exposed as a hoax.[132]

By this time, the actual prospects for cloning had apparently brightened, but it
would not take long for a dispute over another cloning claim to drive a stake through
its heart. "Cloning became a pursuit of those who worked on the edges of science,"
wrote Gina Kolata in *Clone: The Road to Dolly and the Path Ahead.* "It was relegated to
those who worked with farm animals, and whose papers, even when they appeared in
leading journals, often were not read by the scientific elite."[133] The reason for

cloning's stock crash was Karl Illmensee, a highly regarded German biologist who announced at a major scientific conference in 1979 that he had cloned three mice from the cells of early (four-day-old) embryos through nuclear transfer. Illmensee's claim marked the first time any scientist had ever formally claimed to have cloned a mammal (although the Rorvik saga was still unfolding) by nuclear transfer. Illmensee's announcement followed years of failed efforts by scientists around the world to do with mice, on any level, what Briggs and King had done with frogs in 1952. That Illmensee, who worked with Peter Hoppe of the Jackson Laboratory in Bar Harbor, Maine (which remains one of the world's principal repositories of inbred mouse strains), was able to overcome the apparent technical hurdles posed by cloning is what most impressed his professional colleagues. Mouse embryos are truly tiny and their nuclei tinier still, so working with them required a certain technical virtuosity. Illmensee, however, had such a solid reputation in his field that few thought to challenge his ability to get the job done. What astounded everyone, of course, was that the mice (derived from advanced embryo cells—plucked from the inner cell mass, or ICM, of developing embryos) had grown to maturity, according to Illmensee. This left Briggs and King in the dust. It was John Gurdon redux.

When the public first learned of it—from January 1981 news accounts that coincided with Illmensee and Hoppe's technical account of the cloning experiment published in the journal *Cell* that same month—there was the expected hand-wringing over the prospects of human cloning. For Illmensee, the result was virtual stardom in the scientific world. Indeed, the *Cell* paper laid out a very plausible scenario on a technical level.[134] But there were a couple of huge problems.

Most significantly, other well-respected scientists had no luck repeating Illmensee and Hoppe's cloning experiments. Was it simply the case that Illmensee was, as advertised, a laboratory maestro, whereas these other scientists were incompetent? Or had Illmensee faked his results somehow? At about the same time, a group of young scientists who had been attracted to Ilmensee's laboratory in Geneva in hopes of learning his legendary techniques, such as his technique for performing nuclear transfer, became disenchanted with how Illmensee treated them and began to suspect that he was, in fact, faking data in some of his laboratory experiments. Was Illmensee a fraud? If so, what did this say about his much-ballyhooed cloning claim. Eventually, each of these problems converged on Illmensee like violent storm fronts and devastated his career. Hoppe was also ruined. At the same time, the Illmensee winds blew cloning almost completely off the conventional radar screen. Nobody would try it. Nobody would fund it. Nobody believed it was possible to clone mammals from differentiated cells. But what's to be said about the Illmensee saga now? Don't we now know, for a fact, that the cloning of some mammals by nuclear transfer is possible? And from adults, no less. In fact, more than possible, it's routine, right? On the other hand, who are these scientists from Scotland, anyway? Or these other wizards of cloning? Have we been hoodwinked?

None of this is to suggest, however, that a more limited type of mammal cloning research wasn't taking place. In fact, cloning by simply dividing and re-dividing very early embryos (that remain totipotent) to produce identical animals became routine. The first sheep cloned by this method were produced by Danish veterinarian Steen Willadsen at the Agriculture Research Council's Institute of Animal Physiology at Cambridge and announced in 1979. Willadsen would later produce a series of chimeric animals by combining embryo cells of different animals and even different species, among them a goat/sheep mix called a geep. (On a cellular level, chimeras are like animal figures fashioned by squeezing together two different colors of Playdo; some of the cells will be, in the case of a geep, genetically goat, while other cells in the body will be genetically sheep.) In 1985, Willadsen broke additional ground by producing cloned sheep—from early-stage embryos—by nuclear trans-fer.[135] The work by Willadsen and other scientists in the United States on large mam-mals led to a short-lived commercial flirtation with the mass production of elite farm animals using the embryo-splitting cloning technique, but it was costly and failed to catch on.[136]

Eventually, one of the scientists working in Illmensee's laboratory confronted him at a public gathering after Illmensee had just finished presenting some new results to a group of scientists, implying that Illmensee had falsified his results. Subsequently, the young scientist formalized his concerns in a statement to administrators at the University of Geneva, sparking an inquiry. It was scandalous. The University of Geneva quickly appointed a five-member international commission to review the ev-idence against Illmensee. Although the commission found the charges of fraud un-proven, the panel concluded that the questioned experiments were "scientifically worthless." A defeated Illmensee resigned his post in 1985.[137]

By itself, the university scandal was enough to raise serious questions about the authenticity of Illmensee and Hoppe's three cloned mice. Quite independently, how-ever, scientists from outside Illmensee's laboratory tried to reproduce the cloning ex-periments and failed utterly. German embryologist Davor Solter, while working at the Wistar Institute in Philadelphia, teamed up with American postdoctoral student James McGrath to methodically repeat Illmensee's most spectacular work, including the cloning experiment. They were unable to reproduce it. At the conclusion of these efforts, Solter and McGrath submitted two scientific papers that refuted claims made by Illmensee and Hoppe. Despite the enormous difficulties in getting a paper published that outlines only negative results, both papers were accepted for publica-tion. The paper on the inability to clone appeared in *Science* in December 1984. It baldly stated that their conclusions "contrast with the results of Illmensee and Hoppe," ending with the statement that "the cloning of mammals by simple nuclear transfer is biologically impossible." It was a weighty statement, given the respect ac-corded the scientists and the stature of *Science,* so cloning entered its dark ages. Solter would later head a laboratory at the Max-Planck Institute in Freiberg, Germany. McGrath went to medical school and became a geneticist at Yale.

Despite the generally anti-cloning climate, Ian Wilmut—an Englishman who received his Ph.D. in 1971 from Darwin College in Cambridge—began to pursue cloning in earnest in 1986, but only as a means to an end. In 1982 Wilmut received an ultimatum at work to either focus on genetically altering animals or quit his job. He had been working at the Roslin Institute in Roslin, Scotland, near Edinburgh, since 1973 (although it was called the Animal Breeding Research Station when he signed on). After much soul-searching, Wilmut, whose primary interest lay in the physiology of embryo development, decided to stay. The genetic manipulation of one-celled embryos (zygotes) was, however, a time-consuming, difficult, hit-or-miss affair. Charged with developing a protocol for efficiently adding desired DNA to zygotes, Wilmut had a scientific epiphany. He saw that it would be much more efficient to add DNA to a plateful of cultured cells, determine which cells had incorporated the DNA, and then transfer the newly designed genomes into an embryo for transfer into a surrogate mother for gestation. "In other words," Wilmut explains in *The Second Creation: Dolly and the Age of Biological Control* (2000), an account of Dolly's creation written with Keith Campbell and Colin Tudge, "as the 1980s wore on, I began to see that the future of genetic engineering in animals lay in cloning."[138] One of the keys was figuring out how to culture embryonic cells so as to have an inexhaustible supply of them and in such a way that they retain totipotency (or nearly so). This proved tricky.

To work with embryonic stem cells would open up a world of possibilities. Not only could genes be added to animals, they could be knocked out or even taken out, repaired, and put back in. In the early 1980s, genetic engineers had begun creating strains of mice by culturing embryonic stem cells, manipulating them, then installing the changed cells into early embryos. (In other words, they put a designer embryo cell into another early embryo, where it took up residence.) The method produced a kind of chimeric mouse known as a "mosaic" (the creature has two kinds of cells distributed throughout its body, cells with the added gene—or transgene—and unaltered cells). When the engineers got lucky, the transgene ended up in the mosaic mouse's germ cells (sperm or eggs), so its offspring inherited the transgene. To do this with sheep, however, Wilmut saw, would require many years of trial and error, since sheep have much longer gestation and maturation times than mice. Furthermore, embryonic stem cells had not been isolated in sheep. In fact, at the time Wilmut began his work, embryonic stem cells had been found in only two strains of mice (and no other species).

Despite concerted efforts at Roslin to culture embryonic sheep cells so that they would retain totipotency, the cultured cells always began to differentiate, ultimately becoming skinlike (a kind of cell referred to as a fibroblast). Despite these setbacks, Wilmut soon found inspiration in the work of Steen Willadsen, particularly the sheep-cloning feat using early embryos. (When other scientists began using Willadsen's method, cows soon followed, then horses, pigs, rabbits, goats, and, in 1997, rhesus monkeys.[139]) But this was cloning from early embryo subdivision, not from advanced cells. In 1987, at a time when the conventional wisdom held that cloning could only proceed (viable embryos be reconstructed) from embryo cells retrieved within the first few cell divisions, Wilmut learned that Willadsen had succeeded in cloning calves from far more advanced embryonic cells, those at the ICM (inner cell mass) stage taken from a blastocyst (research he never published). Wilmut visited Willadsen in Canada, and Willadsen, who confirmed the rumors, was more than willing to provided technical details and demonstrate his technique for creating viable embryos.[140] Armed with this knowledge, Wilmut and his Roslin team produced their first cloned sheep from an ICM cell in 1989.[141]

Thereafter, Wilmut began to seek the outer limits of totipotency—both for embryonic cells and cultured cells. There had to be a cut-off point, he felt. At the same time, Wilmut and those at Roslin began to sense that another factor came into play, the cell cycle. Perhaps, they began to suspect, the cell cycles of the two cells involved in nuclear transfer (the cell from which the nucleus is derived and the receiving embryo) should be synchronized. In 1991, therefore, postdoctoral researcher Keith Campbell was hired by Roslin for his expertise in the cell cycle. It turned out to be a very fortunate hiring decision. Campbell had a varied background, which included cancer research. Cancer cells, he had observed, defied the conventional wisdom about cell differentiation by returning, to some extent, to a more primordial state: Some tumors produce a wild variety of body cells (albeit in the wrong place), including liver cells, fingernails, bone, and hair. This suggested that the cellular clock can be turned back because cancer cells had somehow figured out how to activate parts of the genome that were previously turned off.[142] Keith Campbell's insight into the cell cycle proved crucial to the genetic engineering research that led to the cloning of Dolly. The pivotal insight came from an understanding of the phases of cell division, during which the cellular machinery marshals its forces to produce two cells from the original one. It is an extremely orderly process. Campbell rejected the emphasis on finding embryonic stem cells to facilitate cloning; the cell cycle was the key, he believed.

The cell cycle has two broad phases: mitosis (or, when referring to germ cells, meiosis) and interphase. For a body cell to duplicate itself, it must go through these two phases. Mitosis is the actual cell division. Interphase has several stages of its own, the most significant of which is the synthesis, or S, phase. Synthesis is the part of the process when the DNA makes copies of itself. Significantly, the synthesis phase produces a whole new set of chromosomes that are identical to the original set contained in the cell nucleus. Since every cell with a nucleus already has two sets of chromosomes, one from the mother and one from the father (this is called being diploid), the synthesis phase ultimately produces four sets of chromosomes (which is called being tetraploid). The mitosis phase sends two sets of chromosomes to one daughter cell and two sets to the other daughter cell (restoring them to the diploid state). Since it is important that each daughter cell receive the correct number of chromosomes, synthesis must be completed before the cell divides (mitosis). Nature has therefore provided intermissions, or Gaps, between the main events. Gap 1 is the phase that precedes the start of the synthesis phase, and Gap 2 comes after the synthesis phase completes DNA replication and before mitosis. Significantly, cells can go into a hibernating, or quiescent, state during the G1 phase, a state known as G0 (for either "G-zero" or "G-nought"). The G0 state, it turns out, not only occurs naturally but can be induced in cells that have been cultured by reducing the nutrients (growth factors) that generally sustain them. It turns out, too, that the G0 resting state provided Campbell with the perfect window of opportunity for resetting the genetic clock and producing clones from more advanced cells.[143]

Differentiated, or specialized, cells use only that portion of the genome that is necessary to the particular cell type's function (i.e., liver or skin cell). The rest of the DNA is masked by proteins that prevent its operation. During cell division, however, the masks come off for a time when the DNA is duplicated, making the DNA amenable to reprogramming. An egg in its resting state, similarly, rearranges the protein masks of the DNA. The trick, then, is to synchronize the resting states of both donor nucleus and egg.[144]

The formation of the sex cells are the key to individuality. A combination of genetic crossing over and the random allocation of the duplicated chromosomes to the various daughter cells allows for an infinite number of genetic combinations. Added to this process is the sexual union of the sperm chromosomes and the egg chromosomes. Every individual that emerges from sexual reproduction (unless he or she is an identical twin or a clone), therefore, is truly unique in a very real genetic sense. And it is this uniqueness that is the foundation of forensic DNA identifications.

During the early 1990s, the scientists at Roslin conducted a series of experiments aimed at determining the optimal time for performing nuclear transfer. What, they wondered, were the best stages of the cell cycle for nuclear transfer, both for the donor nuclei and the receiving oocyte (a diploid egg that has not yet completed meiosis, which turned out to be the preferable stage of egg development for performing nuclear transfer). The experiments at Roslin in the early 1990s differed significantly from those conducted in the 1980s, however. Now the scientists knew exactly which stage of the cell cycle the donor nuclei were in when they were transferred (and this made all the difference). Several stages of the cell cycle were tried on the theory that the cellular biochemistry had to be finessed.

The Roslin team next set out to determine whether differentiation, once it had taken hold, was, in fact, a reversible cellular condition. Campbell felt that by properly coordinating the cell cycles of the donor nucleus and the receiving cytoplasm (of the oocyte), it would be possible to reprogram the donor nucleus to behave as if it were brand new. Beginning with nine-day-old embryos (which showed a distinct inner cell mass—ICM), the Roslin team transferred the ICMs into flasks with ingredients designed to encourage them to multiply. As such cells grow and fill up their original container, they are transferred to additional containers (replated on flat surfaces) ad infinitum. Each such transference is called a passage (pronounced as if it rhymed with collage). They only transferred donor nuclei that were diploid. The easiest way to assure diploidy, the Roslin team found, was to put the donor cells into the resting, or G0, state by "serum starvation." The cells that emerged from their experiments showed clear indications that they were differentiated cells, having acquired, for instance, the skinlike texture of fibroblasts.

With these cells, the team reconstructed embryos and implanted them in surrogate sheep. All of the cells were derived from the same nine-day-old embryo, so, theoretically, all of the embryos were simply additional copies of the original. By the time the team had confirmation of several pregnancies, Campbell recognized that it was the G0 state, in fact, and the G0 state alone, that would allow chromosomes to be reprogrammed. All cells pass through the G0 phase as they differentiate, Campbell reflected, and it is at this stage that nature will briefly release its firm hold on the genome and permit, under the right conditions, reprogramming. In time, 16 surrogate ewes produced five live lambs, although two died immediately and another died ten days later. The remaining two, named Megan and Morag, thrived. It was instantly clear that the experiment was successful because the lambs were Welsh Mountain sheep, a different breed from the surrogate ewes. Nevertheless, DNA analysis targeting microsatellites was later conducted to establish that the cloning procedure had worked. But it revealed not only that Megan and Morag were true clones and genetically identical, but that all of the five lambs born in August 1995 (including the three that died) were genetically identical. The focus of the Scottish team was a major departure from that of earlier cloning experiments, which always emphasized the degree of differentiation of the early embryo cells. "This shift in emphasis," Wilmut and Campbell boast in *The Second Creation,* "from degree of differentiation to cell cycle, is crucial and huge. This indeed is the insight that has made cloning feasible. As such, it is one of the most significant insights of modern biotechnology."[145]

On reporting their accomplishment in *Nature* in March 1996, the Roslin team noted that this was "the first report, to our knowledge, of live mammalian offspring following nuclear transfer from an established cell line." Moreover, they noted that the methods they had devised had achieved, in perhaps a circuitous manner, the goal that Ian Wilmut had been pursuing (at first grudgingly) since 1982, that of genetic engineering.[146] News of the cloning feat did, in fact, circulate around the world (mostly in Europe), but the reception was muted. *The New York Times,* for example, didn't even report it.[147] It would not be until Dolly was born that it would finally sink in for most people, including scientists, that cloning had really arrived.

One significant exception was the response of Davor Solter, whose appraisal of the work of Karl Illmensee in the early 1980s had led him to the conclusion that cloning from differentiated cells would never work in mammals. He had, in fact, abandoned cloning work on the heels of the Illmensee debacle and concentrated his efforts on, among other things, genomic imprinting. Now, however, he was a believer. More than that, he could see what was coming. Writing in the same issue of *Nature* that the Roslin team reported the birth of Megan and Morag, Solter acknowledged the achievement as "a cause for celebration."[148]

Four months later Dolly was born at Roslin Institute. The cells used were derived from the mammary gland tissue of a six-year-old ewe. In February 1997, the Roslin team published its technical account in *Nature* and caused a global sensation.[149] Dolly's birth was a collaborative effort between Roslin Institute and PPL Therapeutics, Inc., a biotechnology company that has its offices on the same site as Roslin, which is itself part of a network of government science institutes located throughout Britain. PPL had provided the mammary gland cells, which they had kept frozen at their facility, not for cloning purposes, but as part of their ongoing efforts to produce pharm animals that secreted human medicines in their milk. This time *The New York Times* not only covered the story, the paper ran it on page one with the headline: "SCIENCE REPORTS FIRST CLONING EVER OF ADULT MAMMAL; RESEARCHERS ASTOUNDED; In Procedure on Sheep, Fiction Becomes True and Dreaded Possibilities Are Raised."[150] Predictably, the public discussion that followed in news accounts focused on the reinvigorated hopes and fears concerning the cloning of humans.

There was, of course, one other principal response: disbelief. A headline in France, for instance, asked, "Clonage: Bluff ou Revolution?"[151] Indeed, many scientists around the world were skeptical of the claim by Wilmut and Campbell, particularly in light of the historical controversies that dogged the science of cloning. Before long, a few scientists stepped forward to publicly challenge the Scottish team directly, raising doubts about Dolly's pedigree and demanding proof that she was what they claimed. Meanwhile, scientists in laboratories around the world were attempting to perform equivalent cloning experiments on a variety of species based on the high-profile cloning recipe of the Roslin team, but nearly a year quickly passed and nobody seemed to be having much luck. The pressure was really on.

On some levels, it should not have been surprising that other laboratories were not having instant success with their cloning attempts. After all, Dolly was the culmination of a great deal of painstaking work, much of which had failed. Wilmut and Campbell—who had been simultaneously running cloning experiments on other cell types at the time they ultimately produced Dolly—had reconstructed a total of 277 embryos from the cultured mammary gland cells (about 400 attempts were made in total), of which only 29 proceeded to develop to an adequate stage (moralae or blastocyst) for implantation into surrogates. These 29 embryos were put into 13 ewes and resulted in one, as Wilmut and Campbell put it, "skin-of-the-teeth success," Dolly.[152] The other experiments involved cells cultured from nine-day-old embryo discs—which resulted in four live lambs, Cedric, Cyril, Cecil, and Tuppence—and fetal fibroblasts (connective-tissue cells)—which resulted in the births of Taffy and Tweed.

The Scottish team had reason to be confident in their results; they had, in fact, anticipated the skepticism and taken steps to verify their claim. For one thing, Dolly was a Finn-Dorset breed just like the owner of the mammary gland cells from which she was derived. The surrogate ewe was a Scottish Blackface; her giving birth to a Finn-Dorset made it reasonably apparent to the Roslin team, at least, that the experiment was a success. Unfortunately, the original Finn-Dorset ewe, who had been pregnant at the time the mammary cells were extracted, was long dead. The mammary cells were made available by PPL and Hannah Research Institute, whose scientists had frozen them as part of their joint work on mammary genes. Dolly and the original model could not be stood side by side for a look-see, to the extent that might have helped anything. To compensate, the Roslin team had delayed announcing Dolly's arrival so that, among other things, DNA tests could be conducted on Dolly and the cultured cell line. (Actually, they were conducted on all the new arrivals, everybody from Cedric to Tweed.) As noted in their Nature article, "DNA microsatellite analysis of the cell populations and the lambs at four polymorphic loci confirmed that each lamb was derived from the cell population used as nuclear donor."[153] Wasn't that sufficient proof? It turned out that it wasn't. The severest critics found these steps wanting.

Ironically, even though the exact experiment that led to Dolly had yet to be repeated, cloning efforts in general had been restored to mainstream science. Indeed, by the anniversary of Dolly's announcement, two teams of scientists had already announced successes in cloning calves from fetal cells obtained at various stages of development (although not from adult cells). One of the teams, a joint effort of the University of Massachusetts and Advanced Cell Technology, used skin cells from fetuses between one and three months old. Like Roslin's efforts, the aim of those experiments was to produce genetically altered livestock that will naturally produce human pharmaceuticals or yield organs for human transplantation.[154]

On this front, however, the Scottish team was also in the lead. In December 1997, while the world was still coming to grips with Dolly, a broader scientific group from Roslin Institute and PPL Therapeutics, including Wilmut and Campbell, published their protocol for producing Polly, a transgenic sheep born the previous summer equipped with the gene for human factor IX, which produces the protein hemophiliacs (suffering from Hemophilia B) require for normal blood clotting. Polly was born along with five sisters (two of whom also acquired the transgene, as established by DNA analysis) and signaled a watershed for the efforts at Roslin since the early 1980s. While the scientists had earlier (1990) produced sheep equipped with a human gene—in this case alpha-a antitrypsin (AAT), which is used in the treatment of cystic fibrosis—they had done so using the standard, and more tiresome and less efficient, methods of genetic engineering (which is random and, at best, produces mosaic animals). Polly and her siblings marked the first time cloning and genetic engineering had successfully worked in unison.[155]

Despite the reported advances in cloning and genetic engineering, some scientists remained deeply suspicious of the claim that Dolly was really and truly cloned from a highly differentiated, adult sheep cell. And as the first anniversary of Dolly's announcement approached (Dolly was about 18 months old by this point, and pregnant with what would become a healthy lamb named Bonnie), no other successful cloning effort was being reported using adult cells. At least three laboratories had tried and failed. Dr. Wilmut exacerbated the situation by announcing at a seminar at the Massachusetts Institute of Technology in early 1998 that he did not intend to personally repeat the feat with Dolly. Given the media hype over cloning, Nobel laureate and MIT biologist Philip Sharp found himself "really leery," he said. Dr. Bruce Alberts, president of the National Academy of Sciences, told Nicholas Wade of *The New York Times*, "[I]t is fair to point out that scientists generally require that a new result be repeated in at least one independent laboratory before they are ready to accept it unambiguously."[156]

With perhaps a certain relish, two leading scientists stepped forward to assume the role of devil's advocates. In a letter to *Science* published in late January 1998, Dr. Norton D. Zinder of Rockefeller University in New York City and Dr. Vittorio Sgaramella of the University of Calabria in Cosenza, Italy, launched a broadside attack on Dolly's origins, citing, among other things, "the scientific weaknesses of the experiment and the possible impact on the societal credibility of science itself." There was, in their view, insufficient proof that Dolly was what they said she was. "The most heinous crime was not saving the parent," Dr. Zinder explained to Nicholas Wade. "I just don't understand that. There could be nothing more exciting than seeing the two twins standing there."[157] Beyond the visual satisfaction of such a photo-op, however, the lack of the original ewe precluded the opportunity of performing a skin graft test; if Dolly was really a clone of the original, a skin graft from one to the other would not have met with tissue rejection (just as is the case with identical twins). Drs. Zinder and Sgaramella castigated Wilmut and Campbell in their *Science* letter for giving "no hint" in their cloning report on Dolly that the original ewe was dead. Referring to the initial number of attempts made to reconstruct embryos (which resulted in 277), Zinder and Sgaramella sneered that "Only one successful attempt out of some 400 is an anecdote, not a result. All kinds of imagined and unimagined experimental error can occur." Quite possibly, they suggested, the mammary gland cell "could have been one of the donor's rare stem cells" or, because the donor ewe was pregnant at the time of her death (in the third trimester, actually), a stray fetal cell circulating in the ewe's bloodstream. "Why was no analysis of the fetus and its father's genotype performed?" they asked. "Given these DNA fingerprints, or even the sex of the fetus, one could have excluded a fetal cell as donor."

As for the DNA tests that were conducted, they felt that these fell short. In fact, they predicted Dolly was headed for court and would lose based on these tests. "The demonstration that the four microsatellite marker DNAs seem the same in Dolly and in the donor mammary cells is good, but not sufficient; it would probably be rejected by a jury called to deliberate on Dolly's origin, not an unlikely event given Dolly's commercial potential." The reason they gave for the inadequacy of the DNA tests was a variation on a statistical argument (regarding isolated subgroups within a population) that had for so long dogged forensic DNA analysis generally: "Sheep are highly inbred," they wrote, "and there are, to our knowledge, no data on gene frequencies in sheep populations; differences in DNA fingerprints can provide exclusion, but similarities are only a statistic." Furthermore, they noted, no analysis of Dolly's mitochondrial DNA had been performed relative to the recipient oocyte and the donor cell.[158]

Not surprisingly, given the stakes, Wilmut, Campbell, and colleagues fought back aggressively. In an immediate response to the letter of Zinder and Sgaramella, published in the same issue of *Science,* they conceded their method was inefficient ("a single birth from 400 attempted fusions is not an efficient system"). Further, they acknowledged that in hindsight it would have been better to deposit original donor tissue with a respected third party. However, they argued, the Dolly experiment, on its face, appeared to have succeeded. Dolly was, after all, a Finn-Dorset who was born to a surrogate of an altogether different breed. Moreover, they insisted, the microsatellite analysis originally conducted should put to rest any suggestion that an errant fetal cell might be responsible for Dolly, particularly since the cultured cells used were predominantly epithelial in nature. Finally, they promised that additional DNA tests, both mitochondrial and nuclear, were being conducted on all available samples.[159] Meanwhile, *Time* magazine reported a month later that Wilmut "and his colleagues are scrambling to track down any other tissue samples taken from Dolly's mom so they can perform the genetic tests that will determine, once and for all, if Dolly's DNA and her mom's DNA are identical."[160]

It is certainly true, as Wilmut and Campbell write in *The Second Creation,* that "Dolly affects the whole sweep of human history."[161] It would not do, consequently, for the Roslin/PPL team to simply conduct another battery of DNA analyses on their own. There had to be more. The DNA results needed, above all, unassailable credibility. Wilmut and Campbell therefore contacted Alec Jeffreys's laboratory at the University of Leicester and Sir Alec agreed to perform his own DNA fingerprint tests. At the same time, the Roslin/PPL team (working under the auspices of a licensed DNA laboratory named Rosgen set up at Roslin Institute) set about conducting their own tests. When it came time to publish these results, the fates smiled upon the cloners of Dolly. Another group of scientists on the far-flung island of

Hawaii, it turned out, had succeeded in cloning mice from adult cells using essentially the same protocol as laid out by the Scottish scientists. It was the vindication they were looking for. In July 1998, all of the studies—the mice cloning studies, the additional DNA tests of the Roslin/PPL scientists, and the DNA fingerprinting results of Sir Alec Jeffreys on the tissues related to Dolly—were all published in the same blockbuster issue of *Nature*. As the headline to Davor Solter's article that also appeared in the issue declared, "Dolly *Is* a Clone—and No Longer Alone."[162]

The Roslin/Rosgen/PPL scientists responded to Zinder and Sgaramella with logic and an expanded set of microsatellite DNA tests. Their new DNA tests ranged over ten microsatellite repeating segments of DNA. Three of the microsatellites used in the original study were reanalyzed and seven new microsatellite markers developed by Perkin-Elmer were also analyzed. The latter seven were developed for paternity testing in cattle but were also found to be polymorphic in sheep. The number of alleles for each targeted site ranged from one to seven, with frequency data derived from a comparative study of 44 Finn-Dorset sheep conducted at the Hannah Research Institute. Comparisons were made on Dolly's DNA (from her blood), the DNA of the other Finn-Dorsets, the DNA of the original mammary tissue, and DNA obtained from the culture cells that were the source of the nuclei for the reconstructed embryos. The team's conclusion: "[T]he alleles present in DNA from Dolly were identical to those in the original mammary tissue, in the cell population prepared from that tissue and in the cells cultured to passage four." The odds that the DNA would match another sheep were astronomical. It was, therefore, "extraordinarily unlikely," they concluded, that Dolly originated from some other source.[163]

Sir Alec's DNA results followed immediately in the pages of *Nature*. Rather than perform a redundant set of DNA tests focused on microsatellites, Jeffreys and a team of researchers at the University of Leicester, working with other scientists at the Hannah Research Institute, called into service the tried-and-true variety of DNA tests that Jeffreys had pioneered in the forensic context, RFLP. Using four multilocus probes targeting variable minisatellite regions of the genome (which were shown to work in a variety of species), the Jeffreys team compared the blood of Dolly (the taking of whose blood was formally witnessed for this test) against the tissue of the original ewe, the cultured cells, and 12 control sheep. Through the steps of electrophoresis and Southern blotting, they produced a set of visually compelling autorads detailing the genetic differences of the test subjects. Dolly matched the DNA of the original ewe and the cultured cells, the team concluded, in every detail, "band number, position and relative intensity." The odds that an unrelated sheep would have the same DNA profile were infinitesimal. Using band-sharing data, they further determined that the possibility of the fetal-cell scenario was remote. To bolster their results, the Jeffreys team conducted a secondary battery of minisatellite analyses (which had been developed for use in humans) and generated, once again, highly variable profiles. These reduced the likelihood of error or fetal contaminant even further. "We therefore conclude," the Jeffreys group wrote, "that Dolly is derived from the nucleus of a cell from the mammary gland of the adult donor."[164] "By throwing the full panoply of forensic DNA-testing methods at the problem of Dolly's origin (including the sworn, unbroken chain of custody in providing samples)," Davor Solter wrote in his *Nature* article, which accompanied the results, the two DNA reports "have shown Dolly is indeed the direct descendant of an udder cell derived from a nameless Finn Dorset ewe."[165] The Hawaiian team conducted PCR-based DNA tests, using a series of three microsatellite markers, to confirm the genetic pedigree of their cloned mice (some of which were clones of clones). Human cloning, cloning expert Dr. George Seidel of Colorado State University told Gina Kolata of *The New York Times,* "is clearly more imminent."[166]

Subsequent mitochondrial DNA tests revealed that Dolly is not an *exact* genetic duplicate of the original ewe. Using a combination of DNA sequencing and RFLP, Wilmut and his colleagues determined in 1999 that, although Dolly's nuclear DNA came exclusively from the original six-year-old ewe, her mitochondrial DNA was derived entirely from the ewe that provided the embryo for the cloning procedure. Dolly was, in this sense, a genetic chimera, after all.[167]

10 Bad Blood
(Part 1)

THE PINNACLE—AT LEAST IN THE PUBLIC EYE—OF COURT CASES USING DNA testing was *People of the State of California v. Orenthal James Simpson*. After months of jury selection and pretrial hearings, the case formally began with opening statements on January 24, 1995. By the time the verdict was rendered more than eight months later, the jury had heard (sometimes ad nauseam) the testimony of 124 expert and lay witnesses (including people called to testify twice), 58 during the prosecution's case-in-chief, 13 more during the prosecution's rebuttal case, and 53 witnesses presented by the defense. Sequestered longer than any other jury in history (which meant spending 266 nights at the Inter-Continental Hotel in Los Angeles cut off from family, friends, and, most significantly, exposure to the media), the jury, presumably, had a lot to think about. The prosecution alone had presented them with 488 exhibits intended to demonstrate that O.J. Simpson, then a month shy of 50, was guilty beyond a reasonable doubt of killing his ex-wife Nicole Brown Simpson, age 35, and her friend Ronald Goldman, 25, on the evening of June 12, 1994.

The case against Simpson was entirely circumstantial, as many murder cases are, but from the prosecution's perspective, it was as strong a circumstantial case as ever there had been. Even though there were no eyewitnesses, the state was convinced there was more than ample physical evidence strewn about that could tie Simpson to the crime. This evidence included state-of-the-art forensic DNA analysis of dozens of bloodstains conducted at three laboratories, as well as sophisticated scientific analyses of hair, fiber, and blood evidence conducted by the FBI Laboratory, and all of it implicated O.J. Simpson. Moreover, the prosecution spent a great deal of time establishing a motive for the crime (even though this was not legally required to prove murder), arguing that the murder of Nicole Brown Simpson was the culmination of years of domestic violence that had punctuated the couple's 17-year relationship and that

Ronald Goldman had simply stumbled upon that ultimate moment at exactly the wrong time only to be overpowered and slain by a man in an uncontrollable rage. Yet when the jury finally received the case on October 3, 1995, they deliberated for less than four hours and, much to the horror of many observers, acquitted O.J. Simpson of both murders and set him free.

O.J. Simpson was as famous as a former football player gets, and one of the most famous black Americans in the country. Right from the start of his football career, as a college player for the University of Southern California Trojans, he distinguished himself as one of the greatest football players ever, chosen All American in both of his seasons at the school and winning the Heisman trophy. Later, as a professional player for the Buffalo Bills, Simpson set new all-time rushing records and ultimately won a place in the National Football League Hall of Fame. After this brilliant sports career, Simpson parlayed his fame into an extremely successful second career as a spokesman for various national companies, most memorably Hertz Rent-a-Car. He also appeared in a number of Hollywood movies and became a sports commentator for two television networks. In 1999, *The New York Daily News* commissioned a panel of football experts to determine the 50 greatest football players of the 20th century: O.J. Simpson was ranked number 15.[1]

For most of his post-football career, O.J. Simpson was either living with or married to Nicole Brown, having divorced his first wife, Marguerite, in 1979, the same year he retired from professional football. O.J. and Nicole had two children together, daughter Sydney and son Justin, both of whom were upstairs sleeping in Nicole's Brentwood condominium at 875 South Bundy Drive the night their mother was killed. After the marriage of O.J. and Nicole had ended in divorce in 1993, O.J. purchased the luxury condominium for Nicole and the children in January 1994; it was located only a few miles from Simpson's sprawling estate on Rockingham, where the entire family had lived prior to the separation and divorce. Although several reconciliations had been attempted over the previous couple of years, by June 1994, according to evidence presented at trial, the relationship between O.J. and Nicole was strained, even hostile. Prosecutor Christopher Darden (one member of a sizeable team assembled to prosecute Simpson), whose principal task was the presentation of evidence about the violence that characterized the relationship, portrayed it in his closing argument to the jury as a time bomb with a long fuse, dramatically replaying a panicky 911 call Nicole placed only eight months before the murders, a tape that clearly reveals an enraged O.J. Simpson ranting at his ex-wife in the background.

On Sunday afternoon June 12, 1994, O.J. and Nicole attended a dance recital at their daughter's junior high school in West Los Angeles, but they were not on amiable terms. O.J., in fact, sat apart from Nicole, Justin, and the members of Nicole's family, who were also in attendance. When the recital ended at 6:00 P.M., Nicole, the children, and the members of the Brown family went to dinner at the Mezzaluna restaurant, where Ron Goldman was working as a waiter, although he was not the group's server that evening. O.J. was not invited, so he returned home where he had several conversations with Kato Kaelin, an aspiring actor who lived in a guest house at the Rockingham estate connected to another guest house occupied by Simpson's adult daughter from his first marriage, Arnelle. Later in the evening, Kaelin and Simpson went together to a nearby MacDonald's restaurant for something to eat, returning, based on the phone records of Kaelin, about 9:35 P.M. O.J. was scheduled to fly to Chicago that evening out of Los Angeles Airport at 11:45 P.M. to play in a charity golf tournament, the Hertz Invitational, so he had arranged for a limousine to pick him up at 10:45 P.M. and take him to the airport.

When Nicole's dinner party left the Mezzaluna between 8:30 and 9:00 P.M., Nicole's mother, Juditha, left behind prescription eyeglasses. At 9:30 P.M. Nicole called the restaurant about the eyeglasses, and Ron Goldman, who had plans to go out later that night with a female friend, volunteered to drop the eyeglasses off at Nicole's en route to his destination. Goldman left the restaurant at about 9:50 P.M. with the eyeglasses in an envelope, walked home to change clothes, and then walked the short distance to Nicole's condominium with the envelope in hand.

At about 10:15 P.M. or so, several of Nicole's neighbors later recalled hearing a dog barking (one neighbor described it as a "plaintive wail," suggesting the dog was in distress) and at about 10:55 P.M., neighbor Steven Schwab found Nicole's dog "Kato" (named by the Simpson children after Kato Kaelin), an Akita, wandering in the street. After the dog followed Schwab home, two of his tenants, Sukru Boztepe and Bettina Rasmussen, allowed the dog to lead them back to where Schwab found him. There, at around 12 midnight, they discovered the bodies of Nicole Brown and Ron Goldman. They had been savagely murdered. Goldman had sustained about 30 stab wounds. Nicole had been nearly decapitated.

A conscientious employee who was worried about being late for his first celebrity pick-up, 24-year-old limousine driver Allan Park arrived early—at 10:22 P.M.—at O.J. Simpson's Rockingham estate. Park later testified to several observations he made while waiting around (he didn't try to announce himself until about the time he was scheduled to be there); most significantly, Park recounted his repeated failed attempts to get a response from the buzzer he pushed outside the gate until about 11:00 P.M. and seeing a darkened figure enter the house just before then, at which time O.J. Simpson answered the call via intercom, said he had overslept, and would be right down. So concerned was Park by the lack of response during his 15 minutes of trying that he had phoned his boss and even his mother, coincidentally creating detailed phone records that established a precise time line. Park and Kaelin, who was rattled by some thumping noises that he had heard behind his room at about 10:38 P.M. and was quite voluble about it, then helped O.J. load his luggage into the limousine and Park then drove Simpson to the airport on time for the flight to Chicago.

The first to arrive at the crime scene, Los Angeles Police Department (LAPD) Officer Robert Riske secured the area and then surveyed the yard and condominium. Riske found Nicole in a pool of blood and curled in a fetal position at the foot of the stairs leading to her doorway. Nearby, in a small enclosure, he found Ron Goldman collapsed against a metal fence. Near Goldman's feet, Riske observed a black wool hat, a white envelope stained with blood, and, he later testified, a single leather glove covered with blood. He also observed bloody shoe prints leading away from the bodies along a 120-foot-long walkway that ended in an alley, noticing further what appeared to be five fresh blood drops to the left of the escaping shoe prints. Inside the condominium Riske found a cup of melting ice cream and candles burning both downstairs and in an upstairs bathroom.

About a dozen police officers and detectives eventually arrived at the scene, some of whom were given a tour by Riske. Among those who responded were West Los Angeles Detective Ronald Phillips, the West Los Angeles homicide unit chief, and his on-call detective, Mark Fuhrman. They arrived together at 2:10 A.M., and Riske took the two detectives on a tour of the crime scene. Fuhrman, who was under the impression that he was to be the lead detective in the case, took careful notes of his observations. The notes reflect that he was shown bloodstains and a possible bloody fingerprint on a back gate on the property. Shortly thereafter, Fuhrman received word that the downtown Robbery-Homicide division was taking over the case and that he was relieved. At 4:05 A.M., Robbery-Homicide Detective Philip Vannatter arrived at the scene, and at 4:30 A.M. his partner, Detective Tom Lange, arrived; these two would now be in charge of the case. By this time, O.J. Simpson had checked into his hotel in Chicago.

Introduced to Phillips and Fuhrman for the first time, Lange and Vannatter were later directed by their superior to immediately notify O.J. Simpson of the death of his ex-wife so that he would not hear about it first in the media. Fuhrman and Phillips accompanied them to the Rockingham estate for the notification after Fuhrman indicated he was familiar with the house, having been there on a domestic-dispute call years earlier. Upon their arrival, the detectives were unable to get a response from anyone in the house, even after getting Simpson's telephone number from his home security company and phoning the main house. After looking around the perimeter of the fenced-in estate, Fuhrman pointed out to his colleagues that there appeared to be a small bloodstain on the driver's door of a Ford Bronco parked on the street adjacent to the property. Fuhrman said he could see O.J.'s name on papers inside the Bronco and a license-plate check came back showing the car listed to Hertz. Fuhrman testified in March 1995 that he further observed additional streaks of blood on the panel below the door as well as items in the car that suggested a crime might be ongoing, such as in a kidnapping or murder-suicide. Ultimately, for reasons that remained contentious throughout the trial, the detectives decided to enter the property without first obtaining a search warrant, something that is legally permissible if officers can establish "exigent circumstances," a term of art that permits an exception to the constitutional requirement to obtain a warrant before conducting a search. This means, in effect, that there exists an emergency of some kind. Among the exceptions recognized under California law, a warrant-less search can be initiated if officers perceive an "imminent and substantial threat to life, health or property," which is what the detectives argued here.

Fuhrman scaled the fence and let the others in to approach the main house. The detectives then walked up the driveway to the front door of the main house, which was at about the halfway point of a crescent-shaped driveway that ended at another gated entrance. They knocked, but got no answer. After several minutes, a groggy Kato Kaelin approached from the side of the house and suggested the detectives speak with Arnelle Simpson. While Phillips, Vannatter, and Lange went to do so, Fuhrman remained behind with Kaelin and learned of the loud thumping noises Kaelin said he had heard the previous evening. The two then joined the three who were talking with Arnelle inside the main house. At about 5 o'clock, Detective Phillips called O.J. Simpson in Chicago about the death of his ex-wife, later testifying that he thought it odd that Simpson never asked how she died. Simpson got on the next available flight to Los Angeles.

Leaving the group in the main house, Fuhrman went alone to investigate the area where Kaelin believed the thumping noises originated, which turned out to be a narrow passageway between the south wall of the guest houses and a six-foot tall cyclone fence. Using his small flashlight, Fuhrman entered the passageway and about 20 feet in, he later testified, he found a dark leather, right-handed glove that appeared to match the glove found at the Bundy crime scene. The glove appeared moist and sticky with blood, he said later. (The defense later claimed this was impossible because of the amount of time that had elapsed since the time of the murders; surely the glove would have had time to dry out completely.) After informing the others of his discovery and pointing it out, Fuhrman, along with Phillips, was dispatched by Vannatter to the Bundy crime scene to determine whether the gloves were in fact a matching pair. At sunrise, Vannatter saw eight drops of blood leading from the Bronco toward the front door of the house. There were also three apparent blood drops in the foyer of the main house and blood inside the Bronco. When Fuhrman reported back that the gloves appeared to match, Vannatter decided to get a search warrant. He later drafted the necessary paperwork, an affidavit, and obtained a search warrant at 10:45 A.M., then returned to the Rockingham location about noon to search the residence.

LAPD criminalists Dennis Fung and Andrea Mazzola, a trainee, did not arrive at the Rockingham location until after dawn, at which time they began the evidence-collection process. After collecting bloodstains outside the Simpson estate, they reported to the Bundy crime scene for evidence collection, then returned again to

Rockingham once the search warrant had been obtained. (The techniques used in evidence collection and the delay in returning evidence to the laboratory later became hotly contested issues at the trial.) Blood samples are collected with swatches of white cotton cloth moistened with distilled water. The swatches must be dried as quickly as possible. The coroner was not called until 6:50 A.M. and, initially told to simply stand by, didn't arrive at Bundy until 9:10 A.M. Prosecutor Marcia Clark—who later admitted having no idea who O.J. Simpson was at the time—was also called to the scene to oversee the activity there on behalf of the Los Angeles District Attorney's Office. As the evidence-collection process proceeded, investigators found themselves increasingly under the glare of the media, which had begun to congregate outside both locations. At one point, officers covered Nicole Brown's body with a blanket obtained from inside her condo. Although that decision was certainly a gesture of respect and concern for the victim and her family, it was an action they would come to regret. Crime-scene evidence had now been compromised by cross-contamination.

In the course of executing the search warrant at the Rockingham estate, criminalists recovered a pair of dark socks on the carpet at the foot of O.J. Simpson's bed. Weeks would pass, however, before it was discovered that the socks contained bloodstains; nobody apparently noticed anything at the time. Blood drops were collected from the foyer, however. Due to the small blood smear on the door of the Bronco, the car was seized and its interior was later processed for evidence. The right-handed glove from Rockingham found by Mark Fuhrman was also collected.

In what would later become a flashpoint in the trial, the criminalists at the Bundy location did not collect any blood from the back gate, the stains that were described during the trial by a number of police officers and investigators, including Fuhrman, as being there on the morning of June 13, 1994. It was not until July 3, 1994, that Dennis Fung returned to the now unsecured crime scene at Bundy and collected bloodstains from the controversial back gate.

When O.J. Simpson returned home at 11:30 A.M. on the morning of June 13, 1994, he was detained outside his house as the search warrant was being executed and briefly handcuffed by an officer on duty outside the house, an image that made news across the country. Although it was an apparent misunderstanding on the part of the officer, and O.J.'s first attorney, Howard Weitzman, quickly persuaded police to uncuff his client, the image was a powerful one. Simpson soon agreed to go to police headquarters for an interview, ignoring Weitzman's advice to remain silent, after which he volunteered a blood sample. The blood was drawn by police nurse Thano Peratis, who testified at a preliminary hearing that he drew eight cubic centimeters of blood from Simpson; this testimony would later undergird defense allegations that some of the blood was unaccountably missing. Nurse Paratis gave O.J.'s blood to Detective Vannatter, who put the blood vial in his pocket rather than log it in at the crime laboratory as evidence. After driving 20 miles across town, Vannatter gave the blood vial to criminalist Fung at the Rockingham estate. The following day, police criminalist Collin Yamauchi began conducting tests on some of the evidence.

On June 15, 1994, Simpson replaced his attorney with Attorney Robert Shapiro, well known in California for handling celebrity cases. Shapiro soon assembled a formidable legal defense team, including famed attorney F. Lee Bailey, dean of University of Santa Clara Law School Gerald Uelmen, Harvard University Law School professor and author Alan Dershowitz; two attorneys well known for their expertise in litigating cases involving DNA evidence, Cardozo Law School professor Barry Scheck and attorney Peter Neufeld; highly regarded California attorney Robert Blasier, who specialized in cases involving serology and DNA; and, finally, attorney Johnnie Cochran, a former prosecutor and now a private attorney famous in Los Angeles for successfully bringing civil rights actions against the LAPD. Cochran would soon assume the lead role in Simpson's legal defense. The so-called Dream Team, as the media soon dubbed it, was also joined by attorney Carl Douglas and Robert Kardashian, a longtime friend of O.J. Simpson's who reactivated his law license to work on the case. Dr. William C. Thompson, an attorney, also appeared on behalf of Simpson to confront the statistical calculations. In the first days of his stewardship of the case, Shapiro retained the services of some of the country's most prominent forensic investigators and began assembling a team of forensic experts to counter what he knew was coming, a flood of forensic evidence.

California authorities responded by assembling an impressive team of seasoned prosecutors from around the state. Altogether, 11 prosecutors participated in the case against O.J. Simpson, with many of them being specifically selected because of their well-known expertise in a particular area of litigation. Deputy District Attorney William Hodgman acted as prosecution team manager, generally coordinating prosecution efforts from behind the scenes. The co-lead prosecutors were Marcia Clark, known for her aggressive courtroom demeanor and popular throughout California for winning the conviction of the murderer of actress Rebecca Schaeffer, and Christopher Darden, a 15-year veteran of the DA's office who specialized in police misconduct cases. The other principal members of the prosecution team included the irascible Deputy DA Rockne Harmon from the DA's office in Alameda County in the San Francisco Bay area, tapped to spearhead the prosecution's DNA case; Deputy DA George "Woody" Clarke, also chosen for his expertise in DNA evidence; Deputy DA Lisa Kahn, another respected DNA expert within the prosecutor's office; and Deputy DAs Hank Goldberg and Brian Kelberg, both of whom specialized in cases involving highly technical medical and forensic evidence.

According to the testimony of an erstwhile friend of O.J. Simpson's, Ronald Shipp, a former police officer with an acknowledged drinking problem who had known O.J. for 26 years, O.J. was himself thinking about the evidence. Shipp was among the many guests at the Rockingham mansion on June 13, 1994, to express sympathy over the death of Nicole. At one point, Shipp testified during the trial, O.J. beckoned Shipp into his bedroom. Concerned about taking a polygraph test, according to Shipp, O.J. "jokingly said, 'To be honest, Shipp, I've had some dreams of killing her.'"[2] It was explosive and controversial testimony, and many legal commentators assailed Judge Lance Ito—the presiding judge throughout the nine-month trial—for permitting it, arguing that the testimony might well constitute reversible error. In the end, Judge Ito instructed the jury to disregard the dream testimony. (Simpson later took and failed—resoundingly—a polygraph test, but since polygraph tests are generally not admissible in court, this never became an issue at the trial.) But Simpson also questioned Shipp about forensic evidence, asking, Shipp testified, "How long does it take for DNA to come back?"

In fact, with the evidence and reference samples from O.J. Simpson and the two victims in hand, the serology unit of the Scientific Investigation Division (SID) of the LAPD began DNA analysis the next day, on June 14, 1994. By the day after that, Collin Yamauchi reported to prosecutors the results of PCR-based HLA DQA1 tests that showed O.J. Simpson's DQA1 type was consistent with the blood drops collected at Bundy and that stains on the glove found at Rockingham revealed an apparent mixture of the blood of O.J. Simpson, Nicole Brown, and Ronald Goldman based on the DQA1 types. O.J. Simpson's DQA1 type was shared by 7 percent of the overall population, so this was far from definitive proof of his guilt.[3] Based on this and other circumstantial evidence, however, an arrest warrant was obtained, and it was arranged that O.J. Simpson would surrender himself on the morning of June 17 after being examined by a number of defense experts and a physician at the home of Robert Kardashian, in Encino, California. (The results of the first round of the more definitive RFLP tests—which tied Simpson to blood drops found along the walkway at the Bundy crime scene—were made public during an evidentiary hearing held on August 22, 1994.)

If the case had not yet monopolized the public's attention, it soon exploded onto the national agenda when Simpson, rather than surrender as agreed, sneaked out of Kardashian's home and became a fugitive, leaving behind an oblique letter to the world that, when it was publicly read by Kardashian later that afternoon, was interpreted by most observers to be a suicide note (because it included a lengthy farewell to his family and friends), maybe even a confession. "I think of my life and feel I've done most of the right things," Simpson wrote, "so why do I end up like this." The tension escalated when a white Ford Bronco being driven by O.J. Simpson's longtime friend A. C. Cowlings was spotted soon after going northbound on the Santa Ana Freeway. In the back of the Bronco, which belonged to Cowlings, was O.J. Simpson. Soon there were 12 police cars in a strange slow-speed pursuit, warned off by Cowlings who frantically explained over his cellular phone that Simpson had a loaded gun pointed at his own head in a suicidal gesture. The perverse convoy was soon joined by seven news helicopters, and at some locations along the Bronco's route crowds gathered and cheered O.J. Simpson. Before the chase finally ended at Simpson's estate with his surrender, 95 million people in America watched some part of the drama unfold on television, with every major network interrupting regular programming to bring the spectacle to its viewers live. By the time Cowlings delivered O.J. home, the estate was surrounded by 75 SWAT team members, whose negotiators gingerly coaxed O.J. out of the Bronco under the glare of dozens of news cameras.[4] Things certainly couldn't have looked worse for O.J. Simpson.

Beginning on June 30 and continuing for six days until July 8, a preliminary hearing was held before a judge to determine whether there existed sufficient evidence to proceed with the prosecution. Since the hearing occurred before more definitive DNA tests could be completed, the prosecutors relied on conventional serology tests to make their case to Municipal Judge Kathleen Kennedy-Powell. This posed some difficulties since O.J. and Nicole both had Type A blood and the same type (Type 1) of a commonly analyzed isoenzyme (ESD). For the PGM isoenzyme, however, Nicole was a PGM sub-type 1+ and O.J. a sub-type 2+2-. Since the blood drops recovered at Bundy that accompanied the bloody shoe prints leading away from the bodies revealed ABO blood Type A, ESD Type 1 and PGM sub-type 2+2-, this combination of serology tests included O.J. Simpson. In making their case to the judge, prosecutors elicited testimony from a serologist that calculated the frequency of that combination of genetic markers at 0.48 percent of the population. Looked at another way, however, there were 40,000 people in Los Angeles with the same profile.[5] At the conclusion of the hearing, which was something of a mini-trial featuring police investigators and witnesses with some knowledge of events on June 12, Judge Kennedy-Powell ruled there was "ample evidence to establish a strong suspicion of the guilt of the accused," but not before she and the defense learned that there were some serious problems with how the LAPD had conducted the investigation. Ultimately, in fact, almost every aspect of the investigation was revealed to be seriously flawed.

Another major issue that surfaced early on was race. The day after Simpson was taken into custody, the LAPD released Simpson's mug shots to the press. *Newsweek* magazine appeared on newsstands that Monday with the unaltered shot on its cover. *Time* magazine, however, ran on its cover a computer-enhanced version depicting O.J. Simpson in a darkened, some argued sinister, light. *Time*'s photo-illustration immediately drew charges of racism and prompted its editors to run a full-page apology the next week, but by this time the national dialogue on the case had turned very much to race, focused on the fact that O.J. Simpson was black and the victims were white. A few weeks later, another racial grenade was thrown into the case. Acting on a tip from Simpson's defense attorneys, author Jeffrey Toobin burrowed into civil court records and discovered that Detective Mark Fuhrman had a well-documented history of problems with racial minorities. As spelled out in the July 18, 1994, issue of *The New Yorker*, Toobin's article, "An Incendiary Defense," portrayed Mark Fuhrman as a police officer who, earlier in his career, had been brought to the breaking point by his animus for the racial minorities he encountered on the job. Assigned as a police officer to East Los Angeles in the late 1970s to contend with street gangs, by the early 1980s Fuhrman was seeking—eventually in court—a disability pension because, his legal papers argued, he had become "substantially incapacitated for the performance of his regular and customary duties as a policeman." In psychiatric interviews contained in the file, Fuhrman referred to beatings he'd inflicted on suspects and spoke of an urge to kill people.[6] The City of Los Angeles fought the lawsuit and portrayed Fuhrman as, on the one hand, competent at his job and, on the other hand, based on the lawsuit, a malingerer. In any event, Fuhrman lost the suit and remained on the police force, later gaining promotion to detective. From this point on, Detective Mark Fuhrman became a central issue in the trial, one that blind-sided many early court watchers because Fuhrman had conducted himself with a high degree of professionalism at the preliminary hearing as he described, among other things, finding the bloody glove on the grounds of O.J. Simpson's Rockingham estate on the morning of June 13, 1994. He was even deluged with fan mail, mostly from single women, after his testimony. But his stardom rapidly faded with the revelations about his past, and at the trial he was subjected to stern and unmerciful cross-examination by F. Lee Bailey, whose questions inevitably turned to the subject of racism. During questioning of Fuhrman on March 15, 1995, Bailey pointedly asked Fuhrman, in what would become one of the most frequently broadcast exchanges of the trial:

"Do you use the word 'nigger' in describing people?"

"No, sir," Fuhrman replied.

"Have you used that word in the past ten years?"

"Not that I recall, no."

"You mean," Bailey persisted, "if you called someone a nigger you have forgotten it?"

"I'm not sure I can answer the question the way you phrased it, sir."

"Are you therefore saying that you have not used that word in the past ten years, Detective Fuhrman?"

"Yes, that is what I'm saying."

"And you say under oath that you have not addressed any black person as a nigger or spoken about black people as niggers in the past ten years, Detective Fuhrman?"

"That's what I'm saying, sir."

"So that anyone who comes to this court and quotes you as using that word in dealing with African-Americans would be a liar, would they not, Detective Fuhrman?"

"Yes, they would."

"All of them, correct?"

"All of them."

"All right," Bailey said, "thank you."[7]

The defense produced a number of witnesses to refute Fuhrman's denials, among them Kathleen Bell. On September 5, 1995, Bell testified that she met Fuhrman in 1985 or 1986, within Bailey's ten-year window, and that on one occasion Fuhrman characterized interracial couples as "disgusting" and went on to say, she recalled, "If I had my way . . . All the niggers would be gathered together and burned."

Most explosively, however, defense attorneys obtained on August 9, 1995, through appellate court action in North Carolina, a collection of audiotapes made by aspiring screenwriter Laura Hart McKinney between April 1985 and July 28, 1994. Having met Fuhrman in 1985, when she was beginning work on a screenplay about the obstacles female police officers face in the LAPD, McKinney enlisted Fuhrman as a consultant and began taping their meetings as Fuhrman gave her his version of life on the streets in the LAPD. Altogether, there were a dozen or more hours of tapes on which Fuhrman is heard disparaging women, blacks, Mexicans, and Jews and describing, even reveling in, encounters with minorities that included police-administered beatings, evidence manipulation, and even the suggestion that suspects had been executed by rogue officers. The tapes included 41 instances of Fuhrman using the word "nigger" and 17 examples of how to plant or fabricate evidence, as well as criticism of Judge Ito's wife, LAPD Captain Margaret York, a former supervisor of Fuhrman's. At the same time, it was unclear from the tenor of the tapes whether Fuhrman was recounting actual events or making it all up as he went along in an effort to provide McKinney with dramatic material for her screenplay. Fuhrman was promised a $10,000 fee if the screenplay was ever produced, but it never was, and he was never paid. But it certainly was clear that Fuhrman had been caught red-handed using the term "nigger" in the previous ten years. It was also clear that Judge Ito had been handed an explosive set of materials to consider. It would take awhile to sort out, he said, and determine which portions, if any, the jury should hear. (A significant issue was resolved when another judge ruled that Judge Ito's wife would not have to testify, which would have meant Judge Ito's replacement by a new judge.) The defense wanted extensive playing of the tapes, proposing that 60 excerpts be introduced as evidence. The prosecution urged that none of it be played to the jury, since, it was argued, Fuhrman was participating in a work of fiction when he made the statements. Judge Ito decided after some soul-searching to allow the tapes to be played in open court (but without the jury present) because they were "of vital public interest." Fred Goldman, the father of Ron Goldman, reacted with ire: "This is now the Fuhrman trial," he exploded before television cameras, "not the trial of O.J. Simpson, accused of killing my son."[8] When they were played on August 29, 1995, the tapes made front-page news across the country. "Voice of Hate" read the *New York Post*'s front page the following day, the paper reprinting, as many papers did, extensive portions of the tapes.[9]

Based on the tapes and Kathleen Bell's testimony, Johnnie Cochran, in his closing argument to the jury, labeled Fuhrman a "lying, perjuring, genocidal racist," and even went so far as to liken him to Hitler. The principal point, however, was that Fuhrman was untrustworthy.

Beyond the evident perjury revealed by the tapes, the defense argued to Judge Ito that major portions of the tapes should be played to the predominantly black jury because Fuhrman's words reveal he is a man capable of planting evidence. In the end, however, Judge Ito disagreed, issuing a ruling that said that the defense had presented no credible evidence that Fuhrman had planted any evidence in this case. "The underlying assumption requires a leap in both law and logic that is too broad to be made based upon the evidence before the jury," Ito wrote. He concluded the jury should hear, and they did, only two statements by Fuhrman related to his denial about using racial epithets. They were:

—"We have no niggers where I grew up."

—"That's where niggers live."

The defense attorneys were livid about Ito's decision, but the snippets were enough to establish that Fuhrman had been untruthful in his testimony and, as permitted by Ito's ruling, Laura Hart McKinney was allowed to tell the jury that Fuhrman had used the epithet 41 times on the tapes. Fuhrman was haled back into court in early September on the heels of these revelations, and defense attorneys posed a series of questions to him, including: "Was the testimony that you gave in the preliminary hearing in this case completely truthful?" "Have you ever falsified a police report?" "Did you plant or manufacture any evidence in this case?"

To all of these questions, Fuhrman—now facing prosecution for perjury—asserted his Fifth Amendment right against self-incrimination and refused to answer. (Legal experts explained at the time that Fuhrman's refusal to answer was more a matter of legal strategy and did not shed any light on whether he had planted or manufactured evidence.) The exchange, however, took place outside the presence of the jury; Fuhrman never appeared before them again. Nevertheless, despite the limitations imposed by the court, the Fuhrman deception provided a springboard for the defense to argue to the jury in closing arguments that Fuhrman was not to be trusted in any respect, that, in fact, he had a motivation to tamper with the evidence in the case. Now, in light of the mean-spirited tapes, the careful recitation of events provided by Fuhrman in his March testimony could be assailed with abandon.

The Fuhrman tapes were the crescendo of a line of argument the defense had been pursuing for more than a year, at least since Toobin's *New Yorker* article had exposed

Fuhrman's past difficulties with minorities. Just before the trial commenced with opening statements in January, Judge Ito ruled that the defense could indeed question Mark Fuhrman about his racial biases, thereby attacking his credibility and suggesting he could not be trusted. Their theory was that Fuhrman, either alone or in concert with others, had fabricated some or all of the evidence against O.J. Simpson. In particular, Fuhrman had been accused by F. Lee Bailey of stealing a bloody glove from the crime scene at Bundy, secretly transporting the second glove to Simpson's Rockingham estate, smearing some of the blood from the glove on the interior of Simpson's Bronco and the door and then depositing it behind Kato Kaelin's guest house. The problem with the theory, however, was that there was no concrete physical evidence to substantiate it, such as hair or fibers traceable to Mark Fuhrman on the glove that criminalists collected at Rockingham (and trace-evidence comparisons were made to rule out a contribution by Fuhrman). Fuhrman denied ever planting any evidence in response to Bailey's questions, but the tapes had seriously damaged his credibility. A little more than a year after the Simpson criminal trial ended, on October 22, 1996, Mark Fuhrman pleaded no contest to the charge of perjury, a felony, and received a small fine and three years of probation. Several investigations into his past, however, failed to uncover any specific wrongdoing, much to the surprise of many observers. While Fuhrman's courtroom lie was arguably not material to the O.J. Simpson case, however, it certainly injected an air of malevolence into the proceedings, casting a long shadow over the entirety of law enforcement involved in the case, investigators and scientists alike. In the end it put the prosecution on the defensive.

Suspicion was not confined, then, to the activities of Mark Fuhrman, and few of the state's witnesses escaped questions that suggested they had participated in, on some level, a frame-up of O.J. Simpson. In addition to Fuhrman, suspicion fell most heavily on Detective Philip Vannatter. For one thing, Vannatter maintained throughout the proceedings that, when the four detectives first went to Simpson's Rockingham estate after the discovery of the bodies at Bundy, O.J. Simpson was not yet a suspect in the murders. They had simply gone to the estate to "notify" O.J. of Nicole's death, he said. "That's the biggest lie we have heard probably in this entire trial," Johnnie Cochran told the jury in his closing argument, brandishing a demonstrative exhibit entitled: "Vannatter Big Lies, the man who carried the blood." In fact, all the responding detectives and their supervisor, Police Commander Keith Bushey, testified to the same thing, but the suggestion that Simpson wasn't a suspect even at this early stage rang hollow. After all, Fuhrman volunteered to accompany Vannatter and Lange because he had been to the Simpson residence on a 1985 domestic-dispute call, during which he encountered the Simpson couple and a Mercedes-Benz (Nicole's) with a broken windshield that Simpson admitted smashing. No arrest was made because Nicole declined to press charges. Fuhrman, too, was aware of Simpson's arrest for battering Nicole in 1989 because he wrote a letter about the 1985 incident in 1989 to assist investigators in the investigation. (Simpson pleaded guilty in that case.)

Vannatter was also assailed for misstatements he had made in the search warrant affidavit he submitted to a judge on June 13, 1994, in order to gain entry to O.J. Simpson's house. Among the questionable items Vannatter included in the affidavit, which is a sworn document the factual accuracy of which a judge necessarily relies upon when deciding to sign it, was the statement that Simpson had left on an "unexpected" flight to Chicago on the night of the killings, suggesting he was fleeing the law. In fact, the flight had been scheduled well in advance, and Vannatter, according to the defense, knew this from talking with Arnelle Simpson and Kato Kaelin. Also, Vannatter wrote that the small blood smear on the Bronco driver's-side door had been "confirmed by Scientific Investigation personnel to be human blood," when in fact no such confirmatory test was ever conducted; only a presumptive test was performed, one that is not definitive. (The test suggested only that the stain could be blood; it was not a test that can specifically identify human blood.) Although these misstatements might simply have been the product of a long night of work and undue haste, Judge Ito considered them sufficiently egregious to personally castigate Detective Vannatter for making them. The most troubling thing Detective Vannatter did, however, was put O.J. Simpson's reference blood sample (drawn at police headquarters on June 13, 1994, right after Simpson was interviewed by detectives) inside his pocket and keep it with him for more than two hours until he handed it off to criminalist Dennis Fung at the Rockingham estate later in the evening. This may simply have been an unthinking breach of protocol on Vannatter's part, but in the context of a case where the integrity of law enforcement officials was so much at issue, this was damning conduct. Vannatter should have immediately logged the vial into evidence while at headquarters. Defense attorneys seized on this to suggest that Vannatter had more than ample opportunity to sprinkle a little of O.J. Simpson's blood around here and there to set him up for the double-murder. But did he in fact have such an opportunity? It was, after all, late in the day when he returned to the Rockingham estate (he did not revisit Bundy), and by this time most of the evidence had already been collected, including the bloody glove found at the estate (which was in Dennis Fung's evidence van; and television cameras showed Vannatter did not visit the truck) and the blood drops in the house and on the driveway. Moreover, the Bronco had already been impounded (blood was collected the next day), and the evidence-gathering process had already concluded at the Bundy crime scene. Furthermore, Simpson readily admitted during his interview with detectives that he had cut his finger (an accident, he said) and was bleeding all over the place at home.[10]

The problem was, according to defense attorneys, blood evidence began turning up in the days, weeks, and even months after the initial crime-scene investigation. As noted earlier, no blood was collected from the back gate at Bundy until July 3, 1994.

It wasn't until August 4, 1994, that police investigators noticed there was blood on the pair of socks retrieved from Simpson's bedroom on June 13, and Dennis Fung conducted a second swabbing of the interior of the Bronco in late August 1994 after finding that initial efforts were inadequate. By this time, the integrity of the blood evidence inside the Bronco could be seriously questioned because the impoundment lot where it was stored, Vertiel's, was shown to be less than secure. In fact, the tow truck driver, John Meraz, who originally brought in the Bronco, eventually confessed to entering the Bronco and stealing two credit card receipts (which he said he later returned) as souvenirs.

A crucial theoretical bridge was established by the defense between Vannatter's retention of Simpson's blood vial and at least some of the blood collected late in the investigation by the testimony (at the grand jury and preliminary hearings) of jail nurse Thano Peratis. In those earlier hearings, Peratis testified that he had drawn between 7.9 and 8.1 cc's of Simpson's blood, which was stored in the vial provided to Vannatter. By the time of trial, however, only 6.5 cc's of blood could be accounted for and, defense attorneys argued, the approximately 1.5-cc discrepancy could account for some of the blood discovered after the initial investigation. Unfortunately, Peratis was recovering from open heart surgery when he was required to testify at the trial, so only his previous testimony was placed into evidence. In response, prosecutors videotaped Peratis at home, where he was not under oath and not subject to cross-examination. On the videotape, which Judge Ito permitted into evidence over the strenuous objections of the defense, Peratis said his earlier testimony was based on an estimate of the amount of blood he had drawn, and now he was convinced he had drawn about 6.5 cc's of Simpson's blood. After the trial, attorney Alan Dershowitz called the original, sworn testimony of Nurse Peratis "a linchpin of the defense" and suggested the jury simply didn't believe the "bit of revisionist history" offered by the prosecution.[11]

At the very heart of the defense case, then, was an accusation that the police had planted evidence. Regrettably, there are well-documented cases of police investigators fabricating evidence against suspects, so the argument has resonance. There is even a word for courtroom testimony of police officers in cases where it is suspected that the evidence has been fabricated: "testilying."

In the end, Johnnie Cochran characterized Detectives Fuhrman and Vannatter as "twin devils of deception" who orchestrated the case against his client. As a practical matter, though, Fuhrman and Vannatter could not have planted all the blood evidence that existed against O. J. Simpson, even if they had wanted to. For that reason, the defense case was double-barreled. Some of the evidence was planted, they argued, and some of it resulted from contamination of the evidence.

At the outset, defense attorneys sought a role in the evaluation and testing of the blood evidence, even proposing that the physical evidence be split equally between the prosecution and defense experts. While these efforts were consistently shot down by Judge Ito (although he did order the state to preserve 10 percent of each evidence sample, wherever possible), the defense experts nevertheless were given sufficient access to the evidence to determine that there had been some potentially serious missteps in how the evidence was collected, packaged, and stored. For one thing, Dennis Fung and Andrea Mazzola did not record the number of swatches they collected with each stain (which ranged from two to seven swatches); they simply placed them in coin envelopes that were themselves marked. Back at the laboratory, the swatches were removed for drying and then placed in paper bindles (folds of paper), but these were found to lack the initials of the criminalists. Most troubling, some of the bindles had transfer stains on them indicating that the swatches were not dry when they were placed inside, which should not have been the case.[12] The extent to which these and other irregularities affected the DNA tests about to be performed would become a central issue in the trial.

Contamination of evidence, of course, can occur at any stage of an investigation, in the field, during transportation, while in storage, and on the laboratory bench in the hands of a careless scientist. Everyone involved in the handling of the physical evidence was grilled at length about the precise steps they took in their respective capacities. In the end, attorneys Barry Scheck and Peter Neufeld effectively dismantled much of the prosecution's blood evidence, piece by painstaking piece, by exposing flawed procedures, outright errors, gaps in record keeping and downright untruthfulness. Emblematic of this was a bruising and sometimes humiliating cross-examination of LAPD criminalist Dennis Fung by Barry Scheck underscoring a series of mistakes and oversights by Fung—who was the supervising criminalist at both crime scenes— that raised serious questions about the integrity of the physical evidence. The cumulative effect was devastating to the prosecution.

On June 13, 1994, Dennis Fung and a criminalist-trainee, Andrea Mazzola (it was her third experience gathering evidence at a crime scene[13]), worked together collecting the physical evidence at both the Bundy crime scene and Simpson's Rockingham estate. During the trial, Fung admitted making some crucial errors on that day, as did Mazzola in later testimony.

Following the direct examination of Dennis Fung by Deputy District Attorney Hank Goldberg, Scheck launched into an excruciating micro-inspection of Fung's every action, inaction, and decision relative to the collection of evidence in the case. Among other things, Fung was forced to concede that his record keeping had been substandard and that certain information was inexplicably omitted from evidence-collection documents he routinely kept (such as exactly when evidence was collected, so that the sequence of events could be re-created). Moreover, Scheck's cross-examination revealed that Fung had failed to adhere to proper evidence-collection procedures, such as wearing disposable and sterile Latex gloves at all times. Scheck, in fact, played a snippet of videotape in court showing Fung grasping a piece of evidence—possibly the envelope that had contained Juditha Brown's eyeglasses—with bare hands, but only after eliciting testimony from Fung in which he insisted he had been wearing Latex gloves when he handled that item. Fung then denied that the paper in question was in fact the envelope, a version that Andrea Mazzola later backed him up on. Fung had also failed to follow other state-of-the-art evidence-collection procedures, his testimony revealed. In particular, he packaged collected blood swatches in *plastic* (as opposed to paper) bags, which he then stored in his police van; these sat for as long as seven hours in the hot sun, a scenario that promoted bacterial growth and led to the degradation of the DNA in the blood samples. Scheck successfully ambushed Fung with evidence-collection guidelines warning against the use of plastic after Fung testified that plastic minimized the risk of bacterial growth. Fung later stated that paper bags invite contamination because blood may leach out. Scheck drove home the point about plastic bags by reading a passage from a well-known text entitled *Techniques of Crime Scene Investigation* by Barry Fisher.[14] The passage read: "It is a certainty that wet or damp bloodstains packaged in airtight plastic bags will be useless as evidence in a matter of days."[15]

On redirect, in an effort to salvage something from the exchange, Deputy DA Goldberg referred to textbooks on evidence collection to underscore that neither bacterial growth nor improperly collected and preserved blood would transform the blood into someone else's, such as the defendant's. But there was more. Fung had to admit he blundered when he didn't intercede to prevent police officers from retrieving a blanket from Nicole's condominium to cover her body, even though the blanket was likely to introduce contaminants into the crime scene by way of what forensic scientists call secondary transfer. This might have introduced O.J. Simpson's hair into the crime scene, hair that perhaps clung to the blanket from a previous visit to the condominium. Perhaps most significantly, Fung acknowledged that he failed to collect enough evidence. A piece of paper seen near the bodies went uncollected, for instance, an item that Johnnie Cochran told the jury in closing arguments showed evidence of footprints.

Ultimately, the prosecution presented DNA evidence to the jury involving 45 bloodstains, all of which were controversial to one degree or another. A few engendered total war. The analyses were conducted at three separate DNA laboratories: Cellmark Diagnostics, the California Department of Justice (DOJ) DNA Laboratory in Berkeley, and the Los Angeles Police Department's laboratory. Eleven of the bloodstains were typed using RFLP. (Five VNTR loci were examined at Cellmark, and 11 at the California DOJ DNA Laboratory; the LAPD laboratory did not conduct any of the RFLP tests.) These stains, and the remaining 34 bloodstains, were also typed with a variety of PCR-based tests (seven tests altogether). Most of the PCR tests targeted sequence-specific loci of certain genes. These tests included the DQA1 system, as well as the series of five sequence-specific markers that are analyzed together and referred to as polymarker. Just as is the case with DQA1, the polymarker system includes prefabricated typing strips that change color to indicate the presence of the various alleles. Unlike the DQA1 gene segment, however, each of the five polymarker loci have only two or three possible variations (alleles). The targets of the examinations are thus called biallelic or triallelic. As such, they don't tell you very much when looked at individually. The power of discrimination derives from combining the five targets together. By comparison, if you are in a room of 100 people and described someone as having brown hair, well, a lot of the people in the room might have brown hair. But if you described the person as a female with brown hair, blue eyes, white skin, and Caucasian features, you will have narrowed things down considerably (although it's likely you'll find several women meeting that description). This is essentially what Polymarker does. The polymarker genes, however, do not correspond to anything as comprehensible as hair or eye color. Instead, the system targets such things as the low-density lipoprotein receptor (LDLR) and hemoglobin G gammaglobin (HBGG). The polymarker system, however, has one singular advantage over DQA1: The interpretation of the typing strips is a great deal more straightforward since the alleles are either present (prompting a color reaction) or not. Polymarker does not require the examiner to infer the presence of certain alleles, as is the case with DQA1.

The other PCR test used in the Simpson case was an AmpFLP (Amplified Fragment Length Polymorphism) test, combining the amplification features of PCR with the greater variation of the length-based analyses typical of RFLP. The specific VNTR site targeted in this test, referred to as D1S80, is on the short side as far as VNTR loci go, containing between 14 and 41 repeats of a core segment (boxcar) that is 16 base pairs long, but it is still an extremely variable locus with a high degree of discriminatory power. In brief, D1S80 combines a short segment (which is thus accessible by PCR) with the higher level of discrimination available in length-based DNA tests. During the DNA-evidence phase of the trial, criminalist Renee Montgomery of the California DOJ laboratory explained the details of the D1S80 testing to the jury.

Because some of the bloodstains were analyzed by all of the DNA systems (RFLP, DQA1, polymarker, and D1S80), some huge numbers, statistically speaking, were generated. Such numbers (one in many billions, for example) result from multiplying together the frequencies of finding the various genotypes exhibited by the individual DNA tests. (RFLP results *times* DQ-alpha results *times* polymarker results *times* D1S80 results *equals* a huge number.)

Significantly, the controversies swirling around the Simpson trial over the blood evidence did not involve a frontal assault on the DNA evidence, at least to the extent that the defense might have tried to block the jury from hearing about it. In the end, the scientific underpinnings of DNA typing were not aggressively challenged by

Simpson's lawyers, even though his principal DNA attorneys, Barry Scheck and Peter Neufeld, had established national reputations for doing, and helping others to do, precisely that. After exploring a number of options, on the eve of trial, the defense waived its right to an admissibility hearing concerning the DNA evidence. (This foreclosed any appeal on this issue.) The decision by the defense to waive the DNA admissibility hearing had more than a hint of pragmatism to it. DNA evidence was entering its tenth year as a forensic tool relied upon by prosecutors and defense attorneys alike, and it had been thoroughly vetted by the adversarial system—and survived. Indeed, vigorous challenges by the defense bar since at least 1989 had in fact strengthened the cause of DNA evidence by forcing forensic scientists to adhere to baseline standards or face humiliation in court. The likelihood that Judge Ito would rule to exclude the DNA evidence from coming before the Simpson jury—most legal analysts observed at the time—approached zero. By the time Simpson was arrested, furthermore, the only enduring controversy over DNA evidence that had any real traction in court involved how to convert evidence of a match (between an evidence sample and a suspect's DNA profile, say) into a reasonably accurate probability calculation that would give meaning to the genetic information. But even this controversy was petering out. In fact, a few months after Simpson's arrest, two of the country's leading adversaries on this point, Eric Lander of the Whitehead Institute for Biomedical Research and Bruce Budowle of the FBI's Forensic Science Research and Training Center, jointly published an article in *Nature* declaring that the "DNA fingerprinting wars are over."[16] Specifically, they concluded that the lingering arguments over statistical calculations were overblown and of little consequence in courtroom proceedings.

Since such a hearing would have involved a protracted legal battle, defense attorneys explained in court papers, the idle jury might be exposed to prejudicial publicity. Moreover, many of their key concerns about DNA typing—its reliability and the manner in which the tests were conducted—could be addressed during the trial itself, and they expressly reserved the right to do so. Instead of challenging *what* was done, then, Simpson's defense team challenged *how* it was done and *by whom*. Challenging *how* the DNA typing was done meant questioning the competence of everyone associated with the handling of the blood evidence, beginning with Dennis Fung, Andrea Mazzola, and Collin Yamauchi. If the handling of the evidence had been "slipshod," as the defense suggested it was in its motion to waive the admissibility hearing, either at the crime scene or in the laboratory, then the DNA results could not be trusted. As Barry Scheck summarized for the jury in his closing argument, "garbage in, garbage out," explaining, "You know, science is no better than the methods employed and the people who employ them. DNA is a sophisticated technology. It is a wonderful technology. But there's a right way to do it and a wrong way to do it." DNA results that implicated their client, the defense asserted, might well be the product of any number of contamination events, occurring at the crime scene or in the hands of laboratory personnel. Due to the exquisite sensitivity of PCR, contamination is always a concern when PCR-based tests are used. And 34 of the bloodstains were only amenable to PCR testing. (Of the 11 stains that were typed with RFLP, however, several produced extremely rare genotypes consistent with O.J. Simpson's blood. And accidental contamination alone could not account for this. Consequently, the defense challenged the investigators themselves, principally Fuhrman and Vannatter, and raised the possibility that these two, alone or in concert with others, planted evidence meant to frame Simpson.)

As part of what was surely meant to be an orderly presentation of the physical evidence, prosecutors started with the criminalists, Fung and Mazzola, who testified about gathering up the individual pieces of evidence that, prosecutors would argue next, pointed directly at Simpson based on multiple DNA tests. What prosecutors did not foresee, apparently, even though the defense had made clear its strategy from the outset (even before jury selection), was the ferocity of the response by Simpson's lawyers. In particular, the street-fighter demeanor of Scheck and Neufeld in their cross-examination of the criminalists managed to kick the teeth out of much of the prosecutors' case even before they got to the most incriminating evidence, the DNA.

Some of the questions posed by Scheck and Neufeld raised the possibility that evidence had been contaminated by such things as the failure of Fung to use gloves at all times, the use of plastic instead of paper evidence bags, and the introduction of the blanket into the crime scene. The mere presence of Andrea Mazzola at such a high-profile and complex crime scene was problematic, the lawyers emphasized, although Hank Goldberg made a point of demonstrating for the jury just how uncomplicated Mazzola's tasks had been on June 13, 1994. Neufeld, however, showed damning videotape of the evidence collection in progress, which showed Mazzola variously dropping cotton swabs to the ground, attempting to clean a pair of tweezers with a soiled hand, and brushing her hand along a dirty knee. She also admitted using the same cotton swab on two different bloodstains taken from Simpson's Bronco.[17]

The contamination theory was bolstered in later testimony by LAPD laboratory analyst Collin Yamauchi, when he explained that a mishap occurred on June 14, 1994, while he was handling the evidence. Just before his handling of the swatches collected at Bundy and the glove discovered at Rockingham by Fuhrman, Yamauchi opened the vial containing Simpson's reference blood sample and spilled some of it. According to defense attorneys, Yamauchi's failure to follow strict clean-up procedures after the spill, as well as the possibility that the blood aerosolized and thus lingered in the air and perhaps settled on the evidence, created the danger of cross-contamination of those items of crime-scene evidence. Also troublesome, according to the defense theory, were instances when the reference samples of Nicole Brown and Ron Goldman were handled at the LAPD laboratory in proximity to the physical evidence, including blood samples taken from the Bronco.

During their case-in-chief some months hence, the defense called microbiologist John Gerdes to more closely examine the possibility of contamination. As clinical director of Immunological Associates of Denver, Dr. Gerdes worked routinely with PCR-based tests relative to organ transplantation and the diagnosis of disease; he conceded, however, that he had no experience whatsoever with DNA analysis in a forensic setting.

In preparation for his testimony, Gerdes had reviewed LAPD laboratory data for 1993 and 1994 and consistently found, he said, a wholesale pattern of contamination, one that escalated up to the time the Simpson case commenced. He testified: "I found the LAPD lab has a substantial contamination problem that is persistent. It is chronic in the sense that it doesn't go away. It's there month after month, and it doesn't go away." In comparison to 23 other DNA laboratories around the country that he had evaluated (Gerdes had been hired as an expert in 30 previous cases since 1990, 29 of them for the defense), he found the LAPD laboratory the worst "by far." Sloppy laboratory practices and incompetent evidence handling in this case, he said, created "a tremendous risk of cross-contamination." In addition, he cited the use of plastic bags for evidence packaging as providing "the perfect environment for bacteria" and condemned the manner in which LAPD laboratory technician Yamauchi stacked test tubes. According to Gerdes, much of the state's blood evidence revealed evidence of cross-contamination, including evidence that indicated Ron Goldman's blood—as revealed by PCR-based tests—had been deposited in O.J. Simpson's Ford Bronco. Gerdes, in fact, rejected as suspect all of the prosecution's blood evidence that was based on PCR, even those tests performed at the DOJ laboratory and Cellmark because the evidence had been funneled through the LAPD facility, where it had been, according to Gerdes, contaminated before distribution. As part of the defense, too, it was suggested that some of the evidence DNA had degraded to the point of vanishing altogether (while stored in plastic bags in a hot truck, for instance). Cross-contamination with Simpson's reference sample, therefore, replaced and overrode whatever original DNA might have been there.

Gerdes, however, specifically excluded from his criticisms the DNA results obtained at DOJ and Cellmark based on RFLP because cross-contamination would not account for the volume of blood necessary to conduct RFLP. The RFLP tests, then, could not be explained as resulting from cross-contamination, and RFLP tests, Gerdes acknowledged, were performed on the socks found in Simpson's bedroom, the bloody glove found at Rockingham, and blood found on the back gate at the Bundy crime scene. (After Gerdes's testimony, additional RFLP test results were introduced by the prosecution that demonstrated a match to Ron Goldman, as well as O.J. and Nicole, from blood discovered in Simpson's Bronco.) Defense attorneys, however, maintained that bloodstains with sufficient DNA to produce RFLP results were planted by law enforcement. As for Collin Yamauchi's mishap with Simpson's blood sample, Gerdes said: "The rules are that you never handle a reference sample at the same time as any evidence." At the least, Yamauchi should have ceased everything and thoroughly cleaned up.[18]

Gerdes's rejection of most of the prosecution's blood evidence went beyond a scathing critique of the LAPD laboratory and its personnel. It was Gerdes's opinion, in fact, that PCR itself was a mistake in a forensic setting, principally because it is "so exquisitely sensitive," he said. While Gerdes succeeded in alienating virtually everyone in the forensic community with his testimony, he did not alienate the Nobel Prize-winning inventor of PCR, Kary Mullis. Mullis, in fact, who was scheduled to be the next witness for the defense, was prepared to give essentially the same opinion, which he outlined in interviews after the trial. In the end, Mullis was not called to testify, for reasons explained elsewhere in this book. His presence at the trial is nevertheless memorable because, when his name was mentioned during Gerdes's testimony, Mullis grinned boyishly and waved at the television camera mounted over the jury box.

Had Mullis actually been called to testify, his testimony might have been a real show stopper, his scientific credentials aside. For one thing, prosecutor Rockne Harman had warned for months that he was prepared to savage the renowned scientist's credibility. Years after the trial, Mullis wrote in his book *Dancing Naked in the Mind Field,* an account of his discovery of PCR and his generally good-natured views on life, that he was disappointed at not being called to testify in the Simpson case. It was for two unrelated reasons, however, that Mullis was anxious to testify, one noble, the other not so. First, he felt he could help the jury understand "just what had gone wrong in the LAPD's lab." Second, he wanted to verbally impale Rockne Harmon in front of the jury while he was being cross-examined. Specifically, Mullis wrote in his book, "I was going to slip in another question of my own—about some outrageous thing he [Harmon] had 'done' in the past." Mullis goes on to suggest a completely false accusation that would get everyone's attention, then continues, "I would be extremely out of order—he would loudly protest—and Ito would slam his gavel down to silence me. But the answer wouldn't matter."[19]

Gerdes singled out three ways contamination might have occurred in the Simpson case. First, of course, was the mishandling of the evidence by evidence collectors (Fung and Mazzola) or laboratory technicians (Yamauchi). Second, contamination could have occurred during DNA extraction procedures (as when Yamauchi spilled Simpson's blood), or perhaps by the unsound use of the chemicals involved. Gerdes fretted, in particular, about a bottle of Chelex that he saw in the LAPD laboratory when he visited. Chelex is a chemical typically used to extract DNA from evidence samples. In some laboratories, the use of Chelex is part of the standard protocol for

PCR because it facilitates DNA retrieval. The Chelex procedure separates the two strands of the double helix because the sample is heated, so only PCR can be performed on the resultant DNA. Gerdes objected that the bottle of Chelex might become a source of contamination from repeated use, but PCR protocols instruct examiners to remove what they need with sterile instruments before an experiment begins. So it wasn't much of a criticism. Third, contamination might have occurred by way of PCR carryover, which can happen if previous amplified DNA is introduced into a separate amplification procedure or finds its way onto other biological evidence.

The problems at LAPD, according to Gerdes, involved the physical layout of the place and the circular nature of the work flow. Gerdes rightly pointed out that high levels of DNA should never be brought back into an area containing low levels of DNA, such as an extraction area, since the higher concentrations of DNA might swamp the lower concentrations and produce erroneous DNA results. The criticism in this case seemed to melt away, however, on the closer scrutiny provided by prosecutor George Clarke's cross-examination. In fact, Gerdes conceded, the PCR product was not returned to the specific area near the extraction room or evidence-handling area but was taken to a completely separate area reserved for gel electrophoresis located a comfortable distance away. Thus, a contamination scenario was highly improbable, based on this criticism.

Unquestionably, the jury in the Simpson case learned far more about the possibilities of DNA contamination than they could ever possibly use over the course of their natural lives, assuming they ever cared to fully grasp the subtleties of such things as aerosolized DNA (or "flying DNA," as it was called during the trial) or whatever other horrors might derail a DNA test. If nothing else, the defense's alternative theory to explain some of the evidence, that it was planted by the LAPD, had the virtue of being instantly understandable. Just how convincing the theory was would no doubt depend on the sum total of one's life experiences.

A certain scorn might well have been deserved for evidence-collection missteps by Dennis Fung and Andrea Mazzola because of the ever present concerns over contamination. But evidence tampering? Was it possible that these two low-key criminalists were drawn into a huddle of nefarious police officers conspiring to frame O.J. Simpson? Well, believable or not, defense attorneys tied them both to the whipping post of cross-examination to explore that possibility.

Barry Scheck suggested in questioning Fung that he was covering up for Vannatter, Lange, and Fuhrman. Fung adamantly denied this, just as he denied that he was tailoring his testimony to favor the prosecution. He further denied pouring off any of Simpson's blood and insisted Vannatter gave him the reference blood vial at Rockingham on the afternoon of June 13, not the next day, as Scheck suggested. But Fung's credibility was damaged considerably when he was forced to admit that his responses under oath to the grand jury and at the preliminary hearing may have given the false impression that he, personally, collected most of the blood evidence at the Bundy and Rockingham crime scenes. In fact, most of the blood was collected by neophyte Mazzola.

Even more disconcerting, Scheck suggested, was Fung's subsequent visit to the Bronco on August 26 to do a second swabbing for blood in the interior. When Scheck characterized the blood that Fung swabbed that day from the console as seeming to "reappear" after Fung's apparent failure to notice it during the initial investigation, an objection choked off the question, but the implication was clear. Blood seemed to be turning up all over the place. The defense's theory on the blood inside the Bronco was somewhat loose. Either Mark Fuhrman planted it when he swiped the bloody Rockingham glove around the interior of the Bronco in the early morning hours of June 13 or someone planted the blood sometime later. The prosecution argued, of course, that the blood originated from the defendant immediately after the crime and was collected both during the early stages of the investigation and, well, also later on. So what? These were matters for the jury to sort out. But they were crucial matters. If the DNA of Ronald Goldman was determined to be in the Bronco, there could only be two explanations (because Goldman never had occasion during his life to be inside the Bronco): Either his blood was planted inside or O.J. Simpson was present at the crime scene. Andrea Mazzola fared little better than Fung during her cross-examination, conducted by Peter Neufeld. She, too, was accused of participating in a conspiracy to frame O.J. Simpson, which she adamantly denied.

Chief among the mistakes Fung made was his failure to notice and collect the bloodstains on the back gate at Bundy, stains whose existence were contemporaneously noted only in the log of Mark Fuhrman. (Several other police officers testified to seeing these stains, however, but they generated no documentation about it. It was, in fact, Officer Riske who pointed them out to Fuhrman, according to Riske's testimony.) It wasn't until Fung revisited the crime scene on July 3, 1994, with Lange that he first observed the blood, at which time it was collected, three weeks after the crime and from a location that was by now a national monument, judging from the number of curiosity seekers. Later, when the DNA results of these stains were carefully scrutinized and compared with those of other bloodstains recovered from Bundy, defense attorneys zeroed in on the molecular hardiness of the gate stains as compared to other bloodstains that were collected immediately after the crime. The molecular weight of the DNA extracted from the gate stains was sufficiently high for RFLP testing, whereas most of the stains collected three weeks earlier had degraded to a point where only PCR tests worked. It was over this discrepancy that some of the fiercest forensic battles would be fought in the Simpson case when the defense began presenting its case. How in the world, defense attorneys wondered, could the DNA from stains left out-of-doors for three weeks at a national shrine be in better shape than the DNA from bloodstains collected immediately after the crime and carefully stored in a forensic laboratory?

Another key controversy involved the dark dress socks Fung testified that he retrieved from the foot of Simpson's bed on June 13, 1994. Ultimately, the socks would prove crucial to the prosecution's case because they contained powerful genetic evidence that indicated O.J. Simpson had killed his wife. The significance of the socks, of course, was recognized from the outset. In her opening statements to the jury, Marcia Clark called the trail of blood leading from the Bundy crime scene to the socks found at the foot of O.J. Simpson's bed "devastating proof of his guilt." By the end of the trial, Barry Scheck, in his closing argument to the jury, also embraced the socks as the "most important piece of evidence," but for entirely different reasons.

Multiple DNA tests on the socks definitively revealed the DNA profiles of two people, O.J. Simpson and Nicole Brown. Most damningly, blood from an ankle area consistently typed to Nicole based on seven PCR tests (DQA1, D1S80, and Polymarker) and 14 RFLP loci, a total of 21 genetic markers. Another area of the socks produced a match to O.J. based on two PCR tests and nine RFLP loci.[20] The odds that the DNA belonged to someone other than O.J. or Nicole were vanishingly low, so the defense did not contest the identifications, only the source.

The principal problem with the socks was that nobody, including Dennis Fung, who collected them, noticed any bloodstains on them until August 4, 1994.[21] A police inventory videotape made on June 13 of Simpson's residence—the kind of videotape made routinely by police to protect themselves from civil litigation after a search— did not capture the socks when LAPD photographer Willie Ford swept Simpson's bedroom with his camera. Possibly the socks had already been picked up by Dennis Fung or possibly Simpson's bed blocked the view that would have revealed their exact location. Or possibly, suggested the defense, the socks were not there. Based on all the confusion surrounding the socks, the prosecution and defense agreed to have them subjected to highly sophisticated chemical tests at the FBI's laboratory, tests that—it was hoped, anyway—would clearly resolve whether the blood originated from the crime scene or, as the defense maintained, it was planted sometime later.

11 Bad Blood (Part 2)

Describing the overall DNA presentation at the O.J. Simpson trial is a little like trying to describe a riot. There was a lot of action, and a lot of punches were thrown, but a lot was unclear, too. For instance, who won? And on what points? Most criminal cases involving DNA evidence are, by comparison, boxing matches. The two sides square off and do battle over one, perhaps two, DNA test results. It's certainly easier to keep score in a boxing match than at a riot, but the Simpson jury and the public witnessed a genetic rumble involving multiple serological and DNA tests of 45 separate bloodstains coming out of three independent laboratories. It was a lot to think about and not a little confusing. Over the course of the entire trial, in fact, the acronym "DNA" was mentioned 10,000 times.[1] During the month it took prosecutors to lay out their DNA evidence members of the jury were reported to be, variously, transfixed, groggy, taking notes, failing to take notes, rubbing their eyes, dozing off, and occasionally even attentive. Even Judge Lance Ito's attention span was severely tested during this part of the trial. While he made some perceptive and rapid-fire evidentiary rulings as the DNA evidence unraveled, he was also observed nodding off occasionally.[2]

Some trial watchers, and perhaps even some jurors, were no doubt expecting the defense to savage DNA typing as a scientific enterprise. Would Barry Scheck and Peter Neufeld, observers seemed to be wondering, so undermine the DNA evidence as to call into question a science that had by this time put tens of thousands of people in jail and kept tens of thousands out? News accounts endeavored to advertise this phase of the trial by anticipating a DNA "holy war," as one reporter put it.[3] But that's not how it turned out. "Three weeks into testimony about DNA tests," reporters for *USA TODAY* lamented, "the war hasn't materialized."[4] And, in fact, it would not. The defense's position on DNA evidence, as it turned out, could be summarized as follows:

DNA typing is a valid, even extraordinary scientific process, but not one that should be entrusted to ordinary mortals, and particularly not those employed by the LAPD.

The real pyrotechnics began, of course, when the DNA evidence began to roll out, although the tediousness of the process created something of a TV ratings trough. Since this was the big forensic contest that everyone had been waiting for, however, news organizations provided what amounted to pre-game analyses of expected testimony, which focused on the professional accomplishments of the experts and lawyers, their personalities, and, wherever applicable, the deep and abiding animus of the participants toward one another. Just as the DNA testimony was getting underway, for example, Kenneth B. Noble of *The New York Times* pointed out that "the lawyers and expert witnesses on both sides are longtime antagonists who have made careers out of debating the complexities of scientific evidence. They also appear to intensely dislike one another."[5] Gale Holland of *USA TODAY* warned of the ferocity of Barry Scheck and Peter Neufeld when confronting experts for the prosecution. "Bad blood runs deep," Holland wrote, noting that Rockne Harmon, while speaking at a fall 1994 DNA conference, had referred to Scheck and Neufeld as the "princes of darkness."[6]

The first DNA expert to testify was Dr. Robin Cotton, director of Cellmark Diagnostics laboratory in Germantown, Maryland. Cotton's testimony began with a methodical, days-long exposition of the science underlying DNA analysis, guided by Prosecutor George "Woody" Clarke, a DNA litigation specialist on loan from the San Diego County District Attorney's Office. She emphasized that although DNA degradation may make it more difficult, even impossible, to detect the various DNA markers, the degradation process does not turn one person's DNA into that of another. "You may lose it altogether," she said, "but you can't just change DNA types."[7] Cotton then unleashed a spate of DNA test results (many of which were backed up by tests performed at the other two laboratories) that appeared to most observers at the time to be a crushing blow to the defense.[8] In particular, Dr. Cotton said Cellmark's tests showed a match between O.J. Simpson's DNA type and the blood drops discovered leading away from the bodies (which the prosecution maintained dripped from the deep cut on O.J. Simpson's left middle finger, a cut sustained during the attack), additional blood drops arrayed around Simpson's Rockingham estate (on the theory the finger wound kept bleeding), blood from the glove found at Simpson's home by Mark Fuhrman, as well as other stains. Just as significantly, Nicole Simpson's DNA type was found on the socks retrieved from Simpson's bedroom. Cotton's testimony was the first time O.J. Simpson had been linked to the double-murder by DNA evidence.[9] "This is the day that everybody paid the price of admission to see," remarked Hastings Law School professor and San Francisco chief deputy public defender Peter Keane.[10]

Whatever the reaction of the jury, the visualizations of DNA concordance were allowed to soak in overnight before Cotton began to assign numerical values to her matches. The tension was in fact so high on the day Cotton began providing probability calculations that Peter Neufeld and George Clarke both drew fines from Judge Ito for incessant arguing over the finer points of the statistical testimony, arguments that forced the judge to halt the proceedings on two occasions and order the jury out of the room.

When the numbers did finally come, the more impressive ones were trumpeted around the country in banner headlines. Four of the drops found leading away from the bodies (Items 47–50) were too degraded for RFLP, so only the seven PCR-based tests were conducted on these (DQA1, polymarker and, except on one of the drops, D1S80). Based on the PCR testing, the probability was one in 240,000 that a person tested at random could have contributed the same DNA profile to the stains of four of the drops, Cotton testified, and a probability of one in 5,200 on the drop that only underwent six tests. (O.J.'s DNA profile matched these stains.) On the last blood drop in the series, however, the drop farthest away from the bodies and found on the driveway (Item 52), there was enough DNA in the stain to conduct both the complete set of PCR tests and also identify five RFLP markers. The odds were only one in 170 million that someone other than O.J. Simpson deposited that blood drop, Cotton said. The probability was the same for a blood drop found in O.J.'s foyer.

While the probability calculations relating to the blood on the socks identified as matching Nicole's DNA were variously stated, all of the estimates were breathtaking. According to Cotton, the chances that the blood was from someone other than Nicole were one in 9.7 billion. And—contrary to defense suggestions—it was Nicole's DNA type, in fact, that was underneath her fingernails, Cotton explained, based on all the available PCR-based tests. The blood under her fingernails did not—as the defense had suggested, based on a confusing battery of serological tests—belong to an unidentified individual (for example, the so-called real killer). The testimony was, according to Southwestern University School of Law professor Myrna Raeder, a "home run" for prosecutors; James Starrs, forensic science professor at George Washington University School of Law, called it "devastating."[11]

This was a defining moment for DNA evidence. During jury selection, a *USA TODAY*/CNN/Gallup Poll found that four in ten people polled considered forensic DNA testing "very reliable."[12] Once some of the DNA results had been presented, however, the number of people who expected a verdict of guilty spurted up dramatically (from 3 percent to 11 percent).[13] Once most of the DNA evidence had been presented, the majority of Americans thought it "likely" that Simpson was guilty, according to another poll.[14] The editors of *The New York Times* captured the significance of the moment when they wrote, just as the DNA testimony was wrapping up, that "the most important development in recent weeks may be the respect accorded to the science of DNA blood testing." The Simpson trial's "lasting importance," they wrote, "might be as a demonstration of the power—and complexities—of DNA technology."[15]

The glow of this moment, however, was seemingly forgotten by the same editors after the defense had presented its case and the jury had rendered its verdict of not guilty. Based on these events, the editors of *The New York Times* concluded that "the bumbling of the scientific evidence in the O.J. Simpson case highlighted glaring deficiencies in how DNA evidence gets collected and tested—a problem, unfortunately, that is not isolated to the astonishingly inept police lab in Los Angeles." In fact, they went on to warn their readers, "many experts believe the poor performance of the police lab in the Simpson case may typify what happens in lower-profile cases nationally."[16]

DNA evidence, of course, wasn't about to slink off into a corner and play dead. It was here to stay. But the message was clear to everyone in the forensic community: Only the most rigorous adherence to professional scientific standards would be acceptable. Anything less risked rejection of the DNA evidence in court and damaging the reputations of the examiners and their laboratories. This was not a new message, but suddenly it had gained universal appeal.

One of the laboratories in the Simpson case, in fact, had long since gotten the message. Cellmark Diagnostics, a private laboratory in Maryland, was one of the earliest players in DNA typing, and, while it had thrived as a laboratory, it had also taken an occasional drubbing in court when its evidentiary work was less than perfect, particularly concerning performance or proficiency tests. As a result, the laboratory strove to meet the highest standards, ultimately submitting itself for accreditation by the American Society of Crime Laboratory Directors/Lab Accreditation Board (ASCLD/LAB), a national accreditation board that establishes exacting standards for staff training and education, laboratory security, testing procedures, and accuracy in testing. In October 1994, just as the Simpson case was heating up, Cellmark won accreditation.[17] The Simpson case prodded other laboratories to follow suit.

New York State, for example, passed a law requiring its crime laboratories to receive accreditation from the ASCLD. To that end, New York City spent $33 million building a new five-story laboratory outfitted with state-of-the-art equipment meant to withstand the scrutiny of the ASCLD, whose accreditation the laboratory ultimately earned.

"Based on the O.J. Simpson case," New York City Police Commissioner Howard Safir said in 1998, "we had to design a lab and procedures that are much less subject to challenge."[18]

After the trial, one commentator, former New York City prosecutor Harlan Levy, called the five Bundy blood drops "the most powerful evidence against Simpson" because they were least aggressively challenged by the defense.[19] What the defense did propose, however, was that Collin Yamauchi had potentially exposed the evidence to contamination from O.J. Simpson's reference vial when he spilled some of the blood on his glove on June 14, 1994. The suggestion was made that Simpson's fresh DNA overrode the completely degraded DNA on the swatches that were tested using PCR;

all of the controls used to test for contamination, however, suggested otherwise and the prosecution steadfastly maintained that the contamination scenario was far-fetched. Microbiologist John Gerdes, however, would testify during the defense case-in-chief that even Item 52 might have been contaminated by Yamauchi during the mishap, something which is far less likely for a stain with enough DNA for RFLP. Gerdes pointed out that the molecular weight of the DNA in that stain was lower (that is, it was more degraded) than any of the other stains subjected to RFLP.[20] The most jarring puzzle related to the Bundy blood drops, however, was the observation by defense experts that there were four bloody transfer stains on the packaging of the swatches of one of the Bundy blood drops (Item 47). Referred to as a paper bindle, this paper packaging was folded, by all accounts, around three blood swatches after they had been allowed to dry overnight. The finding of transfer stains on the paper bindle was, therefore, an incongruous discovery, and not one that lent itself to any easy explanations. The only explanation for how the stains got there at all was that the swatches were still wet when they were wrapped in the bindle. But how? If packaged as the testimony had indicated, the swatches should have been dry by that time. Something wasn't right. Barry Scheck, in his summation, suggested this amounted to clear evidence that the blood swatches had been switched; he noted that Mazzola's initials did not appear on the packaging as it was ultimately received into evidence, even though she had testified that she had initialed each one. This mystery was never solved.

The fiercest battle over the blood evidence, however, would be waged over the blood found on the socks—blood that went long undetected—and the bloodstains collected by Dennis Fung on the back gate at Bundy three weeks after the double-homicide. The next prosecution DNA expert, Gary Sims, the lead forensic analyst at the California Department of Justice laboratory, would have a great deal to say about these bloodstains, as well as many others.

The DOJ laboratory tested blood evidence from the crime scene at Bundy (including the blood drops analyzed by both other laboratories, blood from the back gate, and blood from the victims' clothes), O.J. Simpson's Ford Bronco, the bloody glove found by Mark Fuhrman at Simpson's house, and the socks found in Simpson's bedroom. For the first time in the trial, Sims linked O.J. Simpson to Ronald Goldman through DNA testing. The link was significant because Ronald Goldman and O.J. Simpson were not acquainted prior to the crime. Finding Goldman's DNA type on items of physical evidence or in places tied to O.J. Simpson suggested, according to prosecutors, that the two men had but one moment of contact, a moment that proved to be Goldman's last. Through a combination of PCR and RFLP testing, Sims's laboratory had found DNA markers consistent with Goldman's DNA virtually all over the glove found at Simpson's residence, including inside and outside the glove's wrist notch, inside the various fingers of the glove, and inside the glove where it covers the back of the hand. In several locations where RFLP and PCR were performed, the analysis showed an apparent mixture of the DNA types of Nicole Brown and Ronald Goldman. Altogether, Sims reported on 10 DNA tests his laboratory performed on the glove. Significantly, on a single stain inside the ring finger of the glove, DQA1, D1S80, and eight RFLP loci revealed genetic types consistent only with Goldman. Sims calculated the odds of a random match for that stain (based on only six of the eight RFLP markers, because the frequencies of the other two were unavailable) as approximately one in 2.1 billion for Caucasians.[21] Sims did not factor in the PCR results, which would have made the number even more impressive. Renowned forensic pathologist Cyril Wecht commented after Sims's testimony, "This comes as close as possible to a mortal prosecutorial wound."[22] PCR tests, furthermore, revealed DNA types consistent with O.J. Simpson on several locations on the glove.

The defense had long maintained the glove was planted on Simpson's property by Mark Fuhrman who stole it from the Bundy crime scene, which would explain the DNA types identifying the murder victims. During his testimony, Fuhrman had described the glove as appearing "moist and sticky." The defense argued that, left out in the night air, the glove should have been entirely dry by morning, unless, of course, Fuhrman had kept it hidden somehow overnight, perhaps on his person, only to deposit it for his own convenient "discovery." The defense's explanation for O.J.'s blood on the glove was that either it was contaminated at the laboratory or Fuhrman picked up Simpson's blood when he swiped the glove inside the Bronco, where Simpson had innocently left traces of blood after cutting himself while retrieving his cellular phone on the evening of June 12 (a version of events that O.J. conveyed to one of his forensic experts).

The prosecution sought to demolish this theory through the testimony of other police officers responding to the crime scene who said they only saw one glove near the bodies. In addition, blue-black fibers were found on the Rockingham glove that matched fibers found on the socks and Ron Goldman's shirt; the prosecution alleged the fibers originated from a dark sweat suit that Simpson was believed to have been wearing on the evening of June 12 (and described by Kato Kaelin), but that sweat suit was never produced. (The prosecutors said Simpson must have ditched it after the crime.) Furthermore, no trace evidence on the glove—neither fiber nor hair nor DNA evidence—was ever linked definitively to Mark Fuhrman (although an unaccounted-for Caucasian hair was found on the glove).

The bloody glove found at Simpson's residence was certainly a key piece of evidence. As such, prosecutors struggled mightily to lash that glove securely to O.J. Simpson and at the same time deflect defense arguments that it was tainted. To that end, prosecutors repeatedly called to the stand Richard Rubin, a former executive of Aris Isotoner, the glove manufacturer of the type of gloves found at Bundy and Rockingham. A video clip and photographs had surfaced showing O.J. Simpson working as a sportscaster at a couple of football games in the early 1990s and wearing gloves that

Rubin said he was "100 percent certain" were the same model gloves as those in evidence in the case. He could not be certain, however, that they were the exact same gloves, he testified. The prosecution asserted, based on store receipts, that Nicole had purchased two pair of extra-large Aris Leather Lights from Bloomingdale's department store just weeks before one of the football games. Moreover, the gloves in question were exclusively manufactured for Bloomingdale's. Only a few hundred of that particular model glove (brown and extra-large) were purchased that year by Bloomingdale's, according to Brenda Vemich, a Bloomingdale's buyer who also testified for the prosecution, with only 200 or so pair being sold by the store nationwide that year. They were very rare.

Just at this juncture, however, the gloves took on a life of their own. On the same day in mid-June as the testimony of Rubin and Vemich, prosecutors proposed having O.J. Simpson try on a pristine pair of extra-large gloves similar, but not identical, to the Aris Leather Lights in evidence, but Judge Ito declined to allow it, proposing instead that Simpson try on the actual evidence gloves over a protective pair of Latex gloves. Although the prosecution team had already decided against such a demonstration in front of the jury, Prosecutor Christopher Darden was goaded by defense attorneys into going ahead with it anyway.

During perhaps the most memorable few minutes of the trial, O.J. Simpson put on a pair of Latex gloves and stood before the jury struggling to pull the evidence gloves over his hands, explaining repeatedly "They're too small." Although the gloves did eventually go on, the image of O.J. Simpson grimacing as he began to put the gloves on his hands became a defining moment in the trial. Defense Attorney Johnnie Cochran, in fact, borrowing a phrase suggested by Defense Attorney Gerald Uelmen, would make the glove demonstration the central theme of his closing argument to the jury. "If it doesn't fit, you must acquit," Cochran repeatedly told the jury in his final appearance before them.

Prosecutors scrambled to repair the damage, arguing through Rubin's testimony that the gloves might have shrunk somewhat (15 percent, Rubin said) or that the Latex gloves might have interfered with the demonstration and that they certainly should have fit on Mr. Simpson's hands, except for the defendant's fakery perhaps. Observers saw it with greater clarity. It "changes everything," said UCLA criminal law professor Peter Arenella. It was "an unmitigated disaster" for prosecutors, opined California criminal defense lawyer Gigi Gordon.[23] The following week, prosecutors had Simpson try on a pristine pair of extra-large Aris Leather Lights, the identical style as those in evidence, after having them sent overnight from the Philippines. The gloves fit without difficulty on Simpson's hands. Nevertheless, wrangling over the gloves continued until the last days of the trial. Prosecutors, in fact, opened their rebuttal case in September with a flurry of photographs and some videotape showing O.J. Simpson as a football sportscaster in the early to mid-1990s wearing the gloves that appeared to be the same make and model as the bloody evidence gloves. It was in response to viewing this videotape that Rubin testified that he was "100 percent certain" that the brown gloves Simpson was wearing were the same type of glove in evidence. Photos showed Simpson wearing a black pair of the same type of glove as well, according to Rubin, which would explain Nicole's purchase of two pairs.[24] On one of the very last days of the trial, however, the defense presented testimony from blood-spatter expert Herbert MacDonell to rebut Rubin's earlier suggestion that the evidence gloves might have shrunk by as much as 15 percent due to years of wear and tear, as well as from being splattered with blood. MacDonell told the jury that he had conducted a test on a brand-new pair of Aris Light gloves by smearing them with his own blood. There was no measurable shrinkage, he reported. Moreover, MacDonell said after other tests he performed he came to the conclusion that a 15 percent reduction in size was unimaginable, even after years of wear.

Gary Sims of the California DOJ also testified about DNA analyses of blood found in O.J. Simpson's white Ford Bronco. The DNA evidence from the Bronco, however, came in two distinct waves. The first wave came in May, when the bulk of the DNA testimony was heard by the jury and consisted entirely of PCR-based tests. Initially, PCR tests were all that could be performed because the overall amount of blood in the Bronco—scattered around though it was—was small. Ultimately, the defense argued that since the overall amount of blood found there amounted to a mere 0.07 milliliters, the Bronco contained, not evidence of a brutal and bloody double-murder, but rather evidence of efforts to quickly smear blood inside the car and incriminate O.J. Simpson. This would account for the DNA types of Nicole Brown and Ron Goldman. Simpson's blood was already present, the argument went, because it was his Bronco. Further, one of the defense's experts, Dr. Michael Baden, the former chief medical examiner of New York City, recounted that O.J. had told him about a cut he had sustained on his hand while preparing to go to Chicago on the night of June 12, 1994. This was the injury that occurred when Simpson reached into the Bronco for his cell-phone.[25]

Despite the paucity of blood, a total of 11 bloodstains produced PCR results (most using DQA1 and D1S80), identifying DNA types consistent with O.J. Simpson (ten locations), Nicole Brown (five locations) and, most damningly, Ronald Goldman (four locations). Bloodstains typed to O.J. alone came from the inside of the driver's door, from the driver's-side wall, from the driver's-side carpet, from the instrument panel, and from one of five stains swabbed from the center console. From what appeared to be a partial bloody footprint on the driver's-side carpet, the DNA type of Nicole Brown was obtained. (An FBI analyst could not say, however, that the shoe impression in the Bronco was made by the same type of shoes, size 12 Bruno Magli shoes, as left a series of bloody shoe prints at the Bundy crime scene; he couldn't even state categorically that it was a shoe print.) Several of the stains were apparent mixtures of blood, however, so they posed some interpretive challenges. A stain on the steering wheel (the only stain subjected to polymarker as well as one of the other PCR tests) produced DNA types consistent with O.J. and Nicole. One of the stains collected on the center console produced DNA types consistent with O.J. Simpson and Ron Goldman, based on DQA1 and D1S80. It was computed that there was a one in 1,400 chance that the combined genetic profile in this stain came from two people other than O.J. Simpson and Ron Goldman.[26] Except for the bloody (possible) footprint bloodstain, all of the foregoing were collected on June 14, 1994.

Dennis Fung performed additional, more thorough, swabbings on the center console in late August, however, which produced evidence Item #s 303, 304, and 305. By this time, of course, the Bronco had been burglarized, so the integrity of the evidence could not be assured. In its first wave of testing, the California DOJ conducted DQA1 and D1S80 tests on these late-entry stains. It was this set of stains that created certain interpretive challenges for DNA analysts. The swatches from these stains revealed obvious mixtures of DNA genotypes, but from how many individuals? It was impossible to say for sure. Two of the swatches from the console, for example, produced DQA1 typing strips that clearly showed that the 1.1 and 1.3 alleles were present, as well as some version of type 4. Types 2 and 3 were not present in the mixture. What was unclear was whether subtype 1.2 was present. Reliance on the tie-breaker dots did not, in the end, break the tie. O.J. Simpson's DQA1 genotype is 1.1, 1.2. Nicole Brown's type is 1.1, 1.1. Ronald Goldman's is 1.3, 4. None of their genotypes could be eliminated from the mixtures, but neither could it be said that all three were necessarily present. The D1S80 tests were likewise something of a puzzle, even though the test is by design more straightforward than DQA1. D1S80 differentiates the lengths of a DNA segment composed of a repeating section of DNA 16 base pairs long. In this case, the DNA revealed a mixture of types 18, 24, and 25, but since there was overlap in the genotypes of the principals in the case, the genotypes could not be adequately untangled. It could only be determined that none of the three could be excluded from the mixture.[27]

None of the uncertainties in the PCR testing were, of course, fatal to the prosecution's case. In fact, the tests supported the state's case insofar as nobody could be eliminated as a contributor to the stains. Nevertheless, the prosecution strove for the highest level of proof it could obtain, something particularly important with respect to the presence of DNA in the Bronco that could be linked to Ronald Goldman.

To that end, the DOJ conducted a second wave of DNA tests on the swabbings collected by Dennis Fung in late August 1994 after combining the three pieces of evidence. The objective was to obtain a sufficient amount of blood to enable analysts to perform more definitive RFLP tests, which requires a greater amount of blood at the start. While their efforts succeeded, the tests did not get underway until long after the trial began. Since RFLP can take months to complete, the results themselves were not available to prosecutors until after they had completed their case-in-chief. It was only after the defense had begun presenting its case, and, in fact, right in the middle of the testimony of Dr. John Gerdes, that Prosecutor Rockne Harmon alerted the court to the newly available RFLP results. From the prosecution's point of view, it was a bombshell. For one thing, the new evidence was intended to more definitively place the blood of Ronald Goldman in the Bronco. For another, the ability of analysts to conduct RFLP on the Bronco blood smears undercut the argument that Gerdes was in the process of advancing—namely, that PCR tests are unreliable because of the risk of contamination. A positive RFLP test would, presumably, trump that concern. The defense, naturally, argued for the exclusion of the evidence since it was being offered at a very late stage of the trial and the defense team had been provided no access to the new tests. More generally, the defense assailed the blood evidence from the Bronco because some of it was collected months after the crime, from a vehicle that was stored in an unsecured lot, and from a vehicle that had, in fact, been burglarized. Furthermore, the defense would argue, the detective who had investigated the burglary of the Bronco failed to notice any blood inside.

Although most of the issues surrounding the Bronco blood evidence were matters for the jury to sort out, Judge Ito seriously considered the defense's request to exclude the new RFLP tests because of their lateness. Eventually, though, Ito allowed them into evidence after he was reminded that the delay was attributable to the workload from other cases, believe it or not, of Gary Sims and the California DOJ. The jury did not hear of the evidence until mid-September, however, during the final hectic weeks of the trial at a time when the courtroom was still reeling from the audiotapes of Mark Fuhrman (and the defense was maneuvering, ultimately in vain, to force Mark Fuhrman back on the witness stand in front of the jury) and frenetically dealing with a host of issues as the case lurched toward the finish line. A September headline in *The New York Times* read, aptly, "Nearing Completion, Simpson Trial Is Also Nearing Chaos."[28] By the time Gary Sims retook the witness stand to explain this potentially explosive testimony on September 13, 1995, David Margolick of *The New York Times* described the courtroom as having "degenerated into a cauldron of exhaustion, frustration, rancor and anger."[29] Under these circumstances, it is not hard to imagine that the new DNA evidence—as significant as it was—was simply swallowed up in the din of courtroom pandemonium. When Prosecutor Harmon first advised the court of its existence back in August, he characterized it as "the other shoe" dropping in the case against O.J. Simpson.[30] Quite possibly, few on the jury heard the shoe drop.

The DOJ laboratory managed to obtain results for four RFLP loci on the combined bloodstains (Item #s 303, 304, and 305) from the center console, Sims told the jury. Based on these results, O.J. Simpson and Ronald Goldman could not be excluded as contributors to that blood. Nicole's DNA pattern was not detected by this round of RFLP testing, although her PCR-based DQA1 type was found in the combined blood stains. The RFLP results on the combined stains, the prosecution asserted, corroborated the earlier PCR results on the Bronco blood—which, most significantly, had indicated that Ronald Goldman's DNA was on the center console of O.J.'s car—and undercut the criticisms made by Dr. Gerdes about the dangers of contamination with PCR-based DNA tests. This was powerful evidence, but after a great deal of wrangling over how to present statistical calculations relative to this purposefully mixed stain, the state was content to leave it at that and forgo any testimony about the probability of a match.[31] This may have blunted the impact of the evidence further.

Sims endeavored to make clear the significance of the RFLP findings, however. Asked about John Gerdes's concerns about contamination relative to the PCR results done earlier on the Bronco blood, Sims replied: "I think the RFLP test results undermine that opinion."[32]

There were a multitude of evidence items and issues in the Simpson case—aside from the DNA evidence—that never seemed to provide much in the way of satisfactory proof. Left unresolved in the trial, for example, was whether O.J. Simpson ever owned a pair of shoes consistent with the 30 full and partial bloody shoe prints discovered at the murder scene and leading away into oblivion. According to FBI Special Agent William Bodziak, who examined the evidence for the prosecution, the prints could only have been made by a pair of size 12 Bruno Magli shoes, a rare and expensive Italian shoe. By the conclusion of the criminal trial, however, prosecutors produced no evidence that O.J. Simpson had ever owned such a pair of shoes, let alone the pair involved in the double-murder. (Note that part of the evidence produced in the subsequent civil trial included a sheaf of photographs that surfaced after the criminal trial and showed Simpson wearing a pair of shoes that appeared to be Bruno Maglis.)

For certainty, the prosecution sought refuge in its blood evidence. Sims established the stakes when he reported the DOJ's DNA results obtained from the bloodstains found on the socks recovered from Simpson's bedroom and on the back gate at the Bundy crime scene. In a sense he added insult to injury with respect to the DNA tests on the socks, since the DOJ's tests not only confirmed the analyses by Cellmark, they greatly exceeded them in statistical magnitude. In the course of his testimony, Sims explained how with each additional DNA test, one piled on top of the other, a statistical crescendo is created. The more tests performed, the bigger the numbers (based on the product rule) and the lower the odds that someone else—other than the defendant or the victim, as the case may be—is responsible for the blood deposit. Sims testified that three bloodstains from the ankle areas of the socks produced genetic types consistent with the genotype of Nicole Brown, two by PCR-based tests and one by RFLP and PCR. Altogether, the latter stain was subjected to all seven available PCR tests and a battery of 14 RFLP analyses. All of the tests matched Nicole's DNA type. Based on the 21 genetic markers, Sims recomputed the population statistics and concluded that the probability was one in 21 billion that an individual tested at random would match that DNA profile. In an article published after the trial, the prosecution's statistician, Dr. Bruce S. Weir, wrote that it would be "absurd to claim that such a match could be coincidental" based on the 21-locus match.[33] Southwestern University Law Professor Robert Pugsley commented at the time of Sims's testimony that the socks were "the single most crucial item in the case."[34]

O.J. Simpson's blood was also matched to a stain on the socks based on nine RFLP markers and two PCR tests. It was calculated that the probability that the blood belonged to someone other than O.J. Simpson was a staggering one in 57 billion. (Although the odds were impressive, this particular finding was not, of course, that much of a big deal because the socks were, after all, O.J.'s.) According to the prosecution, O.J.'s blood got there because of his cut finger, Nicole's because Simpson murdered her. The defense argued that the blood on the socks was planted to frame Simpson. And their argument was not entirely insubstantial. The socks, of course, had a lot of baggage arising from the uncertainties over how and when they were collected and the delay in detecting the presence of blood on them. But there was more, and the defense believed they had something of a smoking gun to prove their evidence-tampering theory.

Sims's testimony went beyond just reporting the DNA results. Sims very carefully reviewed for the jury the manner in which the dozens of individual stains (some truly tiny) were distributed on the socks themselves. It was his belief, he told the jury, that the pattern of stains was consistent with someone walking through blood, which created the apparent splatter. (Defense attorneys wondered why there was no blood of Ronald Goldman on the socks, given the allegations, but prosecutors countered that it was Nicole's blood that predominated at the scene.) In anticipation of an argument that defense experts would advance, Sims laid out his theory that the bloodstains on the socks covered only the surface area of the socks. The blood, in other words, had not soaked through to the inside opposite surfaces of the socks, which one might well expect to see if the socks were lying on a flat surface when the blood was applied. The blood on the socks was, in short, not planted.

This conclusion was flatly contradicted by one of the defense's expert witnesses. According to Herbert MacDonell, the stains on the socks found in the ankle area did not result from blood splatter, as Sims testified, but from "direct compression and release," as if someone applied the blood with a finger. Moreover, the stains appeared to be "swiped" onto the outer surface, he said. Furthermore, MacDonell said he observed microscopically "small red balls that appeared to be dried liquid on the inside" surfaces of the socks. These appeared to be blood, he testified, although MacDonell had not performed any serological tests on the substance to make a final determination. This also contradicted Sims and seemed to indicate that blood had seeped through to the inside opposite surfaces of the socks. The question was, how could the blood have gotten through to the inside opposite surfaces if O.J. Simpson's ankles were blocking the way? Wouldn't the blood have been dry by the time Simpson allegedly took them off at home? Prosecutors suggested that perhaps perspiration accounted for the internal wet transfer stains, assuming they were there, and maybe Simpson squeezed the surfaces together somehow. MacDonell didn't buy the perspiration theory (since, he said, there was "no evidence of dilution or diffusion"), but prosecutors also noted that there had been a presumptive blood test, called a phenolphthalein test, performed on the socks, which involves dampening a small area of a sock with a cotton swab or Q-tip. On cross-examination, MacDonell conceded that the internal location of the little red balls seemed to correspond with the outer surface area where the presumptive-test swab was applied. On cross-examination, too, MacDonell agreed with Marcia Clark's suggestion that the apparent swiping might have occurred if Nicole's hand had reached out and brushed the ankle of her assailant or if the socks glanced off her body.

How important was all of this seesawing? It went to the absolute dead center of the case. The prosecution was now, in effect, trying to prove beyond a reasonable doubt that the evidence had not been planted by the police, not just that the defendant was guilty beyond a reasonable doubt. They were fighting on two fronts.

Prosecutors found themselves in the same predicament with respect to the bloodstains that were found by criminalists on the back gate at Bundy. At the same time that he revealed the DNA results obtained by DOJ on the socks, Sims told the jury that three stains from the back gate (Item #s 115, 116, and 117) were found to contain DNA markers consistent with O.J. Simpson's DNA. One of them was found to be very consistent. Early in the trial, the first policeman on the scene at Bundy, Robert Riske, who told the jury he observed the gate with a high-powered flashlight, described the gate this way: "there was blood at the bottom, there was blood on this latch, there was blood at the top and my partner directed my attention to blood on the outside on the grating." Riske went on to explain that he subsequently pointed out the stains to other responding personnel, including Mark Fuhrman. Fuhrman, who was the only person to document these stains contemporaneously, told the jury that he observed what appeared to be a partial bloody fingerprint in amongst the stains. No one, however, ever attempted to lift the print, assuming it was there, so nothing ever came of that observation. Two of the bloodstains on the gate (Item #s 115 and 116) were small, so only PCR tests were conducted, Sims testified, revealing DQA1 and D1S80 matches to Simpson. These translated into modest numbers. The probabilities were one in 520 that someone other than O.J. Simpson deposited that blood. The third stain (Item # 117), on the other hand, produced staggering odds. This larger bloodstain was subjected to DQA1, D1S80, and 9 RFLP tests. Using databases from the FBI and the Orange County Sheriff's Department Forensic Laboratory,[35] DOJ computed the odds that the blood was deposited by someone other than Simpson at one in 57 billion. And unlike the DNA match to Simpson on the socks, this finding was a big deal, or, at least, a bigger deal.

Defense attorneys were quick to remind everyone, and at every opportunity, that these bloodstains were not collected by criminalist Dennis Fung until July 3, 1994, three weeks after the yellow crime-scene tape came down. There was no way that anyone could sanely vouch for the integrity of those bloodstains, they argued. Prosecutors endeavored to compensate for this obvious shortcoming by producing a parade of police investigators to attest to the bloodstains' presence on June 13. Beyond Fuhrman and Riske, these witnesses included Sergeant David Rossi, Detective Ron Phillips, Detective Tom Lange (who said, in fact, that he instructed Fung to collect the gate stains on June 13, although that didn't happen), Lieutenant Frank Spangler, and Detective Philip Vannatter.[36]

On cross-examination of Sims, Barry Scheck established that there was high molecular weight DNA in the blood found on the back gate, just as there was in the blood on the socks. By contrast, most of the blood drops at Bundy, which were collected on the first day of the investigation, contained very little intact DNA. In his closing argument to the jury, Defense Attorney Johnnie Cochran asked: "Why do blood stains with the most DNA not show up until three weeks after the murder, those on the socks, those on the back gate?" In his closing remarks to the jury, Scheck emphasized that the gate was rusty and discolored, stained by berries from the yard, making it anything but clear what the police were seeing in the middle of the night with their flashlights.

To the extent that they had a grand strategy, O.J. Simpson's defense team simply challenged every conceivable aspect of the prosecution's case, from the broadest possible social angles right down to the interpretation of the physical evidence on a chemical level. Before they were through, their microscopic scrutiny would provoke a major confrontation—arguably the most significant—about an analysis of the blood evidence on a *molecular* level. It was a first.

The ultimate question in the O.J. Simpson case was whether the police had fabricated the blood evidence to frame Simpson; the defense had made it clear from the outset that this was one of their principal arguments. Since the prosecution was hopeful that they could demonstrate that this did not happen, they called the FBI Laboratory in late 1994 to see if there was a forensic test for differentiating between untreated fresh blood found at a crime scene and preserved blood taken from a crime victim or criminal suspect as part of an investigation. In particular, prosecutors wanted to know if there was a forensic test for EDTA (ethylenediamine tetraacetic acid), an anti-coagulant, which is routinely mixed with drawn blood samples in test tubes. The chemical additive EDTA prevents the blood from clotting inside the test tube. Prosecutors wanted to test the blood found on the socks in Simpson's bedroom and the blood found on the back gate at Bundy against the preserved blood of O.J. Simpson and Nicole. O.J. Simpson's blood was drawn on June 13, 1994, by nurse Thano Peratis and handed off to Detective Vannatter, who in turn, some hours later, handed it off to Dennis Fung. That vial was not officially logged in as evidence until June 14. Vannatter also handled the blood vials of Nicole Brown and Ron Goldman, which he picked up on June 15 at the coroner's office and delivered to the serology division at the LAPD laboratory. Coupled with questions raised by the defense about a volume of blood from O.J.'s vial that appeared to be unaccounted for (by as much as 1.9 milliliters, by Scheck's estimate), the fact that Vannatter also handled Nicole's blood sample was another problem for prosecutors. The defense argued that this sequence of events provided an opportunity for Vannatter to purloin Nicole's blood and plant it on the socks found in Simpson's bedroom. Naturally, prosecutors countered that this was poppycock, but neither side had any tangible proof to bolster their argument.

The FBI, however, did not have a ready-made test for EDTA in blood. In fact, no one did.

So, just as the Simpson trial was about to get underway in January 1995, Dr. Bruce Budowle and a team of research chemists at the FBI's Forensic Science Research Unit (FSRU) in Quantico, Virginia, set to work on devising a test. About the same time, the chief of the FBI Laboratory's Chemistry-Toxicology Unit (CTU), Roger Martz, also began to tackle the problem. Martz, however, worked on the problem independently at his laboratory in Washington, D.C. The process that emerged involved mass spectrometry. A mass spectrometer detects the presence of electrically charged particles; an ion is an atom or molecule that has become electrified after either gaining or losing at least one electron or proton. In the case of EDTA, three distinct ions confirm its presence, designated the 293 ion, the 160 ion, and the 132 ion. In the first ionization stage of the tandem MS process (MS/MS), so-called parent ions are produced; in the second stage, additional fragments are formed known as daughter ions.

As the FBI was refining its tests for EDTA, Prosecutor Rockne Harmon proposed aloud that some of the disputed blood evidence be sent to the FBI Laboratory for analysis. On February 3, 1995, outside the presence of the jury, he expressed confidence in court that no EDTA would be found in the disputed bloodstains and that the issue of evidence planting would be put to rest. "[T]here will be no question," he said, "no one will have a lingering doubt." He was operating under the assumption that the results would be unequivocal. "There's only two possible outcomes to this test," Harmon explained. "There's either going to be EDTA there or there's not going to be EDTA there. And we're willing to accept the outcome, whatever that is. . . . We agree to accept these results in advance."[37]

About a week and a half later, defense attorneys continued to voice reservations about Harmon's proposal and suggested that such a novel forensic test might not even survive an admissibility hearing. As things would turn out, however, the defense later embraced the EDTA tests as their own and forgot all about earlier suggestions that the testing first overcome an admissibility barricade.

In mid-February, the prosecution sent portions of the sock and gate evidence and the reference vials to Roger Martz for EDTA testing. By sheer happenstance, the fact that Roger Martz was to perform the testing would ensnare the O.J. Simpson trial in a national controversy concerning the integrity of the FBI Laboratory, not that the Simpson litigators needed any additional controversies of national significance to liven things up. The case was pretty lively already. Nevertheless, yet another major subplot was being written into the Simpson drama that played out until the last days of the trial. It was one of the most disturbing.

As it turned out, Rockne Harmon had been unduly optimistic about the EDTA testing as a definitive tie-breaker. In fact, the results of the tests permitted both sides to claim a measure of victory. Once again, the jury would hear completely contradictory expert testimony. Significantly, the testing protocol that Martz had devised was never called into question. It was simply a matter of interpretation. That was more than enough, however, for a good old-fashioned courtroom donnybrook.

Roger Martz first tested the blood from the reference vials and found clear evidence of EDTA in both O.J. Simpson's blood sample and Nicole Brown's blood sample. Computerized printouts produced impressive peaks that reflected the 293 ion, the 160 ion, and the 132 ion. According to his tests, the concentration of EDTA in the preserved blood samples was about 2,000 parts per million. Martz next tested the bloodstains from the socks and the rear gate. These stains also registered signals at locations for the tell-tale ions, most perceptibly the 293 ion and the 160 ion, but at much lower levels and across a field of background noise that made interpretation difficult. Martz concluded that there was no EDTA in the bloodstains from the socks and the rear gate. After these results were reported, Defense Attorney Robert Blasier flew to Washington, D.C., to review the findings with Martz. At the meeting, Martz explained to Blasier that the smaller peaks produced in the testing of the gate and sock blood "could be" EDTA, but at mere trace levels of perhaps 2 parts per million, a far cry from the 2,000 parts per million found in the reference blood. To the extent that there might be EDTA in the crime-scene blood, Martz explained, it could be attributable to the trace amounts of EDTA found in food or detergents, for example.[38]

In fact, Martz had initially been puzzled by these miniature signals. As a control, he tested a sample of blood obtained from the dress Nicole was wearing when she was murdered. There were no questions about how that blood was deposited. He also tested a portion of the dress that was not stained with blood. Contained in both of these he found and reported traces of EDTA, something attributable—if for no other reason—to the fact that laundry detergents contain measurable levels of EDTA. When he tested material from the socks that were clean of blood, however, he detected no EDTA. As an additional check, Martz decided to analyze his own blood, which he knew did not have any EDTA added to it. The graphs generated by this testing produced low peaks similar to those seen in the sock and gate samples. Significantly, however, Martz did not retain the data produced by the tests on his own blood, even though the experiments figured into his conclusions about the evidence bloodstains.

Naturally, defense attorneys sent off Martz's report to another expert. That expert was forensic toxicologist Dr. Fredric Reiders, who founded the National Medical Services laboratory in Willow Grove, Pennsylvania. Dr. Reiders reviewed the Martz report and concluded the FBI expert had completely misinterpreted his own data. In fact, according to Reiders, the amount of EDTA Martz detected in the sock and gate blood went well beyond trace levels or background noise. In his view, the EDTA had to have been artificially introduced into the disputed blood evidence to account for the levels seen. So, who was right? Obviously, there was a courtroom showdown in the offing.

For defense attorneys, EDTA in the sock and gate bloodstains amounted to a smoking gun that those stains were planted by the police. And in their view, Fredric Reiders was a powerful witness to make such a case. In fact, Reiders, a well-known forensic expert, had been on some prosecutors' short lists as an expert, except that the defense had called him first.[39] This did not, however, temper the reception he received by prosecutors when he testified for the defense, any more than Martz's credentials as an FBI agent buffered him from the coming storm.

Despite Rockne Harmon's earlier brinkmanship over EDTA, the prosecution's case-in-chief came to a close without the jury hearing a word about it. As a strategic matter, therefore, the defense needed to call Martz as a foundation witness, his opinion notwithstanding, to verify that he had conducted EDTA tests on the evidence. Oddly, perhaps, they put Reiders on the stand first to explain why he believed the very next witness was entirely wrong.

In the end, it probably mattered little to the jurors who testified first because the testimony of both Reiders and Martz was almost incomprehensible, brought to earth only by an occasional definitive statement, which the jury undoubtedly noted. Although the actual EDTA testing did not involve the theoretical, ten-dimensional world of superstring theory, it may as well have given the clarity of the explanations the witnesses provided on the EDTA testing itself. The testimony was consistently "arcane" and "abstruse," even *The New York Times* complained, as the witnesses crawled through a molasses of scientific terminology. Gone were the halcyon days of the DNA testimony, when everything made perfect sense! With little or no explanation, terms and phrases such as "quadropole," "quasi-molecular ion," "pharmacokinetics" (which Martz couldn't even define), and "full daughter spectrum" ricocheted around the courtroom with dizzying effect. Despite the evident complexity of the technologies used in the EDTA testing, however, both Reiders and Martz appeared to understand completely what they were talking about, insofar as they each were able to draw firm, unshakeable conclusions. Unfortunately for the jury, they drew conclusions that were exactly opposite.[40]

Dr. Fredric Reiders certainly had the bearing of a world-class expert, graying, bespectacled, and graced with the intrigue of a Viennese accent. He appeared in every respect to be completely confident in his assessment of the tests conducted by Agent Martz and was indignant at any suggestion that he might be wrong. Dr. Reiders was, furthermore, categorical about his conclusions. When asked by Attorney Blasier on direct about the EDTA testing on the blood from the back gate at Bundy, Reiders left no wiggle room.

"In my opinion, yes," he said, "it demonstrates there is EDTA present in that stain."

Given the finding by Martz that no EDTA at all was found on an unstained (control) portion of one of the socks, whereas the blood from the sock did produce low-level signals consistent with the EDTA ions, Reiders was equally confident with the sock bloodstain.

"In my opinion, the EDTA came from the blood, not the sock," he said, ruling out the suggestion that laundry detergent might account for the EDTA ions.

Reiders also said he believed that the amount of EDTA found in certain foods, such as mayonnaise, cereals, and dried vegetables, could not account for the EDTA levels seen in the exhibits. Such high levels of EDTA, he said, do not exist in people. "In my opinion," Reiders testified, "it is so unlikely that I would not even consider it."

On this particular point, however, there remains considerable uncertainty, since it is not precisely known how much EDTA, if any, is contained in the blood of healthy people.[41]

It is also a matter of some dispute as to how much EDTA is present in the environment. Prosecutor Marcia Clark thought she had the definitive answer to that question. On cross-examination, she immediately confronted Reiders with a copy of an EPA study—one that Reiders had used in forming his own opinion—suggesting that EDTA is ubiquitous in the environment at levels consistent with the findings of Martz on the crime-scene bloodstains. These EDTA levels also approximated those that Martz detected in his own blood, in experiments that he performed in May and July. Reiders dismissed those passages in the EPA report as "either a typo or a complete absurdity." When Clark's voice rose in exasperation, Judge Ito yanked her to a sidebar conference for a warning. A more restrained Clark, however, had no greater luck.

"Doctor," she began, "how do you account for the readings that came up from Agent Martz's blood?"

"I don't have to account for it," Reiders sneered. "I think he would have to account for it because it's absurd to find that much EDTA in normal blood."

According to Reiders, people would bleed to death with the amount of EDTA in their blood at the level Martz detected in his own blood, which made Martz "an extraordinary, amazing, unusual man," Reiders said.

Reiders's most significant conclusion came in response to a question from Blasier. Could the amount of EDTA Martz detected in the sock and gate bloodstains have come from the reference sample test tubes of blood that Vannatter had handled?

"Yes, of course it could," he said.

The defense knew that Roger Martz was going to put a decidedly different spin on his own data, but they had to call him to the stand for evidentiary and procedural reasons. After visiting Martz in Washington, D.C., Blasier wrote a memo to his colleagues about Martz, calling it "troubling" that Martz was prepared to equate the levels of EDTA he found in his own blood with those of the evidence bloodstains.[42] The strategy the defense team subsequently devised to contend with this was simple: slash and burn.

Within 18 months of his testimony in the Simpson trial, Martz's career at the FBI Laboratory would be in shambles, although for reasons that went well beyond that testimony. Martz's testimony coincided with events at the FBI Laboratory that were about to erupt into one of the most scandalous episodes in the FBI's history. The saga began with internal criticism leveled by Frederic Whitehurst, a Ph.D. in chemistry, who was hired to work in the FBI Laboratory's Materials and Analysis Unit (MAU). When his criticisms went largely unaddressed, Whitehurst became a whistleblower whose complaints eventually triggered a major investigation of the FBI Laboratory by the federal Office of the Inspector General (OIG). After an 18-month-long investigation, during which a team of OIG investigators amassed 50,000 pages of documents and transcribed interviews and a specially appointed team of five independent forensic experts pored over forensic evidence and laboratory reports, the OIG, headed by Inspector General Michael Bromwich, released a 517-page report in April 1997. Entitled *The FBI Laboratory: An Investigation into Laboratory Practices and Alleged Misconduct in Explosives-Related and Other Cases,* the OIG report recommended censure, transfer, or discipline for five FBI agents working at the laboratory. While the report said the OIG was unable to substantiate the worst allegations of Agent Whitehurst, namely that some of his colleagues had committed perjury, fabricated evidence, or obstructed justice, the report nevertheless carefully catalogued a litany of "serious deficiencies" that Whitehurst had originally identified. Among those singled out were Martz.[43]

The looming investigation of the FBI Laboratory became the dark underbelly of the O.J. Simpson case. Roger Martz, by any measure a key witness at the trial relative to the blood evidence, was alleged to have performed substandard work in other high-profile cases. In the end, the OIG would conclude that Martz had made some serious mistakes in the World Trade Center bombing case, the Oklahoma City bombing case, and, to varying degrees, even in the O.J. Simpson case. O.J. Simpson's lawyers noticed the incipient scandal, too, and began aggressive efforts to learn more about Whitehurst's charges, if for no other reason than to find some concrete evidence of forensic wrongdoing on the part of Roger Martz. It was against the backdrop of this burgeoning scandal at the FBI Laboratory that Martz took the stand in the Simpson case to testify about some of the most, if not the most, important evidence in the case, evidence that went to the heart of the defense team's allegations that police had planted evidence to frame their client.

Martz testified over two days in the O.J. Simpson trial, on July 25 and 26, 1995, immediately after Fredric Reiders. Dr. Reiders, of course, thought there was enough EDTA in the bloodstains to have killed everyone involved, the murders aside, including Special Agent Martz. Reiders's testimony was also supposed to be the death blow to the prosecution's case, because if he was right about there being EDTA in the evidence bloodstains, that meant the blood was planted to frame O.J. Simpson.

Martz, however, rejected Reiders's argument that there should be absolutely no EDTA in the evidence samples. Trace amounts might be present due to environmental factors. On the other hand, he also rejected Reiders's conclusion that the third ion indicative of EDTA, the 132 ion, was present in the evidence samples at all, although he did agree with Reiders that the other two, ion 160 and ion 293, were present in tiny amounts in the evidence samples and that these "could be" EDTA. Martz conceded to Robert Blasier on direct that "It responded like EDTA responded."

"Is it consistent with the presence of EDTA?" Blasier asked.

"Yes," Martz said.[44]

But that was early in his testimony. Martz would later clarify that he believed these signals were most likely some other chemical, an artifact in the instrument or "some type of matrix effect with the blood" (whatever that is). Martz testified that the peaks on the graphs that Reiders suggested were the 132 ion were consistent with other small peaks on the graphs that were nothing more than background or "electrical" noise, common to electrospray mass spectrometry (as opposed to conventional mass spectrometry). "It's not signal," he said. To be significant, Martz explained, such peaks needed to be a minimum of three times the size of the noise peaks.

"I was asked to determine whether those bloodstains came from preserved blood. . . . Those bloodstains did not come from preserved blood," Martz testified emphatically. "I was able to prove that."[45]

In addition to testing the evidence stains from the socks and the gate, Martz also ran tests on his own blood to see what would happen. He did this on his own initiative in May and July 1995. In his own blood, Martz also found low levels of the 293 and 160 ions but concluded that the 132 ion did not register. While his testimony on this point bolstered his argument that EDTA was not present in the evidence stains, it got him into trouble back at the FBI Laboratory because Martz had failed to adequately document the testing on his own blood. It also put him further at odds with O.J.'s defense attorneys. While Martz did provide them with graphic readouts of some of the key tests on his own blood, these were provided only weeks prior to his testimony.

"In your opinion, then," asked Marcia Clark, "is his—scientifically speaking, was it scientifically and forensically inappropriate for Dr. Reiders to have interpreted this graph in the manner that he did?"

"In my opinion, it was, yes," Martz replied.[46]

To the defense, there was a sinister pattern to Martz's failure to take notes and generally keep better records of his work in the case. Martz didn't take any specific notes, he said, because he used the exact same procedure he had used on the other blood in the case, so he didn't believe it was necessary. (Martz did eventually generate a report about the testing of his own blood, but only after his meeting with Blasier in Washington, D.C.)

Robert Blasier and the defense team were not alone in their disgust over Martz's forensic records. "We find [the foregoing] record-keeping practices to be unacceptable," the OIG concluded, after a review of Martz's records in the Simpson case. "Martz should have made and retained notes describing his procedures, even if he considered the procedures to be background research and not case work. As a general rule, an examiner should make and retain notes for all work related to any case, but especially work that might be the subject of examination at trial. Further, another examiner should be able to review such notes and have a complete understanding as to all procedures performed in the case. Martz's work in this regard was deficient."

In a practical sense, of course, Martz's deficient(or, at least, minimalist) record-keeping was a boon to the defense because it was arguably a weakness that could be attacked.

"Where is the raw data that you did that formed the basis for all these charts right now?" Blasier asked, referring to the computer-generated digital data produced by the streams of ions moving through the mass spectrometer.

"It no longer exists," Martz replied uncomfortably. "It was erased off the computer when the case was dictated."

"It has been destroyed?" Blasier asked ominously.

"Well, yes," said Martz.[47]

On the basis of this exchange, it certainly would have been fair for a juror to conclude that Martz's erasure of his digital data was suspicious, particularly given the technical dispute that had arisen over the clearly registering ions and the barely registering ones. Martz had, in fact, generated printouts of the ions that registered in any significant way, the ones upon which his opinion was based, but he had decided to flush a voluminous amount of extraneous digital information. Considering the importance of the case, common sense might suggest that every scrap of digital information should have been retained. And isn't that what computers were designed to do, save the tons of information that might have, in a bygone era, become a storage problem? Why hit the "delete" key in such an important case, unless you've got something to hide?

"We only have so much computer space," Martz told Blasier.

Ultimately, the OIG was forced to untangle the episode. The OIG sought to answer two questions. Had Martz lied to the defense attorneys about the digital data? And was Martz wrong to have erased his own digital data? At bottom, both questions were technical. Interestingly, these seemingly nerdy questions provided the OIG investigators with an opportunity to weigh in on Martz's overall conclusions in the Simpson case. The OIG concluded that Martz had not misled O.J. Simpson's defense lawyers about the digital data.

The OIG also sided with Martz, in fact, on his decision to erase. "Martz's decision not to retain digital data," they wrote, "while perhaps subject to criticism for tactical reasons at trial, cannot be criticized from a scientific perspective." In support of this view, the OIG cited ASCLD/LAB guidelines on forensic documentation. In particular, Essential Criterion 1.4.2.14 of these rules provides that "case records such as notes, worksheets, photographs, spectra, printouts, charts and other data or records *which support conclusions* must be generated and kept by the laboratory."[48] The objective of the requirement, according to an explanatory note in the guidelines, is that, in the absence of the person who actually performed the examination, "another competent examiner or supervisor could evaluate what was done and interpret the data."[49] According to the OIG, "Consistent with these guidelines, Martz retained in hard copy from all mass spectra that demonstrated the detection of any significant ions and upon which he based his conclusions. These charts would have enabled a competent examiner to interpret Mart's data and evaluate his conclusions. Martz was not required to retain his digital data. The digital data that Martz allowed to be erased and which was not otherwise reflected in hard copy, was not material to Martz's conclusions." [50] In part, the OIG's conclusions on this point were based on, as they put it, "the limitations of electronic storage" that confronted Martz.[51] The digital data episode, then, was not just another example of poor record-keeping habits by Martz.

But there was a subtext here. Certainly the tenor of Robert Blasier's questions was that the erased data might have held the key to the EDTA controversy. On purely technical grounds, the OIG believed otherwise. In fact, the OIG went so far as to endorse Martz's EDTA conclusions. What Blasier characterized as "missing data" was simply "not material to Martz's conclusions," they said. "In this respect," the OIG wrote in a very telling footnote,

we think that criticism of Martz by Simpson's counsel at trial was especially misplaced. Simpson's counsel criticized Martz for not retaining digital data for trace levels of ions consistent with EDTA at or near the limit of detection of his method. Even if the compound Martz was detecting was EDTA, the concentration was too low to be consistent with EDTA-preserved blood. Moreover, because Martz was detecting such low concentrations, he would not have been able to confirm its identity due to the absence of the 132 daughter ion for that compound. While the defense suggested that Martz could have massaged this data to permit detection of the missing daughter ion, that exercise would not have affected Martz's conclusions. The level of EDTA detected still would not be consistent with that of preserved blood.[52]

However, the OIG found that Martz was cavalier in his preparation for trial, which gave Blasier several opportunities to ambush Martz with reports and studies that Martz had to concede he knew nothing about. In particular, Martz said he had neither seen nor reviewed FSRU's validation study on the EDTA method. He also admitted to Blasier that he was not aware of a recovery study done by FSRU on the extraction method. Consequently, at Blasier's request, Martz was forced to review that study over the lunch break so he could respond to questions about it. Although none of this amounted to perjury, the OIG graded Martz poorly. "Given the importance of this case and the obvious expertise on the defense side," the OIG wrote, "Martz was surprisingly unprepared for his testimony."[53]

Reading the OIG's assessment of his performance in the O.J. Simpson trial in the 1997 report must have been a maddening experience for Martz. One minute there is praise for his testing methodology and even concurrence in his results. The next minute he is fighting for his professional life. This unevenness was especially pronounced in a section of the report that undertook to evaluate Martz's overall presentation of the evidence at trial.

"At the outset," the OIG noted, "we observe that contrary to the suggestion of the defense, Martz's analysis was sound. Martz employed well-established analytical techniques to isolate and identify EDTA in dried blood, and he answered the specific question raised in the case. While Martz came under intense questioning by the defense for not conducting various additional studies, we are not critical of Martz on these grounds. Given an unlimited amount of time and resources, the FBI Laboratory could have conducted all sorts of studies on myriad related and tangential issues. But the reality is that Martz's role was to generate probative information based on the limited samples provided and return the samples for further independent analysis if necessary. He accomplished that task, and it does not appear that any other expert in the case repeated his work and came to any other conclusions."[54]

In the final analysis, however, Martz had been an embarrassment. Martz, the OIG concluded, "poorly represented the Laboratory and the FBI in the case."[55] In January 1997, Roger Martz was reassigned by the FBI to duties outside the FBI Laboratory.

Just as the FBI Laboratory scandal was heating up, the O.J. Simpson case petered out, punctuated only by the drama of closing arguments and the jury's near-instantaneous acquittal of O.J. Simpson.

While the legacy of the O.J. Simpson case can be endlessly debated, one of its clearest effects was to make resoundingly clear to law enforcement generally and forensic laboratory personnel in particular that only the highest professional standards would be tolerated in the future. In August 1999, Executive Director of the National Commission on the Future of DNA Evidence Christopher Asplen explained: "What is different now is the evidence identification, preservation, and collection procedures that are followed. The technology itself really has not changed. One of the lessons from the OJ Simpson case was the need to pay attention to DNA collection issues." The training and education of law enforcement officers was key, he added.[56]

On the heels of the O.J. Simpson case, the FBI Laboratory aggressively pursued accreditation from the American Society of Crime Laboratory Directors/Laboratory Accreditation Board, which establishes stringent national standards for forensic laboratories. To win accreditation is the forensic laboratory equivalent of earning the Good Housekeeping Seal of Approval. After an exhaustive review of the FBI's facilities and documentation over the summer of 1998, which included scrutiny by 28 laboratory inspectors over all of the major laboratory units (including the DNA units), the ASCLD/LAB awarded the FBI Laboratory its accreditation on September 22, 1998.

A few months later, in December 1998, after spending $3.5 million on training and equipment since 1995, the LAPD's crime laboratory, Piper Technical Center, also won ASCLD/LAB accreditation.[57] The laboratory was, in the words of an editorial in *The Los Angeles Times* that ran on December 11, 1998, "completing a long march back to honor," and its achievement was "greater than words imply." The editorial added: "Los Angeles had the nation's first such [crime] lab, and it provided a model. But the lab had never achieved national accreditation and in recent years had a hollow image, one that was destroyed by an aggressive defense in the OJ Simpson double-murder trial. Too many errors, too small a staff, too much outdated equipment and more."[58]

By this time, only about a third of the nation's crime laboratories were accredited. The ASCLD/LAB accreditation process is a difficult, time-consuming, and expensive proposition. Accreditation is no rubber stamp. The New York Police Department's crime laboratory learned this the hard way when it failed to gain accreditation in November 1999, despite New York's expenditure of $33 million on a brand-new facility. It has since been accredited.

Despite the valiant efforts by the LAPD crime laboratory, public relations for a police department can be a fragile thing. In 2000, the image of the LAPD was once again tarnished by a scandal in the LAPD's Rampart Division, where dozens of police officers were accused of police misconduct in an investigation sparked by former LAPD officer Rafael Perez, who pleaded guilty to various counts of stealing cocaine in the course of his duties and who began to tell a tale of wide-ranging corruption in the LAPD. Most disturbing, Perez admitted lying in court about the arrest of Javier Francisco Ovando in 1996, who was shot and paralyzed during his arrest and later sentenced to 23 years in jail for firing on police. Now Perez admits that he planted the gun on Ovando after the shooting. Ovando has since been released from prison and dozens of other convictions have been overturned.

As the Rampart scandal continued to unfold, thoughts naturally turned to the O.J. Simpson case. On February 15, 2000, *Los Angeles Times* contributing editor Robert Scheer wrote: "Now we understand why OJ was acquitted. . . . At the time of the OJ Simpson verdict, it seemed bizarre to many of us that a jury could conclude that officers of the law had conspired to frame an innocent man. If the trial were held today, the once-scorned arguments of Johnnie Cochran would appear far more plausible. . . . The natural outcome [of the Rampart scandal and other instances of police abuse around the country] is the cynicism toward the police that we witnessed in that OJ jury."

O.J. Simpson commented at the time that he felt vindicated.

The O.J. Simpson case continues to spark fierce debate, of course, and opinions about the verdict are often passionately held. Holding a strong opinion about O.J. Simpson's guilt or innocence can be more than an academic exercise, however. It may well chart the course of one's jury service. In a number of subsequent cases around the country, appellate courts have ruled that it is permissible for prosecutors to remove prospective jurors from a jury panel based solely on their belief that O.J. Simpson was innocent or that the verdict in the criminal case was fair. In a Texas murder case, for example, a prosecutor repeatedly bounced such people because of concerns that even "overwhelming scientific evidence" would be insufficient to convince them of the defendant's guilt in his case.[59] The prosecutor persuaded the trial judge that his actions were racially neutral (important because most of the people at issue were African Americans)[60] by explaining, "I believe I struck all people, black and white, who indicated they thought the [O.J.] verdict was fair."[61] The Texas Court of Appeals agreed, at least that it was racially neutral. There have been similar rulings in Alabama and Georgia.[62]

Conclusion: The Future of
Forensic DNA Analysis

WHEN, IN LATE 2001, INVENTOR DEAN KAMEN UNVEILED THE SEGWAY scooter, a supercharged "human transporter" outfitted with five gyroscopes and 10 microprocessors to simulate normal human balance, the machine was consciously promoted as a "disruptive technology," a term that business analysts use to describe a technological advance so profound that it changes the way people live and swamps whatever came before. Kamen predicted that for city-bound transportation the Segway "will be to the car what the car was to the horse and buggy."[1] While some commentators greeted the new invention enthusiastically, others were more circumspect in their assessment of the Segway's eventual impact on the world. Only time will tell.[2]

Whatever the fate of Segway, forensic DNA analysis has clearly proven itself to be a "revolutionary technology." It has fundamentally altered the way criminal cases are investigated and prosecuted and even the way juries are selected. It has solved countless crimes, including cold cases, and revealed serial killings or rapes in situations where only a whiff of suspicion had supported that theory before. It has solved cases that had virtually no chance of being solved without it. It has freed more than 120 men from prison (as of late 2002) following convictions, men who otherwise had no hope of being exonerated. It has inspired many people to approach eyewitness-identification accounts with greater wariness, even some of the eyewitnesses themselves. DNA typing has prompted average people and governmental agencies alike to aggressively pursue the answer to the question, "Who's the daddy?" DNA-matching techniques have allowed disease detectives to pinpoint the source of food-borne illnesses, so that authorities could move more swiftly and confidently to prevent pathogens from spreading. DNA fingerprinting has spawned myriad businesses, with some specializing in only one type of DNA analysis (a kind of DNA-typing boutique, perhaps), such as mtDNA analysis. It has sparked forensic laboratory renovations and building projects coast to coast and around the world. Volumes of state and federal laws have spewed forth from elected bodies to define, finance, and generally facilitate the phenomenon of "DNA fingerprinting." Untold appellate decisions have been written examining DNA evidence. A national DNA database has been established and continues to expand, with an eye toward cobbling together an international DNA database. DNA databases for domestic animals are in the offing, too. Parents are preserving the DNA of their children in commercially available kits designed to help locate missing children; in fact, the child DNA kits have even become a wildly successful giveaway for many U.S companies.[3] Missing persons, both civilian and military, are routinely identified through DNA, as are fugitives from justice. Intriguing questions posed by archeologists, paleontologists, historians, and many other academics are being answered by DNA-identification techniques. DNA typing has even become a standard tool for monitoring the impact of genetically modified organisms in the food supply and for verifying the validity of cloning advances. Forensic DNA analysis has truly changed the world.

A natural byproduct of DNA typing's success is the "creative destruction" that accompanies it, another business term originally coined by economist Joseph Schumpeter to describe how even seemingly minor technological advances can result in the collapse of whole industries.[4] Few, if any, forensic laboratories in the United States perform ABO blood typing anymore, for example, let alone the myriad serological tests that were once deemed so informative in forensic casework. (The ABO blood grouping system was the first useful genetic blood marker. Discovered in 1904, the system lasted an entire century.) The 1990s experienced something of a Cambrian explosion, in fact, of DNA testing methods. Advances in DNA-typing techniques have in fact been so rapid that DNA technology has already eaten its own young. The use of RFLP, for one, is in a precipitous decline in favor of the faster PCR-based STR systems, even though RFLP continues to demonstrate its utility in a few unique applications.

Even when DNA evidence doesn't play a role, it can play a role. When Michael Skakel was convicted on June 7, 2002, of killing teenager Martha Moxley in 1975 in Greenwich, Connecticut, the high-profile case stood out as one of the coldest cases that ever went to trial. The case continued to capture the attention of the American public because Skakel was a nephew of Ethel Kennedy, Robert F. Kennedy's widow, and thus a member of the most famous American political family. Among the other things that kept the investigation alive were the best-selling novel *A Season in Purgatory*[5] by Dominick Dunne, published in 1993 and subsequently the basis of a TV movie, and the 1998 best-selling nonfiction account of the case by former LAPD detective Mark Fuhrman entitled *Murder in Greenwich: Who Killed Martha Moxley?*[6]— which also became a TV movie, in 2002. Skakel, who was 15 at the time of the murder, was convicted of bludgeoning Moxley to death with a golf club near her suburban home. A close neighbor of Moxley's, Skakel had been questioned by authorities over the years about the slaying. In 1992, his story changed dramatically from being at a cousin's house at the time of the killing to actually having gone to the Moxley house to spy on Martha, while maintaining he had no involvement in the murder. According to the 1992 version of events, Skakel climbed a tree in Martha's yard to peer into the house, then masturbated before climbing down and going home. In the years leading up to the trial, Connecticut authorities tried repeatedly to generate usable DNA profiles from the evidence preserved in the case, but came up empty. Even a private laboratory specializing in mtDNA analysis failed to produce anything directly linking the defendant. The three-week trial in 2002, therefore, included no DNA evidence against defendant Skakel. It did, however, include evidence about Skakel's new "alibi," which had been preserved in taped interviews Skakel had done with a writer preparing a manuscript for a book about the Kennedys. In his closing argument to the jury, prosecutor Jonathan Benedict proposed that it was Skakel's *fear of DNA evidence* that had inspired him to revise his story in 1992. It was simply an effort to explain away any possible DNA evidence that might implicate him in the crime, but the strategy backfired. "Rather than spinning a nice tight explanation," Benedict told the jury, "he has spun a web in which he has ultimately entrapped himself."[7] The jury agreed.

The future of DNA profiling has often been charted by the pressing demands of the present, just as it was in the case of Josef Mengele. The unanticipated events and challenges of the future, therefore, will certainly serve to push the technology in directions it might not otherwise go, at least not so quickly. Indeed, this is precisely what occurred in the aftermath of the attacks of September 11, 2001, and over the course of the investigation into the origins of the anthrax that was circulated around the country via mail later that fall. With no warning at all, the forensic science community suddenly found itself confronted with two of the most daunting challenges in the history of criminal investigation. And DNA profiling would play a crucial role in both.

The National Commission on the Future of DNA Evidence expects STRs to remain the predominant technology for forensic investigations at least through 2010, although DNA markers of varying types are likely to be added to the arsenal along the way, including new SNPs from throughout the genome (there are millions of them in everyone's DNA, so the potential is unlimited), additional STR markers (although the huge amount of money expended on establishing the 13 core STR loci for CODIS militates against too much in the way of innovation there), greater use of Alu-repeat sequences, and further exploitation of markers on the Y chromosome (both STRs and SNPs) and from mtDNA. MtDNA, in particular, is expected to play an increasing role in forensic investigations because of its utility with very small or degraded samples, as well as for tracing maternal relationships. Markers along the Y chromosome will also be extremely useful because of their ability to trace paternal lineages and differentiate between multiple male contributors to a mixture of biological sample, which can be crucial in untangling a crime of gang rape, for example.

Analysis of the Y chromosome is already being conducted in criminal cases where it might shed light on a who did what. One New Hampshire case, for example, has for years pitted a client of the Innocence Project against the New Hampshire attorney general as the man, Robert Breest, has continued to battle authorities over the right to get DNA tests performed on biological evidence preserved from his long-ago murder trial, which resulted in a sentence of life imprisonment. Breest was arrested in 1972 for the sex killing of 18-year-old Susan Randall, whose half-naked body was pulled from the icy Merrimack River in Concord, New Hampshire, in the winter of 1971. At Breest's 1973 trial, a considerable amount of evidence seemed to tie him to the crime, including testimony from a forensic chemist who concluded that there was a "high degree of probability" that fibers and paint particles recovered from Randall's coat matched items retrieved from Breest's car and testimony that blood found under Randall's fingernails matched Breest's blood type, based on ABO blood-grouping tests. During the trial, prosecutors argued that Randall had clawed her attacker "to the bone."

Over the years, Breest steadfastly maintained his innocence. Over vigorous objections by the New Hampshire attorney general's office, two batteries of DNA tests were eventually performed in early 2001 by Cellmark Diagnostics (now Orchid Cellmark, a division of Orchid Biosciences, Inc., of Princeton, New Jersey). Both, however, generated only inconclusive results, with the tests of the DNA from underneath the victim's fingernails indicating only a male contribution of unknown origin. More wrangling ensued about whether to go forward with additional DNA tests, but they were finally allowed.

In April 2002, Orchid Cellmark reported the results of additional Y chromosome analysis, which indicated that Robert Breest could not be excluded as a contributor of the blood under Randall's fingernails, although the DNA test results were far from definitive. In fact, the new DNA tests showed only that Breest was among the 10 percent of white males in Orchid Cellmark's Y chromosome DNA database that might have contributed the bloodstain. Added to the fact that the blood under the victim's fingernails also revealed an ABO blood type consistent with that of Breest, however, the odds now seem stacked against Breest's bid for freedom, barring a breakthrough in DNA testing that changes the picture completely.[8]

Differentiating between a male and female contribution to a stain does not necessarily require looking for evidence of Y chromosome markers. Even if Y elements are present, a female contribution may be lurking in the sample as well. Fortunately, a non-STR locus on the X chromosome, the amelogenin locus, has provided a handy way to distinguish between male and female genetic elements. On the X chromosome, a region along the amelogenin locus has a six base-pair deletion that can be isolated with a PCR primer set. In males, on the Y chromosome, the amelogenin locus is

109 base-pairs long, in females 103 with one type of anaylsis. The amelogenin region is normally analyzed during a standard multiplex STR analysis.[9] Y chromosome loci, however, are useful in untangling samples, other than those containing sperm, that have mixtures of male and female DNA, such as with the material under Randall's fingernails in the Breest case. The loci targeted on the Y chromosome are short tandem repeats, Y-STRs, and are specific to that chromosome. Among other things, they can be useful in rape cases where the differential extraction step, as described in the Hammond case, has not been completely successful in separating the male and female DNA.[10] Y-STRs have been used in cases resulting in convictions, as well as cases resulting in the exoneration of someone previously convicted of a crime.[11]

Currently, many forensic science laboratories have the equipment and expertise to examine 20 or more specific DNA loci (including the 13 core STRs). It is difficult to say precisely how many more will be added in the years to come. There is even more uncertainty over the impact of two looming developments in DNA technology. One of those, the use of genetic markers retrieved from a crime-scene sample to identify physical traits, is likely to spark debates over privacy because of fears that the technology may be abused. Nevertheless, the National Commission on the Future of Forensic DNA Evidence expects that a number of such markers will be in use by 2010. The second looming development is the use of portable, miniaturized DNA analyzers that can rapidly analyze multiple STR loci at crime scenes. These are expected to be in use by 2005. By 2010, these mobile devices will readily allow computer-linked remote DNA comparisons to all known DNA profiles contained in the national (and, perhaps, international) DNA database.[12]

Although still in the development phase, the use of portable, microchip-based DNA analyzers appears imminent.[13] Prototypes have been developed by, among others, Nanogen of San Diego, a private biotechnology company that is one of the leaders in developing DNA microchip technology for use in genetics research and medical diagnostics, and the Massachusetts Institute of Technology's Whitehead Institute in Cambridge, one of the principal laboratories involved in the federal human genome project.[14] The development of the hand-held devices is being funded by grants from the National Institute of Justice. The idea behind the technology, of course, is to permit investigators to rapidly identify a suspect, near instantaneously, in fact, and facilitate an arrest. By analyzing biological evidence right at the crime scene, generating a DNA profile based on the 13 core STR loci, and feeding the data into the National DNA Index System through a computer linkup, investigators may actually "solve" a rape or murder case (that is, identify a suspect) within hours, assuming the DNA profile matches a person whose DNA is already part of the national DNA database. This will allow police to focus almost immediately on a likely suspect. The rationale behind the technology is that, statistically, as many as 50 percent of violent crimes are committed by repeat offenders.

The use of such portable DNA analyzers is not intended as a substitute for the more careful and deliberative laboratory work that goes into a criminal prosecution, however. That work will be conducted, as before, to verify or refute the results obtained at the crime scene. But the laboratory, too, could be transformed in the coming years by the miniaturization of the processes that comprise forensic DNA analysis. Every significant step in the process of forensic DNA analysis is proving to be amenable to a miniaturized version of itself, including sample preparation, PCR, capillary electrophoresis, and STR fragment detection and analysis. In the meantime, forensic scientists will continue doing it the old-fashioned, late-20th-century way.

Whether at the crime scene or in the laboratory, the DNA microchip advances are piggybacking on the greater and greater sophistication of microelectronic chips that have been the backbone of the computer revolution in recent decades. Using the techniques of photolithography or chemical etching, for example, the chips will be designed to electronically signal the presence of a particular allele. Biological materials, such as blood or purified DNA, will be directed through the chip by a grid of microchannels, as will the liquid chemicals necessary for each step. Fluids introduced into the microchip canals, which will be composed of glass, silicon, quartz, or plastic, will be propelled through the system by small electrical voltages. Heating and cooling can be likewise regulated. Experimental models of the devices have shown that a complete DNA profile—from PCR to capillary electrophoretic separation of STR fragments and detection—can be performed in mere minutes. It currently takes several hours to get through all of these steps. Significantly, the microchip devices being developed continue to exploit the core STR loci, so any transitional period from current hands-on methods to microchips will not be fundamentally disruptive to the underlying principles of DNA profiling and are expected to be completely compatible to the national DNA database. It will just be faster and less messy.

A Nanogen prototype of the portable DNA analyzer comes in the form of a credit-card–size plastic cartridge that can be inserted into a briefcase-size portable "reader." The reader serves as a reaction chamber, electrical power source, and computer network linkup. (The initial prototypes did not tackle the preparation stage of DNA extraction, so that part of the process still has to be done manually.) Once the STR fragments bind to the analyzer's complementary probes, a signal indicates the particular STR alleles present in the sample.[15]

The attack on America of September 11, 2001, by Osama bin Laden's al Qaeda operatives[16] confronted the United States with one of the greatest challenges in the country's history: how to effectively combat global terrorism aimed at this country and Americans abroad. The carnage left by the attack in New York City would also prove to be the greatest challenge in the history of forensic science. The unprecedented investigation that followed would both demonstrate the extraordinary power of forensic DNA analysis and reveal its current limitations.

At 8:45 A.M., an American Airlines Boeing 767 crashed into the 110-story North Tower of the World Trade Center in New York City. For awhile, at least, it seemed a dreadful accident. At 9:31 A.M., however, a United Airlines Boeing 767 eerily banked and plowed into the sister South Tower of the World Trade Center, confirming that the United States was under attack. The impacts created huge fireballs because the planes were laden with enough jet fuel for cross-country flights out of Boston. In addition to killing hundreds of people immediately upon impact, the fuel began burning inside the towers at more than 2,000 degrees Fahrenheit, rapidly melting the steel structural supports of each building. As people were still scrambling to escape the burning buildings, and firefighters, police officers, and other emergency personnel were working inside to help the people trapped there, the twin towers collapsed in succession, the South Tower at 9:50 A.M. and the North Tower at 10:29 A.M. The final death toll in New York was an estimated 2,792 people (based on adjustments made through December 2002),[17] 343 of whom were firefighters who had responded to the emergency. It was more than the 2,403 civilians and military personnel killed at Pearl Harbor on December 7, 1941. The devastation at the site was beyond words. To make matters worse, within the enormous pile of debris that resulted from collapse of the buildings, fires continued to smolder for weeks.

New York City was not the only target, however. An American Airlines 757 out of Dulles Airport in Washington, D.C., and bound for Los Angeles crashed into the Pentagon at 9:38 A.M., killing the hijackers, 59 passengers and crew on board, and 125 military and civilian workers at the Pentagon. At 10:00 A.M., a United Airlines 757 out of Newark, New Jersey, carrying 38 passengers slammed into a field near Pittsburgh, Pennsylvania, after several passengers boldly attempted to wrest control of the plane from hijackers who were aiming for another suicide run on Washington. At this point, every commercial and private aircraft in the country was ordered to land and the incidents ended.

When the enormity of the task of identifying the dead in the World Trade Center rubble became apparent, the New York City Medical Examiner's office—which boasts one of the largest forensic DNA laboratories in the country—enlisted the help of some of the biggest guns in biotechnology to assist in the process of analyzing the DNA. The task at hand was staggering. DNA had to be rapidly extracted and analyzed not only from recovered bodies and body parts but also from personal items the office requested be provided from families whose relatives were feared lost in the calamity. (Eventually, 15,000 personal items were collected, including 1,400 toothbrushes, 140 razors, and 126 hairbrushes.)[18] As the rubble was being searched (in vain, it turned out) for possible survivors, thousands of body parts and dozens of bodies were being found. The material at the site was loaded onto trucks and transported by truck or barge to the Fresh Kills landfill on Staten Island. At the landfill the material was further sorted and picked over on conveyer belts by local detectives,

FBI agents, and forensic investigators in protective gear searching for human arti-facts, remains, and clues that might aid prosecutors. It was this grim work, and the 175-acre sprawl of the aptly named landfill, that inspired one journalist to describe Fresh Kills as "the world's largest crime lab."[19] The entire forensic operation was co-ordinated by Charles S. Hirsch, New York City's chief medical examiner, who, along with several staff members from his office, was injured when the first tower col-lapsed. Dr. Hirsch and the staffers had gone to the site after the first plane hit to set up a temporary morgue.

Recovered remains were sent to the medical examiner's laboratory for analysis and identification. All of the victims' remains stayed at the New York City medical examiner's facility for ongoing identification efforts by conventional means. Where possible, the DNA was extracted from remains and then sent for analysis to the out-side laboratories, which received frozen packages of carefully labeled test tubes.[20] Within two weeks of the disaster, much of the extracted DNA was being forwarded to Myriad Genetics, Inc., of Salt Lake City, a genomics company most famous for its virtual monopoly of breast-cancer genetics. Myriad is equipped with an assembly-line DNA-sequencing operation for the analysis of patients' blood samples for signs of tell-tale mutations related to breast cancer. By 2001, Myriad was testing 10,000 women a year from blood samples sent by doctors from around the country, but had recently decided to diversify its operation into other areas.[21] Just prior to the World Trade Center disaster, in fact, Myriad had entered into a $16 million contract with New York State to analyze 400,000 DNA samples over several years for criminal casework.[22]

To perform genetic screenings, Myriad uses its expensive automated DNA se-quencers—which were the bedrock technology of the joint efforts to sequence the human genome and are also used for mtDNA sequencing in forensic science labora-tories—which break down the target DNA into sections of varying sizes and tag each of the four nucleotides with a distinct fluorescent dye. The tagged sections are fed through a glass tube filled with a liquid polymer that separates the pieces by size. A laser trained on the glass tube detects the color-coded difference between the nu-cleotides and the results are optically scanned into a computer, which notes the rela-tive positions of each tagged nucleotide and the length of the fragment. Based on the lengths, the computer can then assemble the precise sequence of the entire DNA seg-ment from beginning to end.

With the aid of its powerful DNA testing infrastructure, Myriad was soon analyz-ing hundreds of DNA samples recovered from the World Trade Center, primarily us-ing its DNA sequencing machines to produce STR profiles rather than searching out exact DNA sequences as is necessary in BRCA1/2 analysis. (BRCA1 and BRCA2 are the shorthands for breast cancer gene 1 and breast cancer gene 2.) Simultaneously, the company was daily extracting and analyzing the DNA from a thousand personal items or biological specimens (such as cheek swabs) that were provided by the blood relatives of the presumed victims. (Some of this work was also farmed out to other government laboratories, including the New York State Police Forensic Science Laboratory and the Connecticut State Police Forensic Science Laboratory.) Also under contract with New York City was Bode Technology Group of Springfield, Virginia, and Celera Genomics of Rockville, Maryland, an extremely well-equipped laboratory for DNA sequencing because of its leadership role in sequencing the human genome. Celera, and a second division of its parent company Applera, Applied Biosystems, concentrated on mtDNA in the effort to identify the victims of September 11. By April 2002, Bode had been sent 12,000 bone-derived DNA samples and 5,500 soft tissue-derived DNA samples for DNA analysis, as well as 1,800 samples submitted by families of victims.[23] GeneScreen of Dallas, a division of Orchid Biosciences, later became involved in analyzing samples that had so far yielded no usable DNA.[24]

Despite the early assistance of biotechnology leaders, the New York City medical examiner called an emergency summit on October 3, 2001, of genetics experts from around the country to brainstorm how to convert the increasingly unmanageable flood of DNA data into a comprehensible database that would facilitate the identification process. Although the FBI had produced standardized software for identifying individuals using the 13 core STR loci, which was expected to be the principal means of DNA identification, there were several types of genetic regions being analyzed in the investigation. There needed, in essence, to be efficient computerized cross-talk between the various DNA databases being compiled so that the remains and other pertinent information could be successfully linked to the DNA of the family members or the DNA extracted from the personal items of the victims. Such software programs didn't yet exist.[25] Within weeks, a comprehensive computer system was invented and installed, facilitating the first DNA identification on October 24, 2001.[26]

Thousands of DNA-collection kits were made available so that families could provide authorities with hairbrushes, tooth brushes, combs, dirty laundry, and the like from presumed victims or provide cheek swabs from blood relatives to be used for comparative purposes (reverse-paternity, for example, or mtDNA comparison). The identification process involved other traditional forms of forensic examination, of course, including medical and dental X-ray comparisons, fingerprints, jewelry, and identifying papers or skin markings (such as tattoos or birthmarks), as well as DNA analysis. However, the medical examiner decided early on that DNA analysis would be conducted on all remains, even if the conventional methods had determined an identification.[27]

In one case, the follow-up DNA analysis demonstrated that the wrong body had been buried by a surviving family. The misidentification occurred early in the investigation as the medical examiner's office was still scrambling to contend with a steady stream of human remains. On September 13, 2001, a body lying next to a crushed fire truck from Engine 54 was pulled from the rubble. Although the body was unrecogniz-able, a mask around the victim's neck bore the markings of Engine 54, a fire company from midtown Manhattan that lost 15 firefighters on September 11. On September 11, veteran firefighter Jose Guadalupe, 37, was assigned as chauffeur of the truck. Given the body's proximity to the truck and the logic of Guadalupe's assignment, fire-fighters who assisted in the solemn removal of the remains from ground zero assumed it was Guadalupe. Additionally, the body had a flat gold chain around the neck, which Guadalupe was known to wear. At the morgue, forensic doctors finally relied upon medical X-rays to identify the body because fingerprint identification was impossible and dental X-rays were unavailable. Fire Department X-rays revealed that Jose Guadalupe had a malformation in two adjacent vertebrae of the cervical spine. X-rays taken of the body showed precisely the same anomaly in the neck of the deceased. Af-ter reviewing the set of X-rays, three doctors and a radiologist concurred in the opin-ion that the remains were those of Jose Guadalupe. On October 1, 2001, the body was buried by the Guadalupe family in a cemetery in Queens. Because of the new policy, however, the DNA of the remains was processed anyway.

In a shock to everyone concerned, the DNA results reported back to the medical examiner's office on November 27, 2001, identified the body as that of rookie fire-fighter Christopher Santori, 23, of Engine 54, not Jose Guadalupe, based on the DNA extracted from Santori's toothbrush. When forensic dentists re-examined the case, they concluded that the body was, in fact, that of Santori and pathologists concluded that, in a bizarre coincidence, Santori suffered from the same congenital neck prob-lem that Guadalupe had. It was also determined that Santori commonly wore a gold chain around his neck. The original mix-up, then, resulted from, as the New York Times put it, "a coincidence for which there are undoubtedly no comprehensible odds."[28] Christopher Santori was subsequently disinterred and buried by his parents on Long Island, but the ordeal led to a lawsuit by Guadalupe's widow, distraught over the true whereabouts of her husband's remains. "The peculiar circumstances surrounding the episode," the New York Times wrote when the misidentification first came to light, "seem to touch all that is so raw about the trade center disaster: the extent to which bodies were ruined, the impassioned desire of families to recover remains and the slow but ultimately accurate science that is necessary to sort it all out."[29]

As the recovery and clean-up work at the World Trade Center was coming to an end in May 2002, only 287 intact bodies had been pulled from the rubble. Only ten of these could be visually identified. At the same time, the massive effort had pro-duced nearly 20,000 body parts, all carefully preserved in 18 refrigerated trailers outside the New York City medical examiner's office. However, the initial conflagra-tion that raged through the twin towers and then the fires that smoldered for weeks afterward had reduced many of the body fragments, even bones, to ash. Added to the normal environmental stresses of an outdoor setting and the bacterial assault that accompanies decomposition, the result was a hodgepodge of incomplete DNA profiles from nearly half of the samples tested at the participating DNA laboratories.[30] Not all of the 13 STR loci, in other words, had survived the fires and environmental assaults. Only fragmentary DNA profiles, therefore, could be produced from near totally degraded DNA. For seasoned investigators, none of this was surprising after witnessing the tragedy. Most of the people inside the twin towers had simply been incinerated or pulverized when 1.2 million tons of heavy building materials collapsed to the ground around them. Although the DNA identifications had begun to surge beginning in December 2001, when the recovery effort ended in Manhattan in May 2002, only about 1,018 of the estimated 2,792 victims had been identified, or a little over 36 percent. This was a proportion of the total number that many felt was far beyond expectations, given the sheer devastation found at the World Trade Center site.

For many of the scientists involved in the identification process, however, 36 percent was not good enough, so they pressed on. By the time the work of sifting debris at Fresh Kills concluded on July 15, 2002, the percentage of identifications had risen to 44 percent, or 1,229 of the estimated dead (519 of whom had been identified solely by DNA). They then set their sights even higher, and aimed to identify perhaps as many as 2,000 of the victims, or 70-plus percent. The optimism stems from the successes with using DNA analysis to help identify all of the people lost in recent catastrophic air disasters, including TWA Flight 800 in 1996, the 1998 Swissair crash, and the 1999 EgyptAir crash. Their optimism was tempered, however, by the stark fact that of the 65 people on board the United Airlines 767 that hit the South Tower, only two had been identified. Just as sobering was the fact that roughly 700 DNA profiles that have been generated from partial remains had yet to find a match among the profiles created from items provided by families.[31] At least part of the reason is that the personal items provided by families yielded insufficient DNA for successful DNA extraction or that additional contributions from family members was needed, prompting requests by authorities for additional DNA samples from families.[32] This only caused additional anguish for many surviving family members.[33] At the same time, as many as a quarter of victims' families had declined, at least in the early months of the investigation, to provide any DNA sample at all, generally because they felt it was a futile gesture.[34] To more effectively deal with the confusion and alleviate the anguish of survivors, the medical examiner's office eventually established an actual DNA hot line to quickly answer questions and provide updates on specific cases.[35]

Even in the identification of the initial third or so, however, forensic investigators had found it necessary to rely increasingly on mtDNA. As scientists pressed to account for more victims, the mtDNA data from Celera and Applied Biosystems would become even more important in the identification process because of the greater likelihood that the more plentiful mitochondrial DNA had survived when the nuclear DNA had not, as had been so often demonstrated in other criminal cases and archeological studies. (In some instances, however, mtDNA might not provide a definitive answer, as when two or more people who died in the attack are siblings.[36])

But even with the more practiced hand of STRs, mtDNA, and the other genetic markers with which forensic scientists are familiar, the scientists involved in the World Trade Center investigation recognized that they must push the technology to its outermost boundaries and focus on the tiniest of genetic variables that exist in the human genome. Dr. Robert Shaler, the director of forensic biology at the medical examiner's office, is, for one, determined to make additional positive identifications out of the sea of genetic data in his office.[37] For this reason, the focus is now on single-nucleotide polymorphisms, SNPs, as a method of identification, even though the single-base genetic variations being eyed have never before been used in forensic casework. SNPs (pronounced "snips") are genetic variations at single nucleotide positions and have been detected in DNA fragments as short as 50 base pairs. By the time Dr. Shaler began using them in the World Trade Center investigation, an international research consortium, the SNP Consortium, had identified about 1.4 million of these single-nucleotide variations in humans (there are an estimated three to ten million of them in total) in hopes of zeroing in on the subtle causes of human diseases and untoward pharmacological drug reactions. Typically bi-allelic, however, SNPs are not very discriminating. On the other hand, that last tidbit of genetic information may be just enough to push an unassigned profile into the identified category, especially since the universe of presumed victims is finite. (In the future, scientists hope to develop a number of rapid PCR-based detection systems for SNPs, such as those used for DQA1 analysis or perhaps DNA microchip hybridization arrays.) With the help of Orchid Biosciences, which had a SNP-based paternity test in development that looks at dozens of SNP sites, a specially tailored SNP test was devised for use in making Trade Center identifications, a process expected to continue into 2003.[38]

"Dr. Shaler heads the largest forensic DNA investigation in United States history," wrote *The New York Times* in May 2002., "Now, as the recovery at ground zero winds down, his work is kicking into an intensive stage of identifying victims almost entirely through DNA."[39]

The DNA work continued to pay off. By early November 2002, more than half, or 1,423, of the World Trade Center victims had been identified.[40]

DNA profiling is also playing a key role in the military's efforts to track down members of Osama bin Laden's al Qaeda terrorist organization. Whenever possible, forensic specialists in Afghanistan have gathered DNA samples from abandoned Afghan hideouts or dead al Qaeda or Taliban fighters as part of an overall effort to identify all of the principals involved in waging the terrorist campaign against America. It has been proposed that DNA samples from captured al Qaeda or Taliban fighters in U.S. custody be collected and that a terrorist DNA database be compiled, which federal authorities hope to add to the national DNA database. (The FBI, however, requires congressional approval to expand NDIS and include the DNA of terrorism suspects, such as those being detained in Afghanistan and at Guantanamo Bay in Cuba.) The idea is controversial, however, because the detainees have yet to be convicted of anything. In the meantime, the military and intelligence communities are gathering comprehensive biometric measurements (fingerprints, facial features, voiceprints, iris patterns) of all known or suspected terrorists that they encounter, whether through field interviews in other countries or after suspects are taken into custody. Through a newly developed computer system known as the Biometrics Automated Toolset, or BAT, authorities hope to reliably track the movements and associations of these "people of interest" by way of their digital dossiers.[41]

Bin Laden himself may one day be identified through DNA analysis, should questions arise about a corpse believed to be his remains or even someone in captivity (bin Laden is believed to travel with several doubles). U.S. investigators are actively seeking DNA samples from relatives, including children, of Osama bin Laden through diplomatic channels with Saudi Arabia, bin Laden's home, and Europe, where some of his relatives live. Whether captured or killed, authorities are intent upon quashing inevitable rumors that bin Laden remains at large, because such rumors are certain to energize his followers. Already, authorities are dealing with what they describe as repeated "Elvis sightings."[42] More than a year after the September 11, 2001, attacks, most experts agreed bin Laden remained alive and in hiding, based primarily on several audiotapes that surfaced in Pakistan between September and November 2002. On November 12, 2002, the Arab television network Al Jazeera, operating out of Qatar, broadcast an audiotape purporting to be a message from bin Laden to the United States and other governments in the West. In the message, bin Laden (as confirmed by CIA and National Security Agency experts through linguistics and computer-enhanced voice analysis, but not with 100 percent certainty due to the tape's poor quality) recounts approvingly a series of terrorist attacks around the world over the previous year and aimed at nations allied with the United States. Bin Laden threatens that there is more to come.[43]

Developing a DNA profile for bin Laden is certain to be complicated because he is believed to have no full siblings. Since bin Laden's father, now deceased, had multiple wives, Osama bin Laden has, instead, about 50 half-brothers and half-sisters, so a DNA profile will have to be cobbled together. An mtDNA sample from his mother or her immediate family might prove most telling, however, assuming such a sample can be obtained. Relatives of bin Laden, including four half-brothers and 17 nieces and nephews, lived in the United States prior to September 11, 2001, but most fled the country soon afterward.[44]

At this writing, there is a massive, worldwide search for bin Laden underway reminiscent of the global search for Josef Mengele in 1985. Osama bin Laden is now the most wanted man in the world. This time, however, authorities have a number of technological tools at their disposal that were unimaginable even as recently as the Mengele hunt. In addition to DNA fingerprinting, for example, the search for a live Osama bin Laden has involved the deployment of face-recognition technology inside Afghanistan and at checkpoints on the Pakistan and Uzbekistan borders.[45] Many believe, however, that it will be DNA that eventually certifies that bin Laden has been killed or captured.

So far, DNA analysis has been used principally to identify the victims of terrorism rather than the perpetrators. In addition to those who died on September 11, 2001, the victims include reporter Daniel Pearl, who was the South Asia bureau chief for *The Wall Street Journal* at the time of his death. Pearl, 38, disappeared in Karachi, Pakistan, on January 23, 2002, while researching possible links between Islamic extremists in Pakistan and Richard C. Reid, who was arrested the previous month on a flight from Paris to Miami when it was discovered he had explosives in his shoes. Pakistani authorities believe Pearl was lured to a fake interview and kidnapped by Ahmed Omar Sheikh, a 28-year-old British-born Islamic militant with ties to al Qaeda. E-mails to the *Journal* and other papers contained photographs of Daniel Pearl in chains and with a gun to his head. The e-mails demanded a $2 million ransom and the release of al Qaeda suspects being held at Guantanamo Bay, Cuba. The e-mail account was later traced to Fahad Naseem, whose arrest led to the arrests of Sheikh and two other men. Subsequently, in February, a videotape was sent to U.S. diplomats, which, authorities later announced, confirmed that Pearl had been killed. It was not until May, however, that a decomposed, decapitated body, which authorities believed to be that of Pearl, was recovered from a shallow grave in Karachi. By this time, a trial of Sheikh, Naseem, and the two other men was already underway in Pakistan for Pearl's killing. All four men were convicted in July 2002 of murder, kidnapping, conspiracy to kidnap, and tampering with evidence. Sheikh was sentenced to death by hanging, and the other three men were sentenced to life in prison. Seven other men, three of whom are believed to have actually carried out the murder, were still being sought at the time by U.S. and Pakistani authorities. (At least one of the alleged killers was reportedly picked up by Pakistani authorities in a September 2002 sweep of Karachi.) Shortly after the verdicts, authorities announced that DNA tests conducted by both American and Pakistani forensic scientists on the body found in Karachi established that it was in fact that of Daniel Pearl.[46]

Just as mtDNA is proving its worth in an ever expanding number of forensic cases, it is also experiencing a very serious academic challenge regarding its reliability as a stable indicator of maternal inheritance. This makes the future of mtDNA especially interesting. The controversy dates to the watershed article by Rebecca Cann, Mark Stoneking, and Allan Wilson in 1987 in *Nature* that sought to trace human ancestry back to a hypothetical "mitochondrial Eve" who lived in Africa between 140,00 and 280,000 years ago,[47] which challenged the multiregional view of modern human origins and supported a worldwide replacement model (that is, pre-existing *Homo erectus* and Neanderthal people were completely swept into oblivion). However, a few have raised doubts about both mtDNA and the theory it supports.

The evolutionary mtDNA studies are based upon a number of assumptions. First, mtDNA is assumed to be inherited clonally, meaning it faithfully reproduces itself from the mother's mitochondria when it is transmitted to the next generation. Equally significantly, mtDNA does not—it is assumed, ever—undergo genetic recombination, such as routinely occurs to nuclear DNA during the creation of sperm and egg cells. Moreover, its rapid mutation rate is the basis of the theoretical "molecular clock" that allows scientists to roughly calculate how long ago Eve and her descendants might have lived, given the accumulation of mutations in the mitochondrial genome. (Sections of the noncoding control D-loop are analyzed, just as in forensic casework.)[48]

MtDNA analysis was performed by two teams led by Svante Paabo and Oxford University geneticist Bryan Sykes to determine that Otzi, the famous Iceman found in 1991 in the Tyrolean Alps in Italy (whose age was estimated to be between 5,000 and 5,350 years old based on radio-carbon dating) was a direct ancestor of people now living in Europe north of the Alps. In fact, Sykes and his team found an Irishwoman living in Dorset, England, with an mtDNA sequence on the control region that matched the Iceman's exactly, prompting the headline: "Iceman's relative found in Dorset."[49]

Bryan Sykes, who heads Oxford's Institute of Molecular Medicine, next analyzed the mtDNA of Cheddar Man, the famous 9,000-year-old complete skeleton excavated in 1903 from the caves of Cheddar Gorge (in the tourist village of Cheddar, also famous for lending its name to the popular cheese) in southwest England, and compared the results with living people. Analyzing the cheek swabs of 20 local people, mostly students and teachers, Sykes found, quite remarkably, that two of the young students' mtDNA exactly matched that of Cheddar Man and that a history teacher by the name of Adrian Targett matched exactly except for a single mutation.[50] Such evolutionary "hits" immediately suggested to Sykes that perhaps most people now living in Europe could likewise trace back their maternal ancestry.

The problem was, some of the best minds in human population genetics and evolutionary biology, including Luigi Luca Cavalli-Sforza of Stanford University in California and John Maynard Smith of the University of Sussex in England, were challenging the entire mtDNA enterprise as an untrustworthy mechanism for determining maternal lineages and studying human evolution. And if mtDNA could not be trusted in evolutionary studies, neither could it be trusted in forensic investigations.[51] Questions arose, for example, from studies that reported the "direct" observation of mutations in mtDNA within a couple dozen generations. "Certainly, the 'direct' observation of mitochondrial mutations will have dramatic consequences for the use of mtDNA in forensics," Svante Paabo wrote in 1996. "If 1 person in 25 may carry a DNA sequence different from that of his or her mother, sometimes even at several positions, then the exclusion of a person (as the donor of a sample) on the basis of a different sequence in his or her relatives is questionable."[52] Also, some authorities began to argue that mtDNA was, to some degree at least, subject to the uncertainties of genetic recombination, just like nuclear DNA. If it was true that mtDNA even occasionally underwent the genetic card shuffling of recombination, then the evolutionary studies based on mtDNA were completely wrong.

The competing notions about mtDNA have resulted in a sort of "mtDNA war" still playing out in the scientific literature. Some scientists embrace mtDNA,[53] others are suspicious.[54] (All mtDNA sequences are calibrated against the very first mtDNA sequence ever completed, known as the "Anderson sequence," which was produced by an intrepid research team at Cambridge in 1981.[55] Differences in mtDNA sequence positions are noted by reference to this sequence.) Cavalli-Sforza challenged Sykes and his team about their assumptions concerning the rate at which mtDNA generates mutations. Sykes, however, remains satisfied that the mutation rates that propel the molecular clock are predictable.[56]

The biggest challenge came when John Maynard Smith—a central figure in evolutionary biology—and Erika Hagelberg published studies in 1999 indicating that mtDNA was, at least to some degree, derived from a paternal contribution of mtDNA.[57] Although Hagelberg and her team were forced to retract their study after a flaw in their data was revealed,[58] Maynard Smith and colleagues pressed on.

The crux of Maynard Smith's argument was that there is too much mtDNA variation to have arisen by chance mutation alone. One of the obvious ways to explain an excess of such mutations, according to the Sussex team, is through the recombination of maternal and paternal mtDNA. Additionally, they pointed out, mitochondria contain the precise enzymes necessary for genetic recombination to occur.

This led to a showdown—one that the United States DOJ acknowledged in late 2000 as an unresolved wrinkle in mtDNA research[59]—resulting in dueling scientific journal submissions. In mid-1999, Sykes and colleagues published an article in *The Proceedings of the Royal Society of London B* entitled "Mitochondrial DNA Recombination—No Need to Panic."[60] The article was immediately followed by a response from Maynard Smith and colleagues entitled "Reply to Macaulay et al. (1999): Mitochondrial DNA Recombination—Reasons to Panic."[61] In a subsequent study published in *Science* in late 1999, Maynard Smith and colleagues concluded that the more distant the genetic sites of mitochondria, the less likely it was the sites were being inherited together (in other words, linkage disequilibrium declined). This shouldn't be; the whole mitochondrial genome is supposed to be reproduced exactly. They wrote: "This pattern can be attributed to one mechanism only: recombination."[62] To this position, there have been challenges,[63] more challenges,[64] and defenses.[65]

The mtDNA feud staggered on with no apparent end in sight. Then, suddenly, there was a remarkable DNA study published in *The New England Journal of Medicine* in August 2002 that appears to bear directly on the question of whether there is—at least occasionally—a paternal contribution of mtDNA in offspring, although the paper did not address the issue of recombination. According to Dr. Marianne Schwartz, the laboratory director of the Department of Clinical Genetics of the University Hospital Rigshospitalet in Copenhagen, Denmark, and her colleague from the hospital's Department of Neurology, Dr. John Vissing, a 28-year-old patient of theirs was suffering from a lifelong exercise intolerance related to extreme muscle fatigue. After biopsies and various biochemical analyses revealed a deficiency related to an important mitochondrial enzyme involved in cellular respiration (oxygen intake), and with both parents and a sibling in apparent good health, the doctors decided to take a closer look, at the genetic level. Further testing identified the culprit, a 2-base pair deletion along the crucial ND2 gene that effectively short-circuited the gene; this explained the muscular fatigue, mitochondrial myopathy. But the testing also revealed something quite strange. The mtDNA within the man's skeletal muscle was largely (about 90 percent, it turned out) inherited from the man's father, whereas the mtDNA contained in his blood, hair roots, and other tissues came exclusively from his mother. The Copenhagen team arrived at this conclusion in stages. First, the tests revealed that there was a deletion in the mtDNA of the musculature. Second, the team discovered that the mutation was not present in the blood (lymphocytes) or other tissues they examined in the patient. At the same time, their sequencing data revealed that most of the mtDNA in the muscles was profoundly different from the mtDNA found elsewhere in the man's body. After repeating all the steps involved in the analysis, including new biopsies, and conducting STR-based testing on the patient's nuclear DNA to demonstrate that all the samples derived from the same individual, the researchers were satisfied that there had not been some sort of a laboratory mix-up. Turning to the mtDNA of the parents and the uncle, Drs. Schwartz and Vissing found that the mtDNA profile (or haplotype) of both the man's father and uncle exactly matched that of the mtDNA profile of the patient's muscles, with the exception of the 2-base pair deletion (which may have arisen spontaneously during embryo development or perhaps existed at low levels in the father's mtDNA). The muscle mtDNA profile of the patient did not match his mother's mtDNA profile, however, except at low levels, although the mtDNA profile of all of his other tissues (blood, hair, etc.) did match his mother's mtDNA, as well as that of his sister. Further STR tests confirmed that everyone in the family was biologically related. The Copenhagen patient was, therefore, a mtDNA mosaic, having inherited the distinct mtDNA of both parents. (Nothing in the analysis examined the issue of recombination.) Drs. Schwartz and Vissing can only speculate as to how often this sort of thing happens. They are convinced that it does not happen very often, they explained in their article, because studies indicate that early embryos appear to have an efficient, if little understood, system for ridding the new embryo of any traces of healthy sperm mitochondria that manage—against the odds—to enter the oocyte during fertilization. This has held true, according to studies, even when sperm are injected directly into a woman's egg during the IVF procedure known as intracytoplasmic sperm injection (ICSI). But it clearly happened in their patient, so Drs. Schwartz and Vissing suggested that other medical detectives bear this in mind if confronted with a similar mtDNA anomaly.[66] Writing separately in an editorial about the extraordinary case in the same issue of *The New England Journal of Medicine,* Dr. R. Sanders Williams of Duke University Medical Center in Durham, North Carolina, went a step further. "Likewise," he wrote, "the possibility of paternal inheritance of mtDNA should be accommodated in statistical models that analyze sequence variations in mtDNA in different human or primate populations in order to draw inferences about human evolution or migration. The unusual case described by Schwartz and Vissing is more than a mere curiosity."[67]

The recombination drama remains a work in progress, but Oxford's Bryan Sykes, among others, has claimed total victory in the mtDNA wars,[68] emboldened by corroboration of his mtDNA studies[69] from an altogether different gene system, the Y chromosome, including a study in which Cavalli-Sforza participated.[70] This last was a "crucial endorsement" of his 1996 mtDNA study and his analysis of Cheddar Man's mtDNA, Sykes later wrote.[71] Most (90 percent or more) of the Y chromosome is not subject to the uncertainties of recombination (only the tips of the chromosome correspond to gene segments on the X chromosome). Like mtDNA (so most people believe, anyway), the majority of the chromosome (which is very small to begin with) is faithfully transmitted from father to son over the eons. Its principal function is to prevent growing embryos from becoming girls, which they would otherwise unthinkingly do. The Y chromosome is replete with forensically useful "junk DNA" segments, including dozens of repeating segments.[72] It has long been seen as the obvious complement to mtDNA for evolutionary studies because of its ability to trace paternal lineages.[73] Researchers now began reporting that they had traced the male ancestral line back to an African Adam of antiquity, a finding very much in synch with the "mitochondrial Eve" hypothesis.

The Y chromosome data was lining up with the mtDNA data.[74] "When we can get different lines of evidence that tell the same story," commented senior Stanford research Peter Underhill, a co-author of one of the Y-chromosome studies, "then we feel we are telling the true history of the species."[75]

In 2000, Bryan Sykes and his colleagues at Oxford founded a company called Oxford Ancestors that offers "to harness the power and precision of modern genetics in the service of genealogy," according to the new venture's Web site, and "is the first organisation in the world to offer these DNA-based services in genealogy."[76] Based on the seven clusters of mtDNA Sykes and his colleagues have mapped in Europe, which have been dubbed the Seven Daughters of Eve, as well as another 11 clusters from around the world, and the emerging data from evolutionary studies focusing on Y-chromosome markers, the Oxford Ancestors offers a MatriLine service for placement within the genealogical framework of mtDNA and a MaleMatch service for either surname traces or verifying paternity.[77]

By this time, several intriguing studies had been published that had utilized Y-STR markers, among others, to trace Jewish lineages, primarily in an effort to see if the oral history of Judaism squared with the genetic evidence now available. Two studies sought to determine if the Jewish priesthood, or Cohanim (which is distinct from rabbis), were truly a lineage apart from their lay brethren, as reflected in their genome. The Cohanim are, according to a 3,3000-year-old oral tradition, descended from Aaron, the brother of Moses. To this day, assignment to the priestly class is strictly by patrilineal descent. In the first study, a group of researchers from Israel, the United Kingdom, Canada, and the United States produced Y-chromosome profiles for 188 Jewish men from Israel, North America, and Britain and identified six clusters (haplotypes). The study targeted an STR site, a Y Alu polymorphism (YAP) insert, and a non-Y dinucleotide repeat. The study concluded that the "Y-chromosome haplotype differences confirm a distinct paternal genealogy for Jewish priests."[78] A follow-up study published in 1998 examined the Y chromosomes (at six STR sites and six so-called unique-event polymorphic sites) of 306 Jewish males from Israel, Canada, and the United Kingdom and confirmed that the Cohanim did indeed carry a distinct genetic signature relative to the general flock, which showed diverse Y chromosome profiles. The distinct DNA pattern—or Cohen Modal Haplotype—was found in about 50 percent of the Cohanim of today, but only about 5 percent of Jewish laymen.[79]

A separate DNA investigation by some members of the Cohanim researchers, including Dr. David Goldstein and anthropologist Tudor Parfitt of The Center for Genetic Anthropology at University College, London, also found that Y-chromosome genetic evidence tended to confirm the oral tradition of the black Lemba tribe of southern Africa that claimed Jewish ancestry, even though they are 4,000 miles removed from Jerusalem. The Lemba have long practiced many rituals and traditions that appeared Semitic, but the exact origins of the rites is obscure. Genetic studies comparing members of the tribe with other Jewish male populations demonstrated that the Cohen Modal Haplotype occurred with the same frequency or better among the Lemba (nearly one in ten males) as it did in Jewish males generally. Moreover, a particularly high concentration of the DNA signature (about 50 percent) was found among the Lemba's Buba clan, which oral tradition maintained had played a leadership role in shepherding the Lemba out of Judea to Africa, according to Dr. Goldstein.[80]

It is now possible to trace the ancestry of black Americans back to specific pre-slavery locations in Africa.[81] A large African database is being collected at Howard

University focusing on mtDNA and Y chromosome differences unique to dozens of populations throughout Western Africa,[82] and a commercial venture launched under the auspices of the university by geneticist Rick Kittle, the African Ancestry Project (www.africanancestry.com), hopes eventually to offer genealogical services to the public for a fee.[83] This is the wave of the future. The future of DNA analysis, therefore, will involve a heightened scrutiny of the past.

Occasionally, genetic evidence and oral tradition can intersect, collide, and shoot off fireworks all at once. Never was this more true than when DNA tests were performed on the lineal descendants of the third president of the United States and the author of the Declaration of Independence, Thomas Jefferson, and Sally Hemings, a beautiful and mysterious slave who lived at Jefferson's Monticello estate for much of her life and was long rumored to be Jefferson's mistress and the mother of at least one of his children. For nearly 200 years, the vast majority of scholars discounted the relationship. Then came the age of forensic DNA analysis and an effort was made to examine the issue under the light of modern biology. In the end, the re-examination of the Jefferson/Hemings relationship became one of the most acrimonious episodes of "tabloid history," as examined in Chapter 8. In 1998, the results of a unique DNA study were published that, at first blush, appeared to resolve the controversy in favor of proving the ancestral connection. In the years that followed, however, intense scrutiny of the DNA results and the historical evidence surrounding the relationship between Thomas Jefferson and Sally Hemings has served only to deepen the mystery and polarize the families that descend from Monticello, both black and white. This is no small matter. Beyond the issues of propriety, some claimed the DNA evidence was proof of serious criminal conduct on the part of one of the most revered figures in American history, so the issue remains as pyrotechnic as ever. If nothing else, the Jefferson/Hemings DNA study is the most significant DNA paternity investigation ever. For good or ill, too, the Jefferson study is almost certainly a harbinger of things to come, given the penchant of some scientists to dig up every historical figure they can get their hands on.

According to the oral history of the Hemings line, the Jefferson/Hemings saga begins when an English sea captain by the name of Hemings fathered Betty Hemings, Sally's mother, with an African slave in Williamsburg, Virginia. Betty belonged to slave trader John Wayles, who took Betty as his mistress after the death of his wife. Together Betty Hemings and John Wayles reportedly produced six children, one of whom was Sally Hemings (although there is no direct proof of paternity). One of John Wayles's legitimate children, Martha, would later become the wife of Thomas Jefferson. Martha and Sally, who came to Monticello with her mother and siblings when the wealthy Wayles died, were, therefore, apparently half-sisters.

Very little historical documentation survives about Sally Hemings, but she is known to have had long, straight black hair and was described by contemporaries as "mighty near white," "very handsome," and "decidedly good looking."[84] Sally Hemings, at about age 13 or 14, accompanied Jefferson's young daughter Polly to Paris in 1787, where Jefferson was serving as America's Minister to France, a term which ended in 1789. It was in Paris——at least according to Hollywood (*Jefferson in Paris,* 1995) and some scholars—that a sexual relationship began between Thomas Jefferson and Sally Hemings. By this time, Jefferson was a widower. Sally had her first child, it is believed, in 1790, a child who, according to the oral history of some of her presumed descendants, eventually assumed the name Thomas Woodson. Altogether, Sally Hemings had six or seven children, but only four of them (excluding Thomas) survived to adulthood.

In September 1802, about midway through Jefferson's first term as president, a one-time-ally-turned-enemy of Jefferson, James Thomson Callender, a notorious scandalmonger, published an article in the Richmond, Virginia, *Recorder* accusing Jefferson of having fathered a son named "Tom" with Sally Hemings. The son, though a slave, bore a striking resemblance to the president and was light-skinned, according to Callender, and was estimated to be about ten to twelve years old at the time. Jefferson never publicly answered the charges, but political friends denied the allegations were true, and Jefferson easily won re-election in 1804. The oral history of the descendants of Thomas Woodson asserts that Thomas Woodson was the "Tom" referred to by Callender, but historical records cannot confirm that Sally gave birth to a child named "Tom" in 1790, or even after. There remains uncertainty, in fact, about whether Thomas Woodson was really the son of Sally Hemings at all. (The name Woodson was adopted by Thomas after he was moved to another owner by that name.) Of those children surviving to adulthood, Monticello farm records verify the birth to Sally of son Beverly in 1798, daughter Harriet in 1801, son Madison in 1805, and son Eston, her last known child, in 1808.

In 1873, a newspaper article in the *Pike County Republican* of Ohio written by Samuel F. Wetmore appeared that included the transcribed "memoirs" of Madison Hemings in which Madison claimed that Thomas Jefferson was the father of all of Sally Hemings's children, including himself. Jefferson scholars have pointed out that Madison could not possibly have had firsthand knowledge about either his own paternity or that of his siblings, and that the article refers to no source for Madison's claim (although some scholars attribute it to Sally herself). Many scholars, however, have deemed the account suspect because it was written by Wetmore, an anti-Jeffersonian. Madison's descendants, based on oral history, contend that Thomas Jefferson was Madison's father. Of the three apparent surviving lines of Hemings's descendants, only the oral history of Eston's descendants suggests a different scenario. Thomas Jefferson was not the father of Eston, it was long held, rather Eston's father was an "uncle" or cousin of Jefferson. In the mid-1970s, however, Eston's descendants were persuaded that it was really Thomas Jefferson after all.

In 1974, historian Fawn Brodie published the book *Thomas Jefferson: An Intimate History*,[85] which recounts the long-standing allegations of a Jefferson/Hemings relationship. Brodie's argument convinced the descendants of Eston Hemings Jefferson that Thomas Jefferson himself was their ancestor; moreover, the book led to a series of novels about the alleged affair and several Hollywood movies. In 1997, Annette Gordon-Reed published *Thomas Jefferson and Sally Hemings: An American Controversy*,[86] which painted an even more credible account of a possible Jefferson/Hemings relationship and managed to win over some hitherto doubtful scholars and historians. The stage was now set for actual, tangible proof of some kind.

In 1998 a team of eight scientists led by Dr. Eugene A. Foster, a retired pathologist, published the results of a comprehensive DNA study in *Nature* that sought to throw light on the alleged Jefferson/Hemings affair.[87] They did not dig anyone up for the DNA testing. Instead, the scientists targeted the specific Y chromosome variations (or haplotype) of lineal descendants of Thomas Woodson and Eston Hemings, which they compared to lineal descendants of the Jefferson family, obtained by way of the male-line descendants of Field Jefferson, Thomas Jefferson's paternal uncle. (Thomas Jefferson had no surviving sons, and his brother Randolph's line did not survive past the 1920s or 1930s.) Both Thomas and Field Jefferson would have shared the same Y chromosome from a common ancestor, in this case the grandfather of Thomas Jefferson. Since the majority of the Y chromosome is faithfully passed along the paternal line (except for rare mutations), the Y chromosome can establish whether it is probable or not that individuals are male-line relatives. Because the oral and written history of Thomas Jefferson's descendants indicated that Sally Hemings was sexually involved with Samuel or Peter Carr, who were Thomas Jefferson's nephews (his sister's kids), DNA was also obtained from male-line descendants of John Carr, the grandfather of Samuel and Peter.

Altogether, 19 blood samples were obtained and analyzed at three separate laboratories at Oxford University, England, using different procedures. The study targeted 19 specific variable sites on the Y chromosome, including seven bi-allelic markers, 11 STR sites, and one minisatellite (MSY1). The analysis used both RFLP and PCR-based testing. Five descendants of Field Jefferson were shown to carry the Jefferson haplotype. Four of them had exactly the same patterns, and one showed a single STR deviation (which was still good enough to be considered within the Jefferson haplotype). Similarly, a Carr haplotype was developed from three descendants of John Carr. The Jefferson and Carr haplotypes differed significantly (at nearly half the sites), as would be expected.

Of the five descendants of Thomas Woodson, four showed identical or sufficiently similar patterns to be considered the same haplotype, but one fell completely outside the range of the Woodson haplotype (due, most likely, to a non-paternity event). Neither of the two Woodson types matched the Jefferson haplotype, so Thomas Woodson was excluded as descending from the Jefferson male-line (either as the son of Thomas or another Jefferson). A follow-up DNA test conducted by Dr. Foster confirmed this result in 2000. At the same time, the Carr brothers appeared to be ruled out as responsible for the birth of Thomas Woodson, too, even though the dominant Woodson haplotype differed at only four positions from the predominant Carr haplotype. That many differences would not be expected to arise over so few generations.

Of greatest significance in the study was the finding that the haplotype developed for the one available descendant of Eston Hemings matched precisely at all 19 genetic sites with the dominant Jefferson haplotype. The frequency within the population of the Jefferson haplotype, according to the authors of the study, is less than 0.1 percent.[88] At a minimum, the results strongly indicated that Eston Hemings was sired by someone carrying the Jefferson haplotype, possibly Thomas Jefferson himself, possibly Randolph Jefferson or one of his five sons, or maybe one of about 18 or so other male Jeffersons living at the time. However, the authors of the *Nature* study zeroed in on Thomas Jefferson, based on their understanding of the historical evidence. "The simplest and most probable explanation for our molecular findings," they wrote, "[is] that Thomas Jefferson, rather than one of the Carr brothers, was the father of Eston Hemings Jefferson, and that Thomas Woodson was not Thomas Jefferson's son."[89]

That's where the trouble started. The very title of the *Nature* article by Foster et al., "Jefferson Fathered Slave's Last Child," seemed to announce to the world that conclusive results had been obtained from the genetic evidence establishing Thomas Jefferson's paternity of Eston. In fact, as the scientists involved in the study later acknowledged, the DNA analysis did not confirm paternity by Thomas specifically and was never designed to do so. At best, such a test can rule out paternity (assuming the male line is unbroken), but it could never definitively establish it, at least not with respect to a single individual. There is, however, a definite family connection between Eston's descendants and the Jeffersons. As for Thomas Jefferson's particular involvement, well, that all depends on one's reading of the historical record, which, to say the least, is open to interpretation.

For many, including many in the media, the verdict was in and it was clear-cut. The DNA tests were "confirmation," according to columnist Gwen Daye Richardson, that Thomas Jefferson fathered Eston, if not more of Sally's children.[90] Columnist Ellen Goodman declared: "At last, scientists with DNA tests have proved that the third president of the United States had an 'improper relationship' with Sally Hemings."[91] The cover of *U.S. News & World Report* advertised: "Jefferson and Sally: After two centuries of controversy, new DNA evidence reveals that he fathered a child by one of his slaves." The story inside said the evidence "removes any shadow of a doubt" about Eston's paternity and "[t]hat one can say with certainty that Sally Hemings bore Thomas Jefferson at least one son."[92] Naturally, in this environment, Eston's descendants were convinced. The "DNA match has finally proven long-standing rumors," wrote Dorothy Westerinen, a descendant of Eston Hemings.[93] In the same issue of *Nature* that the DNA results appeared, geneticist Eric Lander and historian Joseph Ellis co-wrote an article entitled "Founding father" (note the lower case "f") that said, "Now, the DNA analysis confirms that Jefferson was indeed the father of at least one of Hemings' children."[94] But others went even further, wondering aloud if the DNA evidence was not, standing alone, evidence of "rape" and even "pedophilia," given Sally's age at the time the relationship is alleged to have begun.[95] "In these days and times," opined the Rev. Thomas Woodson, a modern-day descendant of Thomas Woodson, "Thomas Jefferson would be convicted of rape of a child."[96]

Jefferson defenders were quick to respond that the DNA analysis was in fact not definitive and that up to 25 other Jefferson males in the region might have contributed the Y chromosome to Eston, although the likeliest suspects appear to be Thomas Jefferson's younger brother Randolph or one of his five sons.[97] Dr. Eugene Foster stepped forward on a number of occasions in an effort to clarify the limited nature of the findings, including writing letters to the editor of *The New York Times* and a letter to *Nature* that appeared in early 1999. Dr. Foster and colleagues emphasized that the results were not conclusive (and could never have been), even if the "simplest and most probable" explanation was that Thomas Jefferson fathered Eston, and complained that the headline to their original article selected by the editors at *Nature* was misleading.[98]

Rancor over the DNA findings has continued in the years following the Foster study, pitting various organizations and groups of people that historically have cherished the legacy of Thomas Jefferson against one another. In 1999, a meeting was held by the Monticello Association—which is composed of acknowledged white descendants of Jefferson descending from his daughters Martha and Maria—to consider a proposal to allow the burial of slave descendants at the Monticello cemetery and open the organization's membership to include those descendants. After the meeting erupted into accusations and recriminations, a decision was postponed. The decision was likewise postponed in 2000 and again in 2001. The DNA study ultimately forced a careful review of the historical evidence surrounding the lives of Thomas Jefferson and Sally Hemings.

Shortly after the Foster study was published, the Thomas Jefferson Memorial Foundation (TJMF), which oversees the estate at Monticello, organized a research committee to carefully assess the available evidence. In January 2000, the TJMF released its findings. "Although paternity cannot be established with absolute certainty," wrote TJMF President Daniel P. Jordan, "our evaluation of the best evidence available suggests the strong likelihood that Thomas Jefferson and Sally Hemings had a relationship over time that led to the birth of one, and perhaps all, of the known children of Sally Hemings."[99] In addition to the DNA findings, the committee noted that there was an "observed correlation between Jefferson's presence at Monticello and the conception windows for Hemings's known children," adding: "The committee concludes that convincing evidence does not exist for the hypothesis that another male Jefferson was the father of Sally Hemings's children." The DNA study, the committee felt, indicated a "high probability" that Thomas Jefferson fathered Eston, and that he was "most likely" the father of all of Sally's children. Significantly, however, in an opinion solicited by the foundation and appended to the report, Yale University's Professor of Genetics Kenneth Kidd expressed the view that Eric Lander and Joseph Ellis "over-interpreted the results as proving that Jefferson was the father of Eston," even if it were true, as Foster et al., maintained, that other explanations were "unlikely."[100]

Right after the TJMF report was released, a new organization came into being, the Thomas Jefferson Heritage Society, comprised of admirers of the third president. One of the society's first acts was to ask a group of well-respected Jefferson scholars to re-examine the DNA evidence and the historical record and produce a report on their findings.[101] Thus was formed the Scholars Commission on the Jefferson-Hemings Matter, which included government experts, noted historians, a scientist, and an economics professor from such institutions as Harvard, the University of Virginia, the University of North Carolina School of Medicine, and George Mason University. After about a year of study, the new commission released a voluminous report that disputed the conclusions laid out in the earlier report by the Thomas Jefferson Foundation. According to the Scholars Commission, it was far more likely that Randolph was the father of Eston Hemings than Thomas Jefferson. Among the reasons cited was the reported recollection of Isaac Jefferson, a former Monticello slave, who said Randolph "used to come out among black people [at Monticello], play the fiddle and dance half the night," whereas no similar account exists regarding the President.[102] Moreover, Thomas specifically invited Randolph to visit him—by letter—only days before Eston is believed to have been conceived. Finally, Randolph was commonly referred to as "Uncle Randolph" around Monticello because of the President's daughters, which would explain the long-standing oral history of Eston's descendants (at least until the 1970s) that Eston was fathered by an "uncle" of the President (something not technically true, of course; all of Thomas Jefferson's paternal uncles were, by the time of Eston's birth, long dead).

"Not a single member of our group," the Scholars Commission wrote, "finds the case against Thomas Jefferson to be highly compelling, and the overwhelming majority of us believe it is very unlikely that he fathered any children by Sally Hemings." With respect to the DNA study that ignited the furor, the report explained, "it is our unanimous view that the allegation is by no means proven: and we find it regrettable that public confusion about the 1998 DNA testing and other evidence has misled many people into believing that the issue is closed." In addition to Randolph, they maintained, there remained about two dozen Jefferson males, other than Thomas Jefferson, who was 64 years old at the time Eston was conceived, who might have fathered Eston.[103]

In May 2002, after yet another year of reviewing the mountain of research generated on the heels of the Foster DNA study, the Monticello Association voted 74–6 at its annual meeting to continue to restrict membership of its organization to documented descendants of Thomas Jefferson, frustrating efforts by the descendants of Sally Hemings and some members of the association to open membership to at least some Hemings descendants. The association, in effect, concurred with the judgment of the Scholars Commission that there was insufficient evidence to establish Thomas Jefferson's paternity of any of Sally Hemings' children. The association, which includes more than 700 descendants of Thomas Jefferson by way of his daughters' families, also rejected proposals to create a separate organization honoring the slaves and artisans of Jefferson's Monticello and to authorize a separate cemetery plot at Monticello. The 2002 gathering erupted into a verbal donnybrook, according to media accounts, with opposing sides exchanging heated words and charges of racism. "People got up to speak," Eston Hemings descendant Dorothy Westerinen told *The Washington Post* afterward, "and all the venom just spilled out."[104]

Thomas Jefferson is not the only U.S. president to have his DNA targeted, of course. DNA analysis was performed by the FBI in 1999 on human skin and tissue found adhering to the bullet fragments that resulted from the 1963 assassination of John F. Kennedy in an effort to shed light on the lone gunman theory, but the DNA tests were inconclusive. (DNA tests were also done on the exhumed body of Lee Harvey Oswald to verify that he was, in fact, Lee Harvey Oswald. He was.)[105] DNA analysis may one day be called upon to determine whether another Founding Father, President James Madison, an author of the Constitution and the Bill of Rights, had children with a slave mistress named Coreen. The oral history of lineal descendants of Coreen have long maintained that their family tree descends from the progeny of Madison and Coreen, although there is no immediate clamor for the kind of DNA tests used in the Jefferson/Hemings matter.[106] There is, however, a move afoot to conduct DNA tests similar to those used in the Jefferson case to determine whether a fair-skinned slave who lived at Mt. Vernon was the child of George Washington and a young slave named Venus from the distant plantation of Washington's brother, John, as the oral history of the child's modern-day descendants contends. The claim is being investigated by the same scientist who led the Jefferson DNA inquiry, Dr. Eugene Foster. Even before this, one historian claimed in a biography of Alexander Hamilton that Hamilton was the illegitimate son of George Washington.[107] Abraham Lincoln's DNA has long been eyed as a fit subject of modern genetic evaluation, and has even turned up encased in a Limited Edition Lincoln Pen marketed by Krone Pen of Buffalo Grove, Illinois. (The company extracted Lincoln's DNA from a hair that belonged to the President, then boosted their inventory through PCR.) The principal reason that genetic researchers—such as Dr. Victor A. McKusick of Johns Hopkins University, who confirmed in 2000 that he was going to attempt DNA extraction from Lincoln's skull fragments recovered after the assassination—have coveted Lincoln's DNA is to determine whether the President suffered from Marfan's syndrome, a congenital disease that effects the body's connective tissue and typically produces a physical stature—tall, thin, and with long extremities—not unlike those of Abraham Lincoln. The disease would also have made the President quite ill. For analytic purposes, there are known samples of Lincoln's blood and tissue surviving from the assassination, such as locks of hair and the bloodstained shirt cuffs of one of the doctors who tried to save the President's life after he was shot, as well as those of the physician who performed a partial autopsy. However, scientific curiosity remains unsated because the DNA investigation was stalled over Department of Defense concerns about Lincoln's medical privacy being invaded and, finally, by the uncertainties of accurately diagnosing Marfan's syndrome on a genetic level.[108] And, lastly, President William Jefferson Clinton had nearly the entire spectrum of forensic DNA tests thrown at him to confirm that he had, in fact, had a sexual relationship with White House intern Monica Lewinsky, despite his public statements and sworn testimony to the contrary. The timing of the publication of the results of the Thomas Jefferson study was decried by Clinton's opponents as intentionally designed to create the perception that "everybody does it," at least if they're president. The Jefferson Y chromosome study was published in *Nature* during the (unsuccessful) impeachment proceedings brought against President Clinton arising from the Lewinsky scandal and other curious incidents.

Although there remain many uncertainties about reconstructing the past by way of maternal and paternal lineages, whether for specific family connections or more general ones, the current efforts may—in the long run—prove a great deal more sensible than those traced by population geneticists of the future. Genetic lineages of the future may well turn out to be an impossible tangle. Francis Fukuyama, a professor of international political economy at Johns Hopkins University and a member of President George Bush's Council on Bioethics, warns in *Our Posthuman Future: Consequences of the Biotechnology Revolution*[109] that the genetic engineering technologies of today, if applied to humans, may subvert the very essence of what it means to be human and completely undermine the political and economic systems erected by mankind to guide human affairs. (As implied by its title, a far rosier picture of genetically modified humans in painted by Dr. Gregory Stock, a visiting professor at the School of Public Health of the University of California in Los Angeles, in his book *Redesigning Humans: Our Inevitable Genetic Future*,[110] which, as one reviewer put it, advances "the classic eugenicist's dream."[111]) Fukuyama calls for national and international regulations on reimplantation genetic testing, germ-line genetic engineering in humans, human cloning, and the creation of human chimeras, among other things. Whether or not Fukuyama's predictions turn out to be true, germ-line genetic changes in humans are already under way that will, at a minimum, induce headaches in future population geneticists and forensic scientists alike. Genetic manipulation, of course, can introduce genetic changes that aren't necessarily traceable to one's lineage. Among other things, this can complicate forensic casework.

In recent years, for example, clinics that specialize in assisted reproduction borrowed from the cloning playbook that produced Dolly to help infertile women conceive their own children. In a process called oocyte nuclear transfer, doctors succeeded in removing the nucleus from a woman's egg and injecting it into a donor's egg, from which the nucleus had already been removed. The technique was meant to help older women whose infertility was believed linked to problems with the cytoplasm of her own eggs. The reconstructed egg was then fertilized with the father's sperm and implanted in the patient's womb. (This is not cloning, because an adult cell is not involved, but the nuclear transfer process is the same.)[112] A related procedure, known as ooplasmic transfer, involves transplanting the cytoplasm (or ooplasm) from the donor eggs into the eggs of women with apparent defects in their own eggs' cytoplasm and then fertilizing the recipient egg. In either case, the mtDNA of the egg donor is introduced into the genetic equation of the resulting offspring and the child inherits the DNA of three people.

Oocyte nuclear transfer was first attempted by New York University fertility specialist Dr. Jamie Grifo in 1998, but it is unclear whether any live births resulted. (The FDA subsequently warned fertility clinics to avoid using the techniques related to cloning, but the procedure has gained approval in England, where it is seen as a potential therapy for the diseases related to mtDNA.[113]) The second procedure was first tried at the Institute for Reproductive Medicine and Science of St. Barnabas, in New Jersey, and resulted in the birth of 16 babies there (up to 30 worldwide) before the FDA stepped in to stop the use of ooplasmic transfer in July 2001, pending further study.[114]

Just before the FDA's action, the New Jersey fertility group reported their use of "mtDNA fingerprinting" on two one-year-old children born through ooplasmic transfer to confirm that the children had copies of both the mtDNA of their natural mothers and that of the egg donors. The children were, in other words, heteroplasmic in their mitochondria, just like Czar Nicholas II, only this time it was by design. (Note that mtDNA analysis demonstrated the same thing in Dolly the sheep.) "This report," the New Jersey team claimed, "is the first case of human germline genetic modification resulting in normal healthy children."[115]

Naturally, ooplasmic transplantation is controversial, and several scientists and bioethicists have raised ethical concerns about continuing the practice, at least until there can be an open public debate about it. Some argued, in fact, that the birth of the children was irresponsible human experimentation.[116]

Should the procedure eventually be approved by the FDA and become commonplace, the number of children born this way may quickly balloon to thousands. When this happens, the odds that at least some of them will fall victim to a serious crime, or perhaps commit one, increase dramatically. This will pose significant challenges for DNA investigators. Will all the parents of such children even realize that their children harbor the mtDNA of the egg or cytoplasmic donor? If parents provide a reference sample for a DNA comparison, it will be incomplete. Confirming identity through the maternal lineage using mtDNA will prove very difficult in such cases. So forensic scientists now have some new challenges to ponder. If germline genetic change catches on, some identifications will be very tricky, if not impossible. Moreover, future population geneticists may well wish they had chosen a different profession because of the sheer complexity of tracing genealogical lineages beginning in the late 20th century.

Genetic science is promising modified pets, too, which will probably add a few new wrinkles to forensic DNA work. Following the 1997 announcement of Dolly's cloning, a few enterprising scientists sensed a business opportunity in the cloning of domestic animals. Spurred on by the laments of grieving pet owners who had lost (or were fearful of losing) beloved animals, several companies sprang up around the country focused—at least in part—on resurrecting deceased pets through cloning, including Lazaron BioTechnologies LLC of Baton Rouge, Louisiana; Genetic Savings & Clone of College Station, Texas (in association with Texas A & M); PerPETuate, Inc., of Sturbridge, Massachusetts, and Cyagra, a subsidiary of Advance Cell Technology, Inc., of Worcester, Massachusetts.[117] Researchers at Texas A & M, led by associate professor Mark Westhusin, had by this time already succeeded in cloning a bull calf from an adult cell of a 22-year-old, 2,300-pound Brahman bull named Chance. Born in August 1999, the cloned calf was aptly named Second Chance.[118]

A breakthrough came on December 22, 2001, when researchers at Texas A & M succeeded in bringing the first cloned cat into the world, which they named cc, for carbon copy, and who was, in fact, genetically identical to an adult domestic shorthair named Rainbow. The work was funded by Genetic Savings & Clone's Copy Cat project, which grew out of an earlier effort by millionaire John Sperling to clone his aging dog Missy (the Missiplicity project), which has so far been unsuccessful due to technical obstacles posed by canine reproduction. Interestingly, cc and Rainbow, both calicos, have somewhat distinct fur coloring and patterns. Except for the DNA tests (seven microsatellite loci) that demonstrated they are genetically identical, few would have believed they are actual clones. (Fur markings are determined to a considerable extent in the womb.) Cc, a female, was the only success the researchers had after implanting 87 modified embryos into eight surrogate cats. Marketing research by the companies pursuing the commercialization of cloned pets suggests the market for the new technology may be huge.[119]

In March 2002, a French research team announced that it had successfully cloned four rabbits in early 2001. Although rabbits are often kept as pets, the French team produced their rabbits strictly for research purposes, forgoing even the opportunity to name the animals. Cloning is unlikely to be exploited to any great degree to produce either research or pet rabbits because rabbits quite famously need little help reproducing themselves. Nevertheless, the feat brought the total number of mammalian species successfully cloned to seven.[120] A near-miss—although arguably a success—was reported with an additional species in January 2001. Researchers funded by Advanced Cell Technology were attempting to clone a rare and endangered species of wild ox native to Southeast Asia known as a gaur. The project was intended to show that an extinct or endangered species can be resurrected with the help of readily available materials from a related species. In this case gaur DNA taken from the skin cells of a living male gaur were inserted into cow eggs and then implanted into cows, who acted as surrogates. From dozens of attempts, one baby gaur was eventually born, named Noah, but the calf survived only two days before succumbing to a fatal bacterial infection, *Clostridium perfingens,* which commonly kills newborn calves. The research team involved was sufficiently emboldened by the live birth to immediately make plans to try again with the gaur. The same team also planned a cloning effort using cells from the last known survivor of a species of Spanish mountain goat, known as a bucardo, with the help of more commonly available goats.[121]

While cloning individual pets will do little, if anything, to complicate forensic DNA analysis in the types of criminal cases seen in an earlier chapter, a second line of pet cloning ventures may well do so. Established in 2001, Transgenic Pets LLC of Syracuse, New York, hopes to market a line of allergen-free felines to an estimated 27 million Americans who are allergic to cats. Under contract with Dr. Xiangzhong "Jerry" Yang at the University of Connecticut, who heads UConn's Transgenic Animal Facility, entrepreneur Dr. David Avner of Transgenic Pets envisions using a combination of genetic engineering and cloning to produce the inoffensive cats. Dr. Yang established his cloning credentials in June 1999 with the birth of a Holstein calf named Amy, the first bovine clone produced in the United States from adult cells. UConn's was the first cow cloned from adult cells that were non-reproductive cells.

Six months after Amy's birth, the UConn facility produced six genetically identical calves by cloning cells from the ear of a Japanese bull, only this time the cells had—quite significantly—been cultured for up to three months before they were transferred to the bovine eggs that were implanted in surrogate cows. This was important for two reasons. First, many scientists were convinced that such long-term culturing of cells would compromise the ability of the cells to reprogram themselves. Second, long-term culturing of cells (in other words, time) is exactly what scientists need if they are to succeed at genetically engineering the animals they are cloning.[122] Given that edge, cloning and genetic engineering become a commercially viable combination. The production of a line of allergen-free cats, then, does not seem such a distant prospect.

The trick in ridding cats of their allergy-producing inclinations is to knock out the gene Fel d1, which produces a protein that causes the allergic reaction in humans. According to Dr.Yang, it may be possible to replace the functional Fel d1 gene with a defective Fel d1 gene by manipulating cultured cat cells. Once done, the altered cat DNA can be used to produce, through cloning, male and female cats with the defective protein gene. The cloned cats can then be bred through conventional means to produce a strain of cats that will pose no risk of inducing allergic reactions in prospective cat owners. Imagine if these efforts are successful. Certainly, allergy sufferers might be inclined to have such a cat. But would it not make sense to many people, even to those who are not personally allergic to cats, to own such a cat (assuming the price is reasonable)? Why expose friends or other family members to a potential allergen if it can be avoided? The idea may just catch on. If it does, however, the utility of cat DNA in criminal casework may suffer a setback.

The problem is that, in order to keep the line of allergen-free cats free of the functional Fel d1, they would have to be inbred. The new and improved cats would be emerging from what biologists call a genetic bottleneck, which reduces their genetic diversity. At least initially, the cats would all be closely related, and this would be problematic for forensic scientists attempting to declare that such and such a cat DNA profile is unique among all cats everywhere. In other words, customized cats (or dogs, for that matter, which are also on the drawing board) would add an element of doubt to the relevance of the DNA profile. This is not to suggest that it would be impossible to differentiate between one cloning-derived cat and all the rest. Unless the cats were from the same batch of cloned cats (genetically identical), it would, in fact, be possible to genetically distinguish them, particularly since conventional breeding would be churning the genetic stew. But as a practical matter, it would be a brave prosecutor indeed who sought to put "cloned cat DNA" before a jury and argue that this "cloned cat DNA" is demonstrably different than the DNA of all the other "cloned cats" roaming around town.[123]

Forensic DNA analysis promises to be central to all future investigations involving government-sponsored terror and oppression, as well as international inquiries into allegations of war crimes. In fact, the contributions of forensic DNA advances were felt from its earliest days. Following Argentina's so-called dirty war, for example, which lasted from the late 1970s to 1983, many families demanded to know what had happened to the "disappeared," as the roughly 30,000 missing were known.[124] Among those unaccounted for were the children of alleged dissidents. Before long rumors began circulating that some of the children were living with families connected to the military junta in power during the "dirty war." Beginning in 1984, Dr. Mary-Claire King, a geneticist from the University of California at Berkeley, began using, among other things, specially developed HLA tests for grand-paternity to trace some of the children. Such special tests were necessary because the children's parents were either still missing or known to be dead. Within a few years, however, mtDNA tests were establishing that the children in question were not biologically related to the people who were raising them. Many of the children were later reunited with their biological families.[125]

Forensic DNA testing has been crucial to the investigations of the International Criminal Tribunal for the former Yugoslavia (ICTY) established in 1993 by the United Nations Security Council to investigate and prosecute allegations of war crimes committed throughout the 1990s in the former Yugoslavia, in the series of conflicts that claimed more than 200,000 lives and made the term "ethnic cleansing" familiar to us all.[126] Dozens of indictments have been issued by the court, located in The Hague, Netherlands. One of the most horrific events of the war was the slaughter of as many as 7,500 Muslim men and boys in the UN "safe area" of Srebrenica in July 1995. The massacre has been described as the worst in Europe since the Nazi atrocities of World War II. The Srebrenica investigation brought forensic investigators from the U.S. Naval Criminal Investigations Service (who collected and processed biological evidence) and the U.S. Bureau of Alcohol, Tobacco and Firearms (who collected and processed ballistics evidence).

The Muslim victims of Srebrenica were either summarily executed by automatic weapons fire or shot while attempting to flee to safety. The investigation was seriously complicated by efforts on the part of the Bosnian Serbs to tamper with the evidence by, for example, removing the victims' bodies from mass graves that had been located in U.S. satellite photos and redistributing the corpses—with body parts all mixed together—in dozens of less conspicuous locations.[127] As a result, the identification of the Srebrenica dead initially proved a nearly impossible task. By late 2000, some five years after the mass killing, less than 100 of the victims had been positively identified. Bodies remained warehoused in several makeshift, refrigerated morgues in and around Tuzla's municipal morgue. Finally, in an effort spearheaded by American scientist Dr. Ed Huffine, formerly with the U.S. Armed Forces DNA Identification Laboratory in Rockville, Maryland, forensic investigators in 2001 began a comprehensive DNA testing project to help identify the dead of Srebrenica (as well as those who died elsewhere during the war years) and bring closure to thousands of desperate relatives. Tens of thousands of DNA samples are being collected from relatives and compared with the DNA extracted from the remains. The Srebrenica project is expected to take five to seven years to complete. On smaller scales, the DNA work has been ongoing throughout the former Yugoslavia (still pockmarked by mass graves) since the hostilities began in 1992. "Without DNA technology," one reporter noted, "it could take a century to find and identify Bosnia's estimated 30,000 missing." A year after the DNA testing program began, authorities reported that 1,200 corpses had been identified through DNA analysis from among the dead in Srebrenica and other locations throughout the former Yugoslavia, all of whom were killed in the early to mid-1990s. By this time, of course, the piles of body bags had only increased because of the last stage of the Bosnian conflict.[128] In 1999 the protracted struggle in the former Yugoslavia lurched into a new phase when the Serbs began another wave of killings in Kosovo. By late 2001, forensic investigators had recovered more than 4,000 bodies from 400-plus locations throughout Kosovo.[129]

Similar investigations grew out of the mass slaughter in Rwanda in 1994. The extraordinary events of 1994 prompted the U.N. Security Council to establish the International Criminal Tribunal for Rwanda in Arusha, Tanzania, to prosecute those responsible for the government-sponsored killing of hundreds of thousands of ethnic Tutsis in Rwanda between April and June 1994 as the world inexplicably did nothing. "In ethical terms," *The Boston Globe* editorialized in 2001, "few failures of American leadership in recent years have been as disgraceful as the Clinton administration's refusal to take meaningful actions while it was still possible to stop the 1994 genocide in Rwanda. While top policy makers were marshaling bureaucratic reasons to ignore the genocide after it began on April 6, 1994, 800,000 people, nearly all from the Tutsi ethnic group, were being exterminated with clubs, machetes, and small arms."[130] According to Samantha Power, the executive director of Harvard University's Carr Center of Human Rights, writing for *Atlantic Monthly* in September 2001, the Rwandan carnage "was the fastest, most efficient killing spree of the twentieth century," and very nearly resulted in the total extermination of the Tutsi minority in Rwanda at the hands of the Hutu majority.[131]

U.S. government thinking led to a classic Catch-22. "Pentagon planners understood that stopping the genocide required a military solution," wrote Power. "Neither they nor the White House wanted any part of a military solution. Yet instead of undertaking other forms of intervention that might have at least saved some lives, they justified inaction by arguing that a military solution was required."[132] The U.S. position even led to a policy calling for the withdrawal of most of the UN peacekeeping force already stationed in Rwanda due to fears that a continued presence in the country by UN forces would inevitably lead to the commitment of American reinforcements.

Additionally, a debate broke out within various government agencies over whether to publicly characterize the mass slaughter as technically "genocide" because the United States had become a signatory (in 1986) to the 1948 Genocide Convention. A May 1, 1994, discussion paper drafted in the Office of the Secretary of Defense warned, "Be Careful. Legal at State was worried about this yesterday—Genocide finding would commit [the U.S. government] to actually 'do something.'"[133]

Among the first investigators responding to the Rwandan tragedy was a team of forensic scientists and doctors from Physicians for Human Rights, headquartered in Boston, led by forensic anthropologist William D. Haglund and Dr. Robert H. Kirschner, a medical examiner from Chicago, Illinois. Working on behalf of the United Nations International Criminal Tribunal for Rwanda, the 16 members of

Haglund and Kirschner's team first set their sights on the small town of Kibuye in an effort to document at least a microcosm of the national horror. When they first arrived in January 1996, they discovered that 46 bodies were still out in the open, strewn down a lakeside embankment behind a Catholic church. The specialists' focus soon became four mass graves located around the church containing a total of 506 bodies, mostly women and children. It was, at the time, the largest forensic excavation in history and required the kind of painstaking attention to detail normally associated with an archeological dig. The evidence gathered was destined for use in the criminal prosecutions looming in Arusha.[134]

The task of identifying the dead proved daunting because dental and medical records were extremely rare. In an effort to identify at least some of the dead in Kibuye, Haglund and Kirschner's team turned to mtDNA. When relatives of victims believed that a piece of clothing or some other item appeared to identify one of the bodies, the survivors were asked to provide blood samples for mtDNA comparisons. Even with the mtDNA option, however, the forensic team expected to identify only a small percentage of the victims they found in Kibuye given the time allotted their investigation. Fortunately, mtDNA routinely lasts many years.[135]

The international war-crimes tribunals have certainly broken new ground in bringing war criminals to justice. In 2001, for instance, a panel of three ICTY judges at The Hague established for the first time that rape constituted a crime against humanity, the second-worst category of wartime offenses. Another landmark verdict occurred in Rwanda's war-crimes tribunal in Arusha when Jean-Paul Akayesu was convicted of genocide and crimes against humanity; his was the first conviction in which rape was defined as an act of genocide, the worst of all war crimes. Rape played an enormous role in the 1994 carnage in Rwanda, much of it allegedly orchestrated by Hutu Pauline Nyiramasukuko, who was, at the time, quite perversely, Rwanda's national minister of family and women's affairs. According to U.N. war-crimes prosecutors, Nyiramasukuko, who is being tried for war crimes in Arusha, used her position to incite Hutu soldiers to rape the Tutsi women they were about to execute. A U.N. report estimates that more than 250,000 woman were raped in Rwanda in 1994 as the genocide was unfolding. Although most of the rape victims were killed, tens of thousands survived. However, many of them were forced to bear children from the rapes (5,000 have been documented), and an estimated 70 percent of the rape victims were infected with HIV by their attackers. In 2002, in an interview with a writer for *The New York Times Magazine,* Rwandan President Paul Kagame explained this aspect of the rampage in these terms: "We knew that the [Hutu-led] government was bringing AIDS patients out of the hospitals specifically to form battalions of rapists."[136] Charles B. Strozier, a psychoanalyst and professor of history at John Jay College of Criminal Justice in New York City, characterized the strategy as "biological warfare," one "perpetuating the death unto the generations."[137]

In future war-crimes prosecutions involving rape, the now standard investigative techniques involving the careful preservation of evidence and forensic DNA analysis, as well as DNA paternity testing, will undoubtedly play significant roles in gaining convictions.[138]

In April 2002, a new permanent International Criminal Court came into being after more than the minimum 60 nations ratified the treaty that establishes the court. Whatever the ultimate role of the U.S. government in the proceedings of the International Criminal Court (the United States so far disavows it over fears that U.S. personnel will be the targets of politically motivated prosecutions), the expertise provided by the U.S. forensic experts throughout the 1990s will undoubtedly provide war-crimes investigators with the necessary tools for conducting thorough and meaningful investigations. DNA technology will be central.[139]

Scientists' zeal to understand the genetic past can provide huge payoffs for understanding the present and even anticipating the future. Very often, such research is the basis of a criminal investigation. Consider the study of pathogens. Most viruses and all bacteria are made up of DNA, of course (some viruses are encoded with RNA), and where there's DNA there are DNA signatures. When a dentist in Florida was alleged to have infected some of his patients with HIV, the episode sent shockwaves through the medical community and alarmed countless dental patients around the country. As the story unfolded, teams of scientists scrupulously examined the case in an effort to answer two basic questions: (1) Did it really happen? and (2) If so, how? Careful genetic studies by the CDC, with the help of the PCR, answered the first question. Yes, it really did happen. Dr. David Acer infected six of his patients with an identifiable strain of HIV. The studies were also able to determine that several other patients of Dr. Acer, who came forward as the story broke, were infected with HIV from somewhere else. It was a pioneering study. The second question, however, has never been answered. Was it a series of dreadful accidents? Was it intentional? No definitive answers have been forthcoming.

Two separate studies were launched in 1990 under the auspices of the CDC. A key question was, how can you tell who infected whom? Rather than simply rely on the information gathered from interviews and available records, the researchers trained their sights on the viruses themselves and began looking for genetic evidence. The studies were based on these facts: (1) HIV mutates at an extraordinary rate, and (2) Because number one is true, an infection link between two people can be genetically demonstrated since the genetic sequences will be so similar. In the end, the CDC researchers concluded that the viruses infecting Dr. Acer and six of his patients formed a cluster of genetically similar strains that were distinct from the viruses of other infected AIDS patients from the area. Moreover, they concluded, some of Dr. Acer's patients were infected elsewhere.[140] The groundbreaking study would soon influence other forensic investigations.

A landmark case arose out of these facts: A woman living in Stockholm, Sweden, was raped by a man with a history of intravenous drug abuse. He had tested positive for HIV in 1986. For some reason, the woman did not report the rape on the date of the attack. Just a few weeks later, however, she herself was found to have serocon-verted to HIV. Nevertheless, she did not file a report with the police about the rape until a year after the incident. Consequently, no conventional physical evidence ex-isted for forensic analysis.[141] Based in large part on proof of the victim's subsequent seroconversion to HIV, however, a Stockholm district court convicted the man of rape and of deliberate transmission of HIV. Despite this, the case was very much in jeopardy of being overturned on appeal based on the paucity of hard evidence. Aware of the case of Dr. Acer in South Florida, Swedish investigators seized upon the tech-nology used in that case to perform a genetic analysis of the HIV strains infecting the rape victim and her alleged assailant. Based on the new genetic evidence, which showed a close relationship between the viruses infecting the victim and the defen-dant, the Swedish court of appeal upheld the district court's verdict of guilty.

The first use of the HIV DNA-sequencing evidence in a criminal case in the United States arose out of the 1996 attempted second-degree murder indictment of 48-year-old Dr. Richard Schmidt, a gastroenterologist living in Lafayette, Louisiana, who, authorities believed, infected his one-time, long-term mistress with HIV in 1994, after she had expressed her desire to end their relationship, which did in fact end about a month later.[142] The ten-year relationship had been a stormy one, even abusive, according to the alleged victim Janice Trahan Allen. Both Schmidt and Allen were married to other people throughout the relationship, and the affair had an on-again-off-again quality, with both carrying on other affairs over the course of their tu-multuous romance. They even had a child together.

According to Allen, a nurse, on one of the last occasions the couple was together, on August 4, 1994, Schmidt gave her an injection he led her to believe was Vitamin B-12; he had given her B-12 injections before to combat fatigue. Some months later, Allen began experiencing symptoms of what was subsequently diagnosed, in January 1995, as the earliest manifestations of HIV infection. She suspected Schmidt almost immediately.

How could Allen's charges be proved? As a nurse, wasn't she exposed to the risk of HIV infection through needle-stick injuries at work on a regular basis? Further, might her lifestyle have led to the infection?

The indictment followed a lengthy investigation that suggested that Schmidt had obtained blood from an HIV-infected patient, Donald M., whom he was treating, and used that to inject Allen. Schmidt adamantly denied the charge. His attorney later called Allen's allegations "a crazy shot in the dark."[143] In 1997, after an admissibility hearing, Judge Durwood Conque ruled that HIV sequencing evidence on the issue could be admitted at trial. The judge was guided by the previous work conducted in the CDC's investigation of the Dr. David Acer case. The Louisiana Court of Appeals later upheld Judge Conque's ruling to allow the novel DNA evidence.

The trial was held in 1998. Several genetics experts for the prosecution, including University of Texas biologist David Hillis, testified that a genetic analysis like the one performed in the Dr. David Acer investigation showed a very close genetic relationship between the strain of HIV with which Allen was infected and the strain of HIV infecting Schmidt's patient. Hillis testified that, because of the mutability of HIV, "the viruses from the two individuals were as closely related as viruses from two people could be."[144] Defense experts disagreed, of course, but there was also evidence that Schmidt had kept a secret record book noting blood withdrawals from Donald M.[145] On October 23, 1998, the jury found Dr. Schmidt guilty as charged. He was later sentenced to 50 years of hard labor. In March 2002, the U.S. Supreme Court declined to hear his appeal of the verdict.

Little noticed until recently was a sequencing project targeting the genome of *Yersinia pestis,* the bacterium responsible for three types of plague, pneumatic (in the lungs), septicemic (in the blood), and bubonic (which gets its name because it causes swollen lymph nodes, or "buboes"). Most infamously associated with the Black Death that killed a third of Europe between 1347 and 1350, *Y. pestis* is currently responsible for about 3,000 plague cases a year worldwide. Plague is typically transmitted by fleas carried on small rodents. In the United States, about a dozen people contract the disease each year, mostly in the Southwest, but cases of plague are usually nonfatal because the disease is susceptible to antibiotic treatment. An exception is the airborne version that attacks the lungs with such speed that a victim is typically dead in a few days. Epidemiologists attribute an estimated 200 million deaths to *Y. pestis* throughout recorded history, so plague remains a disease accorded considerable respect.

The plague genome project, funded by the Wellcome Trust, a British medical research philanthropy, and conducted largely at the Sanger Center in Cambridge, England, attracted a considerable amount of attention when the scientists published the completed genome in October 2001 since the announcement followed the terrorist attacks on America of September 11, 2001. Plague has long been considered a possible instrument of bioterrorism because of its deadly airborne variety, so the British sequencing effort targeted an airborne strain.

In 1995, Larry Wayne Harris, a self-described former member of the white-supremacist group Aryan Nations, was twice arrested in the United States for possession of bubonic plague and anthrax. The episode alerted America to the possibility that those bacteria could be used as weapons of terror, even though the case against Harris turned out to be a weak one. Possession of these agents was not at the time, in and of itself, a crime, since research laboratories routinely obtain the germs for study from licensed repositories. What is clear, however, is that Harris had obtained the cultures by deceit (for which he was convicted in 1997), after sending in an order for the germ cultures on a fake letterhead to American Type Culture Collection, a laboratory supply house. (In the 1980s, Iraq had purchased anthrax from the company.) Most of the charges against Harris, who maintained that he was studying the organisms in order to prepare for what he felt was an imminent biological attack, were dropped, and American Type Culture Collection subsequently revamped its procedures for dispensing potentially dangerous organisms.

Even before the completion of the plague genome, professor of microbiology Paul Keim of Northern Arizona of Flagstaff, Arizona, exploited the unfolding sequence data of *Y. pestis* to develop a highly-specific "DNA fingerprinting" system for the rapid detection and diagnosis of plague cases, a system he provided to the U.S. Centers for Disease Control and Prevention in Atlanta, Georgia.[146] Coincidentally, the plague genome project concluded just as the CDC found itself scrambling to defend against the first instance of wide-scale bioterrorism on American soil. The weapon was *Bacillus anthracis*, or anthrax, as it is commonly known, and it was turning up everywhere, even on Capitol Hill and in the mail room that serves the United States Supreme Court. Dr. Keim and the Northern Arizona University laboratory were immediately enlisted in the anthrax investigation because of the university's experience with the bacterium. In fact, Dr. Keim and a colleague at the university, Dr. Kimothy Smith, had used "DNA fingerprinting" techniques (based on RFLP) to, among other things, identify the strain of anthrax that had been released in Tokyo (without harming anyone, fortunately) by the Japanese cult, Aum Shinrikyo, the cult that later gained international infamy when it released sarin gas in 1995 in Tokyo's subway system, killing 12 people.[147] Evidently by mistake, the cult had used a harmless strain of anthrax in its attack, a strain commonly used in vaccines, according to Keim's analysis.

Earlier, Dr. Keim, working with Dr. Martin Hugh-Jones of Louisiana State University School of Veterinary Medicine (another anthrax expert) and some colleagues in Russia, had employed PCR-based DNA testing to analyze tissue samples from victims of a 1979 outbreak of anthrax in Sverdlovsk, U.S.S.R. (now Ekaterinburg, Russia). Officials of the Soviet Union had maintained that the epidemic was caused by contaminated meat, but scientists in the West had long suspected the outbreak—which killed dozens of people despite immediate vaccination efforts—was caused by the release of anthrax spores from a nearby military (perhaps bioweapons) laboratory. The KGB seized all the records related to the incident in an effort to cover it up, but the Russian pathologist who supervised the 1979 autopsies of the victims, Dr. Faina Abramova, and her colleague, Dr. Lev Grinberg, heroically hid tissue samples and autopsy notes from 42 of the 66 victims. In the end, sophisticated DNA analyses conducted by Keim's team on tissue samples from 11 of the fatalities demonstrated that the Sverdlovsk victims were infected by a variety of virulent strains of *Bacillus anthracis*, bolstering the case against the former Soviet Union's military.[148]

In 2000, a group of anthrax experts, including Dr. Keim, published "DNA finger-printing" data useful in differentiating between the subspecies or strains of anthrax.[149] A year earlier, a DNA sequencing project was undertaken at The Institute for Genomic Research (TIGR, pronounced "tiger"), the private, not-for-profit research group originally started by Craig Venter and currently run by his wife, Dr. Claire Fraser, to use the latest sequencing technology to map the genome of the Ames strain of *B. anthracis* obtained from England. (This was part of the Microbial Genome Project launched by the U.S. Department of Energy in 1994.[150]) As things turned out, the work by both groups would become the analytic bedrock of the major criminal investigation launched after the anthrax mailings, an investigation that would be largely guided by DNA typing.

Anthrax, which cannot be transmitted person to person (i.e., is not contagious), is one of three clinical illnesses caused by contact with *Bacillus anthracis*. (The term anthrax is used widely [and here], however, to describe both the bacterial spores and the disease, despite the technical imprecision). Cutaneous anthrax is the skin form of the disease, acquired through direct contact with the bacterium. Gastrointestinal anthrax occurs when the pathogen is ingested. Inhalation anthrax is acquired when the spores enter a person's airways and begin to multiply in the air passages and lungs. Anthrax contains about 5.5 million base pairs of genetic information and has 5,000 to 6,000 genes.

Unlike the devastating attacks of September 11, 2001, the anthrax attacks during the fall of 2001 unfolded gradually, largely unnoticed at first. Just like the September 11 attacks, however, the anthrax mailings seemed to come without warning. On September 18, 2001, letters containing anthrax spores were postmarked in Trenton, New Jersey, destined for the offices of news anchor Tom Brokaw at NBC News in New York and *The New York Post*. Although the envelopes and letters were subsequently discarded, additional letters were almost certainly postmarked in Trenton on that date headed for the offices of news anchor Dan Rather at CBS News in New York, news anchor Peter Jennings at ABC News in New York, and the tabloid publisher, American Media, Inc. (AMI), in Boca Raton, Florida. It is now theorized that many, if not all, of the anthrax-laden letters sent in 2001 were dropped into an outdoor mail box in Princeton, New Jersey, directly across the street from Princeton University. (On August 14, 2002, the FBI found what appeared to be residual anthrax spores inside that mail box. The specifics on the testing of the residue have not been publicized.) Some or all of these letters may have become damp, causing the anthrax spores in the letters to clump together. The letters that survive from this batch of mailings contain a warning to the recipients to take penicillin (misspelled "penacilin"), but did not specify that the contents were *Bacillus anthracis*. Significantly, the anthrax spores appear to have been cautiously contained within the enclosed warning letter, which was folded in a manner consistent with a "pharmaceutical fold," traditionally a method by which druggists package powdered medicines. The message in the letters—all duplicates from a photocopier—gives the impression that Muslim extremists were behind the mailings, but investigators were dubious from the beginning. The envelopes used were a pre-stamped variety, commonly available at most post offices, so there was no need to lick a stamp to mail the letter.

Over the course of the next two weeks or so, a number of people in New York and New Jersey visited their doctors complaining of strange rashes and lesions, including workers at two New Jersey post offices, *The New York Post*, CBS, NBC, and the 7-month old son of an ABC producer who had recently visited the news offices at ABC. (None of the high-profile addressees ever became ill.) Although the ailments were (almost all) mis-diagnosed at first, all of these patients were treated with antibiotics and recover. Tests later conducted by the CDC confirmed these to be cases of cutaneous anthrax. On September 30, photo editor Robert Stevens, 63, of the supermarket tabloid *The Sun* (part of AMI) in Boca Raton, began to feel ill. On October 1, another AMI employee, mail room worker Ernesto Blanco, was hospitalized with apparent pneumonia. The next day, October 2, Bob Stevens was admitted to nearby JFK Medical Center with a 102-degree fever. He had been vomiting and was delirious. On October 3, doctors correctly diagnosed Stevens' illness as inhalation anthrax and he was put on a respirator and administered penicillin intravenously. Florida authorities believed the Stevens case was an isolated one and they reported it as such in a press conference on October 4. The same day the CDC decided to launch an epidemiological investigation. On October 5, as CDC investigators were combing through Stevens' home and the offices at AMI, Stevens succumbed to his illness. His was the first known death from inhalation anthrax in the U.S. since 1976. Over the next couple of days, CDC tests confirmed the presence of anthrax spores in the nasal passages of Ernesto Blanco and on the computer keyboard used at AMI by Bob Stevens, so the building was sealed by the CDC on October 7. (The building was subsequently abandoned by AMI.) Blanco was later confirmed to have contracted inhalation anthrax, but he nevertheless managed to survive the illness. On October 8, the FBI took charge of the investigation—an investigation since dubbed "Amerithrax"—and began to search for the source of the anthrax as part of a criminal probe, although it was not yet suspected that there was a wider problem via the mail system.

On October 9, a second round of anthrax-filled letters were postmarked in Trenton, this time addressed to then-Senate Majority Leader Tom Daschle and Senator Patrick Leahy at their offices in Washington, D.C., the Hart and Russell Senate Office Buildings. These letters, which were subsequently recovered, contain a dry, highly-refined variety of anthrax spore. Like the first batch, the anthrax spores have been placed inside a "pharmaceutical fold," but this time the letters warn specifically that they contain anthrax and the envelope edges were reinforced with common cellophane tape. On this date, also, the FBI reported that no traces of anthrax had been found in the residences recently occupied by the September 11 hijackers (their cars were later declared clean, as well), nor the home of Robert Stevens. On October 12 (it is now believed), the letter addressed to Senator Leahy took a detour through the postal system due to a blurred zip code, passing through a State Department mail facility in Sterling, Virginia, before reaching the senator's office. A worker at the Sterling facility later developed inhalation anthrax, but recovers following treatment.

By this time, the CDC investigation had already enlisted the help of Paul Keim and his colleagues at Northern Arizona University to assist in identifying the specific strain of anthrax used in the attacks through Keim's "DNA fingerprinting" methods. Before long—the announcement is made on October 25—the strain involved in the anthrax attacks was identified through DNA typing at Keim's laboratory to be the "Ames" strain (one of 89 known genetic strains), based on the previously-published VNTR analyss of Keim, Hugh-Jones and colleagues, which targets eight VNTR sites via PCR (so are, in fact, AFLPs).

On October 13, five additional AMI employees tested positive for exposure to anthrax and were treated (none become ill), the letter to Tom Brokaw tested positive for anthrax, and a second employee at NBC reported having symptoms consistent with anthrax. The next day, it was announced that a police officer and two laboratory technicians involved in the NBC investigation had been exposed to anthrax spores. On October 15, the letter to Senator Daschle was opened in his Washington, D.C., office, later testing positive for anthrax, and it was confirmed that the Boca Raton main post office had been contaminated with anthrax spores. On October 16, two postal workers in Trenton reported suspicious symptoms, later confirmed to be inhalation anthrax. Both women were treated and recovered. The U.S. Senate offices were closed and hundreds of workers there were tested for exposure (28 confirmed) and provided antibiotics. It was announced by the FBI that there was a link between the letters to NBC News and Senator Daschle, based on the postmarks, envelopes and similarities in handwriting. It was also announced that the anthrax received by Senator Daschle's office is a highly purified and potent variety, a type effectively "weaponized." Typically, the weaponization of anthrax removes the spores' static charge, which prevents the material from clumping together and permits it to float freely in the air. (Some observers believe the anthrax spores sent in the later mailings were more refined than those sent in the first batch of letters, but this is debatable.) The following day, the House of Representatives was closed. Preliminary tests on the anthrax specimens sent to New York, Washington, D.C., and Florida indicated the anthrax infections were part of a coordinated attack. The CDC had by this time confirmed four cases of anthrax infection using, among other things, PCR-based DNA tests—in a kit only recently vetted—that identified the pathogen as anthrax. These kits were subsequently sent to state health authorities conducting anthrax-related investigations. The DNA-typing protocols developed by Paul Keim and his colleagues indicated that the anthrax found in New York, Washington, D.C., and Florida were from the same strain.

On October 18 and 19, other cases of cutaneous anthrax were diagnosed. On October 20, anthrax spores were found in the mail rooms of the Ford Office Building, which serves the House of Representatives. Subsequently, four postal workers in Washington, D.C., contracted inhalation anthrax, two of whom die, including Thomas Morris, Jr., 55, on October 21 and Joseph Curseen, 47, on October 22. More postal workers reported illness, but none died and authorities conducted tests for exposure on thousands of workers. Small amounts of anthrax were subsequently found in a number of Washington, D.C., area federal buildings, including locations associated with the White House, the U.S. Supreme Court, the USDA, the State Department and the Federal Reserve. The offices of a dozen or more senators' suites were found to be contaminated with anthrax spores stemming from the letter opened in Senator Daschle's office and the volatility of the spores, forcing the closure of several office buildings for decontamination. Anthrax spores were also detected at two additional postal facilities in Washington, D.C., and Virginia.

On October 29, New York City hospital worker Kathy Nguyen, 61, tested positive for inhalation anthrax. Nguyen, who worked in a medical supply room at the Manhattan Eye, Ear and Throat Hospital, died of the disease on October 31. The exact manner of her exposure remains a mystery. However, on November 1, the CDC announced that DNA-typing tests performed on the anthrax specimen that infected Nguyen showed that it was genetically indistinguishable from the anthrax found in Washington, D.C., Florida and elsewhere in New York. All were from a common source. On November 9, the FBI described the anthrax mailer as a home-grown criminal using the September 11 tragedy as a cover. Finally, on November 21, 94-year-old Ottilie Lundgren of rural Oxford, Connecticut, died of inhalation anthrax, the second case where no obvious, direct connection could be made between an anthrax-bearing letter and the victim's infection. Late the same day, the CDC confirmed through DNA tests that the Lundgren anthrax was genetically indistinguishable from the other known cases of infection along the eastern seaboard. According to a CDC report released in September 2002, Lundgren was most likely exposed as a result of cross-contamination in the postal system.[151] Investigators found an anthrax spore on another letter that went to an address in Seymour, Connecticut, four miles away from the house of Ottilie Lundgren. It was determined that the Seymour letter went through a high-speed postal machine in New Jersey 15 seconds after one of the two letters that went to Capitol Hill.[152]

By mid-December 2001, DNA analyses conducted at the University of Northern Arizona (looking now at three dozen VNTRs in samples from several different laboratories) concluded there was a match between the spores contained in the letter to Senator Daschle and an anthrax strain originally developed by the United States Army Medical Research Institute of Infectious Disease at Fort Detrick, Maryland (USAMRIID) and later passed along to several other laboratories. Keim's analyses found that several laboratories, including USAMRIID, had the same genetic strain of Ames as was used in the anthrax mailings, but his tests were not precise enough to say which laboratory possessed *the* anthrax, assuming any of them did. The United States bioweapons program was dismantled under President Richard Nixon in 1969, but USAMRIID continued to experiment with anthrax in its efforts to develop successful vaccines for use by the military. The Ames strain was originally cultured from a dead cow in Texas in 1981 and sent to USAMRIID using a pre-paid, pre-addressed label mistakenly identifying the sender as the USDA's National Veterinary Services Laboratories in Ames, Iowa. A scientific paper published in 1986 by USAMRIID identified the strain as "Ames," it turns out, because that was the return address on the package sent in 1981. Ames really comes from Texas, though. Ames is a particularly virulent strain of anthrax with recognized resistance to some vaccines. Until the late 1990s, when procedures for handling microbes were tightened at USAMRIID, samples of Ames were sent somewhat informally to a small network of scientific and academic researchers working on anthrax, both inside and outside the U.S. Reportedly, USAMRIID sent samples of Ames to a laboratory in Canada (the Canadian Defense Establishment in Suffield, Alberta), the University of New Mexico in Alburquerque, a defense laboratory in England in Porton Down, the military research facility at Dugway Proving Ground in Utah, and the civilian research institute, Battelle Memorial Institute in Columbus, Ohio. Louisiana State University's Special Pathogens Laboratory (where Dr. Smith originally began studying anthrax) subsequently obtained Ames from the English military and later passed it on to Northern Arizona University. Scientists at these facilities may, in turn, have sent Ames samples on to others. Sometimes a genetically defanged version of Ames was sent by USAMRIID to researchers, but other times the virulent form was sent. For example, USAMRIID's counterpart in England, the Centre for Applied Microbiology & Research at Porton Down, was sent the virulent form.

By late November 2001, scientists at TIGR had completed much of the sequencing work on the genome of two specimens of Ames anthrax, the first provided two years earlier by the English military facility at Porton Down and the second obtained from the Florida investigation into the death of Robert Stevens at AMI. Anthrax, like all biological organisms, accumulates mutations as it reproduces, multiplies, and grows. Anthrax, however, undergoes mutational change very slowly, certainly in comparison to HIV. Further, once the bacteria has killed a host, it retreats into its hardened spore stage, where it can remain for perhaps 150 years. It is ths "shelf life" that makes anthrax an attractive bioweapon.

While small genetic changes can be expected each time anthrax is cultured, detecting the minor differences between two cultures requires an exacting scrutiny of the two genomes. If investigators find an actual sample of anthrax in the possession of an individual suspected of sending the killer anthrax through the mail, a genetic analysis might well be able to determine whether the mailed anthrax is from the same batch as that in the hands of the suspect. To date, however, investigators have not found any such evidentiary material, so that DNA option remains merely a possibility. Given the nature of the case, however, there is a second way in which DNA analysis can be useful: It may help narrow the field of suspects. And this is precisely the avenue that investigators have pursued.

Although investigators remain open to the possibility that the terror anthrax originated from a foreign source, the Amerithrax investigation has tilted heavily in favor of the assumption that it originated domestically. The genetic analyses conducted thus far appear to support this assumption. The principal working theory is that a scientist (or someone with a scientific background) who has, or had, access to a laboratory with a supply of anthrax, possibly a military facility, obtained a small anthrax supply by stealth with which he produced the anthrax spores sent through the mail. (It is also possible that the anthrax used in the attacks was obtained in a natural setting within the U.S., although that is less likely.) As the investigation gathered momentum, the likelihood appeared to increase that the anthrax used in the mailings originated from a source within the U.S. military establishment, rather than an academic institution, because the U.S. military conceded that in 1998 it had produced—at Dugway Proving Ground—its own form of highly-refined, or weaponized, anthrax for use in defense tests. Investigators have been retracing the 20-odd-year migratory history of the Ames strain of anthrax in an effort to zero in on the culprit or culprits. The idea is to obtain samples from every laboratory that worked with Ames up until the bioattacks and compare the genetic signatures of each laboratory's batch of Ames with that of the mailed anthrax. The more similar the genomes, the more likely it is that the anthrax sample used in the attacks originated, at some point, from that laboratory. And this would narrow the field of suspects, in theory, to those people affiliated with that laboratory. To aid in these efforts, the DNA sequencing project at TIGR on the Ames sample from Porton Down was expedited following the realization that the anthrax infections in 2001 were part of a bioattack and the Ames sample from the first anthrax death in Florida was hastily added to the effort (funded by an emergency $200,000 grant from the National Science Foundation). Just as in the detailed genetic work conducted in the World Trade Center investigation, scientists at TIGR, led by team leader Dr. Timothy Read, were focusing on identifying SNPs, or single points of departure, in the two Ames samples sequenced, as well as any other polymorphisms. At this point, TIGR and Keim joined forces to conduct the genetic side of the anthrax investigation.[153]

So far as is known, the best evidence developed thus far in the anthrax investigation is the DNA evidence from the anthrax itself. Although the letters sent in September and October 2001 were carefully examined by forensic investigators, the mailer left no fingerprints or telltale DNA of his own and the brands of ink and tape used by the mailer are commonly available. Neither forensic document examination, handwriting analysis, nor language analysis led to a specific suspect. Moreover, scientists appear to be split over the technical difficulty of producing the kind of anthrax recovered from the second batch of letters—highly-purified, volatile and tiny particles (3 to 5 microns). Needless to say, investigators are endeavoring to chemically dissect the bioattack anthrax in an effort to, in effect, reverse engineer it and determine by which methods it was produced.[154]

However arcane or sophisticated the equipment, and however pedestrian the mailing materials, surely the anthrax came from one specific place. Consequently, the forensic DNA analysis may well hold the key to solving the crime. Already, the DNA data generated by TIGR and Keim's laboratory has narrowed down the list of laboratories that might have played an unwitting role in the bioattack episode. Originally, the CDC had a list of 91 laboratories registered to transfer anthrax specimens. The early DNA tests excluded all but about a dozen or so. As a result, the pool of suspects shrank, too, and the FBI was able to develop "short lists" of possible suspects.[155] By February 2002, *The Wall Street Journal* reported that the FBI's Amerithrax investigation was being largely guided by the unfolding DNA-typing data. Nowadays, of course, this is true of many criminal investigations. The FBI, wrote *The Journal*, is "betting that genetic and other scientific analyses will narrow down the number of laboratories that hold anthrax strains closely related to the one used in the terror attacks. As soon as that happens, the agency plans to bear down on those labs with the full force of its investigative powers."[156]

Despite the promise of the DNA-sequencing data, the FBI's investigation was hampered from the beginning by several things. Initially, the ubiquity of the Ames strain of anthrax within the research and defense communities made for a lot of suspects. Here, at least, there was the hope that DNA analysis could narrow the universe of suspects. But there was a related problem: lax security at some of the laboratories under scrutiny made it difficult to develop a complete roster of people affiliated with those laboratories. As detailed in an investigative report by *The Hartford Courant* in January 2002, perhaps the most egregious example of this general lack of security around laboratories working on dangerous organisms was USAMRIID itself. Based on documents related to a 1992 investigation of the biological research facility and interviews with past and present USAMRIID personnel, *The Courant* reported that a 1992 inventory (covering 1991) of the pathology department found 27 sets of pathogens—including anthrax, Ebola, hanta virus, and a couple of "classified" specimens—unaccountably missing from the laboratory (although all but one, an anthrax specimen, was later recovered, according to USAMRIID brass). Typically, pathogens are chemically killed or irradiated when they are being prepared for microscopic study, so the missing samples are not considered dangerous. On the other hand, the investigation also found evidence that an unidentified individual had been conducting unauthorized research at the facility late at night, apparently on anthrax, based on a rolled back counter on a laboratory camera and a misspelled label retained in digital memory reading "Antrax 005." According to the documents obtained by *The Courant*, moreover, USAMRIID was, at the time of these revelations, an organization in a great deal of internal turmoil. An earlier investigative piece by *The Courant* focused on the apparent ease, according to interviews, with which someone might have smuggled a biological specimen, such as anthrax, out of USAMRIID, at least if the person was someone with authorization to be in the facility.[157] It is difficult to say, of course, how many other laboratories working on anthrax had personnel and security problems. Added to these investigative complications was the destruction of anthrax samples by some academic laboratories around the country eager to rid themselves of any specimen that now seemed both a security burden and a threat to the safety of others. Federal subpoenas for anthrax samples went out to laboratories identified as having Ames in late February 2002, according to news accounts, and collected samples were stored by the FBI at USAMRIID, pending DNA analysis.[158]

The FBI's anthrax investigation was hampered finally by the sluggish genome of anthrax itself. Anthrax evolves exceedingly slowly and, as a consequence, it proved impossible to distinguish between one isolate of Ames and another based on the early genetic data. The joint DNA sequencing and genetic analysis project undertaken by TIGR and Keim's laboratory sought to produce an additional assortment of genetic markers that would be useful in sorting out the samples. The original Porton Down material lacked two extra rings of DNA called plasmids, denominated pX01 and pX02. These were removed as part of the process of rendering the bacteria harmless, or nonpathenogenic, because they produce the toxins that cause disease. The researchers, therefore, substituted two plasmids (already sequenced) from other virulent strains (Pasteur and Sterne) to compare with the Florida sample. The normal genetic complement of anthrax includes a single large chromosome and the two virulence plasmids (altogether about 5.5 million base pairs). The comparative full-genome sequencing revealed 60 new genetic locations that differed between the Porton Down and Florida Ames samples, including SNPs, VNTRs of varying sizes (STRs and longer repeats), and genetic inserts and deletions (known as indels, for short). The differences ranged from single nucleotides to a section of DNA that ran for 1,400 base pairs. The results were published in a collaborative article in *Science* in June 2002. The sequencing was done multiple times to insure accuracy in the final data and a statistical model was developed to screen out probable sequencing errors. Excluding seven differences found between two distinct cultures of the Porton Down Ames, the Porton Down version (with its cannibalized plasmids) and the Florida sample differed in the end at 53 sites. Only 11 of these, however, were DNA sequence differences between the main chromosomes. The larger share of differences were between the Florida plasmids and the substitute plasmids used for comparison

(38 SNPs, 8 VNTRs, and 3 large indels, including the 1,400 base pair segment). Only two of the VNTRs previously described by Keim and his colleagues turned up in the data; six new ones were uncovered. Armed with this information, the scientists then screened five Ames isolates that were indistinguishable based on Keim's VNTR profiles, four of which were laboratory isolates obtained before the October 2001 attacks and archived by Keim.

Unfortunately, only one of the variable genetic sites was useful in differentiating between the four samples provided—at least, apparently—by the laboratories under investigation, and the difference was extremely subtle. Except for the one site, these Ames genomes were exactly the same where tested. To protect the information being generated in the case, the *Science* article did not identify which laboratories' samples were analyzed, but instead designated them as A, B, C, and D. The cited difference was a variable string of As (adenines) that occurred in one of the plasmids, pX02. The Florida Ames sample had 35 A's, sample A had 37 A's and sample D 36 A's. Significantly, perhaps, the laboratory isolates designated B and C had the exact same number of adenines as the lethal Florida Ames sample. (The variable site had been previously identified by Keim after examining the published genetic data of the substitute plasmids.) Despite the extra A, strain D was not classified as a distinct group from B, C, and the Florida strain, although strain A was deemed distinct, partly because it had lost its first plasmid, pX01. On a resulting phylogenetic tree, the Florida Ames sample was grouped with B, C, and D. The PCR-based DNA analysis of the laboratory samples was performed at Keim's laboratory in Arizona. DNA sequencing was performed for all of the samples in the study at the 53 variable sites.

According to the magazine *New Scientist*, the two samples that were determined to be identical to the Florida sample originated from the laboratory stocks at USAMRIID. Sample A can likely be ruled out, *New Scientist* deduced, because it is missing a plasmid and is not virulent, and sample D likely came from Dugway Proving Ground. If this is true, Dugway now appears to have a smidgeon of genetic breathing room. Of the two samples that were identical in all respects to the deadly Florida sample, B and C, one appears to have been an isolate in Keim's own collection (which he obtained from Porton Down and they obtained from USAMRIID), wrote *New Scientist*, and the other appears to come directly from USAMRIID's freezers. This is not the end of the story, however, because testing was continuing on samples from other laboratories when the *Science* article appeared.

Assuming the exact match does relate back to USAMRIID, the DNA analysis explains, at least in part, the intense scrutiny that befell Dr. Steven J. Hatfill following the publication of the DNA sequencing results. Dr. Hatfill, who was described by the FBI as a "person of interest," but who vehemently denied any involvement in the the anthrax attacks, apparently fit a psychological profile developed by authorities. Additionally, Dr. Hatfill was employed for several years as an army scientist at

USAMRIID until 1999. Other factors aside, USAMRIID is certainly the "laboratory of interest," based on the DNA analyses to date. After extensive searches by the FBI of Dr. Hatfill's Frederick, Maryland, residence and neighborhood, however, no physical evidence was apparently discovered that linked Dr. Hatfill to the anthrax attacks, so the FBI has continued looking elsewhere.

Whatever the outcome of the anthrax investigation (and it remains unresolved at this writing), new ground was broken in the annals of forensic DNA analysis. Never before had analysts been called upon to sequence the entire genome of an organism in a criminal investigation. Sadly, the ever-present and, perhaps, escalating threat of bioterrorism and biological warfare seem to guarantee that the technique will be called into service again.[159]

The approach is very, very sophisticated. "It is basically like looking for differences in identical twins," TIGR's Timothy Read told USA TODAY.[160] But it should always be remembered, as Keim told the Los Angeles Times in 2002, that "DNA is not the answer to everything. Forensics and eventually convictions still require good police work and supporting evidence."[161]

According to a commentary in Science that accompanied the anthrax sequencing article, the techniques developed by TIGR and Keim for the anthrax investigation constitute an entirely new branch of scientific investigation, one they dub "microbial forensics." "Both criminal investigation of bioterrorism attacks and studies of naturally occurring disease outbreaks will continue to be important applications of this technology," wrote Craig A. Cummings and David A. Relman of the Stanford University School of Medicine, calling for the establishment of a "microbial forensics infrastructure."[162] Drs. Relman and Cummings joined the authors of the sequencing study in urging the creation of a genetic database of major pathogens, including anthrax, that pose either a natural danger to public health or might one day be used as biowar agents. In fact, the sequencing authors note, ominously, that studies such as theirs "may facilitate the identification of genes that have been deliberately introduced into potential biowarfare agents," referring to the hair-raising possibility of genetically engineered bioweapons. At a minimum, they say, "Genome-based analysis will provide a powerful new tool for investigation of unexpected outbreaks of infectious disease, whether these represent biological warfare attacks, emerging agents, or familiar pathogens."[163]

With any luck, however, the future will bring us less interesting times.

Notes

INTRODUCTION

1. Jessie Graham, "Punchy Perp Decks Lawyer," *New York Post,* October 30, 2001, p. 20; Jessie Graham, "Perp Socked with 35 Yrs. for Rape," *New York Post,* November 28, 2001, p. 31.

2. Andy Newman, "Fingerprinting's Reliability Draws Growing Court Challenges," *New York Times,* April 7, 2001, p. A8.

3. Ibid.; Simon Cole, "The Myth of Fingerprints," *New York Times Magazine,* May 13, 2001, p. 13; Andy Newman, "Judge Rules Fingerprints Cannot Be Called a Match," *New York Times,* January 11, 2002, p. A14; Andy Newman, "Judge Who Ruled Out Matching Fingerprints Changes His Mind," *New York Times,* March 14, 2002, p. A27.

4. National Institute of Justice (NIJ), National Commission on the Future of Forensic DNA Evidence (Report), *The Future of Forensic DNA Testing: Predictions of the Research and Development Working Group* (Washington, D.C.: U.S. Department of Justice, Office of Justice Programs, November 2000), p. 7.

5. Fredric U. Dicher, "A 'Gun Control' Law Made in Wonderland," *New York Post,* June 28, 2000, p. 35; "'Ballistic Fingerprint' Legislation Unveiled," *Boston Herald,* November 22, 2000, p. 28

6. Larry Stevens, "Science on the Case," *Popular Science,* October 2002, p. 68.

7. Editorial, "A Way to Trace Bullets," *New York Times,* October 17, 2002, p. A32; Editorial, "Rethinking Ballistic Fingerprints," *New York Times,* November 11, 2002, p. A16; Editorial, "Cross-Hairs Justice," *Boston Globe,* November 11, 2002, p. A16; Editorial, "Today's Debate: Gun Violence: Sniper Attacks Expose Need for Better Gun-Tracking Tools," and, in response, Alan Gottlieb, "System Would Fail," *USA TODAY,* October 8, 2002, p. 20A; Editorial, "Today's Debate: Firearms Regulation: D.C. Sniper Case Highlights NRA's Duplicity on Guns," and, in response, Wayne LaPierre, "Gun-control Logic Fails," *USA*

TODAY, November 12, 2002, p. 22A; Editorial, "Track Guns to Track Killers," *New York Daily News,* October 17, 2002, p. 20; Editorial, "Adopt Ballistic Fingerprinting," *Hartford Courant,* October 14, 2002, p. A8; Jayson Blair, "Sniper Defendant's Bid for Experts Is Rejected," *New York Times,* November 20, 2002, p. A21; Fox Butterfield, "Law Bars a National System for Tracing Bullets and Shells," *New York Times,* October 7, 2002, A12; Steven Milloy (op-ed), "The Fantasy of Gun Fingerprints," *New York Post,* October 28, 2002, p. 25; Gail Gibson and Dennis O'Brien, "Ballistic 'Fingerprint' Database Isn't Foolproof Tool, Experts," *Baltimore Sun,* October 15, 2002 (from www.chicagotribune.com); Fredrick Kunkle (*Washington Post*), "Sniper revives gun debate," *Boston Globe,* October 15, 2002, p. A4; Editorial, "Review & Outlook: The Gun Fear Factor," *Wall Street Journal,* October 17, 2002, p. A18; Fox Butterfield, "Sniper Case Renews Debate Over Firearm Fingerprinting: Backers Reject Criticism of the Technology," *New York Times,* October 18, 2002, p. A26; Susan Milligan and James Geraghty, "Ballistic Fingerprinting: Gun Control a Hot Topic, but Politicians Tread Lightly," *Boston Globe,* October 25, 2002, p. A25; John Ellement, "Call Rises to Create Ballistic Database: Gun 'Fingerprint' Called Police Tool," *Boston Globe,* October 24, 2002, p. B1; Gary Fields and Nicholas Kulish, "Lawmakers Renew Efforts to 'Fingerprint' Firearms: Sniper Attacks Energize Gun-Control Camp," *Wall Street Journal,* October 14, 2002, p. A4; Adam Clymer (op-ed), "Big Brother Joins the Hunt for the Sniper," *New York Times,* October 20, 2002, Section 4, p. 3.

8. Douglas Farah, "Qaeda Linked to Diamond Trade," *Washington Post,* reprinted in *Boston Sunday Globe,* November 4, 2001, p. A22.

9. Bruce Stanley (Associated Press), "'War Diamond' Accord Reached," *Boston Globe,* November 30, 2001, p. A16.

10. Dan Vergano, "Dealing with Rogue Diamonds," *USA TODAY,* January 25, 2001, 8D; Robert Block, "You Can Learn a Lot About a Diamond If You Smash It Up," *Wall Street Journal,* July 12, 2000, p. A1.

11. John McCrone, "Nowhere to Hide," *New Scientist,* June 9, 2001, p. 24; J. R. Romanko, "Truth Extraction," *New York Times Magazine,* November 19, 2000, p. 54.

12. Jeffrey Rosen, "A Watchful State," *New York Times Magazine,* October 7, 2001, p. 38; Dana Canedy, "Tampa Scans the Faces in Its Crowds for Criminals," *New York Times,* July 4, 2001, p. A1; Martin Kasindorf, "'Big Brother' Cameras on Watch for Criminals," *USA TODAY,* August 2, 2001, p. 3A; Reuters, "Street Surveillance Draws Fire in Fla.," *Boston Globe,* July 4, 2001, p. A3; Ross Kerber, "Face Off," *Boston Globe,* August 20, 2001, p. C1; Rob Turner, "Faceprinting," *New York Times Magazine,* August 12, 2001, p. 18. Raphael Lewis and Ross Kerber, "Logan Will Test Face-Data Security," *Boston Globe,* October 25, 2001, p. B1; Simson Garfinkel, "One Face in Six Billion," *Discover,* September 2002, p. 17; Hiawatha Bray, "Reliability of Face-Scan Technology in Dispute," *Boston Globe,* August 5, 2002, p. C1.

13. William Safire (columnist), "The Great Unwatched," *New York Times,* February 18, 2002, p. A15.

14. Peter T. Kilborn, "For Security, Tourists to Be on Other Side of Cameras," *New York Times,* March 23, 2002, p. A12; Associated Press, "D.C. Monuments to Get Cameras," *Newsday,* March 24, 2002, p. A26; Adam Clymer, "Big Brother vs. Terrorist in Spy Camera Debate," *New York Times,* June 19, 2002, p. A16; Ross Kerber, "Getting a Good

Look at the Olympics," *Boston Globe,* February 1, 2002, p. F1; Adam Goodheart (op-ed), "Public Cameras Accost Privacy," *USA TODAY,* July 22, 2002, p. 11A; Corey Kilgannon, "Cameras to Seek Faces of Terror in Visitors to the Statue of Liberty," *New York Times,* May 25, 2002, p. B1.

15. NIJ, *The Future of Forensic DNA Testing,* pp. 53–55.

16. Michael Kranish, "We Are Family," *Boston Globe,* June 24, 2001, p. D1; Elizabeth Bernstein, "Genealogy Gone Haywire," *Wall Street Journal,* June 15, 2001, p. W1

17. Associated Press, "Titanic Victims Exhumed in Effort to Identify Them," *Hartford Courant,* May 18, 2001, p. A27.

18. Colin Nickerson, "Titanic's 'Unknown Child' Now Has a Name," *Boston Globe,* November 7, 2002, p. A8; Michelle Healy, "DNA Identifies Tiny Titanic Victim," *USA TO-DAY,* November 7, 2002, p. 8D; PBS Television documentary (Thirteen/WNET New York), "Secrets of the Dead: *Titanic's* Ghosts," air date November 20, 2002, (www.pbs.org/wnet/secrets/case_titanic/clues)

19. Associated Press, "Hungary: DNA Tests Identify Ex-Prisoner," *New York Times,* October 6, 2000, p. A6; Reuters, "Hungarian Veteran ID'd After 50 years in Russia," *Boston Globe,* September 18, 2000, p. A11; Associated Press, "War Prisoner Close to Having Name Again," *New Haven Register,* September 18, 2000, p. B4; "POW Back in Hungary After Half-Century in Russia," *USA TODAY,* October 6, 2000, p. 18A.

20. Joseph Fried (columnist), "Following Up: DNA Moves Forward, But a Prisoner Stays Put," *New York Times,* July 21, 2002, section 1, p. 27.

21. Richard Willing, "Criminals Try to Outwit DNA," *USA TODAY,* August 28, 2000, pp. 1A, 2A.

22. Ibid.

23. Ibid.

24. Anita Manning, "TB Germ Outlives Victims," *USA TODAY,* January 27, 2000, p. 9D; Timothy R. Sterling, M.D., et al., "Transmission pf Mycobacterium Tuberculosis from a Cadaver to an Embalmer," *New England Journal of Medicine,* vol. 342, no. 4 (January 27, 2000), pp. 246–248.

25. Reuters, "DNA Tests Sniff Out Ownership of Dog," *Boston Globe,* August 16, 2000, p. A12.

26. Mike Cazalas, "DNA evidence: New, Cost Effective Methods Help Charge Man with Burglary," *News Herald (on-line),* January 19, 1998.

27. Kieran Crowley, "DNA Nails Rob Suspect," *New York Post,* June 22, 2001, p. 13.

28. Associated Press, "DNA, in New Twist, Enters Divorce Court," *New York Times,* December 28, 2000, p. A20; "DNA in Divorce Court," Transcript, CNN's *Burden of Proof,* January 2, 2001; "Divorce by DNA: When Bedroom Sheets Become Evidence," abcnews.com, January 24, 2002.

29. Lorie N. Helmuth, "Controversial New Weapons Against Sexual Harassment: Blood Tests and Cheek Swabs," *Virginia Employment Law Letter,* May 2001.

30. Jennifer Ordonez, "Spit Happens: Police in Uniform Are Leery of Fast-Food Places; The Secret Sauce Could Have Something in It for Them, Something Not Very Nice," *Wall Street Journal,* May 23, 2001, p. A1.

31. Abraham McLaughlin, "DNA Goes Mainstream," *Christian Science Monitor,* October 21, 1998.

32. Joe Flint, "TV's Criminal Streak," *Wall Street Journal,* May 18, 2001, p. W9.

33. Bill Keveney, "CBS to Launch TV Crime Wave," *USA TODAY,* July 16, 2002, p. 4D.

34. Frank Vizard, "Inside C.S.I.: Crime Scene Investigation," *Popular Science,* September 2001, pp. 48, 51.

35. New York: Cambridge University Press, 2000.

36. Edward Rothstein, "A Case for Sherlock: The Double Helix of Crime Fiction and Science," *New York Times,* March 4, 2000, p. B11.

37. Ibid.

38. Thomas Doherty, "Why Do We Hunger for Hannibal Lecter?" *Boston Sunday Globe,* September 12, 1999, p. E2.

39. Julia Chaplin, "Outbreak! Suddenly 'Science' Is *Tres* Chic," *New York Post,* February 5, 1998, p. 53.

40. Charles C. Mann, "How Science Got Cool," *Wall Street Journal,* December 10, 1999, p. W11.

41. David Colman, "No Germs on the Runway: Lab Chic," *New York Times,* October 4, 1998, Section 9, p. 1.

42. Gareth Cook, "Worlds Collide: 'Science Couture' in a Magazine," *Boston Globe,* November 13, 2001, p. C1.

43. Albert R. Hunt, "Putting FDR, Michael Jordan in Their Places in History," *Wall Street Journal,* September 16, 1999, p. A12.

44. Ronald Rosenberg, "The Big Wait," *Boston Globe,* February 7, 2001, p. D4.

45. Ann Pasternak, "Biotech Chic," *New York Times Magazine,* July 9, 2000, p. 18.

46. Associated Press, "'DNA Computer' Is Created and Does Complex Calculations," *New York Times,* January 13, 2000, p. A27; Reuters, "Scientists Take DNA Computing Out of the Test Tube," *Yahoo! News,* January 12, 2000.

47. Patricia Reaney (Reuters), "Scientists Build Tiny Computer Using DNA Molecules," *New York Times,* November 27, 2001, p. F3.

48. Catherine Greenman, "Now, Follow the Bouncing Nucleotide," *New York Times,* September 13, 2001, p. G7.

49. D. T. Max, "The Carver Chronicles," *New York Times Magazine,* August 9, 1998, p. 34 at 51.

50. Emily Eakin, "Tilling History with Biology's Tools," *New York Times,* February 10, 2001, p. B7.

51. Robert Aunger, "Culture Vultures," *Sciences,* September/October 1999, p. 36.

52. Ibid., p. 37.

53. Ibid., p. 36.

54. Ibid., p. 38.

55. Ibid., p. 39; Susan Blackmore, *The Meme Machine* (New York: Oxford University Press, 1999).

56. Aunger, "Culture Vultures," pp. 39–42.

57. Naomi Aoki, "Trouble over Genetown," *Boston Globe,* November 8, 2000, p. C1.

58. Roy Furchgot, "Brightening Old Snapshots with PhotoGenetics," *New York Times,* December 9, 1999, p. G3.

59. Advertisement for Hewlett-Packard, *Wall Street Journal,* December 10, 1999, p. B6; Advertisement for Hewlett-Packard, *USA TODAY,* December 15, 1999, p. 5B.

60. Advertisement for Motorola, *Wall Street Journal,* October 31, 2000, p. B26.

61. Mary Behr, "All Eyes Are On You," *Popular Science,* July 2002, p. 49.

62. Ellen Pfeifer, "Triple Helix Explores Beethoven's Mysteries," *Boston Globe,* December 4, 2001, p. D6.

63. Ad for DNA Perfume by Bijan, *New York Times,* July 25, 1993, p. 3.

64. Tim Race, "New Economy: There's Gold in Human DNA, and He Who Maps It First Stands to Win on the Scientific, Software and Business Fronts," *New York Times,* June 19, 2000, p. C4.

65. Dan Vergano and Susan Wloszczyna, "Genetics at Play in Films: Sorry, Godzilla, Your Makeup Is All Wrong," *USA TODAY,* June 18, 2002, p. 8D.

66. Ibid.

67. Chaplin, "Outbreak! Suddenly 'Science' Is *Tres* Chic," p. 53.

68. Jennifer Rewick, "Think Tank Walker Digital to Cut Work Force 80% in Restructuring," *Wall Street Journal,* November 24, 2000, p. A12.

69. Pilar Viladas, "Design," *New York Times Magazine,* March 11, 2001, p. 76 at 77.

70. Sarah Boxer, "Blueprints Built into the Genes," *New York Times,* October 2, 1999, p. B7.

71. Ibid., p. B9.

72. "Capital: The Legal DNA Of Good Economics," *Wall Street Journal,* September 6, 2001, p. A1.

73. David Halberstam, appearing on CNN's *NewsNight with Aaron Brown,* November 6, 2001.

74. Ad for *Piano Today, New York Times Book Review,* September 13, 1998, p. 41.

75. Rob Owen, "History Channel Series Looks at American Icons," *Pittsburgh Post-Gazette,* reprinted in *Boston Herald,* November 26, 2001, p. 36.

76. Maureen Dowd, columnist, *New York Times,* July 19, 2000, p. A25.

77. New York: Henry Holt, 2001.

78. Ellen Clegg, "Solving Mysteries Using DNA of the Written Word," *Boston Globe,* January 5, 2001, p. C9.

CHAPTER 1

1. The facts of the Pitchfork case are derived from the following sources: Joseph Wambaugh, *The Blooding* (New York: William Morrow, 1989); E. Donald Shapiro, "Dangers of DNA: It Ain't Just Fingerprints," *New York Law Journal,* January 23, 1990, p. 1; U.S. Congress, Office of Technology Assessment, *Genetic Witness: Forensic Uses of DNA* (1990), p. 8.

2. National Institute of Justice (NIJ), National Commission on the Future of Forensic DNA Evidence (Report), *The Future of Forensic DNA Testing: Predictions of the Research and*

Development Working Group (Washington, D.C.: U.S. Department of Justice, Office of Justice Programs, November 2000), p. 10.

3. Raja Mishra, "Biotech CEO Says Map Missed Much of Genome," *Boston Globe*, April 9, 2001, p. A1; Walter Kita, "Scientist Seeks Gene Recount," *New Haven Register*, April 11, 2001, p. F4.

4. Becky Beaupre, "Florida Lab Bucks Trend, Keeps Pace," *USA TODAY*, August 21, 1996, p. 9A.

5. NIJ, *Future of Forensic DNA Testing*; A. Jeffreys et al., "Individual-Specific 'Fingerprints' of Human DNA," *Nature*, vol. 316 (1985), p. 76.

6. U.S. Congress, Office of Technology Assessment, *Genetic Witness*, p. 60.

7. William C. Thompson and Simon Ford, "DNA Typing: Acceptance and Weight of the New Genetic Identification Tests," *Virginia Law Review*, vol. 75 (1989), p. 45, at p. 60.

8. National Research Council, *The Evaluation of Forensic DNA Evidence*, Introduction (Washington, D.C.: National Academy Pres, 1996), p. 1.

9. Carol Marie Cropper, Advertising Note, *New York Times*, November 4, 1998, p. C6.

10. Maggie Gallagher, "Who's Daddy? It's Not Just DNA," *New York Post*, August 14, 1999, p. 15.

11. Tamar Levin, "In Genetic Testing for Paternity, Law Often Lags Behind Science," *New York Times*, March 11, 2001, p. 1.

12. "Who's Your Daddy?" ABC-TV's *PrimeTime Thursday*, November 29, 2001.

13. DNA Diagnostics Center, www.dnacenter.com

14. Daniel Costello, "The Perfect Deception: Identical Twins," *Wall Street Journal*, February 12, 1999, p. B1.

15. Maggie Haberman, "Son-of-a-Gun Clinton: DNA Test May Prove He Has Love Child," *New York Post*, January 3, 1999, p. 5.

16. Maggie Haberman, "DNA Shows Ark. Boy Isn't Clinton's Son," *New York Post*, January 10, 1999, p. 2.

17. Associated Press, "Ohio Man Not Son of Lindbergh," *USA TODAY*, October 2, 2000, p. 15A.

18. Robert Frank, "Settled Paternity Suit Makes Junior One Very Rich Kid," *Wall Street Journal*, March 20, 2000, p. A1.

19. Ibid.

20. Claudia Dreifus, "A Math Sleuth Whose Secret Weapon Is Statistics," *New York Times* (Science Times), August 8, 2000, p. F3.

21. Robert Frank, "Settled Paternity Suit Makes Junior One Very Rich Kid," p. A1.

22. "Anybody Can See Mick's Baby," *USA TODAY*, August 5, 1999, p. 2D.

23. Jim Farber, "Concerts by the Numbers," *New York Daily News*, December 20, 1999, p. 24.

24. *People* magazine, February 1, 1999, cover.

25. Associated Press, "'Satisfaction' Is Named No. 1 Rock Song," *Boston Globe*, January 7, 2000, p. C8.

26. Neal Travis, "Mick Faces Paternity Suit," *New York Post*, January 19, 1999, p. 9.

27. Ibid.

28. "'I Admit,' Mick Jagger Says," *New York Times,* March 17, 2000, p. B2.

29. Hallie Levine, "A Chip Off Ol' Rocker," *New York Post,* May 6, 2001, p. 3.

30. "Who's Your Daddy," ABC-TV's *PrimeTime Thursday,* November 29, 2001.

31. Rinker Buck, "Infidelity Silences Jesse Jackson," *Hartford Courant,* January 19, 2001, p. A1.

32. Lynette Clemetson and Flynn McRoberts, "A Confession from Jesse," *Newsweek,* January 29, 2001, p. 38

33. Editorial, *New York Post,* May 6, 2001, p. 52.

34. Pam Belluck, "Questioned About Finances, Jackson Will Amend Taxes," *New York Times,* March 8, 2001, p. A14.

35. Don Terry and Monica Davey, "Jackson pressed on child support," *Chicago Tribune,* May 9, 2001, (reprinted in *Boston Globe*), p. A3.

36. Jill Porter, "Jackson Acts Contrite Simply to Mold Opinion," *Philadelphia Daily News* (reprinted in *New Haven Register*), January 23, 2001, p. A8.

37. Kathleen Parker, syndicated columnist, "Fathers Fight Paternity Fraud," *Hartford Courant,* February 25, 2001, p. A13.

38. Margaret A. Jacobs, "Courts Favor Ancient Paternity Rule Over DNA Tests," *Wall Street Journal,* June 2, 1999, p. B1.

39. Ibid.

40. Angela C. Allen, "He's the Father in Payment Only," *New York Post,* September 4, 1999, p. 11.

41. Levin, "In Genetic Testing for Paternity, Law Often Lags Behind Science," p. 1

42. Kathleen Burge, "SJC Says Fatherhood Goes Past DNA Test," *Boston Globe,* April 25, 2001, p. A1.

43. Kathleen Burge, "7 Years of Child Support Later . . . ," *Boston Globe,* January 28, 2001, p. B1.

44. Adam Peterson, "DNA Tests Emerging As Legal Weapon in Child Support Cases," *Sunday Boston Globe,* July 23, 2000, p. A1.

45. Robert E. Pierre, "Measures Target 'Paternity Fraud,'" *Boston Globe,* October 20, 2002, p. A25.

46. Burge, "SJC Says Fatherhood Goes Past DNA Test," p. A1.

47. Martin Kasindorf, "Men Wage Battle in 'Paternity Fraud,'" *USA TODAY,* December 3, 2002, p. 3A.

48. Levin, "In Genetic Testing for Paternity, Law Often Lags Behind Science," p. 1.

49. Associated Press, "Father Wins Custody of an 8-Year-Old Girl Kidnapped in 1990," *New York Times,* February 11, 1998, p. A11.

50. Adam Pertman, "Measure Aims at Saving Abandoned Babies," *Boston Globe,* May 5, 2000, p. B3.

51. Anne E. Donlan, "State Seeks Guardian for Baby of Comatose Rape Victim," *Boston Herald,* October 27, 1998, p. 1.

52. Ibid.

53. Richard Willing, "Privacy Issue Is the Catch for Police DNA 'Dragnets,'" *USA TODAY,* September 16, 1998, p. 1A.

54. Ibid.

55. William B. Moffitt, appearing in "DNA Dragnet," CBS-TV's *60 Minutes,* May 20, 2001.

56. Anne E. Donlan, "Comatose Rape Victim's Life Mired in Turmoil," *Boston Herald,* November 9, 1998, p. 10.

57. Caroline Louise Cole, "Ex-Nursing Aide Arraigned; Review Set for Lawrence Facility," *Boston Globe,* January 14, 1999, p. B3.

58. Anne E. Donlan, "DNA Test IDs Nursing Home Worker in Coma Patient's Rape," *Boston Herald,* January 19, 1999, p. 1.

59. Anne E. Donlan and Jules Crittenden, "Aide Charged in Rape of Coma Patient," *Boston Herald,* January 13, 1999, p. 1.

60. David Talbot, "Suspect in Lawrence Coma Rape Pleads Innocent," *Boston Herald,* February 12, 1999, p. 17

61. Caroline Louise Cole, "Man Admits Raping Patient in Vegetative State," *Boston Globe,* February 12, 2000, p. B3.

62. Moffitt, appearing in "DNA Dragnet," *60 Minutes.*

63. Caroline Louise Cole, "Coma Patient Impregnated by Worker Dies," *Boston Globe,* May 3, 2000, p. B3.

64. Jim Yardley, "Report Says Embryologist Knew He Erred," *New York Times,* April 17, 1999, p. B1.

65. Dareth Gregorian, "Scrambled Eggs, Busted Hopes," *New York Post,* February 13, 2000, pp. 4–5.

66. Dareth Gregorian, "'Scrambled Egg' Mom Regrets Giving Up Baby," *New York Post,* February 14, 2000, p. 12.

67. Andy Newman, "Visiting Rights Denied in Embryo Mix-Up Case," *New York Times,* October 27, 2000, p. B3.

68. Dareth Gregorian, "Scrambled Eggs Worry Clinic Moms," *New York Post,* September 24, 2000, p. 28; Sarah Lyall, "Whites Have Black Twins in In-Vitro Mix-Up," *New York Times,* July 9, 2002, p. A12; Associated Press, "Judge Says Man Isn't Father of In-Vitro Twins," *Boston Globe,* August 1, 2002, p. A8.

69. Associated Press, "Florida Girl Switched at Birth Renounces Biological Family," *New York Times,* August 3, 1993, p. A13; Associated Press, "Girl Makes Tearful Plea in Court to Stop Biological Family's Visits," *New York Times,* August 6, 1993, p. A12; Carrie Hedges, "For Florida Girl, a Life of Turmoil," *USA TODAY,* August 4, 1998, p. 3A; Malcolm Balfour, "Birth-switched Kimberly: Keep Kids Where They Are," *New York Post,* August 5, 1998, p. 18; Editorial (Review and Outlook), "Suffer the Children," *Wall Street Journal,* October 15, 1999, p. W17; Malcolm Balfour, "Baby-Swap Survivor Gets Own Tot Back," *New York Post,* April 8, 2000, p. 3.

70. Susan Schindehette and Linda Kramer, "Baby Switch," *People* magazine, August 24, 1998, p. 48.

71. T. Trent Gegax, "A Mysterious Baby Mix-Up," *Newsweek,* August 17, 1998.

72. Maria Sanminiatelli, "Switched-Baby Suit Thrown Out," *Boston Globe,* February 11, 2000, p. A9.

73. Associated Press, "Custody Agreement Set in Switched-Babies Case," April 2, 2000, p. 22.

74. "$2M Settlement Reached in Switched-at-Birth Case," *Boston Globe,* April 20, 2001, p. A2.

75. Associated Press, "No Support in Baby-Switch Case," *New York Times,* September 22, 1998, p. A26.

76. Janna Malamud Smith, "Bonds, Genetic and Otherwise," *New York Times,* August 9, 1998, Section 4, p. 19.

77. Grant Wahl and L. Jon Wertheim, "Paternity Ward," *Sports Illustrated,* May 4, 1998, p. 62.

78. Ibid., p. 64.

79. Ibid., p. 65.

80. Ibid., p. 67.

81. Ibid., p. 71.

82. Frankie Edozien, "Home-Run Ball Brawl for Mrs. Mantle," *New York Post,* January 17, 1999, p. 34.

83. Ibid.

84. Michael O'Keefe and Bill Madden, "Wagner's Wild Card," *New York Daily News,* March 25, 2001, pp. 86–87.

85. Richard Wilner, "Mark's About-Face Won't Hurt Andro," *New York Post,* August 6, 1999, p. 82.

86. Dan Shaughnessy, "McGuire's Power Confined to Dugout," *Hartford Courant,* October 16, 2000, p. C1.

87. Allen Barra, "The King of Swing," *Wall Street Journal,* June 9, 2000, p. W11.

88. Murray Chass, "McGuire Fastest in History to Hit 500," *New York Times,* August 6, 1999, p. D1.

89. Arlene Levinson (Associated Press), "Clinton Scandal Overshadows News of '98," *USA TODAY,* December 29, 1998, p. 5A.

90. *Time,* December 28, 1998 through January 4, 1999.

91. Ira Berkow, "McGuire's 70th Homer Still Defying Gravity," *New York Times,* January 13, 1999, p. D2.

92. Ibid.

93. Douglas Martin, "No. 70 Goes After Another Record," *New York Times,* January 4, 1999, p. B1.

94. Anthony Scaduto, "Having a Ball," *Newsday,* February 9, 1999, p. A12.

95. Berkow, "McGuire's 70th Homer Still Defying Gravity."

96. Associated Press, "Fight Goes on Long After Bonds Hit 73," *New York Times,* November 4, 2001, Section 8, p. 2.

97. Gary Smith, "The Ball (An American Story)," *Sports Illustrated,* July 29, 2002, p. 62.

98. Press Release of Collectors Universe, Newport Beach, CA, parent company of PSA/DNA Authenticators, Inc., January 18, 2000.

99. Naomi Aoki, "Block That Fraud: Super Bowl Balls Get DNA Authenticity Tag," *Boston Globe,* January 25, 2001, p. A1.

100. Ibid.

101. Daniel Sorid, "Sydney Olympics Turns to DNA Ink to Mark Official Souvenirs," *New York Times,* June 22, 2000, p. G4.

102. Dan Vergano, "Dab of DNA Helps Keep Counterfeiters at Bay," *USA TODAY,* October 5, 2000, p. 9D.

103. Sorid, "Sydney Olympics Turns to DNA Ink to Mark Official Souvenirs," p. G4.

104. Jere Longman, "Drug Scandal Goes On: Bulgarian Team Is Ousted From Games," *New York Times,* September 23, 2000, p. D8.

105. Dick Patrick, "Drugs Taint Games," *USA TODAY,* September 26, 2000, p. 1A.

106. Larry Tye, "Athletes' Drug Use Outpaces Testing," *Boston Globe,* September 22, 2000, p. A1.

107. Garret Condon and Tom Paleo, "U.S. Poised to Step up Drug Testing of Athletes," *Hartford Courant,* September 28, 2000, p. A1.

108. New York: St. Martin's Press, 2001.

109. Jere Longman, "Just Following Orders, Doctors' Orders," *New York Times,* April 22, 2001, Section 8, p. 11.

110. Roger Cohen, "In German Courthouse: Pain, Doping, Medals," *New York Times,* May 11, 2000, p. D4.

111. Longman, "Just Following Orders, Doctors' Orders."

112. Stephen Moore, "Gene Doping's Olympic Threat," *Wall Street Journal,* September 18, 2000, p. A6.

113. Ibid.

114. Dick Patrick, "U.S. Anti-Doping Agency Willing to Administer Testing for Baseball," *USA TODAY,* June 14, 2002, p. 6C.

115. Jere Longman, "Someday Soon, Athletic Edge May Be from Altered Genes," *New York Times,* May 11, 2001, p. A1.

116. Ibid.

117. Ibid.

118. Ad, American Museum of Natural History, *New York Times,* May 29, 2001, p. F5.

CHAPTER 2

1. Not her real name.

2. *State of Connecticut v. Ricky C. Hammond* (Criminal Docket no. 54057/HHD-CR88-60681), Hartford/New Britain Judicial District, 1990), Trial Transcript; the account that follows is based on the trial proceedings and interviews with many of the participants.

3. Linda Greenhouse, "Justices Agree to Review Case on Issuing Megan's Law Lists," *New York Times,* February 20, 2002, p. A16; Joan Biskupic, "States Push for Sex Offender Laws to Apply Retroactively," *USA TODAY,* February 20, 2002, p. 3A.

4. Linda Greenhouse, "Court Sets Limit on Detaining Sex Offenders After Prison," *New York Times,* January 23, 2002, p. A1; Joan Biskupic, "Court Limits Detaining Sex Predators," *USA TODAY,* January 23, 2002, p. 3A; Lyle Denniston, "Court Limits Sex-Of-

fender Detention," *Boston Globe,* January 23, 2002, p. A2; *Kansas v. Hendricks*, 521 U.S. 346 (1997); *Kansas v. Crane*, 000 U.S. 00-957 (2002).

5. Jo Napolitano, "Oklahoma: Lawmakers Approve Castration Bill," *New York Times,* May 25, 2002, p. A13.

6. Mark Thompson, "DNA's Troubled Debut," *California Lawyer,* June 1988, p. 36 at 44.

7. Felix Frankfurter, *The Case of Sacco and Vanzetti* (Boston: Little, Brown and Company, 1927).

8. Justice William Brennan, *United States v. Wade*, 388 U.S. 218 (1967).

9. Justice Harry Blackmun, *Manson v. Brathwaite*, 432 U.S. 98 (1977).

10. *State of Connecticut v. Christian E. Porter*, 241 Conn. 57 (1997); *Daubert v. Merrill Dow Pharmaceuticals, Inc.,* 509 U.S. 579 (1993).

11. Justice William Brennan, dissenting, *Watkins v. Sowders*, 449 U.S. 341 (1981).

12. *State of Connecticut v. Kemp,* 199 Conn. 473, 507 A.2d 1387 (1986).

13. *State of Connecticut v. Vaughn,* 171 Conn. 454, 370 A.2d 1002 (1976).

14. *State of Connecticut v. Kemp.*

15. *State of Connecticut v. Boscarino,* 204 Conn. 714, 529 A.2d 1260 (1987).

16. *Neil v. Biggers,* U.S. (1972).

17. *Kampshoff v. Smith* (1983).

18. *State of Arizona v. Chapple,* 135 Ariz. 281, 660 P.2d 1208 (1983).

19. *People of California v. McDonald,* 37 Cal. 3d 351, 690 P.2d 709 (1984).

20. Ibid.

21. Elizabeth Loftus, *Eyewitness Testimony* (Cambridge, Mass.: Harvard University Press, 1996).

22. Ibid.

23. *State of New York v. Anthony Lee,* New York Court of Appeals, May 8, 2001; James C. McKinley, Jr., "Court Lets Experts Challenge Witnesses' Accuracy in Trials," *New York Times,* May 9, p. A1.

24. Russell Contreras, "More Courts Let Experts Debunk Witness Accounts," *Wall Street Journal,* August 10, 2001, p. B1.

25. Ann Woolner, "Courts Eye the Problems of Eyewitness Identifications," *USA TODAY,* September 29, 1999, p. 15A.

26. *State of Connecticut v. Barletta,* 238 Conn. 313, 321, 680 A.2d 1284 (1996).

27. *State of Connecticut v. McClendon,* 248 Conn. 572, 730 A.2d 1107 (1999).

28. J. Berdon, dissenting, ibid.

CHAPTER 3

1. U.S. Congress, Office of Technology Assessment, *Genetic Witness: Forensic Uses of DNA* (1990), p. 46.

2. Greg W. Steadman, "Survey of DNA Crime Laboratories, 2001," Bureau of Justice Statistics, Bulletin, U.S. Department of Justice (Office of Justice Programs), January 2002, p. 7.

3. Letter from Peter R. DeForest, D. Crim., of Forensics Consultants, Ardsley, New York, to Attorney Jeffrey Van Kirk, dated June 4, 1990. For Hammond's sentencing hearing, the letter was marked Defendant's Exhibit 1.

4. *Wilson v. United States,* U.S. (1893).

5. Brief of Defendant-Appellant, *State of Connecticut v. Ricky C. Hammond,* A.C. 9381, filed April 12, 1991.

6. Ibid., at 29–30, footnote 28.

7. C. J. Peters, *State of Connecticut v. Ricky C. Hammond* 221 Conn. 264, 275, 604 A.2d 793 (1992).

8. Ibid., p. 275.

9. Ibid., pp. 282–283.

10. Ibid., p. 280.

11. Ibid.

12. Ibid., p. 284.

13. Ibid., p. 280.

14. Ibid., p. 286.

15. Ibid., p. 287.

16. Ibid.

17. Ibid., pp. 287–288.

18. Ibid., p. 288.

19. Ibid., p. 295.

20. Ibid., p. 296.

21. Ibid., pp. 292–293.

22. *State of Connecticut v. Ricky C. Hammond* (Criminal Docket No. 54057/HHD-CR88-60681), Hartford/New Britain Judicial District. Hearing to reconsider Motion for Post-Trial Discovery, April 3, 1992.

23. *State of Connecticut v. Ricky C. Hammond* (Criminal Docket No. 54057/HHD-CR88-60681), Hartford/New Britain Judicial District. Hearing to reconsider Motion for a New Trial, October 2, 1992.

24. *State of Connecticut v. Ricky C. Hammond* (Criminal Docket No. 54057/HHD-CR88-60681), Hartford/New Britain Judicial District. Plea and sentencing hearing, July 15, 1993.

CHAPTER 4

1. *Commonwealth of Pennsylvania v. Pestinikas* (Super. Court, 1986).

2. William F. Allman, "The Amazing Gene Machine: A Revolutionary Process for Rapidly Copying DNA," *U.S. News & World Report,* July 1990, p. 53.

3. Paul Rabinow, *Making PCR: A Story of Biotechnology* (Chicago: University of Chicago Press, 1996), pp. 82–83.

4. Ibid., pp. 92–93; Kary B. Mullis, "The Unusual Origin of Polymerase Chain Reaction," *Scientific American,* April 1990, p. 56 at 61.

5. Rabinow, *Making PCR,* p. 104.

6. Ibid., p. 108.

7. Ibid., pp. 99–109.

8. Ibid., p. 114.

9. Ibid., p. 128.

10. Randall K. Saiki et al., "Enzymatic Amplification of B-Globin Genomic Sequences and Restriction Site Analysis for Diagnosis of Sickle Cell Anemia," *Science,* vol. 230 (December 20, 1985), pp. 1350–1354.

11. Rabinow, *Making PCR,* pp. 111–126.

12. Ibid., p. 126.

13. Ibid., p. 8.

14. Ibid.

15. Ibid., p. 157.

16. "Chiron Buys Cetus: A Tale of Two Companies," *Science,* vol. 253 (August 2, 1991), p. 503.

17. "Looking Beyond the Bloodletting in Biotechnology," *New York Times,* December 30, 1992, p. 5.

18. Kenneth B. Noble, "Unorthodox Expert with a Nobel Prize Prepares for the Simpson Spotlight," *New York Times,* April 5, 1995, p. A18.

19. "AIDS and Drugs," letter by Peter Duesberg in *New York Times Book Review,* May 19, 1996, p. 4; Stephen D. Moore, "Diagnostic Advances Help Doctors Determine Which Drugs to Prescribe," *Wall Street Journal,* June 9, 1997, p. B9B.

20. Dave Margolick, "A Hesitation by the Simpson Defense on Its DNA Expert," *New York Times,* August 9, 1995, p. A20.

21. Ibid.

22. Associated Press, "Four Win Physics, Chemistry Nobels," *Hartford Courant,* October 14, 1993, p. A10.

23. "In Stockholm, a Clean Sweep for North America," *Science,* October 22, 1993, pp. 506–507.

24. Noble, "Unorthodox Expert with a Nobel Prize Prepares for the Simpson Spotlight," p. A18.

25. ABC-TV's *Nightline,* April 4, 1995.

26. *People of the State of California v. Orenthal James Simpson,* No. BA097211, Los Angeles Superior Court, Transcript, March 30, 1995.

27. Michelle Caruso and Jere Hester, "Probe Cops for Lies, Sez O.J. Team," *New York Daily News,* August 9, 1995, p. 3.

28. Rabinow, *Making PCR,* p. 135.

29. Daniel E. Koshland, Jr. (ed.), "The Molecule of the Year," *Science,* vol. 246 (December 22, 1989), no. 4937, p. 1541.

30. Steve Takemoto et al., "Survival of Nationally Shared, HLA-Matched Kidney Transplants from Cadaveric Donors," *New England Journal of Medicine,* vol. 327 (September 17, 1992), no. 12, *New England Journal of Medicine,* pp. 834–839.

31. P. Cros et al., "Oligonucleotide Genotying of HLA Polymorphism on Microtitre Plates," *Lancet,* vol. 340 (October 10, 1992), p. 870.

32. Norah Rudin and Keith Inman, *An Introduction to Forensic DNA Analysis,* 2d edition (Boca Raton: CRC Press, 2002), pp. 48–49.

33. Malcolm W. Browne, "Scientists Study Ancient DNA for Glimpses of Past Worlds," *New York Times,* June 25, 1991, p. C1..

34. Harold M. Schmeck, *New York Times,* April 16, 1985, p. A1.

35. Malcolm W. Browne, *New York Times,* September 25, 1992, p. A14.

36. Malcolm W. Browne, *New York Times,* June 10, 1993, p. A1.

37. John Wilford Noble, *New York Times,* November 18, 1994, p. A26.

38. Malcolme W. Browne, *New York Times,* May 19, 1995, p. A1.

39. Nicholas Wade, *New York Times,* July 11, 1997, p. A1.

40. George O. Poinar, Jr., "Still Life in Amber: Time Capsules of DNA from 40000000 B.C.," *The Sciences,* vol. 33 (March-April 1993), no. 2, p. 38.

41. Ibid.

42. Kathleen McAuliffe, "Resurrecting the Dinosaur," *Omni,* July 1993, p. 16.

43. Michael Crichton, *Jurassic Park* (New York: Alfred A. Knopf, 1990), p. 401 (Acknowledgments).

44. George O. Poinar, Jr., Roberta Hess, "Ultrastructure of 40-Million-Year-Old Insect," *Science,* vol. 215 (March 5, 1982), p. 1241.

45. Allan Wilson, Russell Higuchi, et al., "DNA Sequences from the Quagga, an Extinct Member of the Horse Family," *Nature,* vol. 312 (November 15, 1984), p. 282.

46. Ibid., p. 284.

47. Svante Paabo, "Molecular Cloning of Ancient Egyptian Mummy DNA," *Nature,* vol. 314, (April 18, 1985), p. 644 at 665; Svante Paabo, "Ancient DNA," *Scientific American,* November 1993, p. 86; Robert Shapiro, *The Human Blueprint: The Race to Unlock the Secrets of Our Genetic Script* (New York: St. Martin's Press, 1991), p. 326.

48. Simmett et al., "Alumorphs—Human DNA Polymorphisms Detected by Polymerase Chain Reaction Using Aly-Specific Primers," *Genomics,* vol. 7 (1990), pp. 331–334 at 333.

49. National Institute of Justice, National Commission on the Future of Forensic DNA Evidence (Report), *The Future of Forensic DNA Testing: Predictions of the Research and Development Working Group* (Washington, D.C.: U.S. Department of Justice, Office of Justice Programs, November 2000), pp. 45–46.

50. Svante Paabo, Russell Higuchi, and Allan Wilson, "Ancient DNA and the Polymerase Chain Reaction," *Journal of Biological Chemistry,* vol. 264 (June 15, 1989), no. 17, pp. 9709–9710.

51. Paabo, "Ancient DNA," p. 86 at 88.

52. Ibid., p. 89.

53. Ibid.

54. Russell Higuchi et al., "Mitochondrial DNA of the Extinct Quagga: Relatedness and Extent of Postmortem Change," *Journal of Molecular Evolution,* vol. 25 (1987), pp. 283–287.

55. Ibid., pp. 284–285

56. Ibid.

57. J. Madeleine Nash, "Ultimate Gene Machine: a Method of Multiplying DNA's Revolutionizing Medical Diagnosis, Speeding Forensic Work and Solving Old Mysteries," *Time,* August 12, 1991, p. 54.

58. Ibid.

59. S. L. Vartanyan et al., "Holocene Dwarf Mammoths from Wrangel Island in the Siberian Arctic," *Nature,* vol. 362 (March 25, 1993), p. 337.

60. "Woolly Mammoths Found to Have Survived Ice Age," Knight- Ridder Newspapers wire service, *Hartford Courant,* March 25, 1993, p. A3.

61. Jeremy Cherfas, "Ancient DNA: Still Busy After Death," *Science, 253,* September 20, 1991, p. 1354 at 1356.

62. Ibid.

63. Svante Paabo, Allan Wilson, et al., "Polymerase Chain Reaction Reveals Cloning Artefacts," letter to *Nature,* vol. 334, August 4, 1988.

64. Svante Paabo, "Molecular Genetic Investigations of Ancient Human Remains," *Cold Spring Harbor Symposia on Quantitative Biology,* vol. 51 (1986), p. 441; David A. Lawlor et al., "Ancient HLA Genes from 7,500-Year-Old Archeological Remains," *Nature,* vol. 349 (February 28, 1991), p. 785.

65. Philip Ross, "Eloquent Remains," *Scientific American,* May 1992, p. 114 at 117.

66. Lawlor et al., "Ancient HLA Genes from 7,500-Year-Old Archeological Remains."

67. Svante Paabo, Allan Wilson, "Mitochondrial DNA from a 7000-Year-Old Brain," *Nucleic Acids Research,* vol. 16, no. 20, p. 9775 at 9776.

68. Svante Paabo, Allan Wilson, *Nucleic Acids Research,* vol. 16 (1988), no. 20, p. 9785.

69. Svante Paabo, "Ancient DNA," *Scientific American,* November 1993, p. 86 at 90.

70. Russell Higuchi, Cecilia H. Von Beroldingen, George F. Sensbaugh, Henry A. Erlich, "DNA Typing from Single Hairs," *Nature,* vol. 332 (April 7, 1988), p. 543.

71. Honghua Li et al., "Amplification and Analysis of DNA Sequences in Single Human Sperm and Diploid Cells," *Nature,* vol. 335 (September 29, 1988), p. 414.

72. Scott R. Woodward, Nathan J. Weyand, Mark Bunnell, "DNA Sequence from Cretaceous Period Bone Fragments," *Science,* vol. 266 (November 18, 1994), p. 1229.

73. Wolfgang Saxon, "E. T. Hall, 77, Archeologist Who Debunked Piltown Man," (obituary), *New York Times,* August 21, 2001, p. C14.

74. George and Roberta Poinar, *The Quest for Life in Amber* (Reading, Mass.: Helix Books, Addison-Wesley, 1994); R. DeSalle, M. Barcia and C. Gray, "PCR Jumping in Clones of 30-Million-Year-Old DNA Fragments from Amber Preserved Termites *(Mastotermes electrodominicus),*" *Experimentia,* vol. 49 (1993), p. 906; Virginia Morell, "30-Million-Year-Old DNA Boosts an Emerging Field," *Science,* vol. 257 (September 25, 1992), no. 5078, p. 1860; Raul J. Cano, Hendrik N. Poinar, Norman J. Pieniazek, Aftim Acra, and George O. Poinar, "Amplification and Sequencing of DNA from a 120–135-Million-Year-Old Weevil," *Nature,* vol. 363 (June 10, 1993), p. 536; Hendrik Poinar, Raul J. Cano, and George Poinar, "DNA from an Extinct Plant," *Nature,* vol. 363 (June 24, 1993), p. 677; Charles Pelligrino, "Resurrecting Dinosaurs; Possibility of Cloning Dinosaurs," *Omni,* September 22, 1995, p. 68; Virginia Morell, "Dino DNA: The Hunt and the Hype," *Science,* vol. 261 (September 9, 1993), no. 5118, p. 160; Roger Lewin, "Fact, Fiction and Fossil DNA," *New Scientist,* January 29, 1994, p. 38; Tomas Lindahl, "Instability and Decay of the Primary Structure of DNA," *Nature,* vol. 362 (April 22, 1993), p. 709 at 713; Joshua Fischman, "Have 25-Million-Year-Old Bacteria Returned to Life?" *Science,* vol. 268 (May 19, 1995), p. 977; Robert Lee Hotz, "Prehistoric Bacteria Revived: 25-Million-Year-Old

Microbes Preserved in Insect," *Los Angeles Times,* reprinted in *Hartford Courant,* May 19, 1995, p. A1; Malcolme W. Brown, "30-Million-Year Sleep: Germ Is Declared Alive," *New York Times,* May 19, 1995, p. A1; Woodward, Weyand, Bunnel, "DNA Sequence from Cretaceous Period Bone Fragments," p. 1229; Kathy A. Svatil, "Dinosaur Mine," *Discover,* May 1995, p. 36; Ann Gibbons, "Possible Dino DNA Find Is Greeted with Skepticism," *Science,* vol. 266 (November 18, 1994), p. 1159; Nicholas Wade, "Chemical Traces of Blood Found in Bones of Tyrannosaurus Rex," *New York Times,* June 10, 1997, p. C6; Hans Zischler, Helga Geisert, Arndt von Haeseler, Svante Paabo, "A Nuclear 'Fossil' of the Mitochondrial D-loop and the Origin of Modern Humans," *Nature,* vol. 378 (November 30, 1995), p. 489; Paul G. Taylor, "Reproducibility of Ancient DNA Sequences from Extinct Pleistocene Fauna," *Molecular Biological Evolution,* vol. 13 (1996), no. 1, p. 283; David A. Grimaldi, "Captured in Amber," *Scientific American,* April 1996, p. 84; Hendrik N. Poinar, Matthias Höss, Jeffrey L. Bada, and Svante Paabo, "Amino Acid Racemization and the Preservation of Ancient DNA," *Science,* vol. 272 (May 10, 1996), p. 864; Robert F. Service, "Just How Old Is That DNA, Anyway?" *Science,* vol. 272 (May 10, 1996), p. 810; Jeremy J. Austin, Andrew J. Ross, Andrew B. Smith, Richard A. Fortey, and Richard H. Thomas, "Problems of Reproducibility—Does Geologically Ancient DNA Survive in Amber-Preserved Insects?" *Proceedings of the Royal Society of London B,* vol. 264 (1997), pp. 467–474.

75. Matthias Krings et al., "Neanderthal DNA Sequences and the Origins of Modern Humans," *Cell,* vol. 90 (July 11, 1997), pp. 19–30; Kate Wong, "ANCESTRAL Quandary: Neanderthals Not Our Ancestors? Not So Fast," *Scientific American,* January 1998, p. 30; Richard A. Knox, "Study Finds Neanderthals, Humans Did Not Interbreed," *Boston Globe,* July 11, 1997, p. A1.

76. Tomas Lindahl, *Cell,* July 11, 1997, cited in Nicholas Wade, "Neanderthal DNA Sheds New Light on Human Origins," *New York Times,* July 11, 1997, p. A1 at A14.

77. Michael D. Lemonick, "A Bit of Neanderthal in Us All?" *Time,* May 3, 1999, pp. 68–69; Cidália Duarte et al., "The Early Upper Paleolithic Human Skeleton from the Abrigo do Laga-Velho (Portugal) and Modern Human Emergence in Iberia," *Proceedings of the National Academy of Sciences USA,* vol. 96 (June 1999), pp. 7604–7609.

78. Nicholas Wade, "DNA Tests Cast Doubt on Link Between Neanderthals and Modern Man," *New York Times,* March 29, 2000, p. A18; Igor V. Ovchinnikov et al., "Molecular Analysis of Neanderthal DNA from the Northern Caucasus," *Nature,* vol. 404 (March 30, 2000), pp. 490–493.

79. G. Adcock et al., "Mitochondrial DNA Sequences in Ancient Australians: Implications for Modern Human Origins," *Proceedings of the National Academy of Science USA,* vol. 98 (January 16, 2001), no. 2, pp. 537–542.

80. Joseph D'Aguese, "Not Out of Africa," *Discover,* August 2002, p. 52 at p. 56.

81. Milford H. Wolpoff et al., "Modern Human Ancestry at the Peripheries: A Test of the Replacement Theory," *Science,* vol. 291, January 12, 2001, pp. 293–297; Elizabeth Pennisi, "Skull Study Targets Africa-Only Origins," *Science,* vol. 291 (January 12, 2001), p. 231; Dan Vergano, "Ancestor Theory Falls from Grace," *USA TODAY,* January 15, 2001, p. 6D.

82. Constance Holden, "Oldest Human DNA Reveals Aussie Oddity," *Science,* vol. 291 (January 12, 2001), pp. 230–231.

83. Guy Nolch, "Mungo Man's DNA Shakes the Homo Family Tree," *Australian Science,* March 2001, pp. 29–31 (www.control.comau/~search)

84. Ibid.

85. Vergano, "Ancestor Theory Falls from Grace."

86. L. A. Grivell, "Small, Beautiful and Essential," *Nature,* vol. 341 (October 19, 1989), p. 569.

87. D. C. Wallace et al., "Mitochondrial DNA Associated with Leber's Hereditary Optic Neropathy," *Science,* vol. 242 (December 9, 1988), pp. 1427–1430; Edwin Kiester, Jr., "A Bug in the System," *Discover,* February 1991, p. 70; Douglas C. Wallace, "Mitochondrial DNA in Aging and Disease," *Scientific American,* August 1997, p. 40

88. National Institute of Justice, National Commission on the Future of Forensic DNA Evidence (Report), *The Future of Forensic DNA Testing: Predictions of the Research and Development Working Group* (Washington, D.C.: U.S. Department of Justice, Office of Justice Programs, November 2000), p. 47.

89. Ralph Wetterhahn, "Missing in Action," *Popular Science,* August 1998, p. 46.

90. Keith Inman and Norah Rudin, *Introduction to Forensic DNA Analysis,* 1st edition (Boca Raton, Fla.: CRC Press, 1997), p. 51.

91. Steven Lee Myers, "Pentagon Seeks to Open Unknown's Tomb," *New York Times,* April 28, 1998, p. A1; Andrea Stone, "Mystery in the Tomb: One Mother Wants to Know If It's Her Son; the Other Doesn't," *USA TODAY,* May 6, 1998, p. 1A; Bradley Graham, "Cohen Orders Exhumation at Tomb of the Unknowns," *Washington Post,* May 8, 1998, p. A1; "The Vietnam Unknown Soldier Can Be Identified," U.S. Veteran Dispatch Staff Report, *U.S. Veteran Dispatch* (July 1996 Issue), www.usvetdsp.com; "Background Paper on the Activities of the Department of Defense Senior Working Group on The Vietnam Unknown in the Tomb of the Unknown Soldiers," April 24, 1998, www.defenselink.mil; Jonathan Wright (Reuters), "Pentagon Identifies 'Unknown' Vietnam Warrior," *Yahoo! News,* June 30, 1998; Richard Sisk, "An Unknown Is ID'd: DNA Tests Show Remains Are Mo. Airman's," *New York Daily News,* June 30, 1998, p. 2; William S. Cohen, "Secretary of Defense William S. Cohen's Statement Concerning the Identification of the Vietnam Unknown," News Release, Office of Assistant Secretary of Defense (Public Affairs), June 30, 1998; Thomas M. DeFrank, "U.S. Plans DNA Check of Unknown Military Dead," *New York Daily News,* May 31, 1999, p. 4.

92. Rudin and Inman, *Introduction to Forensic DNA Analysis,* 2d edition, p. 81.

93. Ibid., pp. 51–52 and 74.

94. *State of Connecticut v. Stephen Pappas,* 256 Conn. 854, 776 A.2d 1091 (2001), *Connecticut Law Journal,* July 24, 2001.

95. P. L. Ivanov et al., "Mitochondrial DNA Sequence Heteroplasmy in the Grand Duke of Russia Georgij Romanov Establishes the Authenticity of the Remains of Tsar Nicholas II," *Nature Genetics,* vol. 12 (April 1996), no. 4, pp. 417–420; Editorial, "Romanovs Find Closure in DNA," *Nature Genetics,* vol. 12 (April 1996), no. 4, pp. 339–340; Associated Press, "Czar's Bones for Sure, Russians Say," *New York Daily News,* January 27, 1998, p. 19.

96. *State of Connecticut v. Pappas.*

97. Robert K. Massie, *The Romanovs: The Final Chapter* (New York: Random House, 1995); Rebecca J. Fowler, "Alas, More Fantasia Than Anastasia," *Washington Post* (reprinted in *Hartford Courant*), October 7, 1994, p. A1.

98. "From Hair Cells to a Prison Cell: Mitochondrial DNA Procedure Offers Prosecution Breakthrough," *Law Enforcement News,* November 15, 1996.

99. Report compiled by FBI Laboratory DNA Unit II.

100. Kevin Heldman, "New Test IDs Victim Linked to Green River Case," *APBnews.com,* November 3, 1999.

101. Sam Howe Verhovek, "Suspect Charged in 4 Green River Slayings," *New York Times,* December 6, 2001, p. A32; Gene Johnson (Associated Press), "DNA Links Man to Four 1980s Killings," *Hartford Courant,* December 7, 2001, p. A23; Sam Howe Verhovek, "Suspect Held in Northwest Serial-Killings Case," *New York Times,* December 1, 2001, p. A11; Luis Cabrera (Associated Press), "Suspected of 49 Murders, Wash. Man Finally Arrested," *Boston Sunday Globe,* December 2, 2001, p. A12; Chris Stetkiewicz (Reuters), "Plea Entered in Serial Killings," *Boston Globe,* December 19, 2001, p. A6.

102. Associated Press, "Hair DNA Link in Doc Slaying," *New York Daily News,* January 23, 1999, p. 8; Jose Martinez, "Suspect in Doc's Killing Is Nabbed in France," *Boston Herald,* March 30, 2001, p. 6; James Keaten (Associated Press), "Fugitive Fights Return to US," *Boston Globe,* June 8, 2001, p. A18; David Johnston, "Deal May Spare Life of Abortion Foe Sought in Slaying," *New York Times,* June 2, 2001, p. B5; Associated Press, "Suspect Loses Extradition Appeal," *Boston Sunday Globe,* October 14, 2001, p. A14; Associated Press, "Doctor-Slaying Suspect in U.S.," *Newsday,* June 6, 2002, p. A7; Associated Press, "Tests Ordered on Suspect in Abortion Doctor Death," *New Haven Register,* July 20, 2002, p. A9.

103. Unattributed, "Suspect Tells Newspaper He Killed Abortion Doctor," *New York Times,* November 21, 2002, p. B5.

CHAPTER 5

1. National Institute of Justice (NIJ), U.S. Department of Justice, Office of Justice Programs, *Convicted by Juries, Exonerated by Science: Case Studies in the Use of DNA Evidence to Establish Innocence After Trial* (Washington, D.C.: NIJ, 1996).

2. "DNA Evidence to Set Inmate Free after 18 Years in Prison," *USA TODAY,* October 2, 2001, p. 3A.

3. Editorial, "States Dawdle While Jailed Innocents Languish," *USA TODAY,* June 26, 2001, p. 12A.

4. Richard Willing, "Increasing DNA Exonerations Contradict Predictions," *USA TODAY,* January 18, 2002, p. 8A.

5. Craig Timberg, "Va. Alters Death Penalty," *Washington Post* (reprinted in *The Boston Globe*), May 3, 2001, p. A7; Francis X. Clines, "DNA Clears Virginia Man of 1982 Assault," *New York Times,* December 10, 2001, p. A14.

6. Brooke A. Masters, "Number of Executions Continues a Sharp Decline in U.S," *Washington Post* (reprinted in *Hartford Courant*), September 9, 2001, p. A8; Richard Willing, "Bill Would Expand Access to DNA Tests, *USA TODAY,* June 28, 2001, p. 11A; Eric Lichtblau, "Push Is on for Death Penalty Reform," *Los Angeles Times* (reprinted in *Hartford Courant*), June 25, 2001, p. A5; Susan Milligan, "Support Grows for DNA Testing in Death Row Cases," *Boston Globe,* June 28, 2001, p. A1; Francis X. Clines, "Access by Inmates to Tests for DNA Gains Ground," *New York Times,* December 19, 2000, p. A22.

7. National Institute of Justice (NIJ), National Commission on the Future of DNA Evidence, *Postconviction DNA Testing: Recommendations for Handling Requests* (Washington, D.C.: NIJ, September 1999), p. iii; Neil A. Lewis, "Prosecutors Urged to Allow Appeals on DNA," *New York Times,* September 28, 1999, p. A18.

8. National Institute of Justice (NIJ), *Convicted by Juries, Exonerated by Science: Case Studies in the Use of DNA Evidence to Establish Innocence After Trial* (Washington, D.C.: NIJ, June 1996), p. xxviii.

9. Ibid., p. xxix.

10. Ibid., p. xxx.

11. See e.g., Robert D. McFadden, "Reliability of DNA Testing Challenged by Judge's Ruling," *New York Times,* August 15, 1989, p. B1.

12. Richard Willing, "Increasing DNA Exonerations Contradict Predictions," *USA TODAY,* January 18, 2002, p. 8A.

13. Barry Scheck, Peter Neufeld, and Jim Dwyer, *Actual Innocence: Five Days to Execution and Other Dispatches from the Wrongly Convicted* (New York: Doubleday, 2000), p. 73; Barry Scheck and Peter Neufeld, "Junk Science, Junk Evidence," *New York Times,* May 11, 2001, p. A35; Katherine E. Finkelstein, "When Justice Hinges on What Is Seen, and Believed," *New York Times,* December 4, 2000, p. B1.

14. Ibid.; Michael Rezendes, "A Book with Behind-the-Scenes Splash," *Boston Sunday Globe,* September 17, 2000, p. N4; Joe Beaird, "*Actual Innocence*: Making the Case for DNA," APBnews.com, February 28, 2000; Bess Bezirgan, "Innocence Agenda Needed in Louisiana," NACDL News Release, February 25, 2000.

15. George F. Will (columnist), "Are the Innocent Being Executed?" *Hartford Courant,* April 6, 2000, p. A11.

16. "DNA Testing Frees Man Jailed in Rape: Calling Data Crucial, Virginia Governor Pardons a Man Who Served 6 1/2 Years," *New York Times,* April 25, 1992, p. 29.

17. NIJ, *Convicted by Juries, Exonerated by Science;* Gina Kolata, "DNA Tests Provide Key to Cell Doors for Some Wrongly Convicted Inmates," *New York Times,* August 5, 1994, p. A20.

18. "DNA Tests Clear Man of Rape Nearly 8 Years After Conviction," *New York Times,* January 31, 1995, p. B5.

19. NIJ, *Convicted by Juries, Exonerated by Science,* pp. xxiii–xxiv.

20. Don Terry, "DNA Tests and a Confession Set Three on the Path to Freedom in 1978 Murders," *New York Times,* June 15, 1996; Brian Bregstein (Associated Press), "Students, Professor Provide Key to Freedom for 3 Convicted of Murder," *Hartford Courant,* June 15, 1996, p. A2.

21. David Protess and Rob Warden, *A Promise of Justice* (New York: Hyperion, 1998).

22. Don Terry, "After 18 Years in Prison, 3 Are Cleared of Murders," *New York Times*, July 3, 1996, p. A1.

23. "3 Charged in '78 Double Murder," *Associated Press* article in *New York Times*, July 4, 1996, p. B7.

24. Evelyne Girardet (Associated Press), "Men Exonerated in Two Murders Get $36 Million," *Hartford Courant*, March 6, 1999, p. A14.

25. "What Jennifer Saw," *Frontline*, February 25, 1997, written, produced, and directed by Ben Loeterman.

26. Jennifer Thompson (op-ed), "I Was Certain, but I Was Wrong," *New York Times*, June 18, 2000, Sec. 4, p. 15.

27. Ibid., and NIJ, *Convicted by Juries, Exonerated by Science*, pp. 43–44.

28. Helen O'Neill (Associated Press), "Accused and Accuser Find Forgiveness," *Hartford Courant*, September 28, 2000, p. D1 at D5.

29. Ibid.

30. Ibid.

31. Thompson, "I Was Certain, but I Was Wrong."

32. Peter J. Boyer, "DNA on Trial," *New Yorker*, January 17, 2000, p. 42.

33. "DNA Task Force Wins Freedom for Innocent Inmates," *Champion*, July 1993, p. 51.

34. Ibid.

35. Kevin Krajick, "Genetics in the Courtroom: Controversial DNA Testing Can Clear a Suspect," *Newsweek*, January 11, 1993, p. 64.

36. Kiernan Crowley, "Woman Lives in Terror of Freed Rape Suspect," *New York Post*, April 13, 1996.

37. Ibid.

38. William Glaberson, "Rematch for DNA in a Rape Case: Beneficiary Is Likely to Challenge Test's Reliability," *New York Times*, April 10, 1996, p. B5.

39. Kiernan Crowley, "Rape Suspect to Seek $12M damages," *New York Post*, April 10, 1996, p. 15.

40. David Stout, "Man Cleared of Rape by DNA Is Implicated in a Second Case," *New York Times*, April 9, 1996, p. B5.

41. Crowley, "Woman Lives in Terror of Freed Suspect," p. 3.

42. William Glaberson, "Rematch for DNA in a Rape Case: Beneficiary Is Now Likely to Challenge Test's Reliability," *New York Times*, April 10, 1996, p. B5.

43. Kiernan Crowley, "Lawyer Says Cops Out for Revenge on Rape Suspect," *New York Post*, April 11, 1996, p. 7.

44. Ibid.

45. Kiernan Crowley, "Rape Suspect to Seek $12M Damages," *New York Post*, April 10, 1996, p. 15.

46. Cathy Burke, "DNA Nails L.I. Rapist It Had Previously Freed," *New York Post*, July 19, 1997, p. 12.

47. Kiernan Crowley, "DNA Freed Him from Prison—And May Put Him Back In," *New York Post*, June 26, 1997, p. 14.

48. Boyer, "DNA on Trial," p. 42.

49. Joseph P. Shapiro, "The Wrong Men on Death Row," *U.S. News & World Report*, November 9, 1998, p. 22 at 26; Pam Belluck, "Officials Face Trial in Alleged Plot to Frame a Man for Murder," *New York Times*, March 9, 1999, p. A19.

50. Don Terry, "Ex-Prosecutors and Deputies in Death Row Case Are Charged with Framing Defendant," *New York Times*, December 13, 1996, p. A18.

51. Associated Press, "Officials Face Conspiracy Charges," *Hartford Courant*, December 13, 1996, p. A6.

52. Terry, "Ex-Prosecutors and Deputies in Death Row Case Are Charged with Framing Defendant."

53. NIJ, *Convicted by Juries, Exonerated by Science;* Associated Press, "Officials Face Conspiracy Charges"; Terry, "Ex-Prosecutors and Deputies in Death Row Case Are Charged with Framing Defendant."

54. Belluck, "Officials Face Trial in Alleged Plot to Frame a Man for Murder."

55. Yahoo!News, "Cruz Grilled About Why He Lied to Police in 'Dupage Seven' Trial," April 16, 1999.

56. Associated Press, "Charges Dropped for 2 in Illinois Framing Case," *New York Times*, May 14, 1999, p. A22; Andrew Bluth, "5 Law Officers Acquitted of Plotting to Frame Man in Killing," *New York Times*, June 6, 1999, p. 26.

57. Scheck, Neufeld, and Dwyer, *Actual Innocence*, p. 246.

58. *In the Matter of an INVESTIGATION OF THE WEST VIRGINIA STATE POLICE CRIME LABORATORY, SEROLOGY DIVISION*, 438 S.E. 2d 501, 509 (W. Va. 1993).

59. Scheck and Neufeld (op-ed), "Junk Science, Junk Evidence"; Associated Press, "Mistrial for Police Chemist Accused of Falsifying Tests, *New York Times*, September 19, 2001, p. A18.

60. Jim Yardley, "Inquiry Focuses on Scientist Employed by Prosecutors," *New York Times*, May 2, 2001, p. A14.

61. Jim Yardley, "Flaws in Chemist's Findings Free Man She Helped Convict," *New York Times*, May 8, 2001, p. A1.

62. Reuters, "Oklahoma: Court Overturns Death Sentence," *New York Times*, August 14, 2001, p. A13.

63. Associated Press, "Exonerated Man Cleared of Murder Charges," *APBnews.com*, March 11, 1999.

64. Dirk Johnson, "Illinois, Citing Faulty Verdicts, Bars Executions," *New York Times*, February 1, 2000, p. A1; Editorial, "A Timeout on the Death Penalty," *New York Times*, February 1, 2000, p. A20.

65. Associated Press, "Illinois Grapples with Death Penalty," *New Haven Register*, November 24, 2002, p. A19; Jodi Wilgoren, "G.O.P. Death-Penalty Feud Sinks to First-Name Calling," *New York Times*, September 26, 2002, p. A1; Don Babwin (Associated Press), "Death Row Hearings Revive Old Anguish," *Boston Globe*, October 24, 2002, p. A11; Steve Mills and Christi Parsons, "Tears Send a Message: Hearings' Emotional Impact Surprises Death Penalty Foes," *Chicago Tribune* (online), October 28, 2002; Christopher Wills, "Ryan: 'Pretty Much Ruled Out' Blanket Clemency," *Chicago Tribune* (online), October 22, 2002;

Christi Parsons and Steve Mills, "Ryan Backs off Blanket Clemency," *Chicago Tribune* (online), October 23, 2002.

66. Editorial, "Justice for Death Row," *New York Times,* November 21, 2002, p. A36.

67. John Keilman, "Clemency Hearings a Study in Anguish," *Chicago Tribune* (online), October 28, 2002.

68. Johnson, "Illinois, Citing Faulty Verdicts, Bars Executions"; Dirk Johnson, "Shoddy Defense by Lawyers Puts Innocents on Death Row," *New York Times,* February 5, 2000, p. A1.

69. Claudia Kolker, "Death Penalty Foes Make Gains," *Los Angeles Times,* reprinted in *Hartford Courant,* September 3, 2000, p. A19.

70. Toni Locy, "Push to Reform Death Penalty Growing," *USA TODAY,* February 20, 2001, p. 5A.

71. Jonathan Alter, "The Death Penalty on Trial," *Newsweek,* June 12, 2000, p. 29.

72. Five-Part Series: Tribune Investigative Report. The Failure of the Death Penalty in Illinois: Ken Armstrong and Steve Mills, "Death Row Justice Derailed: Bias, Errors and Incompetence in Capital Cases Have Turned Illinois' Harshest Punishment Into Its Least Credible," *Chicago Tribune,* November 14, 1999, p. 1; Ken Armstrong and Steve Mills, "Inept Defenses Cloud Verdicts: With Their Lives at Stake, Defendants in Illinois Capital Trials Need the Best Attorneys Available. But They Often Get Some of the Worst," *Chicago Tribune,* November 15, 1999, p. 1; Steve Mills and Ken Armstrong, "The Inside Informant," *Chicago Tribune,* November 16, 1999, p. 1; Steve Mills and Ken Armstrong, "A Tortured Path to Death Row," *Chicago Tribune,* November 17, 1999, p. 1; Steve Mills and Ken Armstrong, "Convicted by a Hair," *Chicago Tribune,* November 18, 1999, p. 1.

73. Alter, "The Death Penalty on Trial," cover and p. 27.

74. James S. Liebman et al., *A Broken System: Error Rates in Capital Cases, 1973–1995,* The Justice Project, June 12, 2000.

75. Fox Butterfield, "Death Sentences Being Overturned in 2 of 3 Appeals," *New York Times,* June 12, 2000, p. A1.

76. Paul G. Cassell (op-ed), "We're Not Executing the Innocent," *Wall Street Journal,* June 16, 2000, p. A14.

77. Associated Press, "Condemned Inmate's Plea for DNA Retest Denied," May 31, 2000; Richard A. Oppel Jr. and Frank Bruni, "Bush Expects to Grant His First Reprieve to Killer-Rapist on Texas' Death Row," *New York Times,* June 1, 2000, p. A27; Michael Graczyk (Associated Press), "Bush Grants First Death Row Reprieve," *New York Daily News,* June 2, 2000, p. 7; Richard Willing, "Texas Slows Executions as Election Day Nears," *USA TODAY,* September 8, 2000, p. 6A.

78. Frank Bruni and Richard A. Oppel Jr., "Bush Delays an Execution for the 1st Time in 5 Years," *New York Times,* June 2, 2000, p. A18.

79. John Aloysius Farrell, "DNA Technology Puts Death Penalty to Test," *Boston Globe,* June 2, 2000, p. A1.

80. P. J. McCormick, concurring opinion, *Ex Parte Ricky Nolen McGinn,* Texas Court of Criminal Appeals (No. 35,570–04), published June 14, 2000.

81. Ibid.

82. Ibid.

83. Mark Wrolstad, "Convict's Fate May Hinge on New Tests: Texas Case Joins Debate on DNA Technology Use," *Dallas Morning News,* April 25, 2000.

84. "Defendant's Motion to Authorize Retesting of Physical Evidence by Defense DNA Expert," filed May 15, 2000, in the matter of *Ex Parte Ricky Nolen McGinn,* cited in J. Womack, concurring opinion, *Ex Parte Ricky Nolen McGinn,* Texas Court of Criminal Appeals (No. 35,570–04), published June 14, 2000.

85. P. J. McCormick, concurring opinion.

86. Associated Press, "DNA Test No Help to Reprieved Death Row Inmate," APBnews.com, July 12, 2000; Richard Willing, "DNA Tests Still Point to Texas Inmate," *USA TODAY,* July 12, 2000, p. 1A; Associated Press, "DNA Results Fail to Clear Child Killer," APBnews.com, August 15, 2000; Associated Press, "Rapist-Murderer Executed in Texas," ApB.com, September 28, 2000.

87. Mary Lee Grant (Associated Press), "Prisoner in Texas Freed After DNA Tests," *Hartford Courant,* August 16, 2000, p. A7.

88. Kellie Dworaczyk, "DNA Evidence and Texas' Criminal Justice System," Focus Report, House Research Organization, Texas House of Representatives, November 10, 2000, p. 5.

89. Associated Press, "Judge Says Cons Have Right to DNA Tests," *New Haven Register,* October 1, 2000, p. A7.

90. Laurie P. Cohen, "Reasonable Doubt: Someone Who's Guilty Would Never Want DNA Testing, Right?" *Wall Street Journal,* July 12, 2000, p. A1.

91. Ibid.

92. Ibid., p. A12.

93. Ibid.

94. Scott Turow, *Reversible Errors* (New York: Farrar, Straus, and Giroux, 2002).

95. Donald Lyons, "'Exonerated' Guilty," *New York Post,* October 14, 2002, p. 41.

96. Benjamin Weiser, "Manhattan Judge Finds Federal Death Penalty Law Unconstitutional," *New York Times,* July 2, 2002, p. B1; Jess Bravin, "Death Penalty Is Struck Down by District Judge; Appeal Likely," *Wall Street Journal,* July 2, 2002, p. D5; Fred Kaplan, "Federal Judge Rules Death Penalty Illegal," *Boston Globe,* July 2, 2002, p. A1; Lori Montgomery *(Washington Post),* "Maryland Governor Puts Halt to Death Penalty, Orders Probe," *Boston Globe,* May 10, 2002, p. A2; Francis X. Clines, "Death Penalty Is Suspended in Maryland," *New York Times,* May 10, 2002, p. A20; Richard Willing, "Md. Governor Halts Executions Until Study Done," *USA TODAY,* May 10, 2002, p. 7A.

97. Erika Hagelberg, Bryan Sykes, Robert Hedges, "Ancient Bone DNA Amplified," *Nature,* vol. 342 (November 30, 1989), p. 483.

98. Ibid.

99. Robert Shapiro, *The Human Blueprint: The Race to Unlock the Secrets of Our Genetic Script* (New York: St. Martin's Press, 1991), 331.

100. Paabo, "Ancient DNA," p. 86.

101. Ibid.

102. Erika Hagelberg, Alec Jeffreys, *Nature,* vol. 352 (August 1, 1991), p. 427.

103. Ibid.

104. Ibid.

105. Ibid., p. 428.

106. Ibid.

107. Ibid., pp. 428–429.

108. Bryan Sykes, "The Past Comes Alive," *Nature,* vol. 352 (August 1, 1991), p. 381.

CHAPTER 6

1. Office of Special Investigations, Criminal Division, *In the Matter of Josef Mengele: A Report to the Attorney General of the United States* (U.S. Department of Justice, October 1992), p. 2.

2. Cecilie Rohwedder, "German Companies Approve the Start of Payments to Nazi-Era Slave Laborers," *Wall Street Journal,* May 23, 2001, p. A23; Paul Beckett, "Holocaust Settlement Approved By Judge, But Obstacles Remain," *Wall Street Journal,* July, 27, 2000, p. B15.

3. Gabriel Schoenfeld, "Holocaust Reparations—A Growing Scandal," *Commentary Magazine,* September 2000 (www.commentarymagazine.com/0009/schoenfeld.html).

4. Michael Specter, "The Dangerous Philosopher," *New Yorker,* September 6, 1999, p. 46; Editorial, "Animal Crackers," *Wall Street Journal,* March 30, 2001, p. A14; Don Feder (columnist), "Prof. Death Takes Ideas to Princeton," *Boston Herald,* October 28, 1998, p. 37; Wesley J. Smith, *Culture of Death: The Assault on Medical Ethics in America* (San Francisco: Encounter, 2000); Adam Wolfson, "Are You a Person or a Nonperson? The New Bioethicists Will Decide," *Wall Street Journal,* January 24, 2001, p. A20; Peter Singer, *Writings on an Ethical Life* (New York: Ecco Press/HarperCollins, 2000); William McGurn, "Princeton Defends Its Philosopher of Infanticide," *Wall Street Journal,* November 13, 1998, p. W17; Editorial, "Big Man on Campus," *Wall Street Journal,* October 2, 1998, p. W15.

5. Christopher Joyce and Eric Stover, *Witnesses from the Grave: The Stories Bones Tell* (New York: Ballantine Books, 1992), pp. 143–144.

6. Gerald Astor, *The Last Nazi: The Life and Times of Dr. Josef Mengele* (New York: Donald I. Fine, 1985), p. 223.

7. Joyce and Stover, *Witnesses from the Grave,* p. 161.

8. Ibid., p. 151.

9. Steve Fainaru, "War Crimes Bill Spurs Justice Department Turf Battle," *Boston Globe,* November 6, 1999, p. A4.

10. Office of Special Investigations, *In the Matter of Josef Mengele,* p. 44.

11. Ibid., p. 55.

12. Ibid. (Exhibits), p. 289.

13. Ibid. (Exhibits), p. 290.

14. Ibid., pp. 109–110.

15. Ibid., p. 147; see also Astor, *The Last Nazi*, p. 256.

16. Office of Special Investigations, *In the Matter of Josef Mengele*, p. 147.

17. Ibid. (Exhibits), p. 294.

18. Robert Jay Lifton, *The Nazi Doctors: Medical Killing and the Psychology of Genocide* (New York: Basic Books, 1986), p. 337.

19. Ibid., p. 338; Ernst Schnabel, *Anne Frank: A Portrait of Courage* (New York: Harcourt, Brace, and World, 1958).

20. Rolf Hochhuth, *The Deputy* (New York: Grove Press, 1964 [1963]), pp. 31–32.

21. Ibid., as quoted by Lifton, *The Nazi Doctors,* p. 380.

22. William Goldman, "Butch Cassidy and the Nazi Dentist," *Esquire,* November 1994, p. 126 at 130.

23. Ibid., p. 129.

24. New York: Random House, 1976.

25. Office of Special Investigations, *In the Matter of Josef Mengele*, p. 128.

26. Ibid., p. 128

27. Lifton, *The Nazi Doctors,* p. 358.

28. Office of Special Investigations, *In the Matter of Josef Mengele*, p. 149.

29. Ibid., p. 188.

30. Joyce and Stover, *Witnesses from the Grave*, p. 201.

31. Office of Special Investigations, *In the Matter of Josef Mengele*, p. 158.

32. Ibid., pp. 159–162.

33. Ibid. (Exhibit), p. 378.

34. Joyce and Stover, *Witnesses from the Grave*, p. 164.

35. Alan Wykes, *Himmler* (New York: Ballantine Books, 1972), p. 74.

36. Lifton, *The Nazi Doctors,* p. 43.

37. Ibid.

38. Office of Special Investigations, *In the Matter of Josef Mengele* (Report Exhibits).

39. Astor, *The Last Nazi*, p. 25.

40. Joyce and Stover, *Witnesses from the Grave*, p. 164, citing Gerald Posner and John Ware, *Mengele: The Complete Story* (New York: Dell, 1986), p. 16; Astor, *The Last Nazi*, p. 24.

41. Office of Special Investigations, *In the Matter of Josef Mengele*, p. 152.

42. Ibid., p. 153.

43. Ibid., p. 157.

44. Lifton, *The Nazi Doctors,* p. 31.

45. Joyce and Stover, *Witnesses from the Grave*, p. 148.

46. Ibid., pp. 148–150.

47. Office of Special Investigations, *In the Matter of Josef Mengele*, pp. 139–141.

48. Leonard Flom (Letter to the Editor), "Only Your Own DNA Makes a Perfect Clone," *New York Times,* January 25, 1994, p. A18.

49. Richard Milner, *The Encyclopedia of Evolution: Humanity's Search for Its Origins* (New York: Facts on File, 1990), p. 184.

50. Ibid., p. 156.

51. John Horgan, "Eugenics Revisited," *Scientific American,* June 1993, p. 122 at 123.

52. Robert N. Proctor, *Racial Hygiene: Medicine Under the Nazis* (Cambridge, Mass.: Harvard University Press, 1988), p. 42.

53. Editorial, "Human Sterilization in Germany and the United States," *Journal of the American Medical Association,* vol. 102 (May 5, 1934), no. 18, pp. 1501–1502.

54. *Buck v. Bell,* 274 U.S. 200 (1927).

55. Lifton, *The Nazi Doctors,* p. 23.

56. Daniel J. Kevles, *In the Name of Eugenics: Genetics and Uses of Human Heredity* (Cambridge, Mass.: Harvard University Press, 1985), p. 116.

57. Proctor, *Racial Hygiene,* p. 17.

58. Robert N. Proctor, "Nazi Doctors, Racial Medicine, and Human Experimentation" (Chapter 2) in *The Nazi Doctors and the Nuremberg Code: Human Rights in Human Experimentation* (George J. Annas and Michael A. Grodin, eds.) (New York: Oxford University Press, 1992), pp. 18–20.

59. Ibid., pp. 19–20.

60. Ibid.

61. Proctor, *Racial Hygiene,* pp. 78–79 and 150–151.

62. Proctor, "Nazi Doctors, Racial Medicine, and Human Experimentation," p. 20.

63. Astor, *The Last Nazi,* p. 108.

64. Patrick McMahon, "Ore. Reports 8 Suicides under New Law," *USA TODAY,* August 19, 1998, p. 6A; Sam Howe Verhovek, "Legal Suicide Has Killed 8, Oregon Says," *New York Times,* August 19, 1998, p. A16; (Reuters) "Michigan Law Tests Kervorkian's Mission," *USA TODAY,* August 31, p. 13A.

65. Proctor, "Nazi Doctors, Racial Medicine, and Human Experimentation," p. 20.

66. W. W. Peter, M.D., "Germany's Sterilization Program," *American Journal of Public Health,* vol. 24 (March 1934), no. 3, pp. 187–191.

67. Proctor, "Nazi Doctors, Racial Medicine, and Human Experimentation," p. 21.

68. Peter, "Germany's Sterilization Program," p. 188.

69. Lifton, *The Nazi Doctors,* p. 30.

70. Proctor, *Racial Hygiene,* p. 107.

71. Ibid., p. 140.

72. Proctor, "Nazi Doctors, Racial Medicine, and Human Experimentation," p. 20, fn. 7; Peter, "Germany's Sterilization Program," p. 189.

73. Michael Kimmelman, "When a Glint in the Eye Showed Crime in the Genes," *New York Times,* May 22, 1998, p. E1.

74. Lifton, *The Nazi Doctors,* p. 42.

75. Ibid., p. 29.

76. Ibid. p. 42.

77. Proctor, *Racial Hygiene,* p. 93.

78. Ibid., p. 130.

79. Ibid., p. 92.

80. Lifton, *The Nazi Doctors,* p. 36.

81. Astor, *The Last Nazi,* p. 105.

82. Proctor, *Racial Hygiene,* p. 195.

83. Ibid., pp. 186–189; Astor, *The Last Nazi,* p. 105.

84. Proctor, "Nazi Doctors, Racial Medicine, and Human Experimentation," p. 24.

85. Proctor, *Racial Hygiene,* p. 184.

86. Astor, *The Last Nazi,* p. 106.

87. Proctor, "Nazi Doctors, Racial Medicine, and Human Experimentation," p. 24.

88. Ibid., p. 23.

89. Ibid., p. 191.

90. Astor, *The Last Nazi,* p. 109.

91. Proctor, *Racial Hygiene,* p. 196.

92. Ibid., p. 199.

93. Ibid., p. 205.

94. Ibid., p. 201.

95. Astor, *The Last Nazi,* p. 121.

96. Ibid., p. 120.

97. Proctor, "Nazi Doctors, Racial Medicine, and Human Experimentation," p. 27.

CHAPTER 7

1. Jerry Adler, "The Last Days of Auschwitz," *Newsweek,* January 16, 1995, p. 46; Jack Kelly and Andrea Stone, "50 years After the Liberation of Auschwitz; 'I Hear Crying' in the Silences," *USA TODAY,* January 26, 1995, p. 10A; Gerald Astor, *The Last Nazi: The Life and Times of Dr. Josef Mengele* (New York: Donald I. Fine, 1985), pp. 43–44.

2. Rudolf Hoess, *Commandant of Auschwitz: The Autobiography of Rudolf Hoess* (Cleveland: World, 1959), p. 146.

3. Astor, *The Last Nazi,* p. 92

4. Primo Levi, *Survival in Auschwitz,* translated by Stuart Woolf (New York: Collier Books, 1993), p. 19.

5. Christopher Joyce and Eric Stover, *Witnesses from the Grave: The Stories Bones Tell* (New York: Ballantine, 1992), p. 145.

6. Levi, *Survival in Auschwitz,* p. 20.

7. Robert Jay Lifton, *The Nazi Doctors: Medical Killing and the Psychology of Genocide* (New York: Basic Books, 1986), p. 148.

8. Ibid., p. 165.

9. Ibid., p. 160.

10. Ibid., pp. 160–161.

11. Christopher Rhoads, "Holocaust Lawsuit Shifts Back Focus to German Firms," *Wall Street Journal,* August 25, 1998, p. A11; "Holocaust Survivors Sue German Smelter," *New York Times,* August 22, 1998, p. B4.

12. Lifton, *The Nazi Doctors,* p. 148

13. Ibid., p. 149–150.

14. Ibid., p. 163.

15. Office of Special Investigations, Criminal Division, *In the Matter of Josef Mengele: A Report to the Attorney General of the United States* (U.S. Department of Justice, October 1992), p. 30.

16. Lifton, *The Nazi Doctors,* p. 349,

17. Ibid., p. 355.

18. Lawrence Wright, *Twins: And What They Tell Us About Who We Are* (New York: John Wiley and Sons, 1997), p. 20.

19. Office of Special Investigations, *In the Matter of Josef Mengele,* pp. 97–98.

20. Astor, *The Last Nazi,* pp. 125–126.

21. Ibid., p. 245.

22. Robert N. Proctor, *Racial Hygiene: Medicine Under the Nazis* (Cambridge, Mass.: Harvard University Press, 1988), p. 43.

23. Robert N. Proctor, "Nazi Doctors, Racial Medicine, and Human Experimentation" (Chapter 2) in *The Nazi Doctors and the Nuremberg Code: Human Rights in Human Experimentation* (George J. Annas and Michael A. Grodin, eds.) (New York: Oxford University Press, 1992), p. 20.

24. Ibid.

25. Proctor, *Racial Hygiene,* p. 43.

26. Paul Harris (Associated Press), "Apartheid Testimony Ends," *Hartford Courant,* August 1, 1998, p. A3; Patricia Reaney (Reuters), "Ethnically Targeted Weapons May Not Be Far Off," *Yahoo! News,* January 22, 1999; James Ridgeway, "Biological Weapons Might Not Be Color Blind: Ethnic Warfare," *Village Voice,* February 2, 1999, p. 31; Robert Block, "Doctor Acquitted in South African Germ-War Case," *Wall Street Journal,* April 12, 2002, p. A12; Kurt Shillinger, "Apartheid Figure's Acquittal Reopens Wounds," *Boston Globe,* April 12, 2002, p. A8.

27. Astor, *The Last Nazi,* pp. 97–98.

28. Ibid., p. 101.

29. Ibid., p. 23.

30. Ibid., p. 142.

31. Lifton, *The Nazi Doctors,* pp. 357 and 142.

32. Astor, *The Last Nazi,* p. 142.

33. Lifton, *The Nazi Doctors,* p. 44.

34. Ibid., p. 35.

35. Astor, *The Last Nazi,* p. 133.

36. Ibid., p. 134.

37. Lifton, *The Nazi Doctors,* p. 341.

38. Ibid.

39. Ibid., p. 358.

40. Ibid., p. 377.

41. Dr. Magda V., cited in ibid., p. 377.

42. Astor, *The Last Nazi,* p. 14.

43. Lifton, *The Nazi Doctors,* p. 344.

44. Ibid., p. 344.

45. Eva Mozes-Kor, cited in ibid., pp. 55–57.

46. Astor, *The Last Nazi,* pp. 168–169.

47. Office of Special Investigations, *In the Matter of Josef Mengele,* p. 104.

48. Telford Taylor, "Opening Statement of the Prosecution, December 9, 1946," in *Nazi Doctors and the Nuremberg Code,* p. 87.

49. Historical Documents, "Judgment and Aftermath" (Chapter 6), *Nazi Doctors and the Nuremberg Code,* pp. 94–104.

50. Taylor, *Nazi Doctors and the Nuremberg Code,* p. 71.

51. Ibid., p. 86.

52. Ibid., p. 68.

53. Astor, *The Last Nazi,* p. 227.

54. Office of Special Investigations, *In the Matter of Josef Mengele,* p. 121.

55. Ibid., p. 120.

56. Ibid., p. 30.

57. Ibid., pp. 35–36.

58. Ibid., p. 61.

59. Ibid., p. 43.

60. Ibid., p. 31–32.

61. Ibid., p. 37.

62. Ibid., p. 40.

63. Ibid.

64. Ibid., p. 34.

65. Ibid., p. 47.

66. Ibid., p. 49.

67. Ibid., p. 35.

68. Ibid., p. 45.

69. Ibid., p. 47.

70. Ibid., pp. 48–49.

71. Ibid., p. 35.

72. Ibid., pp. 62–63.

73. Ibid., p. 64.

74. Ibid., p. 65.

75. Ibid., p. 101.

76. Ibid., p. 111.

77. Ibid., p. 103.

78. Ibid., p. 95.

79. Ibid., p. 71.

80. Ibid., p. 74.

81. Ibid., p. 85.

82. Ibid.

83. Ibid., p. 110.

84. Ibid.

85. Ibid., pp. 121–122.

86. Ibid., p. 122.

87. Ibid., p. 123.

88. Ibid., p. 125.

89. Nathaniel C. Nash, "Argentine Files Show Huge Effort to Harbor Nazi," *New York Times,* December 14, 1993, p. A10.

90. Ibid.

91. Simon Wiesenthal, *The Murderers Among Us: The Simon Wiesenthal Memoirs* (New York: McGraw-Hill, 1967), p. 157.

92. Astor, *The Last Nazi,* p. 247.

93. Joyce and Stover, *Witnesses from the Grave,* p. 178.

94. Astor, *The Last Nazi,* p. 220.

95. Joyce and Stover, *Witnesses from the Grave,* p. 181.

96. Ibid., pp. 175–176.

97. Office of Special Investigations, *In the Matter of Josef Mengele,* p. 154, fn 250.

98. Ibid., p. 162.

99. Joyce and Stover, *Witnesses from the Grave,* p. 194.

100. Ibid., p. 195.

101. Office of Special Investigations, *In the Matter of Josef Mengele,* p. 178

102. Ibid., p. 179, fn 289.

103. Ibid., p. 181.

104. Ibid., Exhibit, p. 383.

105. Ibid., p. 185 and Exhibit p. 402.

106. Ibid., Exhibit p. 402.

107. Ibid., Exhibit pp. 419–420.

108. Ibid., p. 191.

109. Ralph Blumenthal, "U.S. Report on Mengele Reaffirms His Death," *New York Times,* October 9, 1992, p. A8.

110. Associated Press, "New Genetic Tests Said to Confirm: It's Martin Bormann," *New York Times,* May 4, 1998, p. A10; Bill Hoffman, "Science Solves Mystery of No. 2 Nazi Martin Bormann," *New York Post,* May 4, 1998, p. 6; Reuters, "Corpse of Hitler Aide Is Quietly Cremated," *New York Post,* August 29, 1999, p. 26; Tara FitzGerald, "Nazi Bormann's Family Welcome DNA Identification," Reuters Limited, May 5, 1998 (www.codoh.com/newsdesk/srnu199822.html).

CHAPTER 8

1. A. Edwards, A. Civitello, H. A. Hammond, and C. T. Caskey, "DNA Typing and Genetic Mapping with Trimeric and Tetrameric Tandem Repeats, *American Journal of Human Genetics,* vol. 49 (1991), pp. 746–756.

2. T. R. Moretti et al., "Validation of Short Tandem Repeats (STRs) for Forensic Usage: Performance Testing of Fluorescent Multiplex STR Systems and Analysis of Authentic and Simulated Forensic Samples," *Journal of Forensic Science,* vol. 46 (2000), pp.

647–660; Norah Rudin and Keith Inman, *An Introduction to Forensic DNA Analysis,* 2nd edition (Boca Raton, Fla.: CRC Press, 2001), pp. 50–51.

3. *Commonwealth of Massachusetts v. Adam Rosier,* 425 Mass. 807 (1997).

4. K. A. Micka et al., "TWGDAM Validation of a Nine-Locus and a Four-Locus Fluorescent STR Multiplex System," *Journal of Forensic Science,* vol. 44 (1999), pp. 1243–1256.

5. *Daubert v. Merrill Dow Pharmaceuticals, Inc.,* 509 U.S 579 (1993); *Commonwealth v. Lanigan,* 419 Mass. 15 (1994).

6. K. A. Micka et al., "Validation of Multiplex Polymorphic STR Amplification Sets Developed for Personal Identification Applications," *Journal of Forensic Science,* vol. 41 (1996), p. 582.

7. Rudin and Inman, *An Introduction to Forensic DNA Analysis,* 2nd edition, pp. 192–194.

8. R. C. Lewontin and D. L. Hartl, "Population Genetics in Forensic DNA Typing," *Science,* vol. 254 (1991), no. 5039, pp. 1745–1750.

9. R. Chakraborty and K. K. Kidd, "The Utility of DNA Typing in Forensic Work," *Science,* vol. 254 (1991), no. 5039, pp. 1735–1739.

10. FBI, "VNTR Population Data: A Worldwide Study," February 1993; Bruce Budowle et al., "The Assessment of Frequency Estimates of Hae III-Generated VNTR Profiles in Various Reference Databases," *Journal of Forensic Science,* vol. 39 (1994), p. 319; Budowle et al., "Evaluation of Hinf I-Generated VNTR Profile Frequencies Determined Using Various Ethnic Databases," *Journal of Forensic Science,* vol. 39 (1994), p. 988.

11. Eric Lander and Bruce Budowle, "DNA Fingerprinting Dispute Laid to Rest," *Nature,* vol. 371 (October 27, 1994), p. 735.

12. B. Budowle, R. Chakaborty, G. Carmody, and K. L. Monson, "Source Attribution of a Forensic DNA Profile," *Journal of Forensic Science Communication,* vol. 2 (2000), no. 3, www.fbi.gov/hq/lab/fsc/backissue/july2000/index; Rudin and Inman, *An Introduction to Forensic DNA Analysis,* 2d edition, p. 151.

13. Richard Willing, "As Police Rely More on DNA, States Take a Closer Look," *USA TODAY,* June 6, 2000, p. 1A.

14. Sacha Pfeiffer, "SJC Upholds Taking DNA from Convicts, Parolees," *Boston Globe,* April 14, 1999, p. B1.

15. Ibid., B5.

16. Rudin and Inman, *An Introduction to Forensic DNA Analysis,* 2d edition, pp. 157–158.

17. Richard Willing, "As Police Rely More on DNA, States Take a Closer Look," *USA TODAY,* June 6, 2002, p. 1A.

18. Helen Bennett Harvey and Tucker McMormack, "DNA Nabs Rape Suspect," *New Haven Register,* March 23, 1999, p. A1.

19. Willing, "As Police Rely More on DNA, States Take a Closer Look,", p. 2A.

20. Richard Willing, "DNA Testing Fails to Live Up to Potential," *USA TODAY,* October 7, 2002, p. 1A; Richard Willing, "Va: DNA Matching Passes 1,000 Milestone," *USA TODAY,* November 18, 2002, p. 3A.

21. Willing, "As Police Rely More on DNA, States Take a Closer Look," p. 1A; "States' DNA Database Laws and 'Qualifying Offenses'" (chart), *USA TODAY,* June 6, 2000, p. 6A.

22. Willing, "As Police Rely More on DNA, States Take a Closer Look,"

23. Richard Willing, "Collection of Prisoner DNA Widens," *USA TODAY*, May 2, 2002, p. 3A; AGID-LAB Working Group (Report), "Forensic DNA Databases and Casework: How Far Have We Come and Where Are We Going?" presented by Smith Alling Lane, P.S., Washington, D.C., March 4, 2002; Editorial, "Today's Debate: Using DNA to Fight Crime: Testing All Felons Is Money Well-Spent," *USA TODAY*, August 1, 2002, p. 10A.

24. Ibid.

25. Rudin and Inman, *An Introduction to Forensic DNA Analysis*, 2d edition, p. 163.

26. Hans C. Chen, "State Crime Labs Drowning in DNA Deluge," APBnews.com, February 18, 2000.

27. Amy Worden, "Backlog of DNA Samples Slows Down Justice," APBnews.com, March 24, 2000.

28. Proceedings, National Commission on the Future of DNA Evidence, April 9, 2000.

29. Rudin and Inman, *An Introduction to Forensic DNA Analysis*, 2d edition, p. 172.

30. Dr. Paul Ferrara, Proceedings, National Commission on the Future of DNA Evidence, April 9, 2000.

31. Hans H. Chen, "National DNA Database Solves Its First Crime," AP, APBnews.com, July 22, 1999.

32. David Doege, "Novel Warrant IDs Suspect Only by DNA; Databank Evidence Used to Charge 'John Doe' in Rape," *Milwaukee Journal Sentinel*, September 2, 1999, p. 1; Bill Dedman, "A Rape Defendant with No Identity, but a DNA Profile," *New York Times*, October 7, 1999, p. A1; Julian E. Barnes, "East Side Rapist, Known Solely by DNA, Is Indicted," *New York Times*, March 16, 2000, p. B1; Maida Cassandra Odom, "Philadelphia DA Targets Sex Abuse," *Boston Globe*, April 28, 2002, p. A9; Erin Hallissy and Charlie Goodyear, "Databank Match Brings Arrest on DNA Warrant; First Such Case Raises Civil Liberty Issues," *San Francisco Chronicle*, October 25, 2000, p. A3; Richard Willing, "Police Expand DNA Use; Charge Man with Rape Using Only Genetic Profile," *USA TODAY*, October 25, 2000, p. 1A; David Doege, "Rapist Identified by DNA Is Convicted; Defendant Is Innocent and Plans Appeal, Attorney Says," *Milwaukee Journal Sentinel*, February 7, 2002, p. 3B; David Doege, "Rapist in DNA Case Sentenced; Appeal Expected in Groundbreaking Prosecution," *Milwaukee Journal Sentinel*, February 21, 2002, p. 1B.

33. National Institute of Justice, National Commission on the Future of DNA Evidence (Report), *The Future of Forensic DNA Testing: Predictions of the Research and Development Working Group* (Washington, D.C.: U.S. Department of Justice, Office of Justice Programs, November 2000), p. 35.

34. Dana Hawkins, "Keeping Secrets: As DNA Banks Quietly Multiply, Who Is Guarding the Safe?" *U.S. News & World Report*, December 2, 2002, p. 58.

35. Michael E. Smith, David H. Kaye, and Edward J. Imwinkelried, "DNA Data from Everyone Would Combat Crime, Racism," *USA TODAY*, July 26, 2001, p. 15A.

36. National Institute of Justice, *The Future of Forensic DNA Testing*, p. 4.

37. Smith, Kaye, and Imwinkelried, "DNA Data from Everyone Would Combat Crime, Racism."

38. Robert Pear, "Bush Accepts Rules to Protect Privacy of Medical Records," *New York Times*, April 13, 2001, p. A1.

39. Darryl Van Duch, "EEOC Goes After Genetic Testing," *National Law Journal,* May 7, 2001, p. B1.

40. Associated Press, "Lab to Pay $2.2 Million to Settle Blood Test Case," *New York Times,* December 12, 1999, p. 47.

41. Darryl Van Duch, "EEOC Goes After Genetic Testing."

42. Hawkins, "Keeping Secrets: As DNA Banks Quietly Multiply, Who Is Guarding the Safe?"

43. Dr. Paul Keim, interview in "Planted Evidence," an episode of Court TV's *Forensic Files,* air date January 29, 2002.

44. "Planted Evidence," episode of Court TV's *Forensic Files,* air date January 29, 2002.

45. *State of Arizona v. Mark Alan Bogan (Court of Appeals of Arizona 1995),* 905 P. 2d 515; Keith Inman and Norah Rudin, *An Introduction to Forensic DNA Analysis,* 1st edition (Boca Raton, Fla.: CRC Press, 1997), pp. 134–136; Carol Kaesuk Yoon, "Botanical Witness for the Prosecution," *Science,* vol. 260 (May 14, 1993), pp. 894–895; Judge Gerald Sheindlin with Catherine Whitney, *Blood Trail* (New York: Ballantine, 1996), pp. 117–122.

46. Nicholas Wade, "For a Noble Grape, Disdained Parentage," *New York Times,* September 3, 1999, p. A16.

47. Michael Hagmann, "A Paternity Case for Wine Lovers," *Science,* vol. 385 (September 3, 1999), p. 1470; John Bowers et al., "Historical Genetics: The Parentage of Chardonnay, Gamay, and Other Wine Grapes of Northeastern France," *Science,* vol. 385 (September 3, 1999), p. 1562; Nicholas Wade, "Vintage Genetics Turns Out to Be Ordinaire," *New York Times, Science Times,* November 23, 1999, p. F1.

48. Frank J. Prial, "WINE TALK: A DNA Match Reveals Zinfandel's Parent," *New York Times,* September 11, 2002, p. F2; Rod Smith (Special to *The Los Angeles Times*), "Age-Old Mystery Solved: Zinfandel Grape Traced to Croatia," *Hartford Courant,* August 15, 2002, p. 61.

49. Stephen Meuse, "DNA Tests: Say Goodbye to the Grape Pretenders," *Boston Globe,* August 28, 2002, p. E4.

50. "Mouse's Genetic Code, Similar to a Human's, Is Sequenced by Group," *Wall Street Journal,* May 9, 2001, p. C7.

51. Erica Goode, "Building a Better Racehorse, from the Genome Up," *New York Times* (Science Times), May 8, 2001, p. F1.

52. Andrew Pollack, "Celera Will Join Rival in Rat Genome Project," *New York Times,* March 1, 2001, p. C10.

53. Dave Longtin and Duane Kraemer, "A Zoo for Endangered Species," *Wall Street Journal,* May 26, 1998, p. A18; Alexander Stille, "New Mission for DNA: Preservation," *New York Times,* February 12, 2000, p. B9; Kenneth Chang, "Creating a Modern Ark of Genetic Samples," *New York Times* (Science Times), May 8, 2001, p. F3.

54. Anthony Faiola, "A Whale of a Breeding Ground," *Washington Post,* reprinted in *Hartford Courant,* September 3, 1998, p. A23.

55. C. S. Baker and S. R. Palumbi, "Which Whales Are Hunted? A Molecular Genetic Approach to Monitoring Whaling," *Science,* vol. 265 (September 9, 1994), pp. 1538–1539; Associated Press, "Illegal Whale Meat Sales Seen," *New York Daily News,* May 12, 1998, p. 10.

56. Frank Cipriano and Stephen R. Palumbi, "Genetic Tracking of a Protected Whale" (Scientific Correspondence), *Nature*, vol. 397, January 28, 1999, pp. 307–308; Reuters (appearing online in *Yahoo! News*), "DNA Samples Link Raw Meat to Protected Whale," January 27, 1999.

57. Mindy Sink, "Genetic Pawprints Are Leading Game Wardens to the Poachers," *New York Times* (Science Times), May 26, 1998, p. F4.

58. Associated Press, "Bear's DNA Used to Convict Hunter," *Hartford Courant*, June 8, 2001, p. B9.

59. Jayne Clark, "High-Tech Tests Help Catch Park's Bad-News Bears," *USA TODAY*, May 4, 2001, p. 9D.

60. John J. Fialka, "Yosemite Bears Prefer Toyotas and Hondas for Late-Night Snacks," *Wall Street Journal*, January 13, 1999, p. A1; Clark, "High-Tech Tests Help Catch Park's Bad-News Bears," p. 9D.

61. Mark Derr, "The Tale of Three Bad News Bears Who Became Killers," *New York Times* (Science Times), August 18, 1998, p. F4.

62. Sarah Chung, "Caviar Becomes Endangered Delicacy," *New York Times*, December 25, 1996, p. C3; Patrick E. Tyler, "Poaching May Kill Fish That Lay the Golden Eggs," *New York Times*, September 24, 2000, p. 3; John Tagliabue, "From Spotted Owls to Caviar," *New York Times*, December 30, 2000, p. C1; Steve LeVine, "What? No Caviar? Caspian States Must Act, or Bye, Bye Beluga," *Wall Street Journal*, June 6, 2001, p. A1; Alexander G. Higgins, "Nations Agree to Reduce Overharvest of Caviar," *Boston Globe*, June 14, 2001, p. A15; Richard Willing, "Wildlife Rules Could Cut off Caviar," *USA TODAY*, December 2, 2002, p. 1A.

63. Rob DeSalle and Vadim Birstein, "PCR Identification of Black Caviar," *Nature*, vol. 381 (May 16, 1996), pp. 197–198; Jerry E. Bishop, "Something Fishy: Genetic Tests Find Caviar from Endangered Sturgeon," *Wall Street Journal*, May 17, 1996, p. B5C.

64. Tim Friend, "Netting Fishy Business in Caviar Trade," *USA TODAY*, March 31, 1998, p. 6D.

65. Molly O'Neill, "Major Importer Is Accused of Smuggling Caviar," *New York Times*, December 19, 1998, p. D1.

66. Andrew C. Revkin, "Stamford: Smuggling Term Lengthened," *New York Times*, January 10, 2002, p. B6.

67. Andrew C. Revkin, "Smuggling Case Brings Guilty Pleas," *New York Times*, July 22, 2000, p. A8; Traci Watson, "Feds: Caviar May Contain Rogue Roe," *USA TODAY*, December 31, 1999, p. 4A; Douglas Frantz, "An Inquiry Finds Roe Impersonating Caviar," *New York Times*, November 19, 1999, p. A22.

68. William Glaberson, "Distributor of Caviar Pleads Guilty in U.S. Court," *New York Times*, November 2, 2002, p. B3; Anthony M. DeStefano, "Fish Czar Reeled In," *Newsday*, November 3, 2002, p. A16; Kati Cornell Smith, "Bad Egg Admits Caviar Con," *New York Post*, November 2, 2002, p. 11.

69. Beth Daley, "Fur Flying in a Fight to Protect Backyard Birds," *Boston Globe*, May 12, 2001, p. A1.

70. M. Menotti-Raymond, V. A. David, J. C. Stephens, L. A. Lyons, and S. J. O'Brien, "Genetic Individualization of Domestic Cats Using Feline STR Loci for Forensic Applications," *Journal of Forensic Sciences,* vol. 42 (1997), pp. 1039–1051.

71. Marilyn A. Menotti-Raymond, Victor A. David, Stephen J. O'Brien, "Pet Cat Hair Implicates Murder Suspect," *Nature,* vol. 386 (April 24, 1997), p. 774; Adam Miller, "New Breed of Forensic Crime Fighting," *New York Post,* September 6, 2000, p. 24.

72. Mark Derr, "To Understand Humans, Geneticists Are Turning to Dogs," *New York Times* (Science Times), April 4, 2000, p. F3; Mark Derr, "It Takes Training and Genes to Make a Mean Dog Mean," *New York Times* (Science Times), February 6, 2001, p. F1; Dan Vergano, "Scientists Dogged by Question of Origin," *USA TODAY,* October 1, 2002, p. 9D.

73. Tim Klass (Associated Press), "Murder Trial Uses Dog DNA," ABCNEWS.com, July 10, 1998; Patrick McMahon, "Dog DNA Contributes to Murder Convictions," *USA TODAY,* September 18, 1998, p. 13A; Associated Press, "Dog DNA Leads to Long Prison Term," APBonline.com, December 9, 1998.

74. Richard Willing, "Prosecutors' Latest Tool: Animal DNA," *USA TODAY,* November 7, 2002, p. 24A; Barbara Whitaker, "Neighbor Guilty of Murder of Girl, 7, in San Diego," *New York Times,* August 22, 2002, p. A14; Valerie Alvord, "Neighbor Guilty in van Dam Killing," *USA TODAY,* August 22, 2002, p. 1A.

75. Steven Greenhouse, "DNA Said to Link a Friend to '75 Hoffa Disappearance," *New York Times,* September 8, 2001, p. A7; David Zeman, David Ashenfelter, and Jim Schaefer (Knight Ridder), "Hoffa Mystery Takes New Turn," *Hartford Courant,* September 8, 2001, p. A3; L. L. Brasier (Knight Ridder), "Report Dims Hope in Hoffa Case," *Boston Globe,* August 31, 2002, p. A5.

76. New York: G. P. Putnam's Sons, 2002.

77. Editorial, "The $6 Million Man," *New York Times,* November 13, 2002, p. A28; Dinitia Smith, "Handy with a Brush and Perhaps a Blade: Book Says a Painter Was Jack the Ripper," *New York Times,* November 11, 2002, p. E1; Richard Willing, "Cornwell Paints 'Portrait' of Jack the Ripper," *USA TODAY,* November 19, 2002, p. 4D; Patricia Cornwell (excerpts from *Portrait of a Killer: Jack the Ripper, Case Closed*), "The Face of Jack the Ripper," *Vanity Fair,* December 2002, p. 342.

78. Smith, "Handy with a Brush and Perhaps a Blade."

79. Ibid.

80. "On the Trail of a Killer," ABC-TV's *PrimeTime Thursday,* air date October 17, 2002; "Tracking the Zodiac," ABCNews.com, October 18, 2002; Scripps Howard, "DNA Clears '60s Zodiac Suspect," *New York Post,* October 16, 2002, p. 26; Fox Butterfield, "The Historical Context: Sniper Cases Prove Hardest for the Authorities to Solve," *New York Times,* October 19, 2002, p. A10.

81. New York: Doubleday, 1999.

82. David Van Biema, "Science and the Shroud," *Time,* April 20, 1998, p. 52; Marcus Baram, "Shroud of Turin Has Human DNA," *New York Daily News,* March 29, 1998, p. 38; Joe Nickell, "Science vs. 'Shroud Science,'" *Skeptical Inquirer,* vol. 22 (July/August 1998), no. 4, p. 20.

83. Daniel Woolls, "Columbus's Brother Disinterred," *Boston Globe*, September 18, 2002, p. A13.

84. David W. Chen, "Rehabilitating Thomas Paine, Bit by Bony Bit," *New York Times*, March 30, 2001, p. B1; Associated Press, "Paine's Admirers Want Home Burial," *New Haven Register*, April 1, 2001, p. A6.

85. Associated Press, "Paine's Admirers Want Home Burial."

86. Philip Weiss, "Beethoven's Hair Tells All!" *New York Times Magazine*, November 29, 1998, p. 108.

87. Ibid.

88. New York: Broadway Books, 2000.

89. www.randomhouse.com/features/beethovenshair/

90. Craig Tomashoff, "Wanted: Answer to Jesse James Mystery," *Boston Globe*, January 17, 2001, p. A1; Associated Press, "Remains Identified As Jesse James," *Hartford Courant*, September 23, 1995, p. A12.

91. Al Goodman (Arts Abroad column), "A Furor for Velazquez: His Art but Also His Bones," *New York Times*, September 7, 1999, p. E2.

92. Reuters, "France: Who Is Buried in Napoleon's Tomb?" *New York Times*, August 17, 2002, p. A6; Associated Press, "French Historian Seeks DNA Test on Remains Said to Be Napoleon's," *New Haven Register*, August 18, 2002, p. A8; Christine Ollivier (Associated Press), "Scientific Sleuths Ponder Clues to Napoleon murder," *Boston Herald*, May 5, 2000, p. 33; Danielle Arnet (Special report), "Napoleon Test Splits Hairs," *USA TODAY*, September 12, 1994, p. 1A; Associated Press, "Tests Fail to Settle Debate on Napoleon's Death," *Hartford Courant*, September 12, 1994, p. A8.

93. Barbara A. Nagy, "DNA from Bones Could Resolve Amelia Earhart Mystery," *Hartford Courant*, December 3, 1998, p. A14; Barbara A. Nagy, "Earhart Mystery May Linger," *Hartford Courant*, December 5, 1998, p. A6; David Morgan (Reuters), "Researchers to Present New Amelia Earhart Claim," *Yahoo! News*, December 4, 1998; Thomas F. King, "Amelia Earhart: Archaeology Joins the Search," *Discovering Archaeology*, January/February 1999, p. 40; Nelson Hernandez *(Washington Post)*, "Sea Sleuths to Look for Earhart Plane," *Hartford Courant*, April 6, 2001, p. B9; Renee Tawa *(Los Angeles Times)*, "New Theories Swirl Around Amelia Earhart's Disappearance," *Hartford Courant*, September 4, 2001, p. D2.

94. Suzanne Daley, "Genetics Offers Denouement to Mystery of Prince's Death," *New York Times*, April 20, 2000, p. A1; Tom Haines, "DNA Solves an Enduring Mystery of France's Revolution," *Boston Globe*, April 20, 2000, p. A2; Associated Press, "Science Cracks Mystery of Louis XVI's Son, *New Haven Register*, April 20, 2000, p. A17; John-Thor Dahlburg *(Los Angeles Times)*, "Dauphin Mystery Laid to Rest: DNA Tests of Boy's Heart Show 10-Year-Old Was French Heir," *Hartford Courant*, April 20, 2000, p. A9; Bernard Edinger (Reuters), "French to Test What Some Call a Royal Heart," *Boston Globe*, December 16, 1999, p. A2; Suzanne Daley, "Geneticists' Latest Probe: The Heart of the Dauphin," *New York Times*, December 21, 1999, p. A3.

CHAPTER 9

1. Lawrence Osborne, "Fuzzy Little Test Tubes," *New York Times Magazine,* July 30, 2000, p. 40.
2. Laura Jonah's, "Protein Clears Alzheimer's Plaque on Brains of Mice, Study Shows," *Wall Street Journal,* March 1, 2001, p. B11.
3. Richard Salts, "Toggling Nature's Auto-erase," *Boston Globe,* March 9, 2001, p. A4.
4. "Glow-in-the-Dark Mice Born," *Hartford Courant,* June 14, 1997, p. A2.
5. Michael D. Lemonick, "Smart Genes?" (cover story), *Time,* September 13, 1999, p. 54.
6. Robert Sapolsky, "Score One for Nature—or Is It Nurture?" *USA TODAY,* June 21, 2000, p. 17A.
7. "Lilly Eyes Antibodies from Altered Mouse," *Boston Globe,* November 11, 2000, p. C4.
8. "Mice in the Pink," *New York Daily News,* May 23, 1999, p. 16.
9. Nicholas Wade, "Scientists Develop a Mouse That Resists Some Cancers," *New York Times,* October 14, 1999, p. A22.
10. Richard Salts, "Of Laboratory Mice and Balding Men," *Boston Globe,* February 15, 2001, p. A1.
11. William McCall (Associated Press), "Scientists Create World's 1st Genetically Modified Primate," *Hartford Courant,* January 12, 2001, p. A19.
12. A. W. S. Chang et al., "Transgenic Monkeys Produced by Retroviral Gene Transfer into Mature Oocytes," *Science,* vol. 291 (January 12, 2001), pp. 309–312.
13. Associated Press, "Company Gets Glow from Sea Genes," *Sunday Boston Herald,* January 9, 2000, p. 34.
14. Gareth Cook, "Cross Hare: Hop and Glow," *Boston Globe,* September 17, 2000, p. A1.
15. Gina Kolata, "Monkey Born with Genetically Engineered Cells," *New York Times,* January 12, 2001, p. A1; Michael D. Lemonick, "Monkey Business," *Time,* January 22, 2001, p. 40; Chet Raymo, "Grappling with Moral Arithmetic," *Boston Globe,* January 30, 2001, p. C2.
16. George F. Will (columnist), "ANDi Is a Threat to Humanity Itself," *Hartford Courant,* January 22, 2001, p. A9.
17. Reuters, "Fertilization Procedure Helps Stop Sickle Cell," *Yahoo! News,* May 11, 1999; M. E. Malone, "A Very Early Checkup: Genetic Screening of Embryo Helps Ease Parents' Fears, but Is It a Step Toward 'Designer Babies'?" *Boston Globe,* December 11, 2001, p. F1.
18. Denise Grady, "Son Conceived to Provide Blood Cells for Daughter," *New York Times,* October 4, 2000, p. A24.
19. Andrew Stern (Reuters), "Ethics of Cell Transfer Debated," *Boston Globe,* October 4, 2000, p. A3.
20. Amy Docks-Marcus, "Ensuring Your Baby Will Be Healthy: Embryo Screening Test Gains in Popularity and Controversy; Choosing a Child's Gender," *Wall Street Journal,* July 25, 2002, p. D1.

21. David Lefer, "An Ad for Smart Eggs Spawns Ethics Uproar," *New York Daily News,* March 7, 1999, p. 38.

22. Kenneth R. Weiss, "Trafficking in Human Eggs: An Issue of Money, Ethics," *Los Angeles Times,* reprinted in *Hartford Courant,* June 8, 2001, p. A12.

23. See www.ronsangels.com.

24. Lenore Skenazy, columnist, "Designer Kids Are Not Such a Bad Concept," *New York Daily News,* June 28, 2000, p. 17.

25. Ad for *Gattaca, USA TODAY,* September 12, 1997, p. 6A.

26. Lee M. Silver, *Remaking Eden: Cloning and Beyond in a Brave New World* (New York: Avon Books, 1997), p. 11.

27. David R. Hamilton, "In Debate on Cloning Humans, Dr. Stock Is One of a Kind," *Wall Street Journal,* June 13, 2002, p. A1.

28. Francis Fukuyama, "A Milestone in the Conquest of Nature," *Wall Street Journal,* June 27, 2000, p. A30.

29. Gina Kolata, "Scientists Brace for Changes in Path of Human Evolution," *New York Times,* March 21, 1998, p. A1.

30. Sharon Begley, "Brave New Monkey," *Newsweek,* January 22, 2001, p. 50 at 52.

31. Ibid.

32. Kolata, "Scientists Brace for Changes in Path of Human Evolution," p. A1.

33. James Watson, "Fixing the Human Embryo Is the Next Step for Science," *Independent,* April 16, 2001 (www.independent.co.uk).

34. Stephen Hawking, "Crystal Ball Drops on a New Century—Revamped Humans," *New York Daily News,* December 29, 1999, p. 9.

35. CNN's "Larry King Live Weekend," Transcript, December 25, 1999, p. 13.

36. Carol Kaesuk Yoon, "If It Walks and Moos Like a Cow, It's a Pharmaceutical Factory," *New York Times,* May 1, 2000, p. A20.

37. Bill Hutchinson, "Supergoat Can Stop Bullets," *New York Daily News,* June 19, 2000, p. 3; Kenneth Chang, "Unraveling Silk's Secrets, One Spider Species at a Time," *New York Times* (Science Times), April 3, 2001, p. F4.

38. "More Testing Is Urged on Animals That Are Genetically Modified," *Wall street Journal,* May 9, 2001, p. C21; Jill Carroll, "Foes of Genetically Engineered Salmon Call for Close FDA Scrutiny of Risks," *Wall Street Journal,* May 10, 2001, p. A20.

39. David L. Chandler, "Greenpeace Protests Seafood Show," *Boston Globe,* March 8, 2001, p. B3.

40. Washington, D.C.: National Academies Press, 2002.

41. Justin Gillis, "Panel Identifies Gene-Altered Animals' Risk," *Washington Post,* August 21, 2002, p. A7; Warren E. Leary, "Panel Urges Caution in Producing Gene-Altered Animals," *New York Times,* August 21, 2002, p. A12; Jill Carroll and Antonio Regaldo, "Gene-Modified Animals Concern Scientists Group," *Wall Street Journal,* August 21, 2002, p. A4; Elizabeth Weise, "FDA Takes First Step Toward Approval: Study Examines Safety Concerns," *USA TODAY,* September 23, 2002, p. 6D.

42. David Barboza, "Suburban Genetics: Scientists Searching for a Perfect Lawn," *New York Times,* July 9, 2000, p. 1.

43. Scott Allen, "Revolution on the Farm: Tinkering with the DNA on Your Dinner Plate," *Boston Globe,* July 11, 1999, p. A1.

44. Karen Hsu, "The Future (of Food) Is Now," *Boston Globe,* July 12, 1999, p. A1.

45. "Harvest of Fear," *Frontline/Nova,* PBS Television, airing April 23, 2001, written, produced, and directed by Jon Palfreman, p. 7.

46. Christopher Marquis, "Ballot Initiatives: Education and Lottery Measures Prove Popular, but Health Care Fares Poorly," *New York Times,* November 7, 2002, p. B3.

47. Anthony Shadid, "The Seeds of Science," *Boston Globe,* March 7, 2001, p. D4.

48. Dean Kleckner, op-ed, "Who's More Moral in Biotech Crop War?" *New York Post,* April 7, 2001, p. 17.

49. "Making New Technologies Work for Human Development," 2001 Human Development Report by the United Nations Development Program, www.undp.org/hdr2001/; "U.N. Urges Research into Genetic Crops for Hungary Nations," *Wall Street Journal,* July 10, 2001, p. A12; Kevin McCoy, "U.N. Agency Backs Biotech Crops," *USA TODAY,* July 10, 2001, p. 6A.

50. Scott Kilman, "Use of Genetically Modified Seed by U.S. Farmers Increases 18%," *Wall Street Journal,* July 2, 2001, p. B2.

51. Kurt Eichenwald, "Biotechnology Food: From the Lab to a Debacle," *New York Times,* January 25, 2001, p. A1.

52. Brendan Mitchener, "Europe Has No Appetite for Modified Food," *Wall Street Journal,* September 23, 2002, p. B3.

53. Editorial, "Friends of ?????" *Wall Street Journal,* February 15, 2001, p. A26; Michael Cooper, "Wave of Eco-Terrorism Appears to Hit Experimental Cornfield," *New York Times,* July 21, 2000, p. B1.

54. Editorial, *Wall Street Journal,* May 2, 2000, p. A26.

55. Andrew Pollack, "Novartis Ended Use of Gene-Altered Foods," *New York Times,* August 4, 2000, p. C4.

56. Scott Kilman, "Food Industry Shuns Bioengineered Sugar," *Wall Street Journal,* April 27, 2001, p. B5.

57. David L. Chandler, "Curbs on Genetically Altered Crops Sought," *Boston Globe,* April 24, 2001, p. D5; Associated Press, "Albany: Genetic Modification Debated," *New York Times,* October 2, 2000, p. B6.

58. Carol Kaesuk Yoon, "Some Biotech Upstarts Fizzle Against Native Plants," *New York Times,* February 20, 2001, p. F2.

59. Anthony Shadid, "Blown Profits," *Boston Globe,* April 8, 2001, p. G1.

60. "Harvest of Fear," *Frontline/Nova.*

61. Shadid, "Blown Profits," p. G7.

62. James P. Miller, "Cargill Agrees to $100 Million Settlement with Pioneer over Genetic Seed Traits," *Wall Street Journal,* May 17, 2000, p. A3.

63. Andrew Pollack, "A Texas-Size Whodunit," *New York Times,* September 30, 2000, p. C1.

64. "Harvest of Fear," *Frontline/Nova,* p. 8.

65. Bruce Ingersoll, "FDA Broadens Test for Unapproved Corn," *Wall Street Journal,* October 3, 2000, p. B6.

66. Greg Winter, "Taco Bell's Core Customers Seem Undaunted by Shell Scare," *New York Times*, October 3, 2000, p. C6.

67. Scott Kilman, "Safeway Recalls Its Taco Shell Brand on Charge It Contains Unapproved Corn," *Wall Street Journal*, October 12, 2000, p. A8.

68. Greg Gatlin, "Shaw's Shucking Corn Taco Shells," *Boston Herald*, October 18, 2000, p. 40.

69. Philip Brasher (Associated Press), "Tortilla Manufacturer Recalls Products," *Hartford Courant*, October 14, 2000, p. A10.

70. Sarah Lueck, "Biotech-Corn Problems Lead to Recall of 300 Products, Disrupt Farm Belt," *Wall Street Journal*, November 2, 2000, p. A2.

71. Associated Press, "Modified Corn in Japan," *New York Times*, December 29, 2000, p. C2.

72. David Barboza, "Negligence Suit Is Filed Over Altered Corn," *New York Times*, December 4, 2000, p. C2.

73. Naomi Aoki, "Biotech's Warrior: Lawyer of Fen-Phen Fame Defending Maker of Engineered Corn," *Boston Globe*, October 16, 2002, p. C1.

74. David Barboza, "Caught in Headlights of the Biotech Debate," *New York Times*, October 11, 2000, p. C1; Holman W. Jenkins, Jr., "Eek! Attack of the Perfectly Harmless Tacos!" *Wall Street Journal*, October 4, 2000, p. A27.

75. Barnaby J. Feder, "Company Says Tracing Problem Corn May Take Weeks," *New York Times*, November 24, 2000, p. C2.

76. David Barboza, "Gene-Altered Corn Changes Dynamics of Grain Industry," *New York Times*, December 11, 2000, p. A1.

77. Reuters, "U.S. to Buy Back Contaminated Corn," *New York Times*, March 18, 2001, p. C9.

78. "Kellogg Orders Tests on Item Due to Traces in It of Modified Corn," *Wall Street Journal*, March 9, 2001, p. B6; "Corn Dogs Said to Contain Banned Corn," *New York Times*, March 9, 2001, p. C2.

79. Anthony Shadid, "EPA: Altered Animal Feed Must Pass Human Standard," *Boston Globe*, March 8, 2001, p. E3.

80. Anthony Shadid, "Bioengineered Corn More Prevalent Than Thought," *Boston Globe*, May 17, 2001, p. C2.

81. Marc Kaufman, "Engineered Corn Found in White Tortilla Chips," *Washington Post*, July 4, 2001, p. A6.

82. Vanessa Fuhrman, "Three Top Officials of Aventis Division in U.S. Are Fired," *Wall Street Journal*, February 12, 2001, p. A17; Mike Glover, "Accord Reached on Biotech Corn That Led to Recall," *Boston Globe*, January 24, 2001, p. A3.

83. David Barboza, "As Biotech Crops Multiply Consumers Get Little Choice," *New York Times*, June 10, 2001, p. A1.

84. Andrew Pollack, "Plan for Use of Bioengineered Corn in Food Is Disputed," *New York Times*, November 29, 2000, p. C4.

85. Mark Kaufman, "Biotech Corn in Test Case for Industry; Engineered Food's Future Hinges on Allergy Study," *Washington Post*, March 19, 2001, p. A1.

86. Andrew Pollack, "U.S. Finds No Allergies to Altered Corn," *New York Times,* June 14, 2001, p. C8.

87. Andrew Pollack, "No Altered Corn Found in Allergy Samples," *New York Times,* July 11, 2001, p. C8.

88. Scott Kilman, "Food, Biotech Industries Feud: Crops Bred to Produce Medicines Raise Contamination Worries," *Wall Street Journal,* November 5, 2002, p. B7; Elizabeth Weise, "FDA Addresses Drugs, Bioengineered Plants," *USA TODAY,* September 9, 2002, p. 6D.

89. Andrew Pollack, "U.S. Investigating Biotech Contamination Case," *New York Times,* November 13, 2002, p. C7; Scott Kilman, "ProdiGene-Modified Corn Plant Nearly Gets Into U.S. Food Supply," *Wall Street Journal,* November 13, 2002, p. A17; Bloomberg News, "Corn Near Gene-Altered Site to Be Destroyed," *New York Times,* November 14, 2002, p. C10; Elizabeth Weise, "Biotech Corn Nearly Escapes," *USA TODAY,* November 14, 2002, p. 10D; Scott Kilman, "Biotech Group Backs off Pledge on Genetically Modified Corn," *Wall Street Journal,* December 4, 2002, p. B3; Andrew Pollack, "Spread of Gene-Altered Pharmaceutical Corn Spurs $3 Million Fine," *New York Times,* December 12, 2002, p. A5; Bloomberg News, "Biotech Company Is Fined $3 Million," *New York Times,* December 7, 2002, p. C2.

90. Patricia Callahan and Scott Kilman, "Seeds of Doubt: Some Ingredients Are Genetically Modified, Despite Labels' Claims," *Wall Street Journal,* April 5, 2001, p. A1.

91. Elizabeth Becker, "Organic Gets an Additive: A U.S.D.A. Seal to Certify It; Green and White Label Makes Debut Today," *New York Times,* October 21, 2002, p. A10; David Hinckley, "Thoughts on Food: Organic, or Maybe Not; New Food-labeling Rules Leave a Lot of Leeway," *New York Daily News* (Lifeline section), November 3, 2002, p. 21.

92. Ibid.; Scott Kilman, "FDA Warns of Misleading Labels on Genetic Modification in Foods," *Wall Street Journal,* December 20, 2001, p. B9.

93. Thomas Hayden, "Bad Seeds in Court," *U.S. News & World Report,* January 28, 2002, p. 34; Jim Carroll, "Gene-Altered Canola Can Spread to Nearby Fields, Risking Lawsuits," *Wall Street Journal,* June 28, 2002, p. B6.

94. Andrew Pollack, "Delay Is Seen for Genetically Modified Wheat," *New York Times,* July 31, 2002, p. C4.

95. Gina Kolata, "Ethics Panel Recommends a Ban on Human Cloning," *New York Times,* June 8, 1997, p. 22.

96. Sheryl Gay Stolberg, "States Pursue Cloning Laws as Congress Debates," *New York Times,* May 26, 2002, p. A1.

97. Robert Davis, "19 Countries Ban Human Cloning," *USA TODAY,* January 13, 1998, p. 1D.

98. Thomas M. Burton, "Plan to Clone People Draws Skepticism Among Experts, Faces Major Hurdles," *Wall Street Journal,* January 8, 1998, p. B6.

99. Margaret Talbot, "A Desire to Duplicate," *New York Times Magazine,* February 4, 2001, pp. 40, 68.

100. www.clonaid.com

101. Ibid.

102. Bill Hoffman, "Cult Boasts: We'll Clone Couple's Dead Daughter," *New York Post*, October 12, 2000, p. 26.

103. Aaron Zitner, "2 Scientists Plan to Clone Humans," *Los Angeles Times* (reprinted in *Hartford Courant*), January 28, 2001, p. A9.

104. Post Wire Services, "Docs Vow Human Cloning," *New York Post*, August 8, 2001, p. 27.

105. Richard Salts, "Testimony Set on Cloning Dangers," *Boston Globe*, March 26, 2001, p. A3.

106. Miriam Falco and Matt Smith (CNN), "House Hearings Turn Skeptical Eye on Cloning," CNN.com/HEALTH, March 28, 2001.

107. Tim Friend, "Human Cloning Project May Have Begun," *USA TODAY*, March 29, 2001, p. 8D.

108. Editorial, "The Cloning Genie," *Boston Globe*, April 4, 2001, p. A20.

109. Jose B. Cibelli, Robert P. Lanza, and Michael D. West, "Rapid Communication: Somatic Cell Nuclear Transfer in Humans: Pronuclear and Early Embryonic Development," *e-biomed: The Journal of Regenerative Medicine*, vol. 12 (November 25, 2001), pp. 25–31 (www.liebertpub.com/ebi); Jose B. Cibelli, Robert P. Lanza, Michael D. West, and Carol Ezzell, "The First Human Cloned Embryo," *Scientific American*, January 2002, p. 44; Michael D. Lemonick, "Just Cloning Around: A Biological Baby Step Sparks a Political Backlash," *Time*, December 10, 2001, p. 75.

110. David P. Hamilton and Antonio Regaldo, "California School Attempted to Clone Human Embryos," *Wall Street Journal*, May 24, 2002, p. A3.

111. Miriam Falco, "Group Says It Will Move Human Cloning Work Offshore," CNN.com/HEALTH, June 29, 2001.

112. Laurie McGinley, "Britain Stirs Stem-Cell Research Debate," *Wall Street Journal*, January 24, 2001, p. B11.

113. Joe Lauria, "W. Va. Lawyer Sought to Develop Clone of Son," *Boston Globe*, August 6, 2001, p. A3.

114. Sheryl Gay Stolberg, "Despite Opposition, Three Vow to Pursue Cloning of Humans," *New York Times*, August 8, 2001, p. A1.

115. Reuters, "Scientists Skeptical About Claim of Cloning," *Los Angeles Times*, April 7, 2002, p. A9.

116. David Humphreys et al., "Abnormal Gene Expression in Cloned Mice Derived from Embryonic Stem Cells and Cumulus Cell Nuclei," *Proceedings of the National Academy of Sciences of the United States*, vol. 99 (October 1, 2002), no. 20, p. 12889.

117. Reuters, "Study Sees Faint Hope on Human Cloning: MIT Scientist Says Existing Methods Are 'Irresponsible,'" *Boston Globe*, September 12, 2002, p. A9.

118. Rick Weiss, "Cloned-Fetus Rumor Stirs Talk," *Washington Post*, April 6, 2002, p. A2; Raja Mishra, "A Human Clone Pregnancy Is Reported," *Boston Globe*, April 6, 2002, p. A6; Sinead O'Hanlon, "Clone Report Galls Scientists," *Boston Globe*, April 7, 2002, p. A5; Reuters, "Scientists Skeptical About Claim of Cloning," *Los Angeles Times*, April 7, 2002, p. A9; Dan Vergano, "Story of Human Cloning Attempt Is Unconfirmed," *USA TODAY*, April 10, 2002, p. 6A; Rick Weiss, "Free to Be Me: Would-Be Cloners Pushing the Debate," *Washington Post*, May 12, 2002, p. A1; Earl Lane, "A Cloned Baby by Next Year; Scientist

Gives House His Prediction," *Newsday,* May 16, 2002, p. A6; David Brown, "Human Clone's Birth Predicted; Delivery Outside U.S. May Come by 2003, Researcher Says," *Washington Post,* May 16, 2002, p. A8; Alan Richman, "Cloning Elvis," *GQ,* May 2002, p. 161; John Crewdson, "Gynecologist Claims Impending Births of 5 Cloned Human Babies," *Chicago Tribune,* June 23, 2002, p. 1; Raja Mishra, "Woman Carries Human Clone, Group Says," *Boston Globe,* July 25, 2002, p. A15; Reuters, "Italian Doctor Says Cloned Baby Due in January," *Yahoo! News,* November 27, 2002; Reuters, "Scientists Doubt Birth of Human Clone," *Yahoo! News,* November 28, 2002; Dana Canedy with Kenneth Chang, "Group Says Human Clone was Born to an American; Cloning Experts Skeptical," *New York Times,* December 28, 2002, p. A16.

119. Ian Wilmut, Keith Campbell, and Colin Tudge, *The Second Creation: Dolly and the Age of Biological Control* (Farrar, Straus, and Giroux, 2000). pp. 66–67.

120. Ibid. p. 68.

121. Ibid.

122. Ibid., pp. 69–73.

123. Ibid., pp. 74–77.

124. Gina Kolata, *Clone: The Road to Dolly and the Path Ahead* (New York: William Morrow, 1998), p. 66.

125. Ibid., pp. 67–69; Wilmut et al., *The Second Creation,* pp. 78–79.

126. Kolata, *Clone: The Road to Dolly* (citing, therein, the following references for Lederberg, Pauling, and Watson), pp. 72–83; Joshua Lederberg, "Unpredictable Variety Still Rules Human Reproduction," *Washington Post,* September 30, 1967, p. A17; Linus Pauling, "Reflections on the New Biology: Foreword," *UCLA Law Review,* vol. 15 (1967–1968), p. 267 at p. 269; James Watson, "Moving Toward Clonal Man," *Atlantic Monthly,* July 1971, p. 26.

127. Kolata *Clone: The Road to Dolly,* pp. 79, 90.

128. Chris Adams, "Bioethics Appointee Says He Is No Indoctrinator," *Wall Street Journal,* August 17, 2001, p. B1.

129. Leon R. Kass and Daniel Callahan, "Cloning's Big Test: Ban Stand," *New Republic,* August 6, 2001, p. 10 at 12.

130. Washington, D.C.: The President's Council on Bioethics, July 2002, available at: www.bioethics.gov/cloningreport/

131. Leon R. Kass (op-ed), "Stop All Cloning of Humans for Four Years," *Wall Street Journal,* July 11, 2002, p. A16.

132. Kolata, *Clone: The Road to Dolly,* Chapter 5, pp. 93–119.

133. Ibid., p. 121.

134. Wilmut et al., *The Second Creation,* p. 117.

135. Steen M. Willadsen, "Nuclear Transplantation in Sheep Embryos," *Nature,* vol. 320 (March 6, 1986), pp. 63–65.

136. Wilmut et al., *The Second Creation,* pp. 111, 129, 131–137.

137. Kolata, *Clone: The Road to Dolly,* pp. 139, 142.

138. Wilmut et al., *The Second Creation,* p. 141.

139. Kolata, *Clone: The Road to Dolly,* p. 186.

140. Wilmut et al., *The Second Creation*, pp. 146, 152–153; Kolata, *Clone: The Road to Dolly*, p. 199.

141. L. C. Smith and I. Wilmut, "Influence of Nuclear and Cytoplasmic Activity on the Development in Vivo of Sheep Embryos After Nuclear Transplantation," *Biology of Reproduction*, vol. 40 (1989), pp. 1027–1035.

142. Wilmut et al., *The Second Creation*, pp. 155, 163.

143. Ibid., pp. 87, 165–169.

144. Kolata, *Clone: The Road to Dolly*, pp. 204–205.

145. Wilmut et al., *The Second Creation*, p. 205.

146. K. H. S. Campbell, J. McWhir, W. A. Ritchie, and I. Wilmut, "Sheep Cloned by Nuclear Transfer from a Cultured Cell Line," *Nature*, vol. 380 (March 7, 1996), pp. 64–66.

147. Kolata, *Clone: The Road to Dolly*, p. 211.

148. Davor Solter, "Lambing by Nuclear Transfer," *Nature*, vol. 30 (March 7, 1996), pp. 24–25.

149. Ian Wilmut et al., "Viable Offspring Derived from Fetal and Adult Mammalian Cells," *Nature*, vol. 385 (February 27, 1997), pp. 810–813.

150. Gina Kolata, "Science Reports First Cloning of Adult Mammal; Researchers Astounded," *New York Times*, February 24, 1997, p. A1.

151. Wilmut et al., *The Second Creation*, p. 222.

152. Ibid., p. 216.

153. Ibid., p. 812.

154. Carey Goldberg and Gina Kolata, "Scientists Announce Births of Cows Cloned in New Way," *New York Times*, January 21, 1998, p. A14.

155. Angelika E. Schnieke et al., "Human Factor IX Transgenic Sheep Produced by Transfer of Nuclei from Transfected Fetal Fibroblasts," *Science*, vol. 278 (December 19, 1997), pp. 2130–2133.

156. Nicholas Wade, "With No Other 'Dollys,' Cloning Report Draws Critics," *New York Times*, January 30, 1998, p. A8.

157. Ibid.

158. Vittorio Sgaramella and Norton D. Zinder, "Dolly Confirmation," *Science*, vol. 279 (January 30, 1998), pp. 635–636.

159. Keith H. S. Campbell, Alan Colman, and Ian Wilmut, "Dolly Confirmation" (response) *Science*, vol. 279 (January 30, 1998), pp. 636–637.

160. J. Madeleine Nash, "Was Dolly a Mistake?" *Time*, March 2, 1998, p. 65.

161. Wilmut et al., *The Second Creation*, p. 224.

162. Davor Solter, "Dolly *Is* a Clone—and No Longer Alone," *Nature*, vol. 394 (July 23, 1998), p. 315.

163. David Ashworth, Matthew Bishop, Keith Campbell, Alan Colman, Alex Kind, Angelika Schnieke, Sarah Blott, Harry Griffin, Chris Haley, Jim McWhir, and Ian Wilmut, "DNA Microsatellite Analysis of Dolly," *Nature*, vol. 394 (July 23, 1998), p. 329.

164. Esther N. Signer, Yuri E. Dubrova, Alec J. Jeffreys, Colin Wilde, Lynn M.B. Finch, Michelle Wells, and Malcolm Peaker, "DNA Fingerprinting Dolly," *Nature*, vol. 394 (July 23, 1998), pp. 329–330.

165. Solter, "Dolly *Is* a Clone—and No Longer Alone," p. 315.

166. Gina Kolata, "In Big Advance in Cloning, Biologists Create 50 Mice," *New York Times,* July 23, 1998, p. 1; T. Wakayama, A. C. F. Perry, M. Zuccotti, K. R. Johnson, and R. Yanagimachi, "Full-Term Development Mice from Enucleated Oocytes Injected with Cumulus Cell Nuclei," *Nature,* vol. 394 (July 23, 1998), pp. 369–374.

167. Matthew J. Evans, Cagan Gurer, John D. Loike, Ian Wilmut, Angelika E. Schnieke, and Eric A. Schon, "Mitochondrial DNA Genotypes in Nuclear Transfer-Derived Cloned Sheep," *Nature Genetics,* vol. 23 (September 23, 1999), pp. 90–93.

CHAPTER 10

1. Gary Myers, "Nifty 50 Blue-Ribbon Panel Picks NFL's Best of the Century," *New York Daily News,* September 12, 1999, special section, p. 14.

2. *People of the State of California v. Orenthal James Simpson,* No. BA097211, Los Angeles Superior Court, Transcript, February 1, 1995 (hereafter Transcript).

3. Jeffrey Toobin, *The Run of His Life: The People v. O. J. Simpson* (New York: Touchstone, 1996), p. 81.

4. Ibid., pp. 104–111.

5. Gerald F. Uelmen, *Lessons from a Trial: The People v. O. J. Simpson* (Kansas City: Andrews and McMeel, 1996), pp. 115–116.

6. Toobin, *The Run of His Life,* p. 149.

7. Transcript, March 15, 1997; Toobin, *The Run of His Life,* pp. 325–326.

8. Ann V. Bollinger and Bill Hoffman, *New York Post,* "Goldman's Family Rips Ito for Playing Tapes," August 30, 1995.

9. Ibid.

10. M. L. Rantala, *O.J. Unmasked: The Trial, the Truth, and the Media* (Chicago: Carus Publishing, 1996), pp. 56–61.

11. Alan M. Dershowitz, *Reasonable Doubts: The Criminal Justice System and the O. J. Simpson Case* (New York: Simon & Schuster, 1996), pp. 78–79.

12. Uelmen, *Lessons from the Trial,* p. 122.

13. Toobin, *The Run of His Life,* p. 338.

14. David Margolick, "Simpson Prosecutors Get Back to the Bedrock Evidence," *New York Times,* April 18, 1995, p. A13.

15. John F. Kelly and Phillip K. Wearne, *Tainting Evidence: Inside the Scandals at the FBI Crime Lab* (New York: Free Press, 1998), p. 242.

16. Eric Lander and Bruce Budowle, "DNA Fingerprinting Dispute Laid to Rest," *Nature,* vol. 371 (October 27, 1994), p. 735.

17. Frank Schmalleger, *The Trial of the Century* (Englewood Cliffs, NJ: Prentice Hall, 1996), p. 160.

18. Transcript, August 2, 3, 4, 7.

19. Kary Mullis, *Dancing Naked in the Mind Field* (New York: Pantheon Books, 1998), pp. 56–58.

20. Bruce Weir, "DNA Statistics in the Simpson matter," *Nature Genetics,* vol. 11 (December 1995), p. 356 at 366.

21. Dershowitz, *Reasonable Doubts,* p. 75.

CHAPTER 11

1. "Simpson Trial Trivia," *U.S. News and World Report,* October 16, 1995, p. 42.

2. Ann Bollinger and Bill Hoffman, "Sleepy Ito Adds Some Zzzz's to DNA," *New York Post,* May 10, 1995, p. 20.

3. Gale Holland, "Curtain Rises on DNA 'Holy War,'" *USA TODAY,* May 9, 1995, p. 3A.

4. Gale Holland and Tom Bradford, "Defense Avoids DNA Showdown," *USA TODAY,* May 22, 1995, p. 3A.

5. Kenneth B. Noble, "DNA Debate in the Simpson Trial, Crucial to Case, Is About to Begin," *New York Times,* May 9, 1995, p. A1.

6. Gale Holland, "Curtain Rises on DNA 'Holy War.'"

7. *People of the State of California v. Orenthal James Simpson,* No. BA097211, Los Angeles Superior Court, Transcript, May 9, 1995 (hereafter Transcript).

8. Kenneth Noble, "Specialist on Testing of DNA Puts Simpson at Crime Scene," *New York Times,* May 11, 1995, p. A1.

9. Gale Holland, "DNA Tests Link OJ to Crime Scene," *USA TODAY,* May 11, 1995, p. 1A.

10. Sally Ann Stewart and Richard Price, "DNA Test Results 'Almost Anticlimactic,'" *USA TODAY,* May 11, 1995, p. 3A.

11. Richard Price, "DNA Evidence 'Powerful,'" *USA TODAY,* May 12, 1995, p. 1A.

12. Sally Ann Stewart, "Credibility 'Low' for DNA Tests," *USA TODAY,* September 23, 1994, p. 4A.

13. Gale Holland, "Simpson Team Attacks DNA Error 'Outbreak,'" *USA TODAY,* May 16, 1995, p. 3A.

14. AP, "Poll: Most Americans Think It Likely Simpson Is Guilty," reported in *Hartford Courant,* May 29, 1995, p. A2

15. Editorial, "The Power of DNA Evidence," *New York Times,* May 28, 1995, Section 4, p. 10.

16. Editorial, "Police Labs Under a Microscope," *New York Times,* October 16, 1995, p. A14.

17. AP story, "DNA Lab Wins Superior Rating," *New York Times,* October 7, 1994, p. A20.

18. John Marzulli, "New Lab Has High Standard," *New York Daily News,* August 16, 1998, p. 8.

19. Harlan Levy, *And the Blood Cried Out* (New York: Basic Books, 1996), p. 171.

20. Ibid., p. 174.

21. Transcript, May 17, 1995.

22. Gale Holland, "DNA Links Bloody Glove to Goldman," *USA TODAY,* May 17, 1995, p. 1A.

23. Kenneth B. Noble, "Prosecution Blundered in Using Gloves, Most Say," *New York Times,* June 17, 1995, p. 7.

24. Transcript, September 12, 1995.

25. Rantala, *O.J. Unmasked,* p. 10.

26. Bruce Weir, "DNA Statistics in the Simpson Matter," *Nature Genetics,* vol. 11 (December 1995), p. 365 at 368.

27. Keith Inman and Norah Rudin, *An Introduction to Forensic DNA Analysis* (Boca Raton, Fla.: CRC Press, 1997), pp. 117–120.

28. David Margolick, *New York Times,* September 19, 1995, p. A15.

29. David Margolick, "Expert Witness Says New DNA Tests Link Simpson to One Victim," *New York Times,* September 14, 1995, p. A20.

30. AP Story, "Simpson Prosecutors Describe New DNA Evidence, but Judge May Rule It Inadmissible," *New York Times,* August 5, 1995, p. 6.

31. Transcript, September 13, 1995.

32. Bill Hoffman, "DNA Test Puts Ron's Blood in O.J.'s Car," *New York Post,* September 14, 1995, p. 4.

33. Weir, "DNA Statistics in the Simpson Matter," p. 367.

34. Richard Price, "O.J.'s Socks May Be 'Most Crucial Item,'" *USA TODAY,* May 18, 1995, p. 3A.

35. Weir, "DNA Statistics in the Simpson Matter," p. 367.

36. Rantala, *O.J. Unmasked,* pp. 72–85.

37. Transcript, February 3, 1995, quoted in Alan M. Dershowitz, *Reasonable Doubts: The Criminal Justice System and the O.J. Simpson Case* (New York: Simon & Schuster, 1996), p. 75.

38. Lawrence Schiller and James Willwerth, *American Tragedy: The Uncensored Story of the Simpson Defense* (New York: Random House, 1996), p. 530.

39. Ibid., p. 529.

40. David Margolick, "FBI Disputes Simpson Defense on Tainted Blood," *New York Times,* July 26, 1995, p. A12; David Margolick, "Simpson Team Says Witness Turns Hostile," *New York Times,* July 27, 1995, p. A24.

41. Rantala, *O.J. Unmasked,* p. 134.

42. Schiller and Willwerth, *American Tragedy,* p. 530.

43. U.S. Department of Justice, Office of the Inspector General (OIG), *The FBI Laboratory: An Investigation into Laboratory Practices and Alleged Misconduct in Explosives-Related and Other Cases* (Washington, D.C.: OIG, 1997).

44. Transcript, Roger Martz testimony, July 25, 1995.

45. Ibid.

46. Ibid.

47. Schiller and Willwerth, *American Tragedy,* p. 531.

48. *ASCLD-LAB Manual* (January 1994), p. 19, with emphasis added by the OIG.

49. Ibid., pp. 30–31.

50. OIG, *The FBI Laboratory,* p. 213.

51. Ibid., p. 214.

388 / *Notes*

52. Ibid., pp. 213–14, footnote 148.
53. Ibid., p. 212.
54. Ibid., p. 215.
55. Ibid., p. 216.
56. Chat Transcript: Christopher Asplen on DNA Evidence, ABCNEWS.com, August 4, 1999.
57. Matt Lait, "Crime Lab at LAPD Wins Accreditation," *Los Angeles Times,* December 10, 1998, p. A12.
58. Editorial, "LAPD Crime Lab Rebounds," Los *Angeles Times,* December 11, 1998, p. A14.
59. *Roy Neal Shelling v. State of Texas,* No. 01-98-01048-CR (Tex. App.–Houston [1st Dist.] 2001), p. 13.
60. *Batson v. Kentucky,* 476 U.S. 79, 106 S.Ct. 1712 (1986).
61. *Roy Neal Shelling,* p. 11.
62. *Harris v. State,* 996 S.W. 2d 232 (Tex. App.–Houston [14th Dist.] 1999, no pet.); *Carroll v. State,* 701 So. 2d 47 (Ala. Crim. App. 1997); *Ridley v. State,* 510 S.E. 2d 113 (Ga. Ct. App. 1999).

CONCLUSION

1. John Heilemann, "Reinventing the Wheel: Here 'It' Is: The Inside Story of the Secret Invention That So Many Are Buzzing About. Could This Thing Really Change the World?" *Time,* December 10, 2001, p. 76.
2. Gareth Cook and Nicholas Thompson, "Ginger's Rough Ride," *Boston Globe,* February 5, 2002, p. C1; Derek Rose, "Future Rolls In," *New York Daily News,* December 3, 2001, p. 8.
3. Richard Willing, "Child DNA Kits Become Businesses' Favorite Freebie," *USA TODAY,* May 13, 2002, p. 1A.
4. Edward Rothstein, "The Unforeseen Disruption of Moving Ahead," *New York Times,* December 23, 2001, p. A19.
5. New York: Ballantine Books, 1993.
6. New York: HarperCollins, 1998.
7. Fred Bayles, "Skakel's Defense Attorneys Plan to Appeal Rulings from His Murder Trial," *USA TODAY,* June 10, 2002, p. 2A.
8. Kay Lazar, "Teen's Killer Wins Bid for DNA Test of Evidence That Convicted Him," *Boston Herald,* December 15, 2000, p. 30; Kay Lazar, "Mass. Man Says DNA Can Set Him Free," *Boston Herald,* April 18, 2000, p. 5; Kay Lazar, "DNA Case to Put N.H. Convictions to Test," *Boston Sunday Herald,* September 24, 2000, p. 1; Kay Lazar, "Jailed Man's Hopes Rest On New DNA Test," *Boston Herald,* March 28, 2001, p. 1; Kay Lazar, "DNA Test Inconclusive in Convict's Bid for Freedom," *Boston Herald,* May 9, 2001, p. 17; Kay Lazar, "BU DNA Expert Says N.H. Convict's Test Wasn't Done Right," *Boston Herald,* April 4, 2001, p.

19; Kay Lazar, "Convicted Murderer May Get Last Chance at DNA Redemption," *Boston Herald,* August 3, 2001, p. 12; Kay Lazar, "N.H. AG Seeks to Block DNA Test on Man Serving Life," *Boston Herald,* August 16, 2001, p. 17; Kay Lazar, "Prosecutors: DNA Test Links Murderer to 1971 Victim," *Boston Herald,* April 18, 2002, p. 26

9. Norah Rudin and Keith Inman, *An Introduction to Forensic DNA Analysis,* 2d edition (Boca Raton, Fla.: CRC Press, 2001), pp. 54–55.

10. Ibid., pp. 51–56.

11. Ibid., pp. 132–133.

12. National Institute of Justice, National Commission on the Future of DNA Evidence (Report), *The Future of Forensic DNA Testing: Predictions of the Research and Development Working Group* (Washington, D.C.: U.S. Department of Justice, Office of Justice Programs, November 2000).

13. D. A. Schmalzing et al., "DNA Sequencing on Microfabricated Electrophoretic Devices," *Analytic Chemistry,* vol. 70 (1998), pp. 2303–2310; D. A. Schmalzing et al., "Two-Color Multiplexed Analysis of Eight Short Tandem Repeat Loci With an Electrophoretic Microdevice," *Analytical Biochemistry,* vol. 270 (1999), pp. 148–152.

14. Eric Convey, "De-Tech-Tive Work: Whitehead Team Targets High-Speed DNA Test," *Boston Herald,* March 8, 1999, p. 28.

15. National Institute of Justice, *The Future of Forensic DNA Testing,* pp. 53–55; Gunjan Sinha, "DNA Detectives," *Popular Science,* August 1999, p. 48, www.popsci.com/context/features/dna_detect/

16. Patrick E. Tyler, "British Detail bin Laden's Link to U.S. Attacks; Release Document; 'Absolutely No Doubt' About Responsibility, Blair Is Declaring," *New York Times,* October 5, 2001, p. A1.

17. Eric Lipton, "Sept. 11 Death Toll Declines As 2 People Are Found Alive," *New York Times,* November 3, 2002, Section 1, p. 17; Joseph P. Fried, "Thousands of Remains Are Still Not Identified," *New York Times,* December 22, 2002, Section 1, p. 43.

18. Dan Barry, "At Morgue, Ceaselessly Sifting 9/11 Traces," *New York Times,* July 14, 2002, p. 1.

19. Fred Kaplan, "They Sift Tons to ID the Lost of Sept. 11," *Boston Globe,* January 15, 2002, p. A12.

20. Andrew Pollack, "Identifying the Dead, 2,000 Miles Away," *New York Times,* September 30, 2001, sec. 3, p. 6.

21. Kimberly Blanton, "Corporate Takeover," *Boston Globe Magazine,* February 24, 2002, p. 10.

22. Andrew Pollack, "Identifying the Dead, 2000 Miles Away," *New York Times,* September 30, 2001, sec. 3, p. 6.

23. Eric Lipton and James Glanz, "DNA Science Pushed to the Limit in Identifying the Dead of Sept. 11," *New York Times,* April 22, 2002, p. A1.

24. Barry, "At Morgue, Ceaselessly Sifting 9/11 Traces."

25. Scott Hensley, "Summit Called to Map Strategy on Victims' DNA," *Wall Street Journal,* October 3, 2001, p. B1.

26. Barry, "At Morgue, Ceaselessly Sifting 9/11 Traces."

27. Amy Waldman, "A Knock at the Door, with the Message of Death," *New York Times,* October 5, 2001, p. A1.

28. N. R. Kleinfield, "Error Puts Body of One Firefighter in Grave of a Firehouse Colleague," *New York Times,* November 28, 2001, p. A1.

29. Ibid.; Dareh Gregorian, "'Wrong Body' Anguish: Widow Sues to Verify ID of Fire Hero She Buried," *New York Post,* January 3, 2002, p. 7; Helen Peterson, "May Sue for Burial Mistake," *New York Daily News,* January 3, 2002, p. 5.

30. Lipton and Glanz, "DNA Science Pushed to the Limit."

31. Ibid.; Barry, "At Morgue, Ceaselessly Sifting 9/11 Traces"; Sara Kugler (Associated Press), "Office of Medical Examiner Still Hopes to Name More WTC Dead," *Boston Globe,* July 13, 2002, p. A8.

32. Emily Gest, "Docs to Seek More DNA," *New York Daily News,* January 10, 2002, p. 9.

33. David W. Chen, "Grim Scavenger Hunt for DNA Drags on for Sept. 11 Families," *New York Times,* February 9, 2002, p. A1.

34. Robert F. Worth, "DNA Matches Surge, Aiding in the Identification of Victims," *New York Times,* December 21, 2001, p. B6.

35. Barry, "At Morgue, Ceaselessly Sifting 9/11 Traces."

36. Lawrence K. Altman, "Now, Doctors Must Identify the Dead," *New York Times* (Science Times), September 25, 2001, p. F1.

37. Lipton and Glanz, "DNA Science Pushed to the Limit."

38. Ibid.; National Institute of Justice, *The Future of Forensic DNA Testing,* pp. 43–44; Scott Hensley, "New Test to ID Sept. 11's Dead," *Wall Street Journal,* August 29, 2002, p. B1.

39. Lynda Richardson, "The Forensic Challenge of a Lifetime, Regrettably," *New York Times,* May 8, 2002, p. B2.

40. Lipton, "Sept. 11 Death Toll Declines as 2 People Are Found Alive."

41. Associated Press, "Biometric Data Being Collected on Terror Suspects," *Hartford Courant,* October 30, 2002, p. A8.

42. Jonathan Weisman, "Intelligence Agencies Put Their Heads Together," *USA TODAY,* April 12, 2002, p. 11A; David Johnston and James Risen, "U.S. Plans to Seek Captives' DNA," *New York Times,* March 3, 2002, p. 1; Associated Press, "U.S. Readying Terror DNA Database," *New Haven Register,* March 3, 2002, p. A6; Bryan Bender, "US Gathers DNA from Al Qaeda Suspects, Sites," *Boston Globe,* May 14, 2002, p. A1.

43. Nicholas Kulish, "New Audiotape Shows bin Laden Might Be Alive," *Wall Street Journal,* November 13, 2002, p. A6; James Risen and Neil MacFarquhar, "New Recording May Be Threat from bin Laden," *New York Times,* November 13, 2002, p. A1; James Risen and Judith Miller, "As New Tape Is Evaluated, Bush Calls Qaeda Threat Real," *New York Times,* November 14, 2002, p. A1; Carlotta Gall, "Trail of Tape Linked to bin Laden Began on Street in Pakistan," *New York Times,* November 15, 2002, p. A18; James Risen, "Experts Conclude That Voice on Tape Belongs to bin Laden," *New York Times,* November 19, 2002, p. A22.

44. Richard Willing and Kevin, "Relatives' DNA May Help Prove Bin Laden Capture," *USA TODAY,* December 17, 2001, p. 2A; Elaine Shannon, "ID'ing a Corpse? Call in the DNA," *Time,* December 3, 2001, p. 17.

45. Jack Sullivan, "Biometrics Ferrets out the Faces of Terrorism," *Boston Sunday Herald*, December 9, 2001, p. 4.

46. Kathy Gannon (Associated Press), "Pearl Death: Four Guilty," *Hartford Courant*, July 15, 2002, p. A1; Kathy Gannon (Associated Press), "Man to Be Hanged for Pearl Slaying," *Boston Globe*, July 16, 2002, p. A8; Zarar Khan (Associated Press), "DNA Test Show Body Was Slain Reporter's," *Boston Globe*, July 20, 2002, p. A7; David Williams, "Kidnapperguy@hotmail.com: How the Killer of Daniel Pearl Spread His E-Mail of Terror," *New York Post*, July 21, 2002, p. 20; Ian Fisher, "DNA Tests Confirm Body in Karachi Is Reporter's," *New York Times*, July 19, 2002, p. A8; Associated Press, "Tests Show Body Found in Karachi Is Pearl's," *Wall Street Journal*, July 22, 2002, p. A13; Associated Press, "DNA Test Positively Identifies Pearl's Body," *New Haven Register*, July 28, 2002, p. A8; Zahid Hussain, Steve LeVine, Gary Fields, and Ian Johnson, "Pakistan Nabs Pearl Murder Suspect," *Wall Street Journal*, September 16, 2002, p. A3; Zahid Hussain, "Pakistani Businessman May Be the Key to Pearl Case," *Wall Street Journal*, October 2, 2002, p. A14.

47. Rebecca L. Cann, Mark Stoneking, Allan Wilson, "Mitochondrial DNA and Human Evolution," *Nature*, vol. 325 (January 1, 1987), p. 31.

48. Bryan Sykes, *The Seven Daughters of Eve* (New York: Norton, 2001), pp. 76–77.

49. Ibid., chap. 1; Oliva Handt et al., "Molecular Genetic Analyses of the Tyrolean Ice Man," *Science*, vol. 264 (June 17, 1994), p. 1775.

50. Sarah Lyall, "Cheddar Journal: Tracing Your Family Tree to Cheddar Man's Mum," *New York Times*, March 24, 1997, p. A4.

51. Svante Paabo, "Mutational Hot Spots in the Mitochondrial Microcosm," *American Journal of Human Genetics*, vol. 59 (1996), pp. 493–496.

52. Ibid.

53. Sykes, *The Seven Daughters of Eve*, p. 148; Martin Richards et al., "Paleolithic and Neolithic Lineages in the European Mitochondrial Gene Pool," *American Journal of Human Genetics*, vol. 59 (1996), pp. 185–203.

54. L. L. Cavalli-Sforza and E. Minch, "Paleolithic and Neolithic Lineages in the European Mitochondrial Gene Pool," (letter to the editor), *American Journal of Human Genetics*, vol. 61 (1997), pp. 247–251.

55. S. Anderson, "Sequence and Organization of the Human Mitochondrial Genome," *Nature*, vol. 290 (1981), pp. 457–465.

56. Sykes, *The Seven Daughters of Eve*, chap. 11.

57. Ibid.

58. E. Hagelberg et al., "Evidence for Mitochondrial DNA Recombination in a Human Population of Island Melanesia: Correction," *Proceedings of the Royal Society of London B*, vol. 267 (2000), pp. 1595–1596.

59. National Institute of Justice, *The Future of Forensic DNA Testing*, p. 72.

60. Vincent Macaulay, Martin Richards, and Bryan Sykes, "Mitochondrial DNA Recombination—No Need to Panic," *Proceedings of the Royal Society of London B*, vol. 266 (1999), pp. 2037–2039.

61. Adam Eyre-Walker, Noel H. Smith, and John Maynard Smith, "Reply to Macaulay *et al.* (1999): Mitochondrial DNA Recombination—Reasons to Panic," *Proceedings of the Royal Society of London B*, vol. 266 (1999), 2041–2042.

62. Philip Awadalla, Adam Eyre-Walker, John Maynard Smith, "Linkage Disequilibrium and Recombination in Hominid Mitochondrial DNA," *Science*, vol. 286 (December 24, 1999), pp. 2524–2525.

63. "Questioning Evidence for Recombination in Human Mitochondrial DNA," Toomas Kivisild and Richard Villems (Technical Comment); L. B. Jorde and M. Bamshad (Technical Comment); Sudhir Kumar, Philip Hedrick, Thomas Dowling and Mark Stoneking (Technical Comment); Thomas J. Parsons and Jodi A. Irwin (Technical Comment), *Science*, vol. 288 (June 16, 2000), p. 1931.

64. J. L. Elson et al., "Analysis of European mtDNAs for Recombination," *American Journal of Human Genetics*, vol. 68 (2001), p. 145.

65. Philip Awadalla, Adam Eyre-Walker, John Maynard Smith, "Questioning Evidence for Recombination in Human Mitochondrial DNA (Response)," *Science*, vol. 288 (June 16, 2000), p. 1931.

66. Marianne Schwartz, Ph.D., and John Vissing, M.D., Ph.D., "Paternal Inheritance of Mitochondrial DNA," *New England Journal of Medicine*, vol. 347 (August 22, 2002), no. 8, pp. 576–580.

67. R. Sanders Williams, M.D., "Another Surprise from the Mitrochondrial Genome," *New England Journal of Medicine*, vol. 347 (August 22, 2002), no. 8, pp. 609–612.

68. Sykes, *The Seven Daughters of Eve*.

69. Martin Richards et al., "Tracing European Founder Lineages in the Near Eastern mtDNA Pool," *American Journal of Human Genetics*, vol. 67 (November 2000), pp. 1251–1276.

70. Ornella Semino et al., "The Genetic Legacy of Paleolithic *Homo sapiens sapiens* in Extant Europeans: A Y Chromosome Perspective," *Science*, vol. 290 (November 10, 2000), p. 1155.

71. Sykes, *The Seven Daughters of Eve*, p. 195.

72. Ibid., chap. 13; Jonathan K. Pritchard et al., "Population Growth of Human Y Chromosomes: A Study of Y Chromosome Microsatellites," *Molecular Biology and Evolution*, vol. 16 (December 1999), no. 12, p. 1791.

73. Svante Paabo, "The Y Chromosome and the Origin of All of Us (Men)," *Science*, vol. 268 (May 26, 1995), p. 1141.

74. Ann Gibbons, "Europeans Trace Ancestry to Paleolithic People," *Science*, vol. 290 (November 10, 2000), p. 1080.

75. Paul Recer (Associated Press), "DNA Study Traces European Ancestors," *Yahoo! News*, November 9, 2000.

76. www.oxfordancestors.com.

77. Ibid.

78. Karl Skorecki et al., "Y Chromosomes of Jewish Priests" (Scientific Correspondence), *Nature*, vol. 385 (January 2, 1997), p. 32.

79. Mark G. Thomas et al., "Origins of Old Testament Priests," *Nature*, vol. 394 (July 9, 1998), p. 138; Jonathan Karp, "Seeking Lost Tribes of Israel in India, Using DNA Testing," *Wall Street Journal*, May 11, 1998, p. A1.

80. "The Lemba, the Black Jews of Southern Africa," www.pbs.org/wgbh/nova/israel/familylemba.html; "The Lost Tribes of Israel," *NOVA*, PBS Television, February 22, 2000; Nicholas Wade, "DNA Backs a Tribe's Tradition of Early Descent from the Jews," *New York Times*, May 9, 1999, p. A1.

81. Susan Saulny, "Genes and Genealogy Lead Back to Africa," *New York Times*, February 26, 2002, p. B1.

82. Tatsha Robertson, "Families Seek Link Across Centuries; DNA May Reunite Blacks, African Kin," *Boston Globe*, September 27, 2000, p. B3.

83. Carey Goldberg, "DNA Offers Link to Black History: Promise in Tracing African-American Ancestry Before Slavery," *New York Times*, August 28, 2000, p. A10.

84. Jefferson-Hemings Scholars Commission, *Final Report of the Scholars Commission on the Jefferson-Hemings Matter*, April 12, 2001.

85. New York: Norton, 1974.

86. Charlottesville: University Press of Virginia, 1997.

87. E. Foster et al., "Jefferson Fathered Slave's Last Child," *Nature*, vol. 396 (November 5, 1998), pp. 27–28.

88. Ibid.

89. Ibid.; Thomas Traut (Individual Views), "Does the DNA Analysis Establish Thomas Jefferson's Paternity of Sally Hemings' Children?" *Final Report of the Scholars Commission on the Jefferson-Hemings Matter*, p. 305.

90. Gwen Daye Richardson, "Never Mind Jefferson: We Deny Children, Too," *USA TODAY*, November 6, 1998, p. 15A.

91. Ellen Goodman, "The Real Jefferson-Hemings Issue: He Owned Her," *Hartford Courant*, November 6, 1998, p. A17.

92. Barbra Murray and Brian Duffy, "Jefferson's Secret Life," *U.S. News and World Report*, November 9, 1998, p. 58.

93. Dorothy Westerinen, "Jefferson Met Sally—The Rest Is My History," *New York Daily News*, November 10, 1998, p. 17.

94. Eric S. Lander and Joseph J. Ellis, "Founding father," *Nature*, vol. 396 (November 5, 1998), p. 13.

95. Ibid.; Robert Hardt, Jr., "B'klyn Pol: Jefferson Was a Perv," *New York Post*, December 20, 2001, p. 20.

96. "Jefferson's Blood," *Frontline*, PBS Television, aired May 2, 2000, Transcript, p. 2.

97. Nicholas Wade, "Defenders of Jefferson Renew Attack on DNA Data Linking Him to Slave Child," *New York Times*, January 7, 1999, p. A20; Dennis Cauchon, "Group Flags Younger Jefferson As Father," *USA TODAY*, January 7, 1999, p. 3A.

98. Eugene Foster (letter to the editor), "Tenable Conclusions," *New York Times*, November 9, 1998, p. A24; E. A. Foster et al., "The Thomas Jefferson Paternity Case," *Nature*, vol. 397 (January 7, 1999), p. 32.

99. Daniel P. Jordan, President, Thomas Jefferson Foundation, Inc., "Statement on the TJMF Research Committee Report on Thomas Jefferson Sally Hemings," *Report of the Research Committee on Thomas Jefferson and Sally Hemings*, Thomas Jefferson Foundation, January

2000 (available at www.monticello.org); Dennis Cauchon, "Foundation Agrees: Jefferson Probably Fathered Slave's Kids," *USA TODAY,* January 27, 2000, p. 1A.

100. Dr. Kenneth K. Kidd, Opinions of Scientists Consulted, *Report of the Research Committee on Thomas Jefferson and Sally Hemings.*

101. *Final Report of the Scholars Commission on the Jefferson-Hemings Matter,* Introduction.

102. Ibid.

103. Ibid.; Robert F. Turner, "The Truth About Jefferson," *Wall Street Journal,* July 3, 2001, p. A14.

104. Leef Smith, "Jefferson Group Bars Kin of Slave," *Washington Post,* May 6, 2002, p. B1.

105. Karen Gullo (Associated Press), "JFK Findings Support Single-Shooter Theory," *Boston Globe,* January 22, 2000, p. A3; Paul G. Labadie, "Advances in Testing Threaten History's Mystique," *USA TODAY,* August 28, 2000, p. 15A.

106. Bettye Kearse, "Our Family Tree Looks for Branches" (op-ed), *Boston Herald,* November 27, 1998, p. 47.

107. Richard Norton Smith, "Founding Fathers in the Dock," *Wall Street Journal,* February 22, 2000, p. A40; Celia McGee, "Getting at George," *New York Daily News,* February 1, 2000, p. 18; Christopher Thorne (Associated Press), "Family Eyes Gene Test of Washington," *Yahoo! News,* November 10, 1998; "Red, White and Blue," *New York Daily News,* July 11, 1996, p. 12.

108. Philip R. Reilly, *Abraham Lincoln's DNA and Other Adventures in Genetics* (Cold Spring Harbor, NY: Cold Spring Harbor Laboratory Press, 2000), chap. 1; Paul G. Labadie, "Advances in Testing Threaten History's Mystique"; Jane E. Brody, "The Lifeline with Marfan's: Its Diagnosis," *New York Times* (Science Times), October 16, 2001, p. F6; Peter Gott, M.D. (Our Health—column), "Marfan's Syndrome Can Affect Aorta," *New Haven Register,* November 27, 1999, p. A10; Richard Saltus, "The Terrorist and the President," *Boston Globe Magazine,* August 22, 1999, p. 4.

109. New York: Farrar, Straus, and Giroux, 2002.

110. Boston: Houghton Mifflin, 2002.

111. Gina Maranto, "Deoxyribonucleic Acid Trip" (book review), *New York Times Book Review,* August 25, 2002, p. 25; David R. Hamilton, "In Debate on Cloning Humans, Dr. Stock Is One of a Kind," *Wall Street Journal,* June 13, 2002, p. A1.

112. Denise Grady, "Doctors Using Hybrid Egg to Tackle Infertility in Older Women," *New York Times,* October 10, 1998, p. A16; Rick Weiss, "Experiments in Human Fertility Move Genes from Egg to Egg," *Washington Post* (reprinted in *Hartford Courant*), October 10, 1998, p. A10.

113. Nell Boyce, "A Mother's Legacy," *U.S. News & World Report,* April 9, 2001, p. 52.

114. Elizabeth Cohen, CNN Correspondent, "FDA Holding Hearing on Controversial Fertility Treatment," *CNN Live Today* (Transcript), May 9, 2002; Leila Abboud, "FDA Interrupts Some Fertility Treatments," *Wall Street Journal,* October 7, 2002, p. A4.

115. Jason A. Barritt et al., "Mitochondria in Human Offspring Derived from Ooplasmic Transplantation," *Human Reproduction,* vol. 16 (March 2001), no. 3, pp. 513–516.

116. Erik Parens and Eric Juengst, "Inadvertently Crossing the Germ Line," *Science,* vol. 292 (April 20, 2001), no. 5516, p. 397; Prof. Joe Cummins and Dr. Mae-Wan Ho, "First GM Humans Already Created," *Institute of Science Society (ISIS) Report*, May 2, 2001.

117. Aaron Zitner, "Cloning: A Cat with More Than 9 Lives?" *Los Angeles Times* (reprinted in *Hartford Courant*), March 30, 2001, p. B7.

118. www.savingsandclone.com; Jeff Franks, "Baby Bull Is Created from World's Oldest Cloned Animal," *Boston Globe,* September 3, 1999, p. A11.

119. Taeyoung Shin et al., "Cell Biology: A Cat Cloned by Nuclear Transplantation," *Nature,* February 21, 2002, p. 859; Antonio Regalado, "Only Nine Lives for Kitty? Not If She Is Cloned," *Wall Street Journal,* February 14, 2002, p. B1; Gina Kolata, "What Is Warm and Fuzzy Forever? With Cloning, Kitty," *New York Times,* February 15, 2002, p. A1; Nell Boyce, "Pets of the Future," *U.S. News & World Report,* March 11, 2002, p. 46; Paul Elias (Associated Press), "Cat Cloning Raises Hopes for Some; Others Predict a Hairball," *Hartford Courant,* February 16, 2002, p. E1; Owen Moritz, "Clones Called Purr-fect," *New York Daily News,* February 16, 2002, p. 4; Susan Ferraro, "Me-wow! Cat Cloned," *New York Daily News,* February 15, 2002, p. 4; William Hathaway, "Frisky Feline Times Two," *Hartford Courant*, February 15, 2002, p. A3.

120. David Brown, "A Big Hop Forward: Rabbits Cloned," *Washington Post,* March 30, 2002, p. A1; Knight Ridder, "Rabbit Becomes 7th Species to Be Cloned," *New Haven Register,* March 31, 2002, p. A12.

121. Gretchen Vogel, "Cloned Gaur a Short-Lived Success," *Science,* vol. 291 (January 19, 2001), p. 409.

122. University of Connecticut News Release by David Bauman, Office of University Communications, "Scientists Use Long-term Cultured Cells for Cloning," January 3, 2000, www.news.uconn.edu.

123. Adam Boyles, "This Is Not Your Father's Agricultural School," *New York Times,* September 3, 2000, Sec. 14, p. 1; Stacy Wong, "A Pet Project at UConn: Gene Researchers May Try to Produce an Allergen-Free Cat," *Hartford Courant,* June 28, 2001, p. A1; Andrew Pollack, "Entrepreneur Envisions a Cat That Doesn't Cause Allergies," *New York Times,* June 27, 2001, p. C1; William Hathaway, "Scientist Undaunted by Setbacks for Cloning Research," *Hartford Courant*, August 1, 2001, p. A8.

124. Christopher Joyce and Eric Stover, *Witnesses for the Grave: The Stories Bones Tell* (New York: Ballantine, 1992), p. 211.

125. Ibid., p. 227.

126. Dragan Primorac et al., "Identification of War Victims from Mass Graves in Croatia, Bosnia, and Herzegovina by the Use of Standard Forensic Methods and DNA Typing," *Journal of Forensic Sciences,* vol. 41 (September 1996), no. 5, pp. 891–894.

127. David J. Lynch, "Solid American Presence in War Crimes Trial," *USA TODAY,* July 28, 2000, p. 14A.

128. David J. Lynch, "'One of the Worst Places on Earth," *USA TODAY,* November 17, 2000, p. 19A; David Rohde, "Warehouse of Death," *New York Times Magazine,* March 11, 2001, p. 46; Daniel Simpson, "Bosnia: Identifying War Victims," *New York Times,* November 16, 2002, p. A5.

129. Sebastian Junger, "The Forensics of War," *Vanity Fair,* October 1999, p. 138; Joshua Hammer, "On the Trail of the Hard Truth," *Newsweek,* July 17, 2000, p. 34; Reuters, "Serbia Exhumes 269 Bodies from Mass Graves," *New York Times,* September 19, 2001, p. A5; Associated Press, "Kosovo Forensic Work 'Very Emotional,'" *Hartford Courant,* September 17, 1999, p. A25.

130. Editorial, "US Cowardice in Rwanda," *Boston Globe,* August 24, 2001, p. A24.

131. Samantha Power, "Bystanders to Genocide," *Atlantic Monthly,* September 2001, p. 84.

132. Ibid., p. 101.

133. Ibid., p. 96; Neil A. Lewis, "Papers Show U.S. Knew of Genocide in Rwanda," *New York Times,* August 22, 2001, p. A5.

134. Richard A. Knox, "Doctors Work to Document Horror—And Prevent It," *Boston Globe,* May 6, 1996, Health and Science section, p. 27.

135. Ibid.

136. Peter Landesman, "A Woman's Work," *New York Times Magazine,* September 15, 2002, pp. 82, 116.

137. Ibid.

138. Marlise Simons, "3 Serbs Convicted in Wartime Rapes," *New York Times,* February 24, 2001, p. A1; Editorial, "A Landmark Ruling on Rape," *New York Times,* February 24, 2001, p. A12; Kevin Cullen, "UN Court Establishes Rape As a War Crime," *Boston Globe,* February 23, 2001, p. A1; Marlise Simons, "An Ex-Bosnian Serb Commander Admits Rape of Muslims in War," *New York Times,* March 10, 1998, p. A10; Janet McBride (Reuters), "U.N. Court Jails Bosnian Croat In Rape Case," *Yahoo! News,* December 10, 1998; Bill Berkeley, "Conspiring Against Humanity" (book review), *New York Times,* January 13, 2002, Section 7, p. 23.

139. Barbara Crossette, "War Crimes Tribunal Becomes Reality, Without U.S. Role," *New York Times,* April 12, 2002, p. A3; Evelyn Leopold (Reuters), "Ratifications Launch Permanent International Criminal Court," *Boston Globe,* April 11, 2002, p. A9; Neil A. Lewis, "U.S. Rejects All Support for New Court on Atrocities," *New York Times,* May 7, 2002, p. A11.

140. C. Ciesliski et al., "Transmission of Human Immunodeficiency Virus in a Dental Practice," *Annals of Internal Medicine,* vol. 116 (1995), pp. 798–805; C.Y. Ou et al., "Molecular Epidemiology of HIV Transmission in a Dental Practice, *Science,* vol. 256 (1992), pp. 1165–1171.

141. J. Albert et al., "Forensic Evidence by DNA Sequencing," *Nature,* vol. 361 (February 18, 1993), p. 595.

142. Gretchen Vogel, "HIV Strain Analysis Debuts in Murder Trial," *Science,* vol. 282 (October 30, 1998), p. 851.

143. Pamela Coyle, "Doctor's Ex-Lover Claims Syringe Had HIV," *Times-Picayune,* October 11, 1998.

144. Vogel, "HIV Strain Analysis Debuts in Murder Trial."

145. Ibid.; *Ruling on the Admissibility of DNA Profiling Evidence, State of Louisiana v. Richard J. Schmidt,* Docket No. 96CR73313, 15th Judicial District Court Div. "G.,"

Lafayette Parish, Louisiana, Judge Durwood Congae presiding, January 26, 1997; Leslie S. King, "Tracking the Chameleon Killer," *Santa Fe Bulletin,* vol. 14, no. 1, online at: www.santafe.edu/sfi/publications/Bulletins/bulletin_wint.../tracking.htm.

146. Nicholas Wade, "DNA Map For Bacterium Of Plague Is Decoded," *The New York Times,* October 4, 2001, page?; Marilyn Chase, "Researchers in Britain Have Determined The Genetic Sequence of Bubonic Plague," *The Wall Street Journal,* October 4, 2001, B2; Joseph B. Verrengia (AP), "Black Death microbe was product of rapid evolution," *USA TODAY,* October 4, 2001, 10D; Adair, D.M., et al., "Diversity in a variable-number tandem repeat from *Yersinia pestis,*" *Journal of Clinical Microbiology,* vol. 38, no. 4, April 2000, pp. 1516-19; Klevystska, A. M., et al., "Identification and characterization of variable-number tandem repeats in the *Yersinia pestis* genome," Journal of Clinical Microbiology, vol. 39, no. 9, September 2001, pp. 3179-85.

147. Paul Keim, et al., "Molecular investigation of the Aum Shinrikyo anthrax release in Kameido, Japan," *Journal of Clinical Microbiology,* vol. 39, No. 12, December 2001, pp. 4566-7; Antonio Regalado and Marilyn Chase, "DNA tests Tie New York and Florida Anthrax Strains, a Vital Clue," *The Wall Street Journal,* October 18, 2001, A8; Tim Larimer, "Why Japan's Terror Cult Still Has Appeal," *Time,* June 10, 2002, p. 8.

148. Paul J. Jackson, et al., "PCR analysis of tissue samples from the 1979 Sverdlovsk anthrax victims: The presence of multiple *Bacillus anthracis* strains in different victims," *Proceedings of the National Academy of Sciences USA,* vol. 95, issue 3, February 3, 1998, pp. 1224-1229; James H. Steele, "Anthrax: The investigation of a deadly outbreak" (book review), *New England Journal of Medicine,* vol. 342, no. 18, May 4, 2000, p. 1373; Jeanne Guillemin (correspondence), "Anthax: The investigation of a deadly outbreak," *New England Journal of Medicine,* vol343, no. 16, October 19, 2000, p. 1198; F.A. Abramova, et al., "Pathology of Inhalational Anthrax in 42 Cases from the Sverdlovsk Outbreak of 1979," (1993) *Proceedings of the National Academy of Sciences USA,* vol. 90, pp. 2291-2294.

149. Andrew C. Revkin, "TRACING THE SPORES: testing Links Anthrax In Florida and at NBC," *The New York Times,* October 18, 2001, B5; Paul Keim, et al., "Multiple-locus variable-number tandem repeat analysis reveals genetic relationships within *Bacillus anthracis,*" *Journal of Bacteriology,* vol. 182, no. 10, May 2000, pp. 2928-36; Paul Keim, et al., "Molecular Evolution and Diversity in *Bacillus anthracis* as detected by AFLP markers," *Journal of Bacteriology* (1997), vol. 179, pp. 818-824.

150. Bob Port, "Anthrax probers follow DNA: Far from deaths, 2 profs finally get to use expertise," *New York Daily News,* November 25, 2001, p. 10; [ADD CITE RE TIGR AND PORTON].

151. Paul Keim, et al., "Multiple-locus variable-number tandem repeat analysis reveals genetic relationships within *Bacillus anthracis,*" Journal of Bacteriology, vol. 182, no. 10, May 2000, pp. 2928-36; "Chronology of anthrax events," *South Florida Sun-Sentinel,* available at www.sun-sentinel.com; Ed Lake, "The Anthrax Cases," available at www.anthraxinvestigation.com; Barbara Hatch Rosenberg, Federation of American Scientists, "Analysis of the Anthrax Attacks," available at www.fas.org; Antonio Regalado, Chad Terhune and Scott Hensley, "Scientists Try to Identify Anthrax Bacteria Strain in Florida," *Wall Street Journal,* October 12, 2001, p. A6; Centers for Disease Control and

Prevention, "Update: Investigation of Anthrax Associated with Intentional Exposure and Interim Public Health Guidelines, October 2001," *Morbidity and Mortality Weekly Report*, 50 (41), October 19, 2001, pp. 889-893; Antonio Regalado and Marilyn Chase, "DNA Tests Tie New York and Florida Anthrax Strains, a Vital Clue," *Wall Street Journal*, October 18, 2001, p. A8; Andrew C. Revkin, "Tracing the Spores: Testing Links Anthrax In Florida and at NBC," *New York Times*, October 18, 2001, p. B5; Kevin Johnson and Laura Parker, "3 anthrax cases linked: N.Y., Fla., D.C. incidents point to a coordinated attack," *USA TODAY*, October 18, 2001, p. 1A; Rick Weiss and Dan Eggen, "U.S. Says Anthrax Germ In Mail Is 'Ames' Strain," *Washington Post*, October 26, 2001, p. A8; Stefan Fatsis and Gary Fields, "Familiar Anthrax Strain Is Tied to Death: DNA Tests on New Yorker Match Other Infections; Spores Keep Spreading," *Wall Street Journal*, November 2, 2001, p. A9; Jennifer Steinhauer, "Familiar Anthrax Strain Is Seen in Woman's Death," *New York Times*, November 2, 2001, p B1; Paul Zielbauer, "The Latest Casualty: An Exhaustive Search for Clues, Down to an Anthrax Victim's Hairpins," *New York Times*, November 23, 2001, p. B6; Dave Altimari and William Hathaway, "Tests Point to Ames Strain: Virulent Anthrax Type Suspected In Oxford," *Hartford Courant*, November 27, 2001, p. A1; Joanne M. Pelton and Tara Stapleton, "Anthrax Test Negative," *Middletown Press*, November 30, 2001, p. A1; Marianne Lippard and Jean Falbo-Sosnovich, "Oxford anthrax linked to other cases," *New Haven Register*, November 30, 2001, p. A1.

152. Dave Altimari, "Anthrax Hits, Misses Traced In CDC Study," *Hartford Courant*, September 18, 2002, p. A1.

153. Steve Fainaru and Joby Warrick (*Washington Post*), "Tracking Anthrax A Formidable Task: Investigation Focuses On Ames Strain," *Hartford Courant*, November 25, 2001, p. A11; Tom Hamburger, Gary Fields and Sarah Lueck, "Anthrax Tests Showing Gene Link Advance Domestic-Terror Theory," *Wall Street Journal*, December 17, 2001, p. B7; Corky Siemaszko, "D.C. anthrax spores match those at Army lab," *New York Daily News*, December 17, 2001, p. 6; Associated Press, "Experts: Anthrax 'match' may not lead to source," *Boston Herald*, December 17, 2001, p. 7; Joby Warrick, "One Anthrax Answer: Ames Strain Not From Iowa," *Washington Post*, January 29, 2002, p. A2; Antonio Regalado and Gary Fields, "New DNA Tests May Find Anthrax Source," *Wall Street Journal*, December 18, 2001, p. A1; William J. Broad, "Genome Offers 'Fingerprint' For Anthrax," *New York Times*, November 28, 2001, p. B1; Antonio Regalado, "The Scientists Probing Terror Anthrax Trace Microbes's Family Tree," *Wall Street Journal*, November 8, 2001, p. A1; Rick Weiss and Susan Schmidt, "Capitol Hill Anthrax Matches Army's Stocks," *Washington Post*, December 16, 2001, p. A1; Laura Parker, "Anthrax probably domestic: Investigators focus on U.S. laboratories," *USA TODAY*, December 18, 2001, p. 1A; William J. Broad, "The Disease: Geographic Gaffe Misguides Anthrax Inquiry," *New York Times*, January 30, 2002, p. A11.

154. Mark Schoofs, Gary Fields and Maureen Tkacik, "Crime Scenes: Anthrax Probe Scours Tips, Trash, Tabletops In Far-Ranging Quest: Screening Labs Serves Only To Expand Possibilities; The Experts Don't Agree," *Wall Street Journal*, December 11, 2001, p. A1; Chad Terhune, "Tests Show Anthrax Spores Don't Stay Put," *Wall Street Journal*, December 11, 2001, B7.

155. Gary Field and Antonio Regalado, "Anthrax Probe Centers on Labs' Staffers," *Wall Street Journal*, January 18, 2002, p. B2.

156. Mark Schoofs and Gary Fields, "Anthrax Search Plans to Target Key Laboratories," *Wall Street Journal*, February 7, 2002, p. B1.

157. Jack Dolan and Dave Altimari, "Anthrax Missing From Army Lab," *Hartford Courant*, January 20, 2002, A1; Michael D. Lemonick, "Anthrax: The Hunt Narrows," *Time*, February 4, 2002, p. 46; Jack Dolan, Dave Altimari and Lynne Tuohy, "Anthrax Easy To Get Out of Lab," *Hartford Courant*, December 20, 2001, p. A3; Jack Dolan, "Army: Most Lab Samples Tracked; Anthrax Specimen Still Among Missing," *Hartford Courant*, February 1, 2002, A3.

158. William J. Broad, "The Anthrax Trail: Labs Are Sent Subpoenas For Samples Of Anthrax," *New York Times*, February 27, 2002, p. A10; Laura Meckler (Associated Press), "Security Examined At University Labs," *Hartford Courant*, March 2, 2002, p. A2.

159. Nicholas Wade, "Analysis May Yield Clues To Origins of Anthrax Mail," *New York Times*, May 10, 2002, p. A32; Debora MacKenzie, "Anthrax attack bug 'identical' to army strain," *New Scientist* (online edition), May 9, 2002; Martin Enserinik, "Useful Data But No Smoking Gun," *Science*, vol. 296, May 10, 2002, p. 1002; Timothy D. Read, et al., "Comparative Genome Sequencing for Discovery of Novel Polymorphisms in *Bacillus anthracis*," *Science*, vol. 296, June 14, 2002, pp. 2028-2033; Marsha Walton, "Gene research may help solve anthrax mystery," *CNN Sci-Tech* (www.cnn.com), May 9, 2002; Eleanor Clift, Michael Isikoff and Mark Miller, "Investigation: Solving the Anthrax Case—With NO Mistakes," *Newsweek*, July 15, 2002, p. 6; Dave Altimari, "Anthrax Killer Outlasting The Hunters; Delays, Lack of Expertise Make It Doubtful That FBI Can Solve Case," *Hartford Courant*, September 7, 2002, p. A1; Dave Altimari and Jack Dolan, "ANTHRAX THEORY EMERGES: Scientists: FBI Questions Suggest Insider Grew Spores At Lab, Refined Them Elsewhere," *Hartford Courant*, June 13, 2002, p. A1.

160. Dan Vergano, "Anthrax was from biodefense lab, study says," *USA TODAY*, May 10, 2002, p. 12A.

161. Rosie Mestel, "The Nation: Study Maps Varied Anthrax Samples," *Los Angeles Times* (online edition), May 10, 2002.

162. Craig A. Cummings and David A. Relman, "Microbial Forensics—'Cross-Examining Pathogens," *Science*, vol. 296, June 14, 2002, pp. 1976-1979.

163. Timothy D. Read, et al., "Comparative Genome Sequencing for Discovery of Novel Polymorphisms in *Bacillus anthracis*," *Science*, Ibid.

Index